Methods in Molecular Biology

For further volumes:
http://www.springer.com/series/7651

Western Blotting

Methods and Protocols

Edited by

Biji T. Kurien

Arthritis and Clinical Immunology, Oklahoma Medical Research Foundation, Oklahoma City, OK, USA; Department of Medicine, University of Oklahoma Health Sciences Center, Oklahoma City, OK, USA; Department of Veterans Affairs Medical Center, Oklahoma City, OK, USA

R. Hal Scofield

Arthritis and Clinical Immunology, Oklahoma Medical Research Foundation, Oklahoma City, OK, USA; Department of Medicine, University of Oklahoma Health Sciences Center, Oklahoma City, OK, USA; Department of Veterans Affairs Medical Center, Oklahoma City, OK, USA

Humana Press

Editors
Biji T. Kurien
Arthritis and Clinical Immunology
Oklahoma Medical Research Foundation
Oklahoma City, OK, USA

Department of Medicine,
University of Oklahoma Health Sciences Center
Oklahoma City, OK, USA

Department of Veterans Affairs Medical Center
Oklahoma City, OK, USA

R. Hal Scofield
Arthritis and Clinical Immunology
Oklahoma Medical Research Foundation
Oklahoma City, OK, USA

Department of Medicine,
University of Oklahoma Health Sciences Center
Oklahoma City, OK, USA

Department of Veterans Affairs Medical Center
Oklahoma City, OK, USA

ISSN 1064-3745 ISSN 1940-6029 (electronic)
Methods in Molecular Biology
ISBN 978-1-4939-2693-0 ISBN 978-1-4939-2694-7 (eBook)
DOI 10.1007/978-1-4939-2694-7

Library of Congress Control Number: 2015938704

Springer New York Heidelberg Dordrecht London

Cover illustration: Coomassie stained urea gel and corresponding amido black stained blot of two-dimensional separations of chicken liver ribosomal proteins. Rearranged and colorized photographs from Towbin, H., Staehelin, T., and Gordon, J. (1979) Electrophoretic transfer of proteins from polyacrylamide gels to nitrocellulose sheets: procedure and some applications. Proc. Natl. Acad. Sci. USA 76, 4350–4354.

Printed on acid-free paper

Humana Press is a brand of Springer
Springer Science+Business Media LLC New York is part of Springer Science+Business Media (www.springer.com)

Preface

Western Blotting

As we noted in a previous volume of protein blotting, western immunoblotting is a fundamental procedure in the modern biomedical research laboratory, as well as in laboratories doing research in other areas. The technique is routinely used now for research purposes, for instance, clinical medical laboratories. The development of this technique occurred within the memory of many of us, including the two of us. But, the variations on the theme nearly 40 years later are quite remarkable, varying from small variations on the original theme to techniques that barely resemble the original idea. But, why does the research enterprise need a compilation of these vast and disparate techniques? It seems to us that virtually everyone who reads these words will have tried to get a new technique started in their laboratory by reading methods found in published papers. We imagine that virtually everyone doing this has found it a daunting task. Similar to cooking recipes, most methods section or even procedure protocols do not tell one exactly what to do. So, the goal of this volume is to give practical methods such that you will be able to open this volume or downloaded chapter on your lab bench, follow the step-by-step instructions, and successfully perform a new procedure. If investigators are able to do this, then we will consider our efforts a success. Since the publication of the last volume, one of the chapter authors has died. Rafael Pont-Lezica, Professor Emeritus of Plant Physiology, Biochemistry, and Biotechnology at the Université de Toulouse, died on September 26, 2011. Despite his emeritus status, he maintained an active engagement in research. He spent much of his career studying plant cell wall. While a Visiting Professor at Washington University in St. Louis, Missouri, USA, he developed the tissue printing technique, a creative alternative to techniques such as in situ hybridization or immunocytochemistry.

Oklahoma City, OK, USA

Biji T. Kurien, Ph.D.
R. Hal Scofield, M.D.

Contents

Preface . . . . *v*
Contributors . . . . *xi*

1 The Early Days of Blotting 1
Edwin Southern

2 Origins of Protein Blotting 5
Harry Towbin

3 Western Blotting: Remembrance of Things Past 9
W. Neal Burnette

4 Simian Virus 40 and Protein Transfer 13
Jaime Renart

5 Western Blotting: An Introduction 17
Biji T. Kurien and R. Hal Scofield

6 From Little Helpers to Automation 31
Biji T. Kurien and Harry Towbin

7 Spectrophotometric Methods to Determine Protein Concentration 41
J.P. Dean Goldring

8 Solubilization of Proteins: The Importance of Lysis Buffer Choice 49
Mandy Peach, Noelle Marsh, Ewa I. Miskiewicz, and Daniel J. MacPhee

9 Simultaneous Immunoblotting Analysis with Activity Gel Electrophoresis and 2-D Gel Electrophoresis 61
Der-Yen Lee and Geen-Dong Chang

10 Diffusion Blotting: A Rapid and Simple Method for Production of Multiple Blots from a Single Gel 73
Ingrid Olsen and Harald G. Wiker

11 Multiple Immunoblots by Passive Diffusion of Proteins from a Single SDS-PAGE Gel 77
Biji T. Kurien and R. Hal Scofield

12 Slice Blotting 87
Graeme Lowe

13 Localizing Proteins by Tissue Printing 93
Rafael F. Pont-Lezica

14 Dot-Immunobinding Assay (Dot-Iba) 105
Sumi Surendran, Annamma Mathai, and Vishnampet Venkataraman Radhakrishnan

15 Analysis of Antibody Clonotype by Affinity Immunoblotting 109
Biji T. Kurien and R. Hal Scofield

16 Glycosaminoglycan Blotting and Detection After Electrophoresis Separation 119
Nicola Volpi and Francesca Maccari

17 A Well-Based Reverse-Phase Protein Array of Formalin-Fixed Paraffin-Embedded Tissue 129
Joon-Yong Chung and Stephen M. Hewitt

18 Quantitative Computerized Western Blotting in Detail 141
Dalit Talmi-Frank, Charles L. Jaffe, and Gad Baneth

19 Cationic Electrophoresis and Eastern Blotting. 151
Engelbert Buxbaum

20 A Miniaturized Blotting System for Simultaneous Detection of Different Autoantibodies 165
Ulrich Canzler, Holger Bartsch, Kai Großmann, Werner Lehmann, Karsten Conrad, Biji T. Kurien, Yaser Dorri, R. Hal Scofield, and Michael P. Bachmann

21 Proteomic Expressional Profiling of a Paraffin-Embedded Tissue by Multiplex Tissue Immunoblotting 175
Joon-Yong Chung and Stephen M. Hewitt

22 Post-Staining Electroblotting for Efficient and Reliable Peptide Blotting. 185
Der-Yen Lee and Geen-Dong Chang

23 Multistrip Western Blotting: A Tool for Comparative Quantitative Analysis of Multiple Proteins. 197
Edita Aksamitiene, Jan B. Hoek, and Anatoly Kiyatkin

24 Western Blotting Using PVDF Membranes and Its Downstream Applications . 227
Setsuko Komatsu

25 Blotting from PhastGel to Membranes by Ultrasound. 237
Joseph Kost and Aharon Azagury

26 Western Blotting of High and Low Molecular Weight Proteins Using Heat. . . 247
Biji T. Kurien and R. Hal Scofield

27 Membrane Strip Affinity Purification of Autoantibodies. 257
Biji T. Kurien

28 Strip Immunoblotting of Multiple Antigenic Peptides 269
Biji T. Kurien

29 Double-Blotting: A Solution to the Problem of Nonspecific Binding of Secondary Antibodies in Immunoblotting Procedures. 277
Françoise Lasne

30 Method for Resolution and Western Blotting of Very Large Proteins Using Agarose Electrophoresis 285
Marion L. Greaser and Chad M. Warren

31 Immunodetection of P-Selectin Using an Antibody to Its C-Terminal Tag . . . 293
Padmaja Mehta-D'souza

32 Improvements and Variants of the Multiple Antigen Blot Assay-MABA: An Immunoenzymatic Technique for Simultaneous Antigen and Antibody Screening 301
Oscar Noya, Sandra Losada, Marilyan Toledo, Adriana Gauna, María Angelita Lorenzo, Henry Bermúdez, and Belkisyolé Alarcón de Noya

33 Blotting from Immobilized pH Gradient Gels: Application to Total Cell Lysates 321
Harry Towbin

34 Immunoprecipitation: Western Blot for Proteins of Low Abundance 327
Edward P. Trieu and Ira N. Targoff

35 Native Electrophoresis and Western Blot Analysis (NEWeB): Methods and Applications 343
Ioannis N. Manoussopoulos and Mina Tsagris

36 Shift-Western Blotting: Separate Analysis of Protein and DNA from Protein–DNA Complexes 355
Matthias Harbers

37 Grid-Immunoblotting 375
Olga Yeliosof and R. Hal Scofield

38 Detection and Quantification of Protein–Protein Interactions by Far-Western Blotting 379
Joshua A. Jadwin, Bruce J. Mayer, and Kazuya Machida

39 Western Blot Analysis of Adhesive Interactions Under Fluid Shear Conditions: The Blot Rolling Assay 399
Robert Sackstein and Robert Fuhlbrigge

40 Centrifuge Blotting 411
Jinny Paul

41 Blotting of Coomassie Blue-Stained Proteins from PAGE Gels to Transparencies 415
Jinny Paul

42 B-Cell ELISPOT: For the Identification of Antigen-Specific Antibody-Secreting Cells 419
Hemangi B. Shah and Kristi A. Koelsch

43 T Cell ELISPOT: For the Identification of Specific Cytokine-Secreting T Cells 427
Kerry M. Leehan and Kristi A. Koelsch

44 Membrane Microplates for One- and Two-Color ELISPOT and FLUOROSPOT Assays 435
Alexander E. Kalyuzhny

45 SDS-PAGE to Immunoblot in One Hour 449
Biji T. Kurien, Debashish Danda, Michael P. Bachmann, and R. Hal Scofield

46 Single-Cell Western Blotting 455
Syed M.S. Quadri

47 Protein Detection by Simple Western™ Analysis 465
Valerie M. Harris

48 Western Blotting Using Microfluidics 469
Pothur R. Srinivas

49 Two-Dimensional Gel-Based Protein Standardization Verified by Western Blot Analysis 473
Hisao Haniu, Daisuke Watanabe, Yusuke Kawashima, and Hiroyuki Matsumoto

50 Fingerprint Deposition on Nitrocellulose and Polyvinylidene Difluoride Membranes Using Alkaline Phosphatase 481
Biji T. Kurien, Debashish Danda, and R. Hal Scofield

51 Other Notable Protein Blotting Methods: A Brief Review 487
Biji T. Kurien and R. Hal Scofield

Index *505*

Contributors

EDITA AKSAMITIENE • *Department of Otolaryngology - Head and Neck Surgery, Department of Pathology, Anatomy and Cell Biology, Thomas Jefferson University, Philadelphia, PA, USA*

AHARON AZAGURY • *Department of Chemical Engineering, Ben Gurion University of the Negev, Beer-Sheva, Israel*

MICHAEL P. BACHMANN • *Helmholtz-Zentrum Dresden-Rossendorf e.V., Institute of Radiopharmaceutical Cancer Research, Dresden, Germany; Department of Radioimmunology, Institute of Radiopharmaceutical Cancer Research, Dresden, Germany; Institut für Immunologie, Carl Gustav Carus Technische Universität Dresden, Dresden, Germany*

GAD BANETH • *School of Veterinary Medicine, Hebrew University, Rehovot, Israel*

HOLGER BARTSCH • *Institut für Immunologie, Carl Gustav Carus Technische Universität Dresden, Dresden, Germany*

HENRY BERMÚDEZ • *Sección de Inmunologia, Instituto de Medicina Tropical, Facultad de Medicina, Universidad Central de Venezuela, Caracas, Venezuela*

W. NEAL BURNETTE • *Molecular Pharmaceutics Corporation, Chapel Hill, NC, USA*

ENGELBERT BUXBAUM • *Ross University School of Medicine, Roseau, Dominica, West Indies*

ULRICH CANZLER • *Institut für Immunologie, Carl Gustav Carus Technische Universität Dresden, Dresden, Germany; Klinik und Polyklinik für Frauenheilkunde und Geburtshilfe, Universitätsklinikum Carl Gustav Carus, Dresden, Germany*

GEEN-DONG CHANG • *Graduate Institute of Biochemical Sciences, National Taiwan University, Taipei, Taiwan*

JOON-YONG CHUNG • *Laboratory of Pathology, Center for Cancer Research, National Cancer Institute, National Institutes of Health, Bethesda, MD, USA*

KARSTEN CONRAD • *Institut für Immunologie, Carl Gustav Carus Technische Universität Dresden, Dresden, Germany*

DEBASHISH DANDA • *Clinical Immunology and Rheumatology, Christian Medical College, Vellore, Tamil Nadu, India*

YASER DORRI • *Arthritis and Clinical Immunology Program, Oklahoma Medical Research Foundation, Oklahoma City, OK, USA*

ROBERT FUHLBRIGGE • *Harvard Skin Disease Research Center, Harvard Medical School, Boston, MA, USA; Department of Dermatology, Brigham and Women's Hospital, Boston, MA, USA*

ADRIANA GAUNA • *Sección de Biohelmintiasis, Instituto de Medicina Tropical, Facultad de Medicina, Universidad Central de Venezuela, Caracas, Venezuela*

J.P. DEAN GOLDRING • *Department of Biochemistry, University of KwaZulu-Natal, Pietermaritzburg, Scottsville, South Africa*

MARION L. GREASER • *Muscle Biology Laboratory, University of Wisconsin-Madison, Madison, WI, USA*

KAI GROßMANN • *Klinik und Polyklinik für Frauenheilkunde und Geburtshilfe, Universitätsklinikum Carl Gustav Carus, Dresden, Germany*
HISAO HANIU • *Institute for Biomedical Sciences, Shinshu University, Matsumoto, Nagano, Japan*
MATTHIAS HARBERS • *RIKEN Center for Life Science Technologies, RIKEN Yokohama Institute, Yokohama, Kanagawa, Japan*
VALERIE M. HARRIS • *University of Oklahoma Health Sciences Center, Oklahoma City, OK, USA*
STEPHEN M. HEWITT • *Laboratory of Pathology, Center for Cancer Research, National Cancer Institute, National Institutes of Health, Bethesda, MD, USA*
JAN B. HOEK • *Department of Pathology, Anatomy and Cell Biology, Thomas Jefferson University, Philadelphia, PA, USA*
JOSHUA A. JADWIN • *Raymond and Beverly Sackler Laboratory of Genetics and Molecular Medicine, Department of Genetics and Genome Sciences, University of Connecticut School of Medicine, Farmington, CT, USA*
CHARLES L. JAFFE • *Department of Microbiology and Molecular Genetics, IMRIC, Hadassah Medical School, Hebrew University, Jerusalem, Israel*
ALEXANDER E. KALYUZHNY • *Department of Immunohistochemistry and Elispot Assays, R&D Systems, Inc., Minneapolis, MN, USA*
YUSUKE KAWASHIMA • *Department of Biochemistry and Molecular Biology, University of Oklahoma Health Sciences Center, Oklahoma City, OK, USA*
ANATOLY KIYATKIN • *Department of Biochemistry and Biophysics, Perelman School of Medicine, University of Pennsylvania, Philadelphia, PA, USA*
KRISTI A. KOELSCH • *Arthritis and Clinical Immunology Program, Oklahoma Medical Research Foundation, Oklahoma City, OK, USA; Department of Medicine, University of Oklahoma Health Sciences Center, Oklahoma City, OK, USA*
SETSUKO KOMATSU • *National Institute of Crop Science, NARO, Kannondai, Tsukuba, Japan*
JOSEPH KOST • *Department of Chemical Engineering, Ben Gurion University of the Negev, Beer-Sheva, Israel*
BIJI T. KURIEN • *Arthritis and Clinical Immunology, Oklahoma Medical Research Foundation, Oklahoma City, OK, USA; Department of Medicine, University of Oklahoma Health Sciences Center, Oklahoma City, OK, USA; Department of Veterans Affairs Medical Center, Oklahoma City, OK, USA*
FRANÇOISE LASNE • *Département des analyses, Agence Française de Lutte contre le Dopage, Châtenay-Malabry, France*
DER-YEN LEE • *Graduate Institute of Biochemical Sciences, National Taiwan University, Taipei, Taiwan*
KERRY M. LEEHAN • *Arthritis and Clinical Immunology Program, Oklahoma Medical Research Foundation, Oklahoma City, OK, USA*
WERNER LEHMANN • *Klinik und Polyklinik für Frauenheilkunde und Geburtshilfe, Universitätsklinikum Carl Gustav Carus, Dresden, Germany*
MARÍA ANGELITA LORENZO • *Sección de Biohelmintiasis, Instituto de Medicina Tropical, Facultad de Medicina, Universidad Central de Venezuela, Caracas, Venezuela*
SANDRA LOSADA • *Sección de Biohelmintiasis, Instituto de Medicina Tropical, Facultad de Medicina, Universidad Central de Venezuela, Caracas, Venezuela*
GRAEME LOWE • *Monell Chemical Senses Center, Philadelphia, PA, USA*
FRANCESCA MACCARI • *Department of Life Sciences, University of Modena and Reggio Emilia, Modena, Italy*

KAZUYA MACHIDA • *Raymond and Beverly Sackler Laboratory of Genetics and Molecular Medicine, Department of Genetics and Genome Sciences, University of Connecticut School of Medicine, Farmington, CT, USA*

DANIEL J. MACPHEE • *Department of Veterinary Biomedical Sciences, One Reproductive Health Research Group, Western College of Veterinary Medicine, University of Saskatchewan, Saskatoon, SK, Canada*

IOANNIS N. MANOUSSOPOULOS • *Laboratory of Virology, Plant Protection Institute of Patras, NEO & Amerikis, Patras, Greece*

NOELLE MARSH • *Division of BioMedical Sciences, Health Sciences Centre, St. John's, NL, Canada*

ANNAMMA MATHAI • *Department of Pathology, Sree Chitra Tirunal Institute for Medical Sciences and Technology (SCTIMST), Thiruvananthapuram, Kerala, India*

HIROYUKI MATSUMOTO • *Department of Biochemistry and Molecular Biology, University of Oklahoma Health Sciences Center, Oklahoma City, OK, USA; Clinical Proteomics and Gene Therapy Laboratory, Kurume University, Kurume City, Fukuoka, Japan*

BRUCE J. MAYER • *Raymond and Beverly Sackler Laboratory of Genetics and Molecular Medicine, Department of Genetics and Genome Sciences, University of Connecticut School of Medicine, Farmington, CT, USA*

PADMAJA MEHTA-D'SOUZA • *Cardiovascular Biology Research Program, Oklahoma Medical Research Foundation, Oklahoma City, OK, USA*

EWA I. MISKIEWICZ • *Department of Veterinary Biomedical Sciences, One Reproductive Health Research Group, Western College of Veterinary Medicine, University of Saskatchewan, Saskatoon, SK, Canada*

BELKISYOLÉ ALARCÓN DE NOYA • *Sección de Inmunologia, Instituto de Medicina Tropical, Facultad de Medicina, Universidad Central de Venezuela, Caracas, Venezuela*

OSCAR NOYA • *Sección de Biohelmintiasis, Instituto de Medicina Tropical, Facultad de Medicina, Universidad Central de Venezuela, Caracas, Venezuela*

INGRID OLSEN • *Section for Immunology, Department of Laboratory Services, Norwegian Veterinary Institute, Oslo, Norway*

JINNY PAUL • *Arthritis and Clinical Immunology, Oklahoma Medical Research Foundation, Oklahoma City, OK, USA*

MANDY PEACH • *Division of BioMedical Sciences, Health Sciences Centre, St. John's, NL, Canada*

RAFAEL F. PONT-LEZICA • *Surfaces Cellulaires et Signalisation chez les Végétaux, UMR 5546 CNRS-Université Paul Sabatier-Toulouse III, Auzeville, Castanet-Tolosan, France*

SYED M.S. QUADRI • *Arthritis and Clinical Immunology Program, Oklahoma Medical Research Foundation, Oklahoma City, OK, USA; Department of Pathology, University of Oklahoma Health Sciences Center, Oklahoma City, OK, USA*

VISHNAMPET VENKATARAMAN RADHAKRISHNAN • *Department of Pathology, Sree Chitra Tirunal Institute for Medical Sciences and Technology (SCTIMST), Thiruvananthapuram, Kerala, India*

JAIME RENART • *Instituto de Investigaciones Biomédicas Alberto Sols, CSIC-UAM, Madrid, Spain*

ROBERT SACKSTEIN • *Program of Excellence in Glycosciences, Harvard Medical School, Boston, MA, USA; Department of Dermatology and Medicine, Brigham and Women's Hospital, Boston, MA, USA*

R. HAL SCOFIELD • *Arthritis and Clinical Immunology, Oklahoma Medical Research Foundation, Oklahoma City, OK, USA; Department of Medicine, University of Oklahoma Health Sciences Center, Oklahoma City, OK, USA; Department of Veterans Affairs Medical Center, Oklahoma City, OK, USA*
HEMANGI B. SHAH • *Arthritis and Clinical Immunology Program, Oklahoma Medical Research Foundation, Oklahoma City, OK, USA*
EDWIN SOUTHERN • *Oxford Gene Technology, Begbroke Business Park, Oxford, UK*
POTHUR R. SRINIVAS • *Division of Cardiovascular Sciences, National Heart, Lung and Blood Institute, National Institutes of Health, Bethesda, MD, USA*
SUMI SURENDRAN • *Department of Pathology, Sree Chitra Tirunal Institute for Medical Sciences and Technology (SCTIMST), Thiruvananthapuram, Kerala, India*
DALIT TALMI-FRANK • *Departments of Biological Regulation, Weizmann institute of Science, Rehovot, Israel*
IRA N. TARGOFF • *Arthritis and Immunology Program, Oklahoma Medical Research Foundation, Oklahoma City, OK, USA*
MARILYAN TOLEDO • *Sección de Biohelmintiasis, Instituto de Medicina Tropical, Facultad de Medicina, Universidad Central de Venezuela, Caracas, Venezuela*
HARRY TOWBIN • *Institute of Pharmaceutical Sciences, Zurich, Switzerland*
EDWARD P. TRIEU • *Arthritis and Immunology Program, Oklahoma Medical Research Foundation, Oklahoma City, OK, USA*
MINA TSAGRIS • *Foundation of Research and Technology Hellas, Institute of Molecular Biology and Biotechnology, Heraklion, Crete, Greece; Department of Biology, University of Crete, Heraklion, Crete, Greece*
NICOLA VOLPI • *Department of Life Sciences, University of Modena and Reggio Emilia, Modena, Italy*
CHAD M. WARREN • *Department of Physiology, University of Illinois at Chicago, Chicago, IL, USA*
DAISUKE WATANABE • *Department of Biochemistry and Molecular Biology, University of Oklahoma Health Sciences Center, Oklahoma City, OK, USA*
HARALD G. WIKER • *Section for Microbiology and Immunology, Gade Institute, University of Bergen, Bergen, Norway; Department of Microbiology and Immunology, Haukeland University Hospital, Bergen, Norway*
OLGA YELIOSOF • *Arthritis and Clinical Immunology Program, Oklahoma Medical Research Foundation, Oklahoma City, OK, USA*

Chapter 1

The Early Days of Blotting

Edwin Southern

Abstract

The history of the development of DNA blotting is described in this chapter. DNA blotting, involving the transfer of electrophoretically separated DNA fragments to a membrane support through capillary action, is also known as Southern blotting. This procedure enables the detection of a specific DNA sequence by hybridization with probes. The term Southern blotting led to a "geographic" naming tradition, with RNA blotting bearing the name Northern blotting and protein transfer to membranes becoming known as Western blotting.

Key words Southern blotting, Electrophoresis, Western blotting, Northern blotting

1 Introduction

It is surprising to me that blotting is still taught and even used for more than 40 years after I did the first one in 1973. At the time I was looking for a way of purifying genes from eukaryotic DNA. The story began with the discovery by Peter Ford that the 5S genes of *Xenopus laevis* are differentially expressed in oocytes and somatic cells: Peter and I sequenced the 5S RNAs and found that their sequences were different [1]. There was a lot of interest at the time in understanding the molecular mechanisms that underpin the control of gene expression, and this seemed to be a case where the genes might be accessible, as they are highly repeated. There were few ways to purify genes in those days—cloning was not available—but the type II restriction endonucleases had recently been described by Tom Kelly and Ham Smith [2]. Combined with agarose gel electrophoresis, these enzymes provided a powerful way of fractionating DNA. The difficult part was finding the bands of 5S genes in the complex smear of fragments that result from running the digest of eukaryotic DNA on a gel. Cutting the gel into slices, eluting the DNA, and analyzing the fractions by filter hybridization were tedious and gave very noisy backgrounds that hid the low signal from the genes. With Julia Thompson, we tried drying the gel to

Biji T. Kurien and R. Hal Scofield (eds.), *Western Blotting: Methods and Protocols*, Methods in Molecular Biology, vol. 1312, DOI 10.1007/978-1-4939-2694-7_1, © Springer Science+Business Media New York 2015

trap the fragments, and hybridizing the dry film with radioactive 5S RNA; this did not work in our hands; and though Oliver Smithies and his colleagues succeeded [3], this method does not seem to have taken off. We shelved the project after Julia left for Canada, but I returned to it when I heard Charlie Thomas describe his way of dissolving agarose gels in concentrated sodium perchlorate [4]. It occurred to me that if we could dissolve the gel while it was laid against a filter, the DNA would stick to the filter, retaining the pattern of bands. The filter-bound DNA could then be hybridized using the well-known Gillespie and Spiegelman method [5]. So I set up a test with a strip of gel floating on a raft of cellulose nitrate filter on a solution of sodium perchlorate. As I sat and watched, I did not see any sign of the gel dissolving, but I did notice a bead of liquid forming on top of the gel. This was the light-bulb moment. I realized that the gel was permeable and that it should be possible to soak out the DNA by a blotting process; it was a matter of putting the filter on top of the gel, with an overlay of filter papers, and substituting SSC for sodium perchlorate. This worked first time and it was a thrill to see the sharp bands on the autoradiograph. Looking back I should have got more quickly to the realization that the gel was permeable. When I was a child I became the favorite of my first teacher at primary school, Mrs Laycock. In those days the way we copied documents was to write a master copy in a thick violet-colored ink on a shiny highly carded paper. This was then laid on a tray of evil-smelling gel so that the ink soaked into the gel and copies could be made by laying blank sheets onto the gel to soak out the impression of the ink. I became very familiar with this method as Mrs Laycock employed me to use it to make copies of examination papers. So I learned a great deal, and I wonder if there was an echo of this experience at the light-bulb moment.

I did not use the method much myself. Peter and I were beaten to the 5S genes by Don Brown, who used density gradient centrifugation to isolate the genes [6]. He went on to identify the region of the genes that control expression. Ironically, the control regions lie within the transcribed part of the gene—the very region that Peter Ford and I had sequenced to show the difference between somatic and oocyte RNAs!

Nevertheless, the method caught on. And it was in widespread use long before it was published in the Journal of Molecular Biology. My first attempt at publication was rejected because it was a "methods paper" and so I had to spend a year or so gathering some more biologically relevant data before it could be accepted. However, it had already been disseminated informally. Late in 1973, we had a visit from Mike Matthews from Cold Spring Harbor Laboratory. I showed him my early results and he asked if he could have the protocol, which I sketched on a scrap of paper. Visitors to CSHL saw his work and in turn asked for the protocol. Mike asked my permission to pass it on, which I agreed provided that they were told the origin of the method. This probably explains why my name became so firmly a part of the method's label. I was pleased and amused by the Northern (*see* Chapter 4) and Western blots (*see* Chapters 2 and 3) which quickly followed.

References

1. Ford PJ, Southern EM (1973) Different sequences for 5S RNA in kidney cells and ovaries of Xenopus laevis. Nat New Biol 241:7–12
2. Kelly TJ, Smith HO (1970) A restriction enzyme from Hemophilus influenzae. II. J Mol Biol 51: 393–409
3. Shinnick TM, Lund E, Smithies O, Blattner FR (1975) Hybridization of labeled RNA to DNA in agarose gels. Nucleic Acids Res 10: 1911–1929
4. Chen CW, Thomas CA Jr (1980) A rapid two-dimensional fractionation of DNA restriction fragments. Anal Biochem 101:339–341
5. Gillespie D, Spiegelman S (1965) A quantitative assay for DNA-RNA hybrids with DNA immobilized on a membrane. J Mol Biol 12:829–842
6. Sakonju S, Bogenhagen DF, Brown DD (1980) A control region in the center of the 5S RNA gene directs specific initiation of transcription: I. The 5′ border of the region. Cell 19:13–25

Chapter 2

Origins of Protein Blotting

Harry Towbin

Abstract

The development of protein blotting in its early days is recounted as arising from the need to tackle a specific analytical problem. Combining diverse elements of common methods and simple lab equipment resulted in a procedure of general utility. The expansion of the idea of carrying out immunoassays on membranes as predecessors of microarrays is briefly touched upon.

Key words Blotting, Western, History, Protein array

1 Introduction

Easily persuaded by the editors' invitation to write about the origins of protein blotting, I begin the story in the laboratory of Julian Gordon at the Friedrich-Miescher-Institute in Basel. In need of a postdoctoral fellow, Julian had hired me in 1978 for the task of creating antibodies against ribosomal proteins. With some experience in ribosomes and none in immunology, I set out to purify our proteins from chicken liver. We believed that it was necessary to get really fresh samples. So, together with a courageous colleague, who later turned to running a motorbike shop, we ventured to a slaughterhouse in the countryside. While we watched their grisly machinery, the friendly workers were captivated by the fog running over the brim of the Dewars as the freshest ever chicken liver dropped into the liquid nitrogen.

Back in the new and spacious laboratory, we struggled with ultracentrifuges, ample amounts of urea, and countless column fractions. We wondered whether the animals we were going to immunize would raise the antibodies we so eagerly sought. How could we ever be sure we would not get antibodies to some contaminants? We were lucky to be in contact with Theo Staehelin at Roche, a pioneer in the field of initiation factors for mammalian protein synthesis who had ample expertise in ribosomes. Theo also told us about the new hybridoma technique. The idea of immunizing

Biji T. Kurien and R. Hal Scofield (eds.), *Western Blotting: Methods and Protocols*, Methods in Molecular Biology, vol. 1312, DOI 10.1007/978-1-4939-2694-7_2, © Springer Science+Business Media New York 2015

with mixtures of proteins and still getting an eternal source of a specific antibody was irresistible. After learning the secrets of the trade from Theo, we happily switched to mice and spared the rabbits and goats we had already injected to familiarize ourselves with time-honored immunological techniques. I was fascinated by precipitation arcs of the Ouchterlony double diffusion test and by the sensitivity of solid-phase immunoassays achieved with remarkably simple equipment. Still, the problem of assuring specificity of hybridoma antibodies remained.

A common way of characterizing ribosomal proteins was by electrophoresis on two-dimensional gels. Could one recover the proteins from the gel as the literature described and use them in these sensitive immunoassays? The extraction worked, in principle, but the bulk of homogenized polyacrylamide was deterring. Even today, few researchers take that approach for purifying proteins from gels. We also discarded the idea of letting the antibodies react with proteins in the gel because of the impeded diffusion within the polyacrylamide matrix.

As the three of us later realized, the idea of preparing a replica of a protein gel on a membrane, in close analogy to Southern's DNA blotting (*see* Chapter 1), was in many people's mind [1]. But how could we copy the proteins to a membrane? It was a lucky coincidence that Julian had an electrophoretic destainer in use. This apparatus, now rarely seen in the laboratory, served to remove excess stain from gels, simply by placing the gel between two grids and applying current at a right angle to the plane of the sheet. The electrically charged dye molecules quickly cleared off the gel. Well, proteins were also charged, would they behave like the dye? The basic setup was quickly put together. Meticulously cleaning the destainer from residual amido black I remember as being the most tedious part of the chore. Pipette tip holders, Scotch Brite scouring pads, and rubber strings were all what was needed to build the sandwich that is still popular for protein blotting. A series of straightforward experiments showed us that the nitrocellulose sheets reliably captured the proteins as they were leaving the gel. From my wife, Marion, I knew about the art of immunohistochemical staining. One could easily test staining procedures by placing little dots of proteins directly on the membrane and run series of dilutions. Developing the first blot from a gel with antibodies thrilled me with bands that darkened within seconds. I felt like a child that reveals secret messages written in invisible ink by holding a sheet of paper over a flame.

After publication of the method [2], it dawned on us that the blotting procedure might have some commercial value, after all. We learned from the lawyers that we could still claim protection in some countries. We also learned how hard it was to define what was really new in an invention and also that there needed to be an unforeseen element in it. In some way, almost everything in our procedure had some precedence! We stand on the shoulders of giants, as every Google Scholar user knows. Finally, the patent application was written; the fact that Theo Staehelin was at Roche and Julian and myself at Ciba-Geigy, though unusual, was no impediment.

The idea of placing proteins on nitrocellulose sheets by direct spotting, as trivial as it appeared, proved to be stimulating [3]. You could easily probe little dots of protein on nitrocellulose or create sandwich tests, for example for determining antibody subtypes in hybridoma supernatants. The potential of carrying out assays on arrayed protein spots, a bit awkwardly named dot immunobinding, was most clearly recognized by Julian. With the advent of spotting devices, always in the footsteps of DNA technologies, protein arraying is only now gaining popularity. These efforts might all be viewed as aiming at eliminating the cumbersome gel electrophoresis step from Western blotting. Also, for those weary of running gels and handling membranes, there is relief in sight with new systems that automatically resolve proteins in capillaries ([4, 5]; *see* Chapters 6 and 47). Still, as the contributions to this volume attest, membranes remain attractive supports, giving room for countless variations, unforeseen applications, and an expanding nomenclature inspired from the compass set by E. M. Southern (*see* Chapter 1).

References

1. Rocco RM (2006) Western blots [commentary]. In: Rocco RM (ed) Landmark papers in clinical chemistry. Elsevier, Amsterdam, pp 381–383
2. Towbin H, Staehelin T, Gordon J (1979) Electrophoretic transfer of proteins from polyacrylamide gels to nitrocellulose sheets: procedure and some applications. Proc Natl Acad Sci U S A 76:4350–4354
3. Hawkes R, Niday E, Gordon J (1982) A dot-immunobinding assay for monoclonal and other antibodies. Anal Biochem 119:142–147
4. O'Neill RA, Bhamidipati A, Bi X, Deb-Basu D, Cahill L, Ferrante J et al (2006) Isoelectric focusing technology quantifies protein signaling in 25 cells. Proc Natl Acad Sci U S A 103:16153–16158
5. Kurien BT, Towbin H (2015) From little helpers to automation. Methods Mol Biol. (this volume, Chapter 6)

Chapter 3

Western Blotting: Remembrance of Things Past

W. Neal Burnette

Abstract

Western blotting sprung from the need to develop a sensitive visual assay for the antigen specificity of monoclonal antibodies. The technique employed SDS-PAGE of protein antigens, electrophoretic replica transfer of gel-resolved proteins to unmodified nitrocellulose sheets, probing the immobilized antigens with hybridomas, and detection of antibody–antigen complexes with radiolabeled staphylococcal protein A and autoradiography. The simplicity and relevance of the method have led to its expansive application as an immunodiagnostic and a ubiquitous research tool in biology and medicine.

Key words Western blotting, Immunoblotting, Electroblotting, Antibodies, Antigens, SDS-PAGE, Immunodetection, Protein A

1 Introduction

Paraphrasing Plato, Jonathan Swift once famously observed that "Necessity is the Mother of Invention" [1]. Such necessity was the antecedent of Western blotting. The fact that similar techniques arose within the same time frame indicates the temporal pressure of an unfilled demand in biology and medicine—a common exigency to provide a tool by which to visualize specific antigens.

The requirement that impelled the development of Western blotting [2] in my laboratory came to light in 1977, when I moved to Robert Nowinski's RNA tumor virus group at the Fred Hutchinson Cancer Research Center. This was just at the time when monoclonal antibodies were first described by Köhler and Milstein [3], and Bob's group was developing monoclonal reagents as probes to assess the structural and immunologic nature of retrovirus proteins [4]. It quickly became clear that there was no simple, objectively visual way to easily screen the vast numbers of generated clones for their specificity toward individual structural polypeptides comprising the retrovirus envelope and core.

Although the main focus of my work at the time was in other areas of retroviral research, I had a methodological background in

Biji T. Kurien and R. Hal Scofield (eds.), *Western Blotting: Methods and Protocols*, Methods in Molecular Biology, vol. 1312, DOI 10.1007/978-1-4939-2694-7_3, © Springer Science+Business Media New York 2015

electrophoretic antigen assessment; therefore, I agreed to undertake the effort in the Nowinski group to develop new and streamlined techniques to facilitate screening of the hybridomas for antigen specificity. Having been trained as a postdoc in Tom August's lab at Albert Einstein College of Medicine in radioimmunoassays, immunoprecipitation, and SDS-polyacrylamide gel electrophoresis (SDS-PAGE), I attempted to conceive of ways in which these methods might be combined. RIAs had great sensitivity, but lacked the ability to give a simple picture of specificity, especially in complex protein mixtures. Conversely, immunoprecipitation required radiolabeling of diverse antigen species and, while it provided reasonable sensitivity and definition of specificity when linked to SDS-PAGE and autoradiography, it was plagued by significant background that led to substantial uncertainty and was not easily adaptable to high-throughput screening.

Launching into this project, essentially on my own and without benefit of knowledge of others who might be engaged in similar work, I attempted a wide array of techniques, hoping that I would stumble upon something useful or, at least, something that might light the pathway to proceed further. Here, I was trying through trial and error to fulfill another Swiftian dictum: "Discovery consists of seeing what everybody has seen and thinking what nobody else has thought" [1]. In retrospect, some of the things I tried verged on the laughable. Nevertheless, the early work furnished me with the recognition that purified, radiolabeled (in this case, radioiodinated) staphylococcal protein A [5] provided a more functionally stable and "universal" imaging agent for detection of antigen–antibody complexes than did "second-antibody" reagents.

As incongruous as it might seem in the hindsight of more than three decades, I struggled with how to apply the monoclonal antibodies (as well as monospecific antisera) to gel-separated antigens. The "Eureka" moment occurred while I was concomitantly performing other experiments that employed "Northern" blots [6], an effulgent clarity of vision that an immobilized "replica" of the PAGE-resolved proteins was to be an intrinsic element. Initially, I attempted passive transfer by placing gels in direct contact with derivatized, and later unmodified, nitrocellulose sheets. After overcoming problems associated with nonspecific binding of immunoglobulin and protein A reagents to the nitrocellulose by the use of a blocking agent (I employed immunoglobulin-depleted, purified bovine serum albumin), it became apparent that capillary transfer was slow, inefficient, and resulted in unacceptable diffusional band-spreading of the gel-resolved antigens.

A second Archimedean moment occurred at this point, when I came across an old electrophoretic gel destainer that I had not used in years. Perhaps, I reasoned, if I could work fast enough or keep temperatures low enough to minimize band diffusion *within* the parent gel, and find electrophoretic conditions and nitrocellulose pore size to prevent driving the proteins out of the gel and *through* the paper, I might be able to make better "replicas" of the gel-resolved antigens.

It only took about a week from this point to work out the "final" parameters of the basic electroblotting technique, and another few weeks to work on adaptations that could increase resolution and sensitivity in complex mixtures (e.g., cell culture, blood, tissue, and other clinical samples) using isotachophoresis in a first dimension, and then applying such cylindrical gels to the SDS-PAGE slab gels. During this period, a manuscript was prepared and a discussion with Bob Nowinski ensued wherein the name "Western blotting" was conceived. It was just at this time that the publication of Towbin et al. [7] appeared. While the basic technique described by these investigators was similar, I believed that many of the simplifying and "universalizing" aspects of Western blotting (e.g., unmodified nitrocellulose, radiolabeled protein A detection, 2-D separations) were sufficiently important to warrant submission of my manuscript. I also became aware at this time of the publication by Renart et al. [8] (*see* Chapter 4); however, the technique described in their paper employed conditions with which I had experimented (e.g., derivatized paper, passive capillary transfer, second antibody) and found wanting from the perspectives of simplicity, ease of use, resolution, sensitivity, and specificity.

The manuscript was submitted to *Analytical Biochemistry* and was rejected without, it seemed, any recourse to resubmission. It was interesting to note that the rejection appeared to me to be based not on any technical criticisms or its ostensible similarity to the methods of Towbin et al. [7] but rather on the reviewers' sentiment of the pedestrian nature of the contribution and, particularly, the flippant and frivolous whimsy in the name "Western blotting."

As previously documented [9], preprints of the rejected manuscript had been sent to colleagues, who subsequently provided them to others, and they to others until, eventually (even in this pre-electronic era of written communications), it seemed as though this unpublished article had received wider distribution than many published ones. I only became aware of this subsequent to my move to the Salk Institute at the end of 1979. It was there that I was tracked down and spent a good part of every work day fielding telephonic questions about the technique and providing readable copies of the preprint—the original I had sent to a few colleagues had undergone many cycles of photocopy replication as it wended its way from lab to lab, the later generations being difficult to read. After about a half year of operating this private "journal club," I called the editor-in-chief of *Analytical Biochemistry*; he agreed that the situation was untenable, that the general immunoblotting technique (as well as the name "Western blotting") was becoming widely accepted, and that the initial rejection of my manuscript was probably unfortunate. Therefore, I re-submitted the paper (with only very minor changes); it was accepted immediately, and finally published a few months later [2].

For those who have felt the sting of journal rejection, it is worth noting that this paper has entered a small pantheon of the most highly cited scientific articles, all of which were initially rejected for publication [10]. Humility is an oft-reinforced virtue in science; it is humbling to realize that this little paper on Western blotting far transcended the sum of journal citations for all of my other published research efforts. Nevertheless, it is a source of immense satisfaction to have made—along with Towbin et al. [7] (*see* Chapter 2)—a lasting contribution to the methodological armamentarium of biological and medical scientists.

To complete the analogy hinted in the title of this review, I wish to thank the editors of this volume for providing me, like the proffered "madeleine" in Proust's *À la recherche du temps perdu* [11], the occasion for this reminiscence.

References

1. Swift J (1726) Travels into several remote nations of the world, in four parts. By Lemuel Gulliver, first a surgeon, and then a captain of several ships. London: Benjamin Motte
2. Burnette WN (1981) "Western blotting": electrophoretic transfer of proteins from sodium dodecyl sulfate-polyacrylamide gels to unmodified nitrocellulose and radiographic detection with antibody and radioiodinated protein A. Anal Biochem 112:195–203
3. Köhler G, Milstein C (1975) Continuous cultures of fused cells secreting antibody of predefined specificity. Nature 256:495–497
4. Nowinski RC, Stone MR, Tam M, Lostrom ME, Burnette WN et al (1980) Mapping of viral proteins with monoclonal antibodies: analysis of the envelope proteins of murine leukemia viruses. In: Kennett RH, McKearn TJ, Bechtol KB (eds) Monoclonal antibodies: hybridomas: a new dimension in biological analyses. Plenum Publishing Corp, New York, NY, pp 295–316
5. Langone JJ (1978) [^{125}I]protein A: a tracer for general use in immunoassay. J Immunol Methods 24:269–285
6. Alwine JC, Kemp DJ, Stark GR (1977) Method for detection of specific RNAs in agarose gels by transfer to diazobenzyloxymethyl-paper and hybridization with DNA probes. Proc Natl Acad Sci U S A 74:5350–5354
7. Towbin H, Staehelin T, Gordon J (1979) Electrophoretic transfer of proteins from polyacrylamide gels to nitrocellulose sheets: procedure and some applications. Proc Natl Acad Sci U S A 76:4350–4354
8. Renart J, Reiser J, Stark GR (1979) Transfer of proteins from gels to diazobenzyloxymethyl-paper and detection with antisera: a method for studying antibody specificity and antigen structure. Proc Natl Acad Sci U S A 76: 3116–3120
9. Burnette WN (1991) Western blotting (citation classic). Curr Cont 34:8
10. Campanario JM (1996) Have referees rejected some of the most-cited articles of all times? J Am Soc Inf Sci 47:302–310
11. Proust M (1913) À la recherche du temps perdu, vol 1. Bernard Grasset, Paris

Chapter 4

Simian Virus 40 and Protein Transfer

Jaime Renart

Abstract

Protein transfer to solid supports after polyacrylamide gel electrophoresis, and subsequent probing with specific antibodies, is one of the most important tools in modern molecular and cellular biology. Since its development in 1979, the improvement of the technique has been impressive, from new apparatus to streamline the electrophoresis step to different modalities of the transfer step or solid supports for the transfer. Perhaps most impressive has been the explosion of the production and availability of antibodies. In this chapter, I describe the environment and conditions that led to the development of this technique in George Stark's laboratory.

Key words SDS-PAGE, Capillary protein transfer, Cleavable cross-linkers, Simian virus 40, Diazo-paper, SV40 T antigen, Antibody specificity

In the summer of 1975 I joined George Stark's laboratory in the Biochemistry Department, Stanford University, Palo Alto, California, as a postdoctoral fellow. George's laboratory had at that moment three research lines.

One was the continuation of the successful studies on aspartate transcarbamylase, the regulatory enzyme in pyrimidine nucleotide biosynthesis. After seminal studies on the mechanism of the bacterial enzyme, another postdoc there, Patrick Coleman, was purifying the mammalian enzyme, which ended up being a very large polypeptide (CAD), which accounted for the three first activities of the pathway, namely carbamoyl-phosphate synthetase 2, aspartate transcarbamylase, and dihydroorotase. At almost the same time as when I came to the laboratory, D. Parker Suttle joined this line of research [1].

The second line of research was the identification of an immunogenic material present in hamster cell lines [2], which ended up being lymphocytic choriomeningitis virus.

Biji T. Kurien and R. Hal Scofield (eds.), *Western Blotting: Methods and Protocols*, Methods in Molecular Biology, vol. 1312, DOI 10.1007/978-1-4939-2694-7_4, © Springer Science+Business Media New York 2015

The third line of research was on the regulation of the early phase of Simian virus 40 (SV40) infection, and was being worked out by James C. Alwine (postdoc) and Steven I. Reed, a graduate student. Their studies were challenging, given the paucity of SV40 early mRNA. Two strategies were taken; one was the use of SV40 *ts* mutants to increase the amount of early mRNA [3]. Another was to increase the specific activity of [^{32}P]-labeled mRNA, to get a detectable amount of material. To this end, we took advantage of the tools available in the laboratory for the aspartate transcarbamylase project. George and Kim Collins had developed a transition-state analog, *N*-(phosphonacetyl)-L-aspartate (PALA), that effectively inhibited pyrimidine biosynthesis, and therefore permitted us to increase the specific activity of the pool of labeled pyrimidines. The other tool we used relied on the work of another member of the group, Barbara Noyes, who had developed a cellulose-powder derivatized with a diazobenzyloxymethyl arm that could bind nucleic acids covalently [4]. In addition to this laboratory background, other groups in the Biochemistry Department were actively developing molecular biology ideas and methods.

There were two important earlier references for nucleic acid work; the first was the realization by Nygaard and Hall that ^{32}P-labeled RNA was not retained by nitrocellulose filters [5]. The other was the recently developed method to study DNAs separated by gel electrophoresis [6]. We were, then, in an inverse position working with SV40: DNA was easy to obtain, but specific RNAs were not, so using the Southern technique was not possible.

The stage, then, was ready for new developments. The first one came from the work of Jim Alwine and David Kemp. Instead of using the derivatized cellulose powder developed by Barbara Noyes, they derivatized sheets of paper. In this way RNA separated by electrophoresis in agarose gels could be studied much in the same way as Southern had done for DNA [7]. This technique soon became known as the "Northern," as opposed to the "Southern" technique.

At around that time I started to collaborate with a new postdoc, Jakob Reiser, from Switzerland. Jakob was mainly interested in the function of SV40 T antigen during infection, and we worked together characterizing the interaction of SV40 T antigen with SV40 chromatin in infected cells. In fact, this work is one of the first examples of the now widely used ChIP technique [8]. At the same time we started to work in extensions of the method developed by Alwine and Kemp to other molecules of interest, namely proteins and small fragments of DNA that were not retained in nitrocellulose in the Southern technique [9].

As we were trying to find some way to extend the available techniques, we used composite acrylamide/agarose gels, in which bis-acrylamide was substituted for a cleavable cross-linker. After cleaving the acrylamide matrix, the agarose gel remained and proteins transferred from them by upward capillary transfer, much in the same way as in the Southern or Northern techniques. Proteins could be detected by incubating the paper first with antibodies and then with iodinated-protein A from *Staphylococcus aureus*. The first cleavable cross-linker we used was *N,N'*-diallyltartardiamide (DATD). This compound has a *cis*-diol and could therefore be cleaved by treatment with sodium periodate.

We had at our disposal sera prepared against T antigen, SV40 virions and against the SV40 virion proteins VP2 and VP3. The first results were very positive. I always remember coming back to the laboratory from the University of California at San Francisco hospital, where my second child had just been born (27 December, 1977), and rejoicing at seeing those bands on the radiographic film!

We started to look for other antibodies in the laboratories around us, and (as mentioned in our paper) [10] we could detect proteins from *Escherichia coli*, *Myxococcus xanthus*, and the G protein of vesicular stomatitis virus. Of course we also tried T antigen, for which we had two different sera. Surprisingly, the periodate treatment of the gel worked fine with one serum but not with the other. We had already found another cleavable cross-linker, ethylene diacrylate (EDA), an ester that can be degraded with alkali. Using EDA-cross-linked gels, both sera detected T antigen perfectly. Further experiments were done to confirm this difference in the behavior of the two anti-T antigen sera, demonstrating clearly that it was a property of T antigen. In fact, we stressed this fact in the title of our paper, namely that the method could be used for studying the structure of antigens.

In retrospect, any breakthrough in science, and this technique clearly was one, is dependent on where it was developed and on the interests in the particular laboratory and around it. Both Towbin and Burnette, in their historical perspectives in these series [11, 12] (*see* Chapters 2 and 3), mention the presence in their laboratories of an electrophoretic destainer as an important hint for their work. In our case, molecular biology problems and techniques, namely detection of low-abundance RNAs, directed our work. Of course, electrophoretic transfer of protein to diazo-paper was soon demonstrated to be feasible [13]. The ease of the electrophoretic transfer, together with the development of new supports in addition to nitrocellulose, is now the basis of most cellular and molecular biology. However, the usefulness of covalent transfer to diazo-paper has been largely forgotten. In this regard, it is worth mentioning the work of Olmsted, using diazo-paper to affinity purify specific antibodies from complex mixtures [14].

Acknowledgement

I am indebted to George Stark and Jakob Reiser for helpful comments and advice.

References

1. Coleman PF, Suttle DP, Stark GR (1977) Purification from hamster cells of the multifunctional protein that initiates de novo synthesis of pyrimidine nucleotides. J Biol Chem 252:6379–6385
2. Mirault ME, Reed SI, Stark GR (1975) An antigen associated with messenger RNA in a transformed hamster cell line. Cold Spring Harb Symp Quant Biol 39(Pt 1):295–303
3. Reed SI, Stark GR, Alwine JC (1976) Autoregulation of simian virus 40 gene A by T antigen. Proc Natl Acad Sci U S A 73: 3083–3087
4. Noyes BE, Stark GR (1975) Nucleic acid hybridization using DNA covalently coupled to cellulose. Cell 5:301–310
5. Nygaard AP, Hall BD (1963) A method for the detection of RNA-DNA complexes. Biochem Biophys Res Commun 12:98–104
6. Southern EM (1975) Detection of specific sequences among DNA fragments separated by gel electrophoresis. J Mol Biol 98: 503–517
7. Alwine JC, Kemp DJ, Stark GR (1977) Method for detection of specific RNAs in agarose gels by transfer to diazobenzyloxymethyl-paper and hybridization with DNA probes. Proc Natl Acad Sci U S A 74:5350–5354
8. Reiser J, Renart J, Crawford LV, Stark GR (1980) Specific association of simian virus 40 tumor antigen with simian virus 40 chromatin. J Virol 33:78–87
9. Reiser J, Renart J, Stark GR (1978) Transfer of small DNA fragments from polyacrylamide gels to diazobenzyloxymethyl-paper and detection by hybridization with DNA probes. Biochem Biophys Res Commun 85:1104–1112
10. Renart J, Reiser J, Stark GR (1979) Transfer of proteins from gels to diazobenzyloxymethyl-paper and detection with antisera: a method for studying antibody specificity and antigen structure. Proc Natl Acad Sci U S A 76:3116–3120
11. Towbin H (2009) Origins of protein blotting. Methods Mol Biol 536:1–3
12. Burnette WN (2009) Western blotting: remembrance of past things. Methods Mol Biol 536:5–8
13. Reiser J, Wardale J (1981) Immunological detection of specific proteins in total cell extracts by fractionation in gels and transfer to diazophenylthioether paper. Eur J Biochem 114:569–575
14. Olmsted JB (1981) Affinity purification of antibodies from diazotized paper blots of heterogeneous protein samples. J Biol Chem 256:11955–11957

Chapter 5

Western Blotting: An Introduction

Biji T. Kurien and R. Hal Scofield

Abstract

Western blotting is an important procedure for the immunodetection of proteins, particularly proteins that are of low abundance. This process involves the transfer of protein patterns from gel to microporous membrane. Electrophoretic as well as non-electrophoretic transfer of proteins to membranes was first described in 1979. Protein blotting has evolved greatly since the inception of this protocol, allowing protein transfer to be accomplished in a variety of ways.

Key words Western blotting, Sodium dodecyl sulfate-polyacrylamide gel electrophoresis, Nitrocellulose membrane, Polyvinylidene difluoride membrane

1 Introduction

The transfer of proteins or nucleic acids to microporous membranes is referred to as "blotting" and this term encompasses both "spotting" (manual sample deposition) and transfer from planar gels. Proteins that are resolved on sodium dodecyl sulfate-polyacrylamide gel electrophoresis (SDS-PAGE) gels are typically transferred to adsorbent membrane supports under the influence of an electric current in a procedure that is known as western blotting (WB) or protein blotting [1, 2] (*see* chapters 2 and 3). Nucleic acids are routinely transferred from agarose gels to a membrane support, through capillary action (Southern blotting). Protein blotting evolved from DNA (Southern) blotting [3] (*see* Chapter 1) and RNA (northern) blotting [4]. The term "western blotting" was coined to describe [5] this procedure to retain the "geographic" naming tradition initiated by Southern's paper [3] (*see* Chapter 3). The blotted proteins form an exact replica of the gel and have proved to be the starting step for a variety of experiments. The subsequent employment of antibody probes directed against the membrane-bound proteins (immunoblotting) has revolutionized the field of immunology

Biji T. Kurien and R. Hal Scofield (eds.), *Western Blotting: Methods and Protocols*, Methods in Molecular Biology, vol. 1312, DOI 10.1007/978-1-4939-2694-7_5,

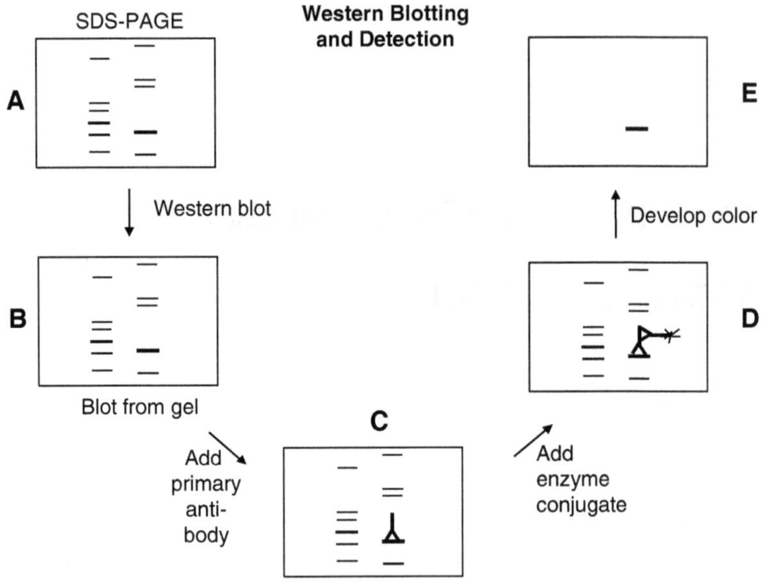

Fig. 1 Schematic representation of Western blotting and detection procedure. (*A*) Unstained SDS-PAGE gel prior to Western blot. The bands shown are hypothetical. (*B*) Exact replica of SDS-PAGE gel obtained as a blot following Western transfer. (*C*) Primary antibody binding to a specific band on the blot. (*D*) Secondary antibody conjugated to an enzyme (alkaline phosphatase or horseradish peroxidase) binding to primary antibody. (*E*) Color development of specific band (reproduced from ref. 10 with permission from Elsevier)

(Fig. 1). Dot blotting refers to the analysis of proteins applied directly to the membrane rather than after transfer from a gel.

The utility of the high resolving power of SDS-PAGE [6] was limited in purpose, owing to the fact that the separated proteins in the gel matrix were difficult to access with molecular probes, until the advent of protein blotting. Protein transfer with subsequent immunodetection has found wide application in the fields of life sciences and biochemistry. This procedure [1, 2] is a powerful tool to detect and characterize a multitude of proteins, especially those proteins that are of low abundance. It offers the following specific advantages: (a) wet membranes are pliable and are easy to handle compared to gels; (b) easy accessibility of the proteins immobilized on the membrane to different ligands; (c) only small amount of reagents are required for transfer analysis; (d) multiple replicas of a gel are possible; (e) prolonged storage of transferred patterns, prior to use, becomes possible; and (f) the same protein transfer can be used for multiple successive analyses [7–9].

Protein blotting has been evolving constantly, since its inception, and now the scientific community is faced with a multitude of ways and means of transferring proteins [10]. Nonetheless, western blot sensitivity is dependent on efficiency of blotting or transfer, retention of antigen during processing, and the final detection/amplification system used. Results are compromised if there are deficiencies at any of these steps [11].

1.1 Blotting Efficiency

The efficient transfer of proteins from a gel to a solid membrane support depends greatly on the nature of the gel, the molecular mass of the proteins being transferred, and the membrane used. Running the softest gel, in terms of acrylamide and cross-linker that yields the required resolution, is the best option. Transfer becomes more complete and faster with the use of thinner gels. However, the use of ultrathin gels may cause handling problems and a 0.4 mm thickness represents the lower practical limit [12]. Proteins with a high molecular mass blot poorly following SDS-PAGE, resulting in low levels of detection on immunoblots. However, the efficiency of transfer of such proteins has been facilitated with heat, special buffers, and partial proteolytic digestion of the proteins prior to transfer [11, 13–17].

2 Immobilizing Supports for Protein Transfer

A wide range of solid phases is available for immobilization, ranging from the truly solid such as glass and plastic to latex and cellulose that are porous. The most common phases used for blotting comprise microporous surfaces and membranes like cellulose, nitrocellulose (NC), polyvinylidene difluoride (PVDF), cellulose acetate, Polyethersulfone, and nylon. The unique properties of microporous surfaces that make them suitable for traditional assays such as "western blotting" are (a) large volume-to-surface area ratio, (b) high binding capacity, (c) short- and long-term storage of immobilized molecules, (d) ease of processing by allowing a solution phase to interact with the immobilized molecule, (e) lack of interference with the detection strategy, and (f) reproducibility. These properties are useful for the high-throughput assays used in the post-genomic era as well [2, 4, 14, 18, 19].

Typically, these microporous surfaces are used in the form of membranes or sheets with a thickness of 100 μm and possessing an average pore size that ranges from 0.05 to 10 μm in diameter. The interaction of biomolecules with each of these membranes is not completely understood, except for the fact that it is generally known to interact in a non-covalent fashion [20, 21].

Regardless of the type of membrane used, it must be borne in mind that exceeding the protein binding capacity of the membrane used tends to reduce the signal obtained in immunoblotting. Excess protein, weakly associated with the membrane, is readily accessible to react with the primary antibody or any other ligand in solution (e.g., lectin). However, the resulting antibody-protein complexes will easily wash off during further processing of the membrane. Such a scenario would not have prevailed if the protein had initially made good contact with the membrane [18].

2.1 Nitrocellulose Membranes

Nitrocellulose (NC) is perhaps the most versatile of all the surfaces mentioned earlier for the immobilization of proteins, glycoproteins, or nucleic acids [3, 4, 19]. In addition to traditional blotting, NC is used in high-throughput array, immunodiagnostic, and mass-spectrometry-coupled proteomic applications, filtration/concentration, ion exchange, and amino acid sequencing in addition to traditional blotting procedures. It was Southern who first demonstrated (in 1975) the usefulness of NC to capture nucleic acids. Towbin in 1979 [4] and Burnette in 1981 [1] showed that they could also be used for proteins.

This unique polymer derived from cellulose has been used as the most common immobilization surface in biological research for over 65 years. Since high-throughput methodologies for proteomics and genomics rely heavily on traditional concepts of molecular immobilization followed by hybridization binding or analysis, NC continues to be useful in the post-genomic era technology [19].

2.1.1 Synthesis of Nitrocellulose from Cellulose

Treatment of cellulose with nitric acid results in the hydroxyl moieties on each sugar unit of cellulose being substituted by nitrate groups, resulting in NC. Organic solvents readily dissolve dry NC resulting in the formation of a lacquer. When the solvents are evaporated the polymer is deposited as a thin film. By including a non-solvent such as water in the lacquer pores, nonsolvent can be introduced into the film to create a microporous membrane. Pore formation is a consequence of differential evaporation of the non-solvent and the solvent. Therefore pore size and porosity can be readily controlled by the amount of the non-solvent in the lacquer [2]. The pore size of 0.45 μm refers to the average effective diameter of the irregular long and tortuous channels that traverse the membrane. The pores of 0.45 μm in NC membranes account for about 80 % of the filters' volume reaching an average density of $450 \times 10^6/\text{cm}^2$ [18]. In the blotting process, the membrane needs to be porous to allow it to be saturated with buffer and will permit the required flow of current or liquid for electro- and convection blotting.

2.1.2 Immobilization Mechanism

Even though the exact mechanism by which biomolecules interact with NC is unknown, several lines of evidence suggest that the interaction is non-covalent and hydrophobic. One evidence favoring hydrophobic interaction is the fact that since most proteins at pH values above 7 are negatively charged it is surprising that NC which is also negatively charged can bind proteins efficiently. An additional fact is that nonionic detergents, like Triton X-100, are effective in removing bound antigens from NC [8].

High concentrations of salt and low concentrations of methanol increase immobilization efficiency [22]. NC is unique, when compared to other microporous membranes, in its ability to distinguish between single- and double-stranded nucleic acids, small and large proteins, short and long nucleic acids, and complexed versus uncomplexed molecules [22].

It can be stained with amido black [4], Coomassie brilliant blue (CBB) [1], aniline blue black, Ponceau S, fast green, or toluidine blue. Amido black staining can detect a 25 ng dot of bovine serum albumin readily with acceptable background staining. The background staining tends to be higher with CBB while Ponceau S gives a very clean pattern but with slightly less sensitivity than amido black.

2.1.3 Disadvantages of NC

One clear disadvantage of NC is the fact that it cannot be stripped and reprobed multiple times owing to its fragile nature. It also has a tendency to become brittle when dry. In addition, small proteins tend to move through NC membranes and only a small fraction of the total amount actually binds. Using membranes with smaller pores can obviate this [12]. Gelatin-coated NC have been used for quantitative retention [10, 23]. In supported NC (e.g., Hybond-C Extra), the mechanical strength of the membrane has been improved by incorporating a polyester support web, thereby making handling easier.

2.2 Polyvinylidene Difluoride

PVDF is a linear polymer with repeating –(CF_2–CH_2)– units. The use of "di" in polyvinylidene *di*fluoride is redundant (including its use here) and its use needs to be discouraged [2]. Polyvinylidene fluoride or polyvinylidene difluoride refer to the same membrane first made available for protein blotting by Millipore in June of 1986. The product was renamed as Immobilon-P™ Transfer Membrane after being initially referred to as Immobilon™ PVDF transfer membrane to differentiate it from other PVDF and non-PVDF-based blotting membranes referred to collectively as Immobilon family and marketed by Millipore. Immobilon-P^{SQ} membrane with a 0.2 μm pore size suitable for proteins with a molecular weight less than 20 kDa (to prevent blow through) and Immobilon-FL membrane optimized for all fluorescence applications also form part of the Immobilon family of PVDF membranes, added recently. Sequelon [24], a PVDF-based sequencing membrane, sold by Milligen/BioSearch, a Millipore subsidiary, is advantageous because of high protein binding capacity, physical strength, and chemical stability.

2.2.1 Immobilization Mechanism

Proteins transferred to the Immobilon-P membrane during western transfer are retained well on its surface throughout the immunodetection process via a combination of dipole and hydrophobic interactions. The antigen binding capacity of the membrane is 170 μg/cm^2 for bovine serum albumin and this is proportionate with the binding capacity of NC. In addition the Immobilon-P membrane has very good mechanical strength and like Teflon™ (a related fluorocarbon polymer) is compatible with a range of chemicals and organic solvents (acetonitrile, trifluoroacetic acid, hexane, ethylacetate, and trimethylamine) [2, 25].

Blotting mechanics are not different from those seen with NC, except for the fact that it is necessary to pre-wet the membrane in either methanol or ethanol before using with aqueous buffers. This is because PVDF is highly hydrophobic and there is no added surfactant in PVDF.

2.2.2 Advantages of PVDF

One of the advantages of electroblotting proteins onto PVDF membranes is that replicate lanes from a single gel can be used for various purposes such as N-terminal sequencing and proteolysis/peptide separation/internal sequencing along with western analysis. Proteins blotted to PVDF membranes can be stained with amido black, India ink, and silver nitrate [26]. These membranes are also amenable to staining with CBB, thus allowing excision of proteins for N-terminal protein sequencing, a procedure first demonstrated by Matsudaira in 1987 [25].

2.3 Activated Paper

Activated paper (diazo groups) binds proteins covalently, but is disadvantageous in that the coupling method is incompatible with many gel electrophoresis systems. Linkage is through primary amines and therefore systems that use gel buffers without free amino groups must be used with this paper. In addition, the paper is expensive and the reactive groups have a limited half-life once the paper is activated.

2.4 Nylon Membranes

Nylon-based membranes are thin and smooth surfaced as NC, but with much better durability. Two kinds of membranes are available commercially, Gene Screen and Zetabind (ZB). ZB is a nylon matrix (polyhexamethylene adipamine or Nylon 66) modified by the addition of numerous tertiary amino groups during the manufacturing process (extensive cationization). It has excellent mechanical strength and also offers the potential of very significant (yet reversible) electrostatic interactions between the membrane and polyanions. Nylon shows a greater protein binding capacity compared to NC (480 μg versus 80 μg BSA bound/cm^2). In addition, nylon offers the advantages of more consistent transfer results and a significantly increased sensitivity compared to other membranes [7, 18]. This effect is possible owing to the extra potential difference created by the positive charge of ZB.

2.4.1 Disadvantages of Nylon

The high binding capacity of these membranes however produces higher nonspecific binding. Another problem with using nylon membranes is the fact that they bind strongly to the commonly used anionic dyes like Coomassie blue, amido black 10B [18], aniline blue black, Ponceau S, fast green, or toluidine blue. SDS, dodecyl trimethylammonium bromide, or Triton X-100 at low concentrations (0.1 % in water) remove the dyes from the membrane while simultaneously destaining the transferred proteins, with SDS being the best. Destaining of this membrane is thus not possible, unlike NC, and therefore the background remains as high as the signal [8]. On account of these problems, NC membranes have remained the best compromise for most situations. However, an immunological stain and India ink have been used to detect proteins on ZB [27, 28] and NC membranes.

Nylon membranes have been found very useful in binding the negatively charged DNA, especially the positively charged ZB membranes. As a consequence it has been used more for DNA blotting than for protein blotting.

3 Buffers Used in Transfer Protocols

Commonly used buffers for western blotting are (a) Towbin system buffer (25 mM Tris, 192 mM glycine, 20 % methanol (v/v), none to 0.01 % SDS) [1] and (b) CAPS buffer system (CAPS: 10 mM 3-(cyclohexyl-amino)-1-propanesulfonic acid, 10 % methanol (v/v), pH 11) for transfer to PVDF popularized by Matsudaira [24] for use prior to in situ blot sequencing. Transfer buffers without SDS are better, in general, when using Immobilon-P, since proteins have been reported to pass through the plane of the membrane in the presence of SDS [29, 30]. However, for proteins that have a tendency to precipitate, it would be better to have SDS in the buffer (<0.01 %) during the transfer and fine-tune transfer time, current, etc. The Towbin system is used widely for applications that require immuno-development while the low-ionic-strength buffer system of Matsudaira [25] allows rapid transfer (ca. 10 min) and prevents introduction of additional Tris and glycine that is detrimental to sequence analysis using PVDF membranes.

Methanol, introduced originally by Towbin, is typically present in the transfer buffer and aids in stripping SDS from proteins transferred from denaturing SDS-containing polyacrylamide gels. It stabilizes the geometry of the gel during the transfer process, and tends to increase the binding capacity of NC for protein as well as helps proteins to bind better to NC membrane [5, 8, 18, 31]. It is possible to eliminate methanol completely from transfer buffer when using Immobilon-P membranes as well as NC. Ten to fifteen percentage methanol is suggested for general protein transfer (standard Towbin buffer used 20 % methanol). Methanol shrinks the gel and therefore when transferring high-molecular-weight proteins (>150 kDa) best results are obtained without added methanol. Non-methanolic transfer is also advised when enzyme activity needs to be preserved as well as when transferring conformation-sensitive antibodies. PAGE gels tend to swell in low-ionic-strength buffers in the absence of methanol. The "bands" may become distorted if this swelling is allowed to occur during protein transfer. Pre-swelling of the gel by incubating it in transfer buffer for 30 min to 1 h prior to transfer has been shown to obviate this problem [5, 8].

4 Settings (Current/Voltage) for Protein Transfer

Some of the issues to be considered before electrotransfer include deciding on whether to use constant voltage or constant current and the use of tank or semidry electroblotting units. The use of constant voltage provides the best driving force (that is, potential difference) during transfer [2]. The buffer composition changes as salts are eluted from the gels, resulting in an increase in current and a drop in resistance [8, 18]. However, joule heating can cause an accompanying rise in current. Ohm's law states that voltage (V) = current (I) × resistance (R). A transfer using constant voltage leads to an increase in current and a decrease in resistance while in a transfer using constant current there is decrease in voltage as well as resistance ($I = V/R$). When current reaches over 500 mA heating can be problematic in tank buffer systems and the use of cooling elements has been recommended in such a scenario. However, it has been shown recently that transfer can be efficiently carried out using heated buffer, from which methanol was omitted, to transfer high-molecular-weight proteins [17, 32]. Semidry blotters have been used to rapidly transfer proteins electrophoretically without excessive heat and using small volumes of buffer, short electrode distances, and planar electrodes that also serve as heat sinks [33].

Low-molecular-weight proteins are preferentially eluted out of the gel into the plane of the blotting membrane when a planar gel having electrophoretically resolved protein is exposed to a current perpendicular to its surface. As a result, large-molecular-weight proteins will be under-transferred under conditions optimized for transfer of low-molecular-weight polypeptides. On the other hand, a prolonged transfer will help the movement of large-molecular-weight species with accompanying loss of smaller species consequent to "blow through." A second sheet of membrane as a "backup" is useful to capture proteins that span a large molecular weight range. The use of gradient electric fields to reduce overall current use and allow the quantitative transfer of a wide range of proteins has been suggested [18]. Another approach involves a two-step electrotransfer beginning with elution of low-molecular-weight proteins at low current (1 mA/cm^2) for an hour followed by transfer at high current density (3.5–7.5 mA/cm^2) which aids the elution of high-molecular-weight proteins [34]. Recent work has shown the utility of heated buffer to transfer high-molecular-weight proteins rapidly [17, 32].

5 Techniques to Transfer Proteins from Gel to Membrane

Transfer of proteins from SDS-PAGE or native gels to nitrocellulose or PVDF membranes has been achieved by (a) simple diffusion, (b) vacuum-assisted solvent flow, and (c) "western" blotting or electrophoretic elution [4, 12, 35–39].

5.1 Simple Diffusion

Diffusion blotting was originally developed for transferring proteins separated by isoelectric focusing on thin gels to membranes and this was later expanded to other gel systems [32, 40–46]. In this procedure a membrane is placed on the gel surface with a stack of dry filter papers on top of the membrane. A glass plate and an object with a certain weight are usually placed on this assembly to enable the diffusion process. However, since there was no quantitative transfer of protein this protocol had not gained widespread acceptance. A waning interest in diffusion transfer was resuscitated when it was demonstrated that it was possible to obtain up to 12 blots from a single gel by sandwiching it between two membranes sequentially (*see* Chapter 11) [31] (Fig. 2).

Non-electrophoretic membrane lifts from SDS-PAGE gels for immunoblotting, obtained by using this method, are very useful for identification of proteins by mass spectrometry [47, 48].

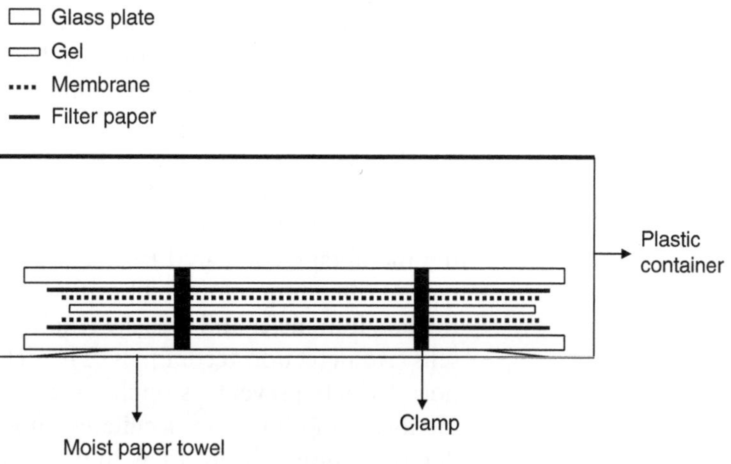

Fig. 2 Bidirectional, non-electrophoretic transfer of proteins from SDS-PAGE gels to NC membranes to obtain up to 12 blots. The PAGE gel is sandwiched between two membranes, filter paper, and glass plates and incubated at 37 °C for varying periods of time to obtain up to 12 blots (reproduced from ref. 10 with permission from Elsevier)

The gel can be stained with Coomassie following diffusion blotting. The antigens on the blot are detected by immunostaining and the immunoblotted target band can be compared with the Coomassie-stained gel by superimposing the blot and the stained gel, allowing the identification of the band to be excised for tryptic digestion and subsequent matrix-assisted laser desorption time-of-flight mass spectrometric analysis. The main advantage of diffusion blotting compared to electroblotting is that several transfers or imprints can be obtained from the same gel and different antisera can be tested on identical imprints.

Subsequently, quantitative information regarding protein transfer during diffusion blotting was obtained using ^{14}C-labeled proteins. A 3-min diffusion blotting was shown to allow a transfer of 10 % compared to electroblotting. Diffusion blotting of the same gels carried out multiple times for prolonged periods at 37 °C causes the gel to shrink. This can be overcome by using gels cast on plastic supports [44, 45].

Zymography or activity gel electrophoresis has also been studied with regard to the utility of diffusion. This involves the electrophoresis of enzymes (either nucleases or proteases) through discontinuous polyacrylamide gels containing enzyme substrate (either type III gelatin or β-casein). Following electrophoresis, SDS is removed from the gel by washing in 2.5 % Triton X-100. This allows the enzyme to renature, and the substrate to be degraded. Staining of the gel with CBB (in the case of proteins) allows the bands of enzyme activity to be detected as clear bands of lysis against a blue background [49]. An additional immunoblotting analysis using another gel is often required in this procedure to examine a particular band that is involved. Diffusion blotting has been used to circumvent the use of a second gel for this purpose [45]. The activity gel was blotted onto PVDF for immunostaining and the remaining gel after blotting was used for routine "activity staining." Since the blot and the activity staining are derived from the same gel, the signal localization in the gel and the replica can be easily aligned for comparison.

Diffusion blotting transfers 25–50 % of the [45] proteins to the membrane compared to electroblotting. However, the advantage of obtaining multiple blots from the same gel could outweigh the loss in transfer and actually could be compensated for by using sensitive detection techniques. The gel remains on its plastic support, which prevents stretching and compression; this ensures identical imprints and facilitates more reliable molecular mass determination. If only a few imprints are made, sufficient protein remains within the gel for general protein staining. These advantages make diffusion blotting the method of choice when quantitative protein transfer is not required.

5.2 Vacuum Blotting

This method was developed [50, 51] as an alternative to diffusion blotting and electroblotting. The suction power of a pump connected to a slab gel dryer system was used to drive the separated polypeptides from the gel to the nitrocellulose membrane. Both low- and high-molecular-weight proteins could be transferred using this method. Since small-molecular-weight proteins (±14,000 Da) are not well adsorbed by the 0.45 μm membrane nitrocellulose, membranes with a small pore size (0.2 or 0.1 μm) should be used when using low-molecular-weight proteins.

The gel can dry out if the procedure is carried out over 45 min and in such a scenario enough buffer should be used. In some instances low-concentration polyacrylamide gels stuck to the membrane following transfer. Rehydrating the gel helps detaching the nitrocellulose membrane from the gel remnants in such a scenario.

5.3 Electroblotting

Electroblotting is the most commonly used procedure to transfer proteins from a gel to a membrane. The main advantages are the speed and the completeness of transfer compared to diffusion or vacuum blotting. Electroelution can be achieved either by (a) complete immersion of a gel-membrane sandwich (Fig. 3) in a buffer (wet transfer) or (b) placing the gel-membrane sandwich between absorbent paper soaked in transfer buffer (semidry transfer).

The transfer conditions as such are dependent on the gel type, immobilization membrane, transfer apparatus used, as well as protein themselves. SDS gels, urea gels [4], lithium dodecyl sulfate-containing gels, nondenaturing gels, two-dimensional gels, and agarose gels have been used for protein blotting (electrophoretic) [18]. The electric charge of the protein should be determined and the membrane should be placed on the appropriate side of the gel. When using urea gels the membrane should be placed on the cathode side of the gel [4]. Proteins from SDS-PAGE gels are eluted as anions and therefore the filter should be placed on the anode side of the gel.

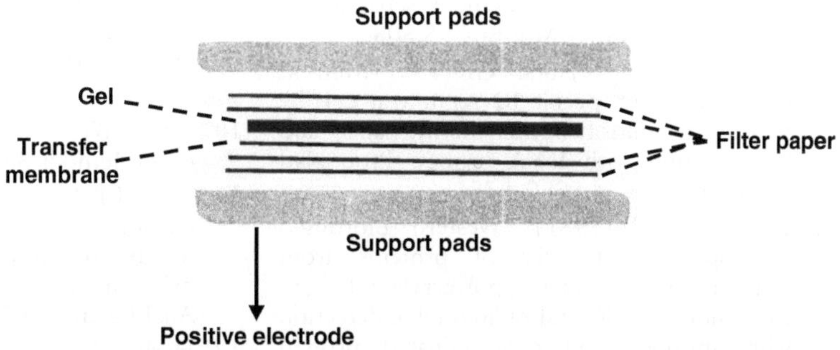

Fig. 3 The Western blot transfer assembly (reproduced from ref. 10 with permission from Elsevier)

5.3.1 Wet Transfer

In this procedure, the sandwich is placed in a buffer tank with platinum wire electrodes. A large number of different apparatuses are available to efficiently transfer proteins (or other macromolecules) transversely from gel to membrane. Most of these, however, are based on the design of Towbin et al. [1]: that is, they have vertical stainless steel/platinum electrodes in a large tank.

5.3.2 "Semidry" Transfer

In "semidry" transfer, the gel-membrane sandwich is placed between carbon plate electrodes. "Semidry" or "horizontal" blotting uses two plate electrodes (stainless steel or graphite/carbon) for uniform electrical field over a short distance, and sandwiches between these up to six gel/membrane/filter paper assemblies, all well soaked in transfer buffer. The assembly is clamped or otherwise secured on its side, and electrophoretic transfer effected in this position, using as transfer buffer only the liquid contained in the gel and filter papers or other pads in the assembly.

The advantages to this procedure over the conventional upright protocol are that(a) gels can be blotted simultaneously, (b) electrodes can be cheap carbon blocks, and (c) less power is required for transfer (and therefore a simpler power pack).

As will be seen in the following chapters, protein blotting has been evolving constantly and now the scientific community is faced with a plethora of ways and means of transferring and detecting proteins.

References

1. Towbin H, Staehelin T, Gordon J (1979) Electrophoretic transfer of proteins from polyacrylamide gels to NC sheets: procedure and applications. Proc Natl Acad Sci U S A 76:4350–4354
2. LeGendre N (1990) Immobilon-P transfer membrane: applications and utility in protein biochemical analysis. Biotechniques 9(6 Suppl):788–805, Review
3. Southern EM (1975) Detection of specific sequences among DNA fragments separated by gel electrophoresis. J Mol Biol 98:503–517
4. Alwine JC, Kemp DJ, Stark GR (1977) Method for detection of specific RNAs in agar gels by transfer to diazobenzyloxymethyl-paper and hybridization with DNA probes. Proc Natl Acad Sci U S A 74:5350–5354
5. Burnette WN (1981) "Western Blotting": electrophoretic transfer of proteins from sodium dodecyl sulfate—polyacrylamide gels to unmodified NC and radiographic detection with antibody and radioiodinated protein A. Anal Biochem 112:195–203
6. Laemmli UK (1970) Cleavage of structural proteins during assembly of the head of bacteriophage T4. Nature 227:680–685
7. Kost J, Liu L-S, Ferreira J, Langer R (1994) Enhanced protein blotting from PhastGel media to membranes by irradiation of low-intensity. Anal Biochem 216:27–32
8. Gershoni JM, Palade GE (1982) Electrophoretic transfer of proteins from sodium dodecyl sulfate-polyacrylamide gels to a positively charged membrane filter. Anal Biochem 124:396–405
9. Gershoni JM (1988) Protein blotting: a manual. Methods Biochem Anal 33:1–58, Review
10. Kurien BT, Scofield RH (2006) Western blotting. Methods 38:283–293
11. Karey KP, Sirbasku DA (1989) Glutaraldehyde fixation increases retention of low molecular weight proteins (growth factors) transferred to nylon membranes for Western blot analysis. Anal Biochem 178:255–259
12. Harlow E, Lane D (1988) Immunoblotting. In: Harlow E, Lane D (eds) Antibodies.

A laboratory manual. Cold Spring Harbor Laboratory, NY, p 485
13. Renart J, Reiser J, Stark GR (1979) Transfer of proteins from gels to diazobenzyloxymethyl paper and detection with anti-sera: a method for studying antibody specificity and antigen structure. Proc Natl Acad Sci U S A 76: 3116–3120
14. Elkon KB, Jankowski PW, Chu JL (1984) Blotting intact immunoglobulins and other high-molecular-weight proteins after composite agarose-polyacrylamide gel electrophoresis. Anal Biochem 140:208–213
15. Gibson W (1981) Protease-facilitated transfer of high-molecular-weight proteins during electrotransfer to NC. Anal Biochem 118:1–3
16. Bolt MW, Mahoney PA (1997) High efficiency blotting of proteins of diverse sizes following sodium dodecyl sulfate-polyacrylamide gel electrophoresis. Anal Biochem 247:185–192
17. Kurien BT, Scofield RH (2002) Heat mediated, ultra-rapid electrophoretic transfer of high and low molecular weight proteins to NC membranes. J Immunol Methods 266:127–133
18. Gershoni JM, Palade GE (1983) Protein blotting: principles and applications. Anal Biochem 131:1–15
19. Thornton DJ, Carlstedt I, Sheehan JK (1996) Identification of glycoproteins on nitrocellulose membranes and gels. Mol Biotechnol 5:171–176
20. Tonkinson JL, Stillman B (2002) NC: a tried and true polymer finds utility as a post-genomic substrate. Front Biosci 7:c1–c12, Review
21. Lauritzen E, Masson M, Rubin I, Bjerrum OJ, Holm A (1993) Peptide dot immunoassay and immunoblotting: electroblotting from aluminum thin-layer chromatography plates and isoelectric focusing gels to activated NC. Electrophoresis 14:852–859
22. Masson M, Lauritzen E, Holm A (1993) Chemical activation of NC membranes for peptide antigen-antibody binding studies: direct substitution of the nitrate group with diaminoalkane. Electrophoresis 14:860–865
23. Too CK, Murphy PR, Croll RP (1994) Western blotting of formaldehyde-fixed neuropeptides as small as 400 daltons on gelatin-coated NC paper. Anal Biochem 219:341–348
24. Coull JM, Dixon JD, Laursen RA, Koester H, Pappin DJC (1989) Development of membrane supports for the solid-phase sequence analysis of proteins and peptides. In: Witmann-Liebold B (ed) Methods in protein sequence analysis. Springer, Berlin, pp 69–78
25. Matsudaira P (1987) Sequence from picomole quantities of proteins electroblotted onto polyvinylidene difluoride membranes. J Biol Chem 262:10035–10038
26. Pluskal MF, Przekop MB, Kavonian MR, Vecoli C, Hick DA (1986) Biotechniques 4:272–282
27. Kittler JM, Meisler NT, Viceps-Madore D, Cidlowski JA, Thanassi JW (1984) A general immunochemical method for detecting proteins on blots. Anal Biochem 137:210–216
28. Hughes JH, Mack K, Hamparian VV (1988) India ink staining of proteins on nylon and hydrophobic membranes. Anal Biochem 173:18–25
29. Tovey ER, Baldo BA (1989) Protein binding to NC, nylon and PVDF membranes in immunoassays and electroblotting. J Biochem Biophys Methods 19:169–183
30. Xu QY, Shively JE (1988) Microsequence analysis of peptides and proteins. VIII. Improved electroblotting of proteins onto membranes and derivatized glass-fiber sheets. Anal Biochem 170:19–30
31. Nielsen PJ (1982) The phosphorylation of ribosomal protein S6 in rat tissues following cycloheximide injection, in diabetes, and after denervation of diaphragm. A simple immunological determination of the extent of S6 phosphorylation on protein blots. J Biol Chem 257:12316–12321
32. Kurien BT, Scofield RH (1997) Multiple immunoblots after non-electrophoretic bidirectional transfer of a single SDS-PAGE gel with multiple antigens. J Immunol Methods 205:91–94
33. Kyhse-Andersen J (1984) Electroblotting of multiple gels: a simple apparatus without buffer tank for rapid transfer of proteins from polyacrylamide to nitrocellulose. J Biochem Biophys Methods 10:203–209
34. Otter T, King SM, Witman GB (1987) A two-step procedure for efficient electro transfer of both high-molecular weight (greater than 400,000) and low-molecular weight (less than 20,000) proteins. Anal Biochem 162:370–377
35. Harper DR, Kit ML, Kangro HO (1990) Protein blotting: ten years on. J Virol Methods 30:25–39, Review
36. Egger D, Bienz K (1994) Protein (western) blotting. Mol Biotechnol 1:289–305
37. Wisdom GB (1994) Protein blotting. Methods Mol Biol 32:207–213
38. Kurien BT, Scofield RH (2003) Protein blotting: a review. J Immunol Methods 274:1–15, Review

39. Kurien BT, Scofield RH (2005) Blotting techniques. In: Worsfold PJ, Townshend A, Poole CF (eds) Encyclopedia of analytical science, 2nd edn. Elsevier, Oxford, p 425
40. Reinhart MP, Malamud D (1982) Protein transfer from isoelectric focusing gels: the native blot. Anal Biochem 123:229–235
41. Jagersten C, Edstrom A, Olsson B, Jacobson G (1988) Blotting from PhastGel media after horizontal sodium dodecyl sulfate-polyacrylamide gel electrophoresis. Electrophoresis 9:662–665
42. Kazemi M, Finkelstein RA (1990) Checkerboard immunoblotting (CBIB): an efficient, rapid, and sensitive method of assaying multiple antigen/antibody cross-reactivities. J Immunol Methods 128:143–146
43. Heukeshoven J, Dernick R (1995) Effective blotting of ultrathin polyacrylamide gels anchored to a solid matrix. Electrophoresis 16:748–756
44. Olsen I, Wiker HG (1998) Diffusion blotting for rapid production of multiple identical imprints from sodium dodecyl sulfate polyacrylamide gel electrophoresis on a solid support. J Immunol Methods 220:77–84
45. Chen H, Chang GD (2001) Simultaneous immunoblotting analysis with activity gel electrophoresis in a single polyacrylamide gel. Electrophoresis 22:1894–1899
46. Bowen B, Steinberg J, Laemmli UK, Weintraub H (1980) The detection of DNA-binding proteins by protein blotting. Nucleic Acids Res 8:1–20
47. Kurien BT, Scofield RH (2000) Association of neutropenia in systemic lupus erythematosus with anti-Ro and binding of an immunologically cross-reactive neutrophil membrane antigen. Clin Exp Immunol 120:209–217
48. Kurien BT, Matsumoto H, Scofield RH (2001) Purification of tryptic peptides for mass spectrometry using polyvinylidene fluoride membrane. Indian J Biochem Biophys 38:274–276
49. Bischoff KM, Shi L, Kennelly PJ (1998) The detection of enzyme activity following sodium dodecyl sulfate-polyacrylamide gel electrophoresis. Anal Biochem 260:1–17, Review
50. Peferoen M, Huybrechts R, De Loof A (1982) Vacuum-blotting: a new simple and efficient transfer of proteins from sodium dodecyl sulfate-polyacrylamide gels to NC. FEBS Lett 145:369–372
51. Garribba F, Turi S, Corpetti G, Khiabani A, la Torre XD, Botrè F (2014) A modified procedure based on a vacuum-driven blotting system for the detection of erythropoietin and its analogs. Bioanalysis 6:1605–1615

Chapter 6

From Little Helpers to Automation

Biji T. Kurien and Harry Towbin

Abstract

Western blot technology has continually evolved to enhance sensitivity, speed, and ease of operation. For enhancing awareness to these developments, this brief review outlines a representative selection of methods and devices, many of which are commercial products. In particular, the steps taken towards partial and full automation of western blotting are addressed.

Key words Western blot, Automated western, Microwestern, Capillary electrophoresis, Diagnostic western blot

1 Introduction

Whoever engages in a western blot experiment will become aware of the time it takes to complete the procedure. One will also realize that it demands a little practice and dexterity to handle wobbly gels and to push out air bubbles. In many laboratories, especially at universities, most of the steps are often still done manually, starting with preparation of buffers and casting gels and ending with a densitometric analysis in front of the computer screen. When a laboratory engages in longer term studies that require western blotting, operators will be glad to reach for gadgets that make work easier. In this chapter, we shall focus on items pertinent to blotting while only cursorily referring to those facilitating gel electrophoresis. In recent years, innovative devices have reached the market, which promise to ease the workload of protein blotting or even allow complete automation. One can expect significant changes in these technologies in the years to follow—this chapter is but a snapshot of these ongoing trends. Rather than trying to comprehensively cover all technologies, we hope that interested readers will be able to spot relevant literature and company websites from keywords in the text and from Table 1.

Biji T. Kurien and R. Hal Scofield (eds.), *Western Blotting: Methods and Protocols*, Methods in Molecular Biology, vol. 1312, DOI 10.1007/978-1-4939-2694-7_6, © Springer Science+Business Media New York 2015

Table 1
Western blotting tools

Description	Points to consider/keywords/suppliers
Precut membranes and filter papers	Filter paper sandwiches, precut blotting membrane, PVDF, preassembled nitrocellulose membranes/many suppliers
Buffer-soaked membranes and filter stacks, devices for fast blotting	Transfer stack, transfer pads, fast transfer blotting/Trans-Blot® Turbo™ (BioRad Laboratories, Hercules, CA, USA), iBlot® (Invitrogen, Carlsbad, CA, USA), eBlot® (GenScript, Piscataway, NJ, USA), G2 Fast blotter (Pierce, Rockford, IL)
Total protein stain for normalization	Total protein stain, reversible protein stain/Stain-Free™ precast gels (Biorad), many suppliers
Automated western blot processors	Recovery of solutions, number of membranes processed simultaneously, different antibody solutions used simultaneously/automated western blot processing instruments/many suppliers
Reagent buffer solutions that enhance reactivity of blotted antigens	Antibody Extender, Signal Enhancer (Pierce), Signalboost (EMD Millipore), MaxBlot (MBL Medical and Biological Laboratories Co. Ltd)
Processing of diagnostic western blot strips	Strip, immunoblot, line blot, serology, diagnostic, western blot processor/many suppliers
Software for automated evaluation of diagnostic strips	BLOTrix Software B4C (BioScitec GmbH), Virascan, recomScan (Mikrogen GmbH)
Software for western blot analysis	Automatic band and lane recognition, correction of distortions, background subtraction, signal saturation, molecular weight determination, relative and absolute quantification/many suppliers

Appropriately combined terms may be helpful for finding similar products or improvements when used as terms in general search engines or at specialized websites (e.g., www.biocompare.com)

1.1 Cutting Time on Sandwich Preparation

Although it does not look like much work to cut membranes and filter paper by oneself, it requires some discipline to prepare them in advance. Hence, taking precut membranes and filter papers out of the box are convenient. Many suppliers offer such packages. They can be used for both tank and semidry blotting. Ideally, membrane dimensions should match the size of the gel cassettes in use. Of course, ready-made membrane-filter paper sandwiches come at a higher price than membranes on rolls or large sheets.

Most vendors of blotting devices emphasize the gain of speed and efficiency of protein transfer to membranes achievable with their products. Several measures contribute to faster transfer, mostly in equipment with a horizontal setup for electrophoretic transfer. Firstly, pre-soaked membranes do away with methanol or ethanol conditioning of PVDF membranes; filter papers soaked with buffer ease handling further. Secondly, high-ionic-strength buffer formulations generate a strong electric field across the gel while the distance between the electrodes is also kept small, similar to a principle described earlier [1]. Some systems (eBlot® and iBlot®) integrate these features in dedicated devices that combine power supply and transfer unit.

1.2 Washers for Speedy Staining

With the proteins faithfully translocated to the membrane, quite a few incubation steps with blocking buffers, antibody solutions, and washings need to be done. An early step towards developing western blots automatically has used the phastsystem [2]. This apparatus contains a Coomassie blue staining unit that the authors transmuted into a blot processor. Several companies have engineered instruments dedicated to completing the time-consuming development steps automatically. Depending on the needs for varying protocols and numbers of membranes to be processed, researchers can choose a suitable apparatus. Some instruments are able to recover fluids, which helps saving solutions of precious antibodies. While most systems use pumps and rockers to move liquids around, a device called iBind™ (Life Technologies Corp.) works with sequential lateral flow. In this non-electric piece of equipment, the blot is placed on a porous support. The antibody and wash solutions, ingeniously designed to be released sequentially, flow across the membrane and are drawn into a stack of absorbent material at the opposite side [3].

How, in general, is it possible to reduce the time needed for incubation steps? Among the factors that obviously contribute are temperature and concentration of antibodies. Also, stirring to avoid transient depletion at binding sites and diffusion into the pores of the membranes are important. It is less obvious that even the intrinsic kinetic properties of an antibody-antigen reaction (on and off rates) may be changed by microwave irradiation [4]. Owing to their relatively high thickness in the range of 100 μm and tortuous pores, diffusion of molecules into and out of the membrane may take time. It is therefore intuitively useful to speed up these processes by inducing a bulk flow through the membrane. This principle is realized in the manually controlled SNAP i.d.® (Millipore) setup. Here, solutions are drawn through the membrane by vacuum and, in particular, allow rapid washing. The ONE-HOUR Western™ kit (GenScript, Piscataway, NJ) reduces the number of incubation and washing steps by allowing the primary antibody to form a soluble complex with a labeled secondary reagent, which may be peroxidase conjugated or fluorescent. Lastly, for increasing sensitivity of detection and saving on antibody consumption, reagents are available that rely on briefly treating membranes after protein transfer with one or two solutions. The underlying principle has remained undisclosed.

1.3 Processing Diagnostic Western Blot Strips

In a clinical environment, automation is particularly relevant. The strips used are often secondary tests for diagnosing a disease, such as a bacterial or viral infection and autoimmune diseases. The strips consist of separated antigens, either resolved by electrophoresis or deposited as lines of purified proteins. They are used in serology with the goal of determining whether a patient's serum contains an antibody pattern characteristic for a disease. Thus, strip tests are still widely used as confirmatory tests in the diagnosis of Lyme disease (borreliosis) [5], of syphilis [6, 7], and autoimmunity [8]. Often it is important to differentiate between IgM and IgG class antibodies and two assays are run in parallel. Most instruments can process around 40 strips per session. For the visualization step of commercial strip assays, traditional color development reagents are usually employed. Benefits we expect from automation are standardized incubation conditions, including temperature, time, volume, and also improved safety when working with potentially infectious samples. Important additional features of processors are software control for various protocols, automated interpretation, and transmission of the data to a laboratory information management system (LIMS). Advanced image processing software allows shape recognition of the strips in their incubation trays, followed by virtual alignment and automated interpretation of the staining pattern. Binnicker et al. [9] compared two commercial automated interpretative systems. The authors concluded that both systems produced results that agreed well with visual interpretation.

1.4 Software and Strategies for Hard Western Data

In the course of a western blotting experiment, it is comforting to make sure that the gel has run properly and that the transfer has been completed successfully. For some time already, prestained molecular weight markers have proven helpful to this end. Now, control at every step is possible with gels that turn proteins fluorescent. This feat is achieved after electrophoresis in polyacrylamide gels that contain a trihalocompound (Stain-Free™ precast gels, Bio-rad). A UV-light-induced photochemical reaction renders tryptophan-containing proteins fluorescent [10]. At this stage, the quality of separation can be assessed in the gel, and, importantly, also after transfer to the membrane [11]. Since the fluorescence intensity integrated along a lane is a direct measure of the total protein load, normalization for western blot quantification is possible [12]. The benefits of normalizing to total protein and thus replacing separate determinations of housekeeping proteins, which are at times not as stable as their name implies, have recently been discussed [13].

A common goal of a western blotting experiment is determining molecular weights of protein antigens. Since the separation step in western blotting is identical to conventional SDS-gel electrophoresis, reporting molecular weights of antigens is mostly unproblematic and is a task that appropriate image analysis software can do automatically on the basis of molecular weight standards on the membrane. Quantifying band intensities is more demanding. Unavoidably, the more steps involved the larger the errors that accrue. Image processing software helps to objectively determine band intensities and is very useful if normalization to internal standards, be these housekeeping or total protein, is intended. Besides the popular ImageJ software [14], which is freely available, many programs for western blot quantification have been developed. Often they are provided by vendors of imaging equipment and come with varying degrees of sophistication. Some of them offer automatic recognition of bands and lanes, allow for correction of distortions of the gel, and perform background subtraction. Many software implementations also detect saturated areas and are capable of inferring molecular weights from lanes of molecular weight standards. The overall goal is to report integrated density values, and the "volumes" of a band, for relative or absolute quantification. In spite of all the electronic features available, it is useful to be aware of potential limitations of the quantification process that arise from factors such as limited linearity of the signal-to-protein concentration, uneven background, or even vignetting in CCD cameras [15, 16]. Because the number of slots on a gel is quite limited, extensive dilutions for constructing standard curves or replicates, which are customary in microtiter plate assays, are usually not done. Lastly, for heavy users of western blotting, electronic gel documentation systems help archiving and retrieving images and annotations.

1.5 Micro and Automated Westerns

A lot of thought has been devoted to miniaturization, multiplexing, and finally full automation of western blotting. An example of a multiplexing manual system, termed microfluidic western blot, is described by Pan et al. [17]. It resolves proteins by conventional blotting after SDS-PAGE. The trick lies in gaskets that divide up each protein lane into as many as ten narrow parallel incubation slots. The gaskets providing the channels are crafted by soft lithography and are pressed against the PVDF membrane. With a mere 150 μm width and a length of 3.5 cm, it takes only 1 μl to fill a channel. After incubation, the gasket with the "microfluidic network" is lifted off to allow washing and access to fluorescent secondary antibodies. Likely due to dilution of antibodies into the membrane or depletion around reacting sites, a higher antibody concentration is needed than for normal-sized blots; the saving of precious primary antibody is still by a factor of about 100. An equivalent technology with increased sensitivity attributed to a chemiluminescent readout has also been devised [18].

Similar to the progress achieved by automation in DNA sequencing, remarkable steps towards miniaturization and full automation have been made by carrying out the separation step in capillary systems. These can be classical capillary electrophoresis (CE) or scaled-down microchip versions. A number of approaches have been taken—some retaining the deposition on membranes, and others culminating in full integration and avoiding membranes altogether.

Western blotting without the need of gels has been realized with capillary electrophoresis for both conventional CE [19] and microchip [20] electrophoresis. The capillary is filled with a semi-liquid medium of entangled polymers that are well known to provide SDS-PAGE-like separations. The tip of the capillary touches a moving membrane on an X–Y translational stage and as the eluting proteins reach the end of the tubing, they are immediately immobilized on a PVDF membrane. A sheath flow of buffer at the exit of the capillary ensures firm electrical contact during this step. This system thus achieves immobilization without the transfer step characteristic of conventional western blotting. Taking only 2 min on a 2 cm capillary length for the chip version, separations are fast. It is unproblematic to deposit the trace of many separations on one membrane in any suitable pattern. The membrane is now ready for incubation steps with primary and secondary fluorescent antibodies for which the authors chose a rapid vacuum-assisted technique.

Complete automation of western blotting has been accomplished with a range of instruments recently developed by Protein Simple (cf. Fig. 1a) (*see* Chapter 47). Here, the capillary itself serves as protein-immobilizing support. After the electrophoretic run, which requires only small volumes of sample, the entire capillary is irradiated with UV light to induce a photochemical reaction that covalently ties the resolved proteins to the capillary wall. Immobilization is possible because the inner surface is pre-coated with a benzophenone derivative [21], a moiety known for photoreactions with proteins. As the sieving gel is semifluid, it can be expelled under pressure. All developing antibody and wash solutions can now be pushed through the capillary without dislodging the covalently bound proteins. Detection is by sensitive enzymatic chemiluminescence combined with a CCD camera for recording

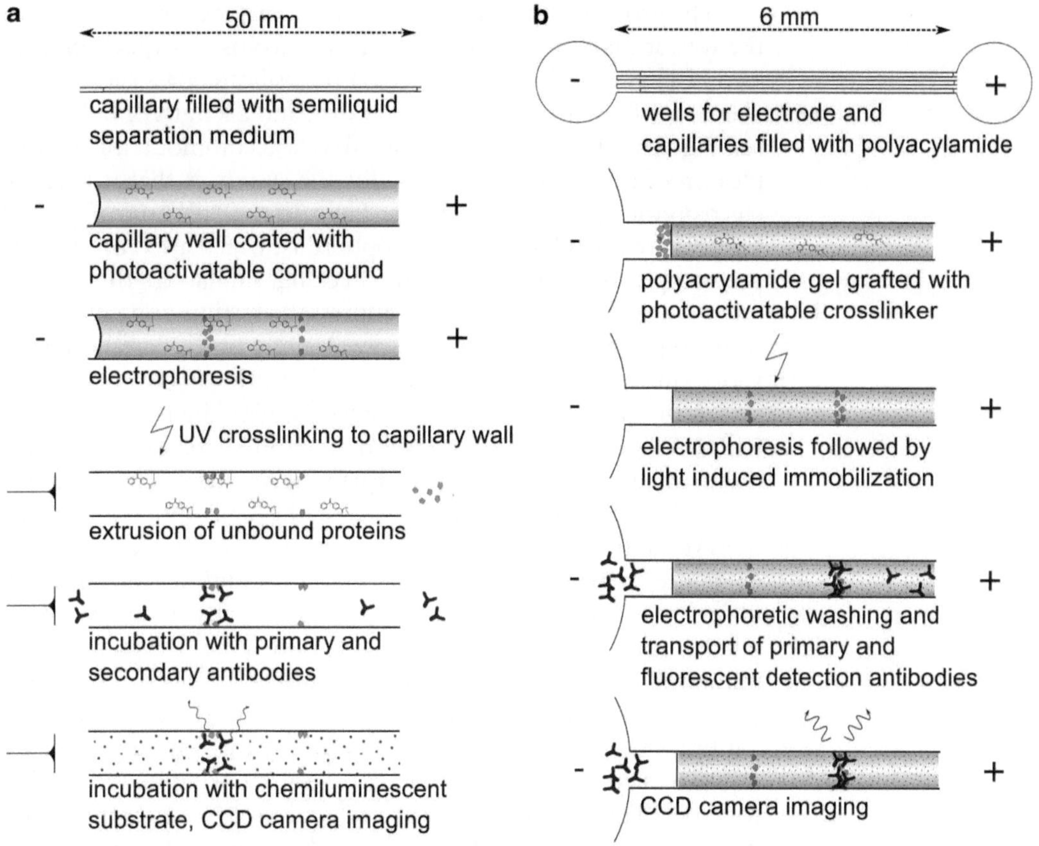

Fig. 1 Schematics of two automated western capillary electrophoretic systems. (**a**) Commercial automated western system [21]. Separation and detection by antibodies take place in individually manipulated capillaries; the device processes up to 96 capillaries in an unattended mode. (**b**) Microfluidic system on microscope slide-sized chips according to Hughes and Herr, 2012 [24]; the chip accommodates 48 separation channels per slide (shown on top is one unit, each containing three replicate channels, for the close-ups, only one channel is shown). Both devices have been developed for isoelectric focusing and sizing applications

the light emitted from the capillary. The instrument has been first described for isoelectric focusing applications [21], which are ideally suited to assess changes in phosphorylation or other charge-modifying modifications on proteins such as sialylation and acetylation. The system has later been extended to a sieving-type separation resulting in SDS-gel-like profiles. Each of the 12 capillaries, which are run in parallel, contains fluorescent molecular weight markers that assure consistent running conditions. Up to 96 samples can be processed unattendedly. Besides the obvious benefit of complete automation the question of reliable quantification is of special interest and has been addressed (e.g., [22, 23]). Reproducibility was always found to be superior to conventional manual western blotting.

The group of Amy Herr has devoted a major effort in rebuilding the western blot at microscales that also avoids an explicit blotting step [24]. Separations take place on borosilicate glass slides in tiny channels of 10 μm depth, 70 μm width, and a length of only 6 mm (cf. Fig. 1b). This gives room for a device accommodating 48 samples on one slide. Characteristic for the system is that it relies on electrophoresis at multiple steps. Thus, in the "μWestern" protein samples are introduced by electrophoresis, which also concentrates the sample by isotachophoretic stacking similar to the original Laemmli gel technique. The compressed starting zone allows for short separation distances (below 3 mm) and correspondingly short run times. The separation gel is made of benzophenone-derivatized polyacrylamide. Immediately after the run, irradiation with UV light, as in the application mentioned previously, captures the proteins in situ with nearly quantitative yield. This high efficiency is likely due to the density of the benzophenone (about 3 mM) and the short distances a protein needs to diffuse to encounter a reactive site in the gel. In addition, the inventors observed that the denatured state of the proteins enhanced immobilization. The washing and incubation steps of conventional immunoblotting on membranes are replaced by electrophoresis of the antibody reagents, including fluorescence-tagged secondary antibodies, along the capillary. Following this staining step, the capillary is imaged by a CCD camera in several sections and stitched together. The authors emphasize as the four key achievements the high immobilization efficiency, speed of the entire procedure, economy with respect to reagent consumption, and lastly small footprint of the device.

Immunoprecipitation of radiolabeled proteins is the traditional method complementary to western blotting that informs about quantity and molecular weights of proteins present in complex mixtures. Since the immunoprecipitation purifies a protein it becomes amenable to further biochemical analysis. Of special interest is probing for posttranslational modifications or interacting protein or DNA and RNA components. For the less demanding task of molecular weight determination and quantification, immunoprecipitation has been reintroduced as an alternative to western blotting [25]. In this method, radioactive labeling of the total mixture of proteins is replaced by fluorescence tagging and conventional electrophoresis is substituted by a chip equivalent (an Agilent 2100 Bioanalyzer). A quite similar technique is the "SWAP" methodology of Caliper Life Sciences offered as a "Microfluidic Alternative to Western Blotting" (*see* Chapter 48). In a nutshell, both systems consist of an immunoprecipitation of fluorescence-labeled proteins. The workflow calls for a labeling step with an amine-reactive fluorescent dye carried out at elevated temperature. Protein A beads pre-loaded with antibody then capture the protein of interest directly from the labeling mixture. All of these steps are carried out in microtiter plates, including appropriate washing and elution. Finally, microfluidic electrophoresis reveals the trace of released proteins, which might, for example, consist of a single peak. In order to see the profile of all the proteins in the sample, one simply analyzes the labeling mixture before bead addition. A further related methodology, mostly intended for applications in clinical chemistry, is immunoaffinity capillary electrophoresis (reviewed in [26]). The potential of such methods has been demonstrated in a study of cytokines extracted from microdissected tissue [27]. On this microfluidic chip, affinity purification, fluorescence tagging, elution, and electrophoretic separation were all integrated in a single unit.

2 Conclusion

The methods mentioned here and elsewhere in this volume show us that there are now many options to complete western experiments faster and make them more reliable. With increasing automation and other creative solutions to combine separation and immunodetection, the days of cutting membranes and reassembling broken gels may soon be a thing of the past.

References

1. Dalton RG, Lasham A, Savidge GF (1988) A new rapid semi-dry blotting technique for multimeric sizing of von Willebrand factor. Thromb Res 50:345–349
2. Prieur B, Russo-Marie F (1988) An automated western blot analysis using the phastsystem. Anal Biochem 172:338–343
3. Semenov E, Gargir A (2013) Sequential lateral flow capillary device for analyte determination. US Patent US20130164193 A1
4. Liu Y-T, Toyokuni S (2009) Intermittent microwave irradiation facilitates antigen-antibody reaction in Western blot analysis. Methods Mol Biol 536:307–312
5. Aguero-Rosenfeld ME, Wang G, Schwartz I, Wormser GP (2005) Diagnosis of Lyme borreliosis. Clin Microbiol Rev 18:484–509
6. Hagedorn H-J, Kraminer-Hagedorn A, De Bosschere K, Hulstaert F, Pottel H, Zrein M (2002) Evaluation of INNO-LIA syphilis assay as a confirmatory test for syphilis. J Clin Microbiol 40:973–978
7. Lam TK, Lau HY, Lee YP, Fung SM, Leung WL, Kam KM (2010) Comparative evaluation of the INNO-LIA syphilis score and the MarDx Treponema pallidum immunoglobulin G Marblot test assays for the serological diagnosis of syphilis. Int J STD AIDS 21:110–113
8. Mietzner B, Tsuiji M, Scheid J, Velinzon K, Tiller T, Abraham K et al (2008) Autoreactive IgG memory antibodies in patients with systemic lupus erythematosus arise from nonreactive and polyreactive precursors. Proc Natl Acad Sci U S A 105:9727–9732
9. Binnicker MJ, Jespersen DJ, Harring JA, Rollins LO, Bryant SC, Beito EM (2008) Evaluation of two commercial systems for automated processing, reading, and interpretation of Lyme borreliosis Western blots. J Clin Microbiol 46:2216–2221
10. Ladner CL, Turner RJ, Edwards RA (2007) Development of indole chemistry to label tryptophan residues in protein for determination of tryptophan surface accessibility. Protein Sci 16:1204–1213
11. Colella AD, Chegenii N, Tea MN, Gibbins IL, Williams KA, Chataway TK (2012) Comparison of Stain-Free gels with traditional immunoblot loading control methodology. Anal Biochem 430:108–110
12. Gürtler A, Kunz N, Gomolka M, Hornhardt S, Friedl AA, McDonald K et al (2013) Stain-Free technology as a normalization tool in western blot analysis. Anal Biochem 433:105–111
13. Li R, Shen Y (2013) An old method facing a new challenge: re-visiting housekeeping proteins

as internal reference control for neuroscience research. Life Sci 92:747–751

14. Rasband W (2014) ImageJ, 1997–2014, Bethesda, MD: U.S. National Institutes of Health. Available at: imagej.nih.gov/ij/
15. Heidebrecht F, Heidebrecht A, Schulz I, Behrens S-E, Bader A (2009) Improved semi-quantitative Western blot technique with increased quantification range. J Immunol Methods 345:40–48
16. Charette SJ, Lambert H, Nadeau PJ, Landry J (2010) Protein quantification by chemiluminescent Western blotting: elimination of the antibody factor by dilution series and calibration curve. J Immunol Methods 353:148–150
17. Pan W, Chen W, Jiang X (2010) Microfluidic Western blot. Anal Chem 82:3974–3976
18. Chang H-N, Leroueil PR, Selwa K, Gasper CJ, Tsuchida RE, Wang JJ et al (2013) Profiling inflammatory responses with microfluidic immunoblotting. PLoS One 8:e81889
19. Anderson GJ, Cipolla CM, Kennedy RT (2011) Western blotting using capillary electrophoresis. Anal Chem 83:1350–1355
20. Jin S, Anderson GJ, Kennedy RT (2013) Western blotting using microchip electrophoresis interfaced to a protein capture membrane. Anal Chem 85:6073–6079
21. O'Neill RA, Bhamidipati A, Bi X, Deb-Basu D, Cahill L, Ferrante J et al (2006) Isoelectric focusing technology quantifies protein signaling in 25 cells. Proc Natl Acad Sci U S A 103: 16153–16158
22. Rustandi RR, Loughney JW, Hamm M, Hamm C, Lancaster C, Mach A et al (2012) Qualitative and quantitative evaluation of Simon™, a new CE-based automated western blot system as applied to vaccine development. Electrophoresis 33:2790–2797
23. Chen J-Q, Heldman MR, Herrmann MA, Kedei N, Woo W, Blumberg PM et al (2013) Absolute quantitation of endogenous proteins with precision and accuracy using a capillary Western system. Anal Biochem 442:97–103
24. Hughes AJ, Herr AE (2012) Microfluidic western blotting. Proc Natl Acad Sci U S A 109:21450–21455
25. Wenz C, Rüfer A (2009) Microchip CGE linked to immunoprecipitation as an alternative to western blotting. Electrophoresis 30: 4264–4269
26. Guzman NA, Phillips TM (2011) Immunoaffinity capillary electrophoresis: a new versatile tool for determining protein biomarkers in inflammatory processes. Electrophoresis 32: 1565–1578
27. Phillips TM, Wellner EF (2007) Analysis of inflammatory biomarkers from tissue biopsies by chip-based immunoaffinity CE. Electrophoresis 28:3041–3048

Chapter 7

Spectrophotometric Methods to Determine Protein Concentration

J.P. Dean Goldring

Abstract

Measuring the concentration of proteins is an essential part of enzyme evaluations or to monitor protein yields during protein isolation procedures. Decisions on the usefulness of any isolation procedure depend on knowing the relative concentrations of a particular protein or enzyme in relation to the concentrations of all the proteins present. Protein concentration in solution is generally measured with spectrophotometry in the UV range or in the presence of dyes or copper interacting with the protein. This review describes protein determination methods for measuring protein concentration in solution.

Key words Protein staining, Coomassie, Bicinchoninic acid, Lowry, Absorbance, Bradford, Protein concentration

1 Introduction

During protein isolation procedures from plant, animal, insect, yeast, or bacterial material it is important to determine the concentration of protein at each step in the purification procedure. Measuring protein concentration is also essential for enzyme activity assays. Protein quantification methods detect particular amino acids based on absorbance by aromatic ring structures or interaction of dyes or copper with charged amino acid residues or peptide bonds. The choice of method depends on the amount of protein available, detection limit of the method, ease of use, time required to complete the method, and whether one can afford to lose some protein for the measurement. The methods described here are limited to those that detect 10–50 μg protein per sample.

1.1 Detecting Proteins with Absorbance at 280 nm

Proteins comprising aromatic rings in their primary sequence absorb light at 280 nm. The absorbance at 280 nm is primarily due to the presence of the amino acids tryptophan (λ max 279.8 nm) and tyrosine (λ max 274.6 nm) which have extinction coefficients, ε, of 5.6 M^{-1} cm^{-1} and 1.42 M^{-1} cm^{-1}, respectively.

Biji T. Kurien and R. Hal Scofield (eds.), *Western Blotting: Methods and Protocols*, Methods in Molecular Biology, vol. 1312, DOI 10.1007/978-1-4939-2694-7_7, © Springer Science+Business Media New York 2015

Phenylalanine (λ max 257 nm, ε 0.197 M^{-1} cm^{-1}) makes a minor contribution [1]. The absorbance measured is directly proportional to the concentration of the protein solution and the pathlength, i.e., follows the Beer-Lambert law. Since the method is based on absorbance by tryptophan and tyrosine residues in a protein, the absorbance values will differ in relation to the number of tryptophan and tyrosine residues present (*see* **Note 1**). The method can be used to detect protein in the 20–3,000 μg range. The method is particularly useful for the detection of proteins eluting from gel filtration, ion-exchange, affinity, hydrophobic, and chromatofocusing chromatography columns as there is no loss of protein (*see* **Note 2**). Nucleic acids are often present in protein solutions and contribute to absorbance values at 280 nm. A compensation for the presence of nucleic acids should be made (*see* **Note 3**).

1.2 Detecting Proteins with the Lowry Method

The biuret reaction detects a cupric complex formed between copper and four nitrogen atoms of peptides. Lowry in 1951 modified the biuret reaction producing a more sensitive method of protein detection. Lowry [2] added phosphomolybdic/phosphotungstic acid (Folin-Ciocalteu reagent) which stains the protein with a blue-green color after interaction with the cuprous ions and the side chains of tryptophan, tyrosine, and cysteine residues. The blue-green color can be detected between 650 and 750 nm [3]. The protein detection range is 2–100 μg. For some 25 years the Lowry assay was the dominant method to quantify proteins until being replaced with the Bradford (Coomassie G-250) assay and the introduction of bicinchoninic acid to the biuret reaction by Smith [4, 5]. The Lowry assay has a number of steps and many common laboratory reagents interfere with the method (*see* **Notes 1, 4, 5, 10–13**).

1.3 Detecting Proteins with the Smith Bicinchoninic Acid Method

The biuret reaction is also the basis of the bicinchoninic acid assay (BCA) developed in 1985 by Smith [5]. The cuprous complex from the biuret reaction reacts with the sodium salt of bicinchoninic acid to produce a deep blue color with an absorbance detected at 562 nm. The method detects proteins in the range of 0.2–50 μg [5]. The BCA reagent is stable under alkali conditions and so can be included in the biuret alkaline copper solution. The assay can be run at room temperature or sensitivity can be increased at 60 °C. The assay is compatible with detergents, giving it an advantage over both the Lowry and Coomassie dye assays (*see* **Notes 1, 6, 7, 10–13**).

1.4 Detecting Proteins with the Bradford Coomassie Blue G-250 Method

The Coomassie Blue G-250 dye is reddish/brown with an absorbance maximum of 465 nm and interacts with proteins in the Bradford method producing a change in color [4]. The blue color produced under acidic conditions is the result of the dye reacting with arginine and to a lesser extent with lysine, histidine, tyrosine, tryptophan, and phenylalanine residues in proteins. The absorbance maximum is 595 nm (the absorption range is between 575 and 615 nm) [4, 6] and 0.2–20 μg of protein can be detected.

The method is the easiest and fastest of the protein determination methods currently in use (*see* **Notes 1, 8–13**).

2 Materials

2.1 Detecting Proteins by Measuring Absorbance at 280 nm

1. One milliliter Quartz cuvettes are recommended. Some plastic cuvettes are now available where the plastic does not absorb light at 280 nm.
2. A UV spectrophotometer.

2.2 Detecting Proteins with the Lowry Method

1. Spectrophotometer absorbing light in the visible range and 1 mL plastic or glass cuvettes.
2. Copper reagent: Dissolve 20 g of Na_2CO_3 in 260 mL water, 0.4 g $CuSO_4 \cdot 5\ H_2O$ in 20 mL water, and 0.2 g sodium potassium tartrate in 20 mL water and mix.
3. SDS solution: 1.0 g sodium dodecyl sulfate in 100 mL water (1 % w/v).
4. NaOH solution: Dissolve 4 g NaOH in 100 mL water to make a 1 M solution.
5. 2× Lowry working reagent: Mix three parts copper reagent with one part SDS solution and one part NaOH solution. This reagent is stable for 3 weeks.
6. 0.2 N Folin reagent: Mix 10 mL Folin reagent with 90 mL water. The reagent is stable for several months if stored in an amber bottle.

2.3 Detecting Proteins with the Smith Bicinchoninic Acid Method

1. Spectrophotometer absorbing in the visible range and 1 mL plastic or glass cuvettes.
2. Reagent A: 1 g sodium bicinchoninate, 2 g Na_2CO_3, 0.16 g sodium tartrate, 0.4 g NaOH, and 0.95 g $NaHCO_3$ made up to 100 mL with water. Adjust the pH to 11.25 with NaOH.
3. Reagent B: 0.4 g $CuSO_4 \cdot 5H_2O$ in 10 mL water. Reagents A and B are stable.
4. Mix 100 volumes of reagent A with 2 volumes reagent B for a working solution. The solution is stable for a week at room temperature.

2.4 Detecting Proteins with the Bradford Coomassie Blue G-250 Method

1. Spectrophotometer absorbing in the visible range and 1 mL plastic or glass cuvettes.
2. Dissolve 100 mg Coomassie Brilliant Blue G250 in 100 mL of 85 % (v/v) phosphoric acid and 50 mL 95 % (v/v) ethanol. After the dye is dissolved adjust the volume to 1 L with water. The reagent is stable for several months. Should a precipitate form, filter through filter paper and determine the standard curve again.

3 Methods

3.1 Detecting Proteins by Measuring Absorbance at 280 nm

1. Set the UV spectrophotometer to read at 280 nm allowing 15 min for the instrument to equilibrate.
2. Set the absorbance reading to zero with a solution of the buffer and all components except the protein present. Alternately take a reading of solution without protein and subtract the value from each reading of solution containing protein.
3. Place the protein solution in the 1 mL cuvette and determine the absorbance. This step should be repeated to obtain duplicate readings. If a reading is obtained with an absorbance value greater than 2, then dilute the protein sample with the parent buffer and determine the absorbance value. An initial 1 in 10 dilution is suggested. The dilution and readings of samples should be performed in duplicate. It is advisable to use the same cuvette or matched cuvettes for samples and controls.
4. If the extinction coefficient of the protein is known, then the following equation can be employed. Absorbance = Extinction coefficient × concentration of protein × path length (1 cm) to determine the concentration of the protein.

3.2 Detecting Proteins with the Lowry Method

1. Set the spectrophotometer to read at 750 nm allowing 15 min for the instrument to equilibrate.
2. Prepare a bovine serum albumin (BSA) protein standard at 1 mg/mL concentration in duplicate. Dilute the protein standard in a volume of 0.4 mL to give five concentrations over the range of 10–100 μg protein. Perform the dilution and readings for both standard solutions to give four readings at each protein concentration (*see* **Note 1**).
3. To 0.4 mL protein sample add 0.4 mL of the Lowry working reagent and incubate for 10 min at room temperature.
4. Add 0.2 mL of the 0.2 N Folin reagent, vortex, leave for 30 min, and read the absorbance values at 750 nm in the spectrophotometer. Readings can be done between 650 and 750 nm, but are more sensitive at 750 nm.

3.3 Detecting Proteins with the Smith Bicinchoninic Acid Method

1. Set the spectrophotometer to read at 562 nm, allowing 15 min for the instrument to equilibrate.
2. Prepare a BSA protein standard at 1 mg/mL concentration in duplicate. Dilute the protein standard in a volume of 20 μL to give five concentrations over a range of 10–50 μg protein. Each dilution of each of the two standards should be done in duplicate to produce four readings per protein concentration.
3. Add 20 μL of protein sample to 1 mL of working solution and incubate for 30 min at 37 or 60 °C for increased sensitivity.

Cool the sample and place in a 1 mL cuvette and read the absorbance.

3.4 Detecting Proteins with the Bradford Coomassie Blue G-250 Method

1. Set the spectrophotometer to read at 595 nm, allowing 15 min for the instrument to equilibrate.
2. Prepare a BSA protein standard at 1 mg/mL concentration in duplicate. Dilute the protein standard in a volume of 20 μL to give five concentrations over a range of 10–50 μg protein. Each dilution of each of the two standards should be done in duplicate to produce four readings per protein concentration.
3. Add 20 μL of protein solution to 1 mL of dye reagent, mix, incubate for 2 min at room temperature, and measure the absorbance in a cuvette.

4 Notes

1. The recommended calibration protein for protein determination is BSA due to its availability and common use. One should be aware that each protein contains unique amounts of each amino acid. For example BSA (66.3 kDa) in serum contains 2 tryptophan, 20 tyrosine, 23 arginine, and 60 lysine; mature ovalbumin (44.3 kDa) contains 3 tryptophan, 10 tyrosine, 15 arginine, and 20 lysine amino acid residues. The two proteins will thus provide different standard curves in all the assays described above.
2. Advantages of using absorbance at 280 nm: The method (a) is fast, (b) easily automated, (c) reasonably sensitive, (d) does not destroy protein, and (e) can be automated, and (f) most proteins contain tryptophan and tyrosine residues.
3. Disadvantages of using the absorbance at 280 nm to detect proteins: (a) It should be noted that DNA and RNA have absorbance maxima at 260 nm, but still absorb at 280 nm and have tenfold higher absorbance values at 280 nm compared to the equivalent concentration of protein. Groves et al. developed an elegant method to correct for the presence of nucleic acid in a protein solution [7]. A pure protein solution has a ratio of absorbance values (A_{280}/A_{260}) of greater than 1.7 and pure nucleic acid of less than 0.5. (b) Cuvettes should be handled carefully as fingerprints on the cuvette will distort readings.
4. Advantages of the Lowry method: (a) The method is 100-fold more sensitive than the original biuret reagent, and (b) the method is more sensitive than determining absorbance at 280 nm.
5. Disadvantages of the Lowry method: (a) Many common substances (K^{+}, Mg^{2+}, NH_4^{+}, EDTA, Tris, carbohydrates, and

reducing agents) interfere with the method; (b) the Folin reagent is reactive for only a short period of time after addition; (c) the method is complicated and requires more steps and reagents than the BCA or Bradford assays, and (d) the method is destructive to proteins; that is, once the protein sample has reacted with the dye, the protein cannot be used for other assays.

6. Advantages of the Smith bicinchoninic acid assay: (a) The method is less complicated than the Lowry assay; (b) the method is more sensitive than the Lowry assay; (c) the method has less protein:protein variability than the Bradford assay; (d) the volume of reagents can be reduced and a micro-assay can be performed in 96-well plates; (e) detergents are compatible with the assay (up to 5 %); and (f) the method can be automated.
7. Disadvantages of the Smith bicinchoninic acid assay: (a) Reducing reagents (e.g., ascorbic acid) that reduce copper interfere with the assay; (b) chelating agents (e.g., EDTA) that chelate copper interfere with the assay; (c) the method is destructive to proteins; that is, once the protein sample has reacted with the copper, the protein cannot be used for other assays.
8. Advantages of the Bradford Coomassie Blue G-250 assay: (a) The method is fast. (b) The method is simple. (c) The reagents are stable for long periods of time. (d) The volume of reagents can be reduced and the assay performed in a 96-well microtiter plate. (e) The method is sensitive and (f) can be automated.
9. Disadvantages of the Bradford Coomassie Blue G-250 assay: (a) The dye stains cuvettes. Cuvettes can be cleaned by washing with a dilute sodium dodecyl sulfate solution. (b) Dye binding depends on the basic amino acid content that can vary between proteins. (c) Concentrated protein solutions can form a precipitate upon contact with the dye reagent. If this is observed, the protein solution should be diluted to determine protein concentration, and (d) the method is destructive to proteins; that is, once the protein sample has reacted with the dye, the protein cannot be used for other assays.
10. It is advisable to dissolve the protein standard in duplicate. If a protein standard is weighed or solubilized incorrectly in a single preparation, then all protein determinations will be inaccurate.
11. Protein determinations with the Lowry method, copper (BCA), or Coomassie Blue reagents should always be considered as a relative concentration of protein. It is accepted practice to express the relative concentrations of proteins during protein purification steps, provided that all protein concentrations are determined with the same method.

12. The Lowry, copper (BCA), and Coomassie Blue protein determination methods all produce a nonlinear curve, whilst absorbance at 280 nm produces a straight line as protein concentration increases.
13. The Lowry, copper (BCA), and Coomassie Blue protein curves all proceed from the origin of the graph, provided that the correct "blank" or "control" has been used.

References

1. Fasman GD (1990) Practical handbook of biochemistry and molecular biology. CRC Press, Boston, MA
2. Lowry OH, Rosebrough NJ, Farr AL, Randall RJ (1951) Protein measurement with the Folin phenol reagent. J Biol Chem 193:265–275
3. Stoscheck CM (1990) Quantitation of protein. Methods Enzymol 182:50–68
4. Bradford MM (1976) Rapid and sensitive method for quantitation of microgram quantities of protein utilizing principle of protein-dye binding. Anal Biochem 72:248–254
5. Smith PK, Krohn RI, Hermanson GT, Mallia AK, Gartner FH, Provenzano MD, Fujimoto EK, Goeke NM, Olson BJ, Klenk DC (1985) Measurement of protein using bicinchoninic acid. Anal Biochem 150:76–85
6. Compton SJ, Jones CG (1985) Mechanism of dye response and interference in the Bradford protein assay. Anal Biochem 151:369–374
7. Groves WE, Davis FC, Sells BH (1968) Spectrophotometric determination of microgram quantities of protein without nucleic acid interference. Anal Biochem 22:195–210

Chapter 8

Solubilization of Proteins: The Importance of Lysis Buffer Choice

Mandy Peach, Noelle Marsh, Ewa I. Miskiewicz, and Daniel J. MacPhee

Abstract

The efficient extraction of proteins of interest from cells and tissues is not always straightforward. Here we demonstrate the differences in extraction of the focal adhesion protein Kindlin-2 from choriocarcinoma cells using NP-40 and RIPA lysis buffer. Furthermore, we demonstrate the use of a more denaturing urea/thiourea lysis buffer for solubilization, by comparing its effectiveness for solubilization of small heat-shock proteins from smooth muscle with the often utilized RIPA lysis buffer. Overall, the results demonstrate the importance of establishing the optimal lysis buffer for specific protein solubilization within the experimental workflow.

Key words Lysis buffer, NP-40 lysis buffer, RIPA lysis buffer, Urea/thiourea lysis buffer, Protein extraction, Protein solubilization, Small heat-shock proteins, Focal adhesion proteins

1 Introduction

There are many obstacles to the appropriate extraction of proteins from cells and tissues. For example, it has been estimated that intracellular proteins may represent only a tiny fraction of total cellular protein. Thus, it would be difficult to extract these intracellular proteins in reasonable quantities without proper lysis and extraction protocols [1]. The hydrophobicity of membrane proteins is also a significant stumbling block, particularly to 2D gel electrophoresis [2]. The advent of proteomics has accelerated the development and modification of strategies for cell lysis as well as protein extraction at both small- and large-scale levels. The result is that despite more traditionally known cell and tissue lysis buffers [3, 4] there are now quite a number of commercially available reagents for use in such procedures [1]. Nonetheless, one must consider if a particular lysis buffer is optimally designed to extract the protein(s) of interest, particularly the use of the appropriate

Biji T. Kurien and R. Hal Scofield (eds.), *Western Blotting: Methods and Protocols*, Methods in Molecular Biology, vol. 1312, DOI 10.1007/978-1-4939-2694-7_8, © Springer Science+Business Media New York 2015

denaturants and detergent(s) for protein extraction and solubilization.

Several basic criteria are considered for the production of an effective lysis buffer for protein extraction: (1) pH, (2) ionic strength, (3) choice of detergents and denaturants, and (4) constituents to improve protein stability or prevent proteolysis. Specific details on these parameters are described in detail elsewhere (*see* **Note 1**) [1, 5–8]. Although Gromov et al. [9] have reported the development of a lysis buffer that could be used for several different proteomic strategies, it has so far proven difficult to design a lysis buffer effective for isolation of a broad array of proteins. This has necessitated the tailoring of buffers and protocols for individual use.

The *s*mall *h*eat-*s*hock *p*roteins (sHsp) are a family of highly conserved stress proteins ranging in size from 12 to 43 kDa in mass [10, 11]. These proteins have the ability to form low- and high-molecular-mass (>300 kDa) complexes with other heat-shock proteins [12–14] and possess a dynamic quaternary structure. In periods of stress, sHsp can bind denatured proteins and prevent their irreversible aggregation [15], thus aiding in the assembly, disassembly, stabilization, and internal transport of intracellular proteins [16]. Some sHSP, such as HSPB1, can be found in high quantities (e.g., 2 mg/g of tissue protein) within specific tissues such as different types of muscle [15]. sHSPs like HspB1 also appear to interact with integrins and the actin cytoskeleton at focal adhesions [10]. These characteristics and interactions present challenges to the effective extraction and solubilization of such proteins for SDS-PAGE and immunoblot analysis. Clearly, the determination of optimal lysis buffer conditions would be especially important if relative abundance of a specific protein(s) was going to be calculated between control and experimental conditions with downstream immunoblot analysis. Efficient immunoprecipitation of proteins could also be affected by suboptimal lysis conditions.

The use of more standard tissue lysis buffers such as NP-40 and RIPA lysis buffers is quite common for extraction of proteins from tissues, such as muscle, or established mammalian cell lines [17–22]. Using the focal adhesion protein Kindlin-2 as an example, differences in extraction of this protein from BeWo choriocarcinoma cells using NP-40 and RIPA lysis buffer were demonstrated. Tyson and colleagues [23] demonstrated the very effective extraction of HSPB6 from uterine smooth muscle with a urea/thiourea lysis buffer (*see* **Note 2**) and subsequent SDS-PAGE and immunoblot analysis. Thus, using sHSPs as an example, the ability of RIPA lysis buffer and urea/thiourea lysis buffer to solubilize such proteins from uterine smooth muscle was assessed. Overall, the data shown here demonstrate the importance of lysis buffer choice within the experimental workflow.

2 Materials

All aqueous solutions utilized deionized ddH_2O or Milli-Q-grade water (ultrapure; ~18 MΩ cm^2) and all reagents used were of electrophoresis or analytical grade.

1. Phosphate-buffered saline (10×): Dissolve 40 g NaCl, 1 g KCL, 7.2 g Na_2HPO_4, and 1.2 g KH_2PO_4 in 500 mL of ultrapure water and adjust the pH to 7.4 with HCl. Filter-sterilize and store at room temperature. Create 1× working solutions by diluting with ultrapure water.
2. Modified RIPA lysis buffer: 50 mM Tris–HCl (pH 7.5), 150 mM NaCl, 1 % (vol/vol) Triton X-100, 1 % (wt/vol) sodium deoxycholate, 0.1 % (wt/vol) SDS. Dissolve 3.02 g of Tris base, 4.38 g of NaCl, 5 g of deoxycholic acid, and 0.5 g of SDS in 400 mL of ultrapure water. Add 5 mL of Triton X-100, mix, and bring up the volume to 500 mL following adjustment of pH to 7.5 with HCl. Filter-sterilize and store at 4 C (*see* **Note 3**).
3. NP-40 lysis buffer: 50 mM Tris (pH 8.0), 150 mM NaCl, 1 % Nonidet P-40 [3]. Dissolve 0.88 g of Tris base and 0.61 g NaCl in 80 mL of ultrapure water. Add 10 mL of NP-40 detergent, mix, and bring up the volume to 100 mL following adjustment of pH to 8.0 with HCl (*see* **Note 3**).
4. Urea/thiourea lysis buffer: 7 M Urea, 2 M thiourea, 4 % (w/v) CHAPS, in 30 mM Tris–HCl solution (pH 8.5) [23]. Combine 4.2 g of urea, 1.52 g of thiourea, and 0.4 g CHAPS in 10 mL of 30 mM Tris–HCl (pH 8.5). Dissolve one tablet each of Mini EDTA-free Protease and PhosSTOP phosphatase inhibitor tablets and store the buffer in 1 mL aliquots at −80 °C (*see* **Note 4**).
5. 1.5 M Tris–HCl resolving gel buffer: Dissolve 18.2 g of Tris base in ultrapure water to a final volume of 100 mL following adjustment of pH to 8.8 with HCl. Store at 4 °C.
6. 0.5 M Tris–HCl stacking gel buffer: Dissolve 6.1 g of Tris base in ultrapure water to a final volume of 100 mL following adjustment of pH to 6.8 using HCl. Store at 4 °C.
7. SDS-PAGE 15 % resolving gel: Combine 7.5 mL 30 % acrylamide mix (29:1), 3.8 mL of 1.5 M Tris–Cl (pH 8.8), 0.15 mL of 10 % SDS, 0.15 mL of freshly made 10 % ammonium persulfate (w/v; *see* **Note 5**), and 0.006 mL of TEMED in ultrapure water to a final volume of 15 mL.
8. SDS-PAGE 10 % resolving gel: Combine 5 mL 30 % acrylamide mix (29:1), 3.8 mL of 1.5 M Tris–HCl (pH 8.8), 0.15 mL of 10 % SDS, 0.15 mL of freshly made 10 % ammonium persulfate (w/v), and 0.006 mL of TEMED in ultrapure water to a final volume of 15 mL.

9. SDS-PAGE 4 % stacking gel: Combine 1 mL 30 % acrylamide mix (29:1), 2 mL of 0.5 M Tris–HCl (pH 6.8), 0.08 mL of 10 % SDS, 0.04 mL of freshly prepared 10 % ammonium persulfate (*see* **Note 5**), and 0.008 mL of TEMED in ultrapure water to a final volume of 8 mL.
10. SDS-PAGE running buffer (5×): 15.1 g of Tris base, 94 g glycine, and 50 mL of 10 % SDS made up to 1 L in deionized ddH_2O water with a final pH of 8.3. Filter-sterilize and refrigerate until use. Dilute to 1× with deionized ddH_2O when required.
11. SDS-PAGE loading dye (2×): 4 % (w/v) SDS, 0.2 % (w/v) bromophenol blue, 20 % (v/v) glycerol in 100 mM Tris–HCl, pH 8.0. Dissolve 0.4 g of SDS and 0.02 g of bromophenol blue in 8 mL of 100 mM Tris–HCl, pH 8.0 and add 2 mL of glycerol (20 % final concentration v/v). Divide into 1 mL aliquots and add β-mercaptoethanol, at a final concentration of 10 % (v/v), to a 1 mL working solution. Store at room temperature.
12. Transfer membranes: 0.22 μm nitrocellulose membranes.
13. Gel transfer buffer (10× stock): Dissolve 29 g glycine, 58 g Tris base, and 3.7 g SDS in 1 L of deionized ddH_2O water. Adjust pH to 8.3 and filter-sterilize. Store in the refrigerator and produce 1× working solutions by diluting with deionized ddH_2O water.
14. TBST: Dissolve 2.42 g Tris base and 8.0 g of NaCl in deionized water and adjust the pH to 7.6 with HCl. Add 1.0 mL of Tween-20 and bring the volume to 1 L. Buffer can be stored in the refrigerator.
15. Immunoblot blocking buffer: Dissolve 5 g of fat-free skim milk powder in TBST buffer and mix by vigorous shaking.
16. Coomassie brilliant blue (CBB) protein stain: Add 0.25 g of CBB R-250 to 100 mL of methanol:acetic acid solution (45 % methanol (v/v), 10 % acetic acid (v/v) in deionized water). Filter through Whatman paper to remove particulates.
17. Methanol:acetic acid destaining solution: Mix 25 mL of methanol and 7.5 mL of acetic acid in deionized water.
18. Electrophoresis and transfer system: Mini-PROTEAN 3 system.
19. Whatman Paper: 3 mm chromatography.
20. Molecular weight protein markers: Precision plus dual-color protein standards.
21. Pierce reversible protein stain kit for nitrocellulose membranes.
22. Antisera: Rabbit polyclonal anti-HspB8 or anti-HspB6 (Abcam, Inc, Cambridge, MA, USA). Mouse monoclonal anti-Kindlin-1 (EMD-Millipore, Etobicoke, Canada). Mouse monoclonal

anti-smooth muscle calponin and rabbit polyclonal anti-Kindlin-2 (Sigma Chemical Co, St. Louis, MO, USA).

23. Tissue: Rat uterine smooth muscle tissue from d15 of pregnancy.
24. Cells: BeWo choriocarcinoma cells (American Type Culture Collection, Manassas, VA, USA).

3 Methods

3.1 Tissue Collection

Isolate all tissue samples rapidly (*see* **Note 6**) and place in ice-cooled PBS for rapid washing. Subsequently, place tissues in polypropylene vials and freeze in liquid nitrogen.

3.2 Protein Solubilization

1. For uterine smooth muscle samples, chip off pieces of frozen tissues in a fume hood, quickly weigh the fragments (~100–250 mg) in precooled weigh boats, and place in a precooled mortar on dry ice. Grind the samples into a fine powder with a pestle under liquid nitrogen. The use of a fume hood is recommended to avoid inhalation of vapors from the liquid nitrogen. For BeWo cells, wash cells with PBS in a biological safety cabinet, then add 0.2 mL of lysis buffer to the cells, and collect with a plastic cell scraper. Place the mixture in a prechilled 0.5 mL lysis tube (PreCellys CK14 lysis kit) containing ceramic beads and homogenize cells with a Minilys Bead Mill using a 10-s burst. Centrifuge the sample at full speed for 15 min in a Beckman Microfuge-18 microcentrifuge, and collect supernatants for protein analysis.
2. For uterine smooth muscle samples, transfer the powdered samples to 1 mL of urea/thiourea lysis buffer (*see* **Notes 2** and **4**) or 1 mL RIPA lysis buffer in 15 mL polypropylene tubes and homogenize for up to 1 min on ice with a Polytron PT10-35 homogenizer (*see* **Note 7**).
3. For samples homogenized in urea/thiourea lysis buffer, allow the lysates to settle at room temperature for 30 min while RIPA tissue lysates are kept on ice. Subsequently, transfer all sample lysates to appropriately labeled microcentrifuge tubes, centrifuge at full speed for 15 min in a Beckman Microfuge-18 microcentrifuge, and collect supernatants for protein analysis.
4. Determine sample protein concentrations using the Bradford Assay [24].

3.3 SDS-PAGE and Electroblotting

1. Prepare the polyacrylamide gel casting module according to instructions provided by the appropriate manufacturer (*see* **Note 8**).
2. Immediately after the addition of TEMED to the resolving gel mixture, add the mixture to the prepared gel cassette with a

Pasteur pipet (*see* **Notes 9** and **10**). Add isopropanol over the top of the resolving gel to ensure that gel polymerization is not inhibited. After 45 min, remove the overlaid isopropanol by tipping the gel molds to pour off the solvent and soak up residual isopropanol with Kimwipes. Pour the stacking gel in the same manner as the resolving gel. Insert the appropriate gel comb into the stacking gels ensuring that no air bubbles are trapped under the teeth of the comb.

3. Once gel polymerization is complete and the gel assembled in the electrophoresis tank, incubate protein samples (e.g., 50 μg volumes for uterine muscle, 20 μg volumes for BeWo cells) with 2× SDS-PAGE loading dye at 95 °C for 5 min prior to gel loading (*see* **Note 11**). Run the gel at 60 V until samples and the prestained molecular mass standards enter the resolving gel, and then separate proteins at 100 V until the dye front reaches the bottom of the gel (*see* **Note 12**).
4. Following electrophoresis, gently pry open the gel plates with a plastic wedge to recover the gel.
5. To help assess the effective solubilization of sample proteins with the different lysis buffers, stain the polyacrylamide gel with Coomasie blue for at least 0.5 h with gentle agitation on a shaker. Destain the gel with methanol:acetic acid solution using gentle agitation until bands are finely resolved (~4–8 h). The use of small pieces of foam added to the destain solution aids the wicking up of dye from the destain solution. Photograph the gel with a gel documentation system.
6. For immunoblot analysis skip **step 5** and place the gel in transfer buffer. Cut a nitrocellulose membrane to the same size of the gel and also place it in transfer buffer. Assemble the gel for electroblotting as has been described in detail elsewhere [25, 26] (*see* **Note 13**) and conduct electroblotting for 1 h at 300 mA in transfer buffer with constant buffer stirring in an ice bucket or use an ice pack in the transfer apparatus.
7. To help assess effective solubilization and transfer of sample proteins in an alternative way to **step 5** above, reversibly stain the immunoblot using a Pierce Reversible Protein Stain Kit according to the manufacturer's instructions and photograph the blot with a digital immunoblot imaging system. Following erasure of the protein staining, proceed to Subheading 3.4, **step 1**.

3.4 Immunoblot Analysis

Unless otherwise stated, all incubations should be conducted at room temperature and with constant agitation.

1. Rinse the membrane with TBST for 5 min.
2. Block the blot in 5 % skim milk powder/TBST for 30 min (*see* **Note 14**).

3. Incubate the membrane in appropriate antisera, diluted in blocking solution, for 1 h.
4. Rinse the blot 1 × 15 min in TBST, followed by 2 × 5 min in TBST.
5. Incubate membranes for 1 h in HRP-conjugated goat anti-rabbit IgG (H+L) or HRP-conjugated goat anti-mouse IgG (H+L) antisera (1:10,000 and 1:10,000 dilutions, respectively) diluted in blocking solution.
6. Wash the blot 1 × 15 min in TBST, and then 4 × 5 min in TBST.
7. Detect proteins on the immunoblot using the Pierce SuperSignal West Pico Chemiluminescent Substrate (MJS BioLynx, Inc., Brockville, Ontario, Canada) detection system (Figs. 1 and 2). Generate multiple exposures on ECL X-ray

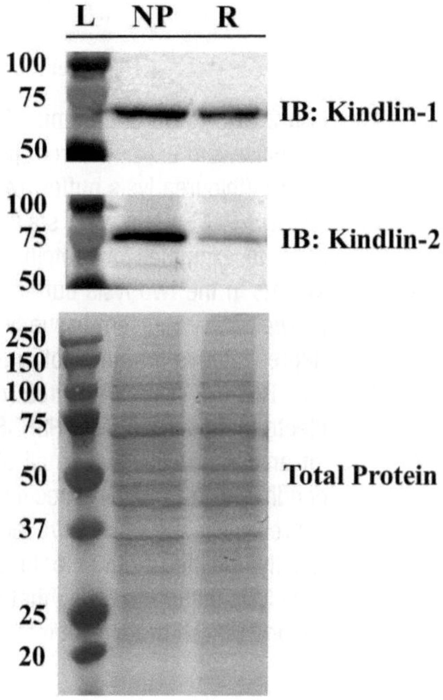

Fig. 1 Immunoblot analysis of Kindlin expression using NP-40 and RIPA lysis buffers. Proteins were extracted from BeWo choriocarcinoma cells with NP-40 or RIPA lysis buffer followed by SDS-PAGE and immunoblot analysis. Prior to immunoblot analysis, the blot was stained with a reversible protein staining kit. Kindlin-1 and particularly Kindlin-2 were solubilized to a greater degree in NP-40 rather than RIPA lysis buffer even though the overall extent of total proteins extracted was comparable. These results indicate the care that must be taken to predetermine the optimal lysis buffer prior to an experimental workflow as proteins of interest could remain in the insoluble fraction. *L* protein molecular mass ladder listed in kDa, *NP* NP-40 lysis buffer extracts, *R* RIPA lysis buffer extracts. Shown are representative immunoblots ($n = 4$)

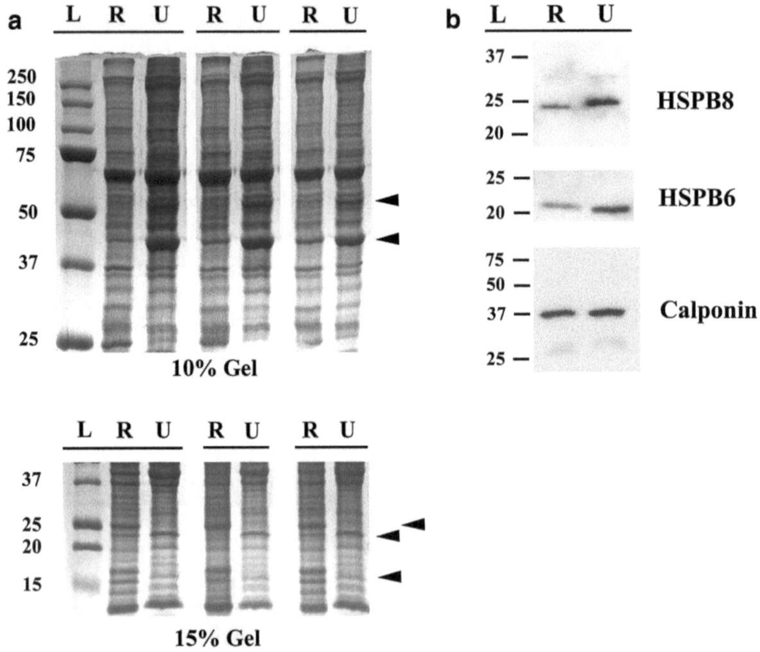

Fig. 2 SDS-PAGE and immunoblot analysis using urea/thiourea lysis buffer. (**a**) Proteins were extracted from pregnant rat uterine smooth muscle with RIPA (*R*) or urea/thiourea lysis buffer (*U*) and loaded on 10 or 15 % polyacrylamide gels for SDS-PAGE. Gels were stained with Coomassie blue and destained. Wedges indicate examples of protein species that appeared to solubilize to different extents in the two lysis buffers. *L* protein molecular mass ladder listed in kDa. *Paired lanes* represent tissue lysates obtained from different animals (*n*= 3). (**b**) Representative immunoblot analyses of HSPB6 and HSPB8. Proteins extracted with RIPA (*R*) or urea/thiourea (*U*) lysis buffer were separated by SDS-PAGE and electroblotted. HSPB6- or HSPB8-specific antisera were then used for immunoblot analysis of the two small stress proteins obtained from the different lysis conditions. h-Calponin-specific antiserum was also used to assess calponin expression in the same extraction conditions. Both HSPB6 and HSPB8 were more readily detected from smooth muscle tissue lysates prepared in the urea/thiourea lysis buffer indicating that it may be a more optimal lysis buffer for solubilization of these proteins of interest

films to ensure the linearity of the film exposures or capture chemiluminescence signal using a digital imaging documentation system.

4 Notes

1. Some researchers like to modify or develop their own lysis buffers in which case Harlow and Lane [3] have recommended considering the following range of variables to optimize the lysis buffer for downstream western blot analyses: salt concentrations

0–1 M, nonionic detergents 0.1–2 %, ionic detergents 0.01–0.5 %, divalent cation concentrations 0–10 mM, EDTA concentrations 0–5 mM, and pH 6–9.

2. The urea/thiourea lysis buffer appears to have been originally used for extraction of skeletal muscle-specific proteins in 1983 by Yates and Greaser [27, 28]. Their utilization of thiourea was based on the report of Pace and Marshall [29] indicating that thiourea was a potent protein denaturant. The use of this type of buffer, particularly in 2D gel electrophoresis, has been described in detail elsewhere [30, 31]. The combined use of urea and thiourea increases protein solubility since urea is effective at disrupting hydrogen bonds aiding protein unfolding and denaturation while thiourea is much better at reducing hydrophobic interactions between proteins [32]. The volumes of lysis buffers utilized are also very important for lysis efficiency. Gorg and colleagues [31] have previously reported the use of 1 mL of this lysis buffer with ~50–100 mg of mammalian tissue. In our hands, it has proven reasonable to use up to a maximum of 250 mg of tissue with a 1 mL volume. It is recommended that the appropriate volume be determined by the investigator on a case-by-case basis.
3. When required for lysis, take 10 mL aliquots and completely dissolve one tablet each of Mini EDTA-free Protease and PhosSTOP phosphatase inhibitor tablets prior to use. Any unused buffer can be frozen at −20 °C for future use. As with just about every lysis buffer, care should be taken to ensure that all components are in solution prior to use as cold storage can lead to precipitation of some constituents (e.g., SDS).
4. It is recommended that the urea/thiourea lysis buffer be prepared fresh whenever possible, but it can also be aliquoted (1 mL) and stored at −80 °C for up to several months. It has also been reported that once the buffer is thawed, it should not be refrozen [30, 31].
5. The use of 10 % ammonium persulfate in the preparation of the gel mixtures eliminates the need to degas the solution and remove dissolved oxygen.
6. Cells and tissues should be frozen rapidly with liquid nitrogen to avoid protease degradation of proteins in the sample or collected and lysed quickly, preferably while chilled. Since proteases as well as phosphatases can be released during lysis and act on your target protein(s), protease and phosphatase inhibitors should be included in the lysis buffers. Many of these are produced as cocktails in tablet form for easy purchase and their use is as simple as dissolving the tablet in the lysis buffer prior to utilization.
7. Methods utilized for tissue disruption clearly depend on the tissue origin. There are a large number of other means to lyse

cells/tissues and readers are directed to Simpson [33] for specific details and discussion of these protocols. When using mechanical homogenization, it is important to avoid the production of excessive amounts of foam as this could decrease your recovery volume (i.e., becomes difficult to recover from the homogenizer). Short bursts of mechanical homogenization, while the sample(s) is cooled with ice, are usually best.

8. Place 2–3 folded Kimwipes under the thermoplastic rubber gaskets of the casting module and a strip of Parafilm on top of each gasket. This prevents leaks by increasing the thickness of the rubber gaskets upon which the glass plates are held against with a spring-loaded lever. Leakage can be a problem as the gaskets age and lose their flexibility and overall thickness from constant use.

9. It is critical to use high-quality SDS from a single source and polyacrylamide that is free of contaminating metal ions. Sambrook and Russell [25] have reported that the migration pattern of polypeptides can change significantly when SDS from different manufacturers are interchanged. Purchase of premade acrylamide from reputable companies is becoming the normal procedure. Acrylamide solutions with a 1:29 bisacrylamide:acrylamide ratio are usually employed as they are capable of resolving polypeptides differing in size as little as 3 % [25], but solutions can be purchased with different ratios if necessary to vary the pore size of the gel [26]. In addition, Tris base should always be used for the preparation of gel buffers to avoid production of diffuse protein bands and even improper polypeptide migration [26].

10. Leave approximately 1 cm of space below the eventual bottom of the combs for the later addition of the stacking gel mixture.

11. It has been noted that heating of samples containing urea for 2D gel electrophoresis can result in some decomposition of urea and release of isocyanate leading to protein carbamylation and charge heterogeneities of the samples. However, in this instance there is no need to worry about protein carbamylation during heating of the samples in 2× SDS loading dye at 95 °C as the samples are not being used for isoelectric focusing. This heating step is necessary to produce SDS-polypeptide complexes for subsequent SDS-polyacrylamide electrophoresis.

12. The system used here is a discontinuous buffer system. As a result, the SDS-polypeptide complexes in the 4 % stacking gel become deposited and concentrated on the surface of the resolving gel. The SDS-polypeptide complexes are then separated in the resolving gel according to size by molecular sieving in a zone of uniform voltage and pH. Greater details on the

mechanism of polyacrylamide gel electrophoresis are found elsewhere [25].

13. There are now many options for transfer of polypeptides to membranes and the reader is directed to a review of these procedures [26]. It is also imperative that no air bubbles be trapped between the nitrocellulose membrane and the polyacrylamide gel as this will result in the lack of polypeptide transfer to the membrane in these regions. Use a blot roller to remove any bubbles between the gel and membrane.
14. The researcher should consider the blocking buffer that is most appropriate for the specific antiserum (e.g., skim milk powder vs. BSA). Blocking a blot serves two important purposes. The first is well known in that it can help mask any potential nonspecific binding sites on the membrane itself. The second purpose, being less known and even less understood, is that blocking a membrane can promote renaturation of antigenic sites [34]. However, it has been reported that prolonged blocking times (>24 h) can actually remove antigens [35].

Acknowledgement

This work was supported by a Natural Sciences and Engineering Research Council Discovery Grant (#250218), an Establishment Grant from the Saskatchewan Health Research Foundation (SHRF; #2695), and a regional partnership program grant from SHRF (#2776) and the Canadian Institutes of Health Research (#ROP-101051) to D.J.M. M.P. and N.M. were holders of Alexander Graham Bell NSERC postgraduate fellowships.

References

1. Grabski AC (2009) Advances in preparation of biological extracts for protein purification. Methods Enzymol 463:285–305
2. Cordwell SJ (2008) Sequential extraction of proteins by chemical reagents. 2D PAGE: sample preparation and fractionation. In: Posch A (ed) Methods in molecular biology, vol 424. Humana, Totowa, NJ, pp 139–146
3. Harlow E, Lane E (1988) Immunoprecipitation. In: Harlow E, Lane D (eds) Antibodies: a laboratory manual. Cold Spring Harbor Laboratory, New York, NY, p 231
4. Rosenberg IM (2005) Protein analysis and purification, 2nd edn. Birkhauser, Boston, p 37
5. Helenius A, Simons K (1975) Solubilization of membranes by detergents. Biochim Biophys Acta 415:29–79
6. Helenius A, McCaslin DR, Fries E, Tanford C (1979) Properties of detergents. Methods Enzymol 56:734–749
7. Dawson RMC, Elliot DC, Elliot WH, Jones KM (1986) pH, buffers, and physiological media. Data for biochemical research. Oxford University Press, New York, In, pp 417–448
8. Linke D (2009) Detergents: an overview. Methods Enzymol 463:603–617
9. Gromov P, Celis JE, Gromova I, Rank F, Timmermans-Wielenga V, Moreira JMA (2008) A single lysis solution for the analysis of tissue samples by different proteomic technologies. Mol Oncol 2:368–379
10. Salinthone S, Gerhoffer WT (2008) Small heat shock proteins in smooth muscle. Pharmacol Ther 119:44–54

11. Laskowska E, Matuszewska E, Kuczynska-Wisnik D (2010) Small heat shock proteins and protein misfolding diseases. Curr Pharm Biotechnol 11:146–157
12. Kato K, Goto S, Inaguma Y, Hasegawa K, Morishita R, Asano T (1994) Purification and characterization of a 20 kDa protein that is highly homologous to alpha B crystallin. J Biol Chem 269:15302–15309
13. Pipkin W, Johnson JA, Creazzo TL, Burch J, Komalavilas P, Brophy CM (2003) Localization, macromolecular associations, and function of the small heat shock related protein HSP20 in rat heart. Circulation 107:469–476
14. Sun X, Fontaine J-M, Rest JS, Sheldon EA, Welsh MJ, Benndorf R (2004) Interaction of human hsp22 (HSPB8) with other small heat shock proteins. J Biol Chem 279:2394–2402
15. Gusev NB, Bogatcheva NV, Marston SB (2002) Structure and properties of small heat shock proteins and their interaction with cytoskeleton proteins. Biochemistry (Moscow) 67:511–519
16. Drieza CM, Komalavilas P, Furnish EJ, Flynn CR, Sheller MR, Smoke CC, Lopes LB, Brophy CM (2010) The small heat shock protein, HSPB6, in muscle function and disease. Cell Stress Chaperones 15:1–11
17. Williams SJ, Shynlova O, Lye SJ, MacPhee DJ (2009) Spatiotemporal expression of α1, α3, and β1 integrin subunits is altered in rat myometrium during pregnancy and labour. Reprod Fertil Dev 22:718–732
18. Palliser HK, Zakar T, Symonds IM, Hirst JJ (2010) Progesterone receptor isoform expression in the Guinea Pig myometrium from normal and growth restricted pregnancies. Reprod Sci 17:776–782
19. Shynlova O, Dorogin A, Lye SJ (2010) Stretch-induced uterine myocyte differentiation during rat pregnancy: involvement of caspase activation. Biol Reprod 82:1248–1255
20. Huo P, Zhao L, Li Y, Luo F, Wang S, Song J, Bai J (2014) Comparative expression of thioredoxin-1 in uterine leiomyomas and myometrium. Mol Hum Reprod 20:148–154
21. Elustondo PA, Hannigan GE, Caniggia I, MacPhee DJ (2006) Integrin-linked kinase (ILK) is highly expressed in first trimester human chorionic villi and regulates migration of a human cytotrophoblast-derived cell line. Biol Reprod 74:959–968
22. Butler TM, Elustondo PA, Hannigan GE, MacPhee DJ (2009) Integrin-linked kinase can facilitate syncytialization and hormonal differentiation of the human trophoblast-derived BeWo cell line. Reprod Biol Endocrinol 7:51
23. Tyson EK, MacIntyre DA, Smith R, Chan E-C, Read M (2008) Evidence that a protein kinase A substrate, small heat shock protein 20, modulates myometrial relaxation in human pregnancy. Endocrinology 149:6157–6165
24. Bradford MM (1976) A rapid and sensitive method for the quantification of microgram quantities of protein utilizing the principle of protein-dye binding. Anal Biochem 72: 248–254
25. Sambrook J, Russell D (2001) Molecular cloning: a laboratory manual. Cold Spring Harbor Laboratory Press, Cold Spring Harbor, NY. Pp, 2344
26. MacPhee DJ (2010) Methodological considerations for improving western blot analysis. J Pharmacol Toxicol Meth 61:171–177
27. Yates LD, Greaser ML (1983) Quantitative determination of myosin and actin in rabbit skeletal muscle. J Mol Biol 168:123–141
28. Yates LD, Greaser ML (1983) Troponin subunit stoichiometry and content in rabbit skeletal muscle and myofibrils. J Biol Chem 258: 5770–5774
29. Pace CN, Marshall HF Jr (1980) A comparison of the effectiveness of protein denaturants for β-lactoglobulin and ribonuclease. Arch Biochem Biophys 199:270–276
30. Weiss W, Gorg A (2008) Sample solubilization buffers for two-dimensional electrophoresis. 2D PAGE: Sample preparation and fractionation. In: Posch A (ed) Methods in molecular biology, vol 424. Humana Press, New Jersey, pp 35–42
31. Gorg A, Drews O, Weiss W (2006) Extraction and solubilization of total protein from mammalian tissue samples. Cold Spring Harb Protoc . pp 3. doi:10.1101/pdb.prot4226
32. Rabilloud T (1998) Use of thiourea to increase the solubility of membrane proteins in two dimensional electrophoresis. Electrophoresis 19:755–760
33. Simpson RJ (2009) Preparation of cellular and subcellular extracts. In: Simpson RJ, Adams PD, Golemis EA (eds) Basic methods in protein purification and analysis. Cold Spring Harbor Laboratory Press, New York, pp 39–78
34. Towbin H, Gordon J (1984) Immunoblotting and dot immunobinding – current status and outlook. J Immunol Methods 72:313–340
35. DenHollander N, Befus D (1989) Loss of antigens from immunoblotting membranes. J Immunol Methods 122:129–135

Chapter 9

Simultaneous Immunoblotting Analysis with Activity Gel Electrophoresis and 2-D Gel Electrophoresis

Der-Yen Lee and Geen-Dong Chang

Abstract

Diffusion blotting method can couple immunoblotting analysis with another biochemical technique in a single polyacrylamide gel, however, with lower transfer efficiency as compared to the conventional electroblotting method. Thus, with diffusion blotting, protein blots can be obtained from an SDS polyacrylamide gel for zymography assay, from a native polyacrylamide gel for electrophoretic mobility shift assay (EMSA) or from a 2-D polyacrylamide gel for large-scale screening and identification of a protein marker. Thereafter, a particular signal in zymography, electrophoretic mobility shift assay, and 2-dimensional gel can be confirmed or identified by simultaneous immunoblotting analysis with a corresponding antiserum. These advantages make diffusion blotting desirable when partial loss of transfer efficiency can be tolerated or be compensated by a more sensitive immunodetection reaction using enhanced chemiluminescence detection.

Key words Diffusion blotting, Zymography, Electrophoretic mobility shift assay, Autoantigen, Posttranslational modification

1 Introduction

Diffusion blotting is originally developed for isoelectric focusing (IEF) gels or ultrathin gels [1–4], which requires laying a blotting membrane on the gel surface and a stack of dry filter paper on top of the blotting membrane. Usually a glass plate and an object carrying certain weight are further stacked on the filter to facilitate the diffusion process. It is more suitable using diffusion blotting than the electrophoretic blotting for particular applications such as electrophoresis on gels bound on plastic sheets [3, 4], multiple blotting from a single gel [1, 4, 5] (*see* Chapters 10 and 11), and simultaneous immunoblotting analysis with activity gel electrophoresis such as proteolytic zymography and electrophoretic mobility shift assay (EMSA) [6].

We have previously demonstrated several applications of diffusion blotting [6]. They are multiple blotting [1], combined immunoblot-

Biji T. Kurien and R. Hal Scofield (eds.), *Western Blotting: Methods and Protocols*, Methods in Molecular Biology, vol. 1312, DOI 10.1007/978-1-4939-2694-7_9, © Springer Science+Business Media New York 2015

ting analysis with gel staining [2], combined immunoblotting analysis with proteolytic zymography [3], and combined immunoblotting analysis with EMSA from a single gel [4]. In activity gel electrophoresis, several positive signals are frequently observed, which requires an additional immunoblotting analysis in another gel to examine whether a particular protein is involved. A method of simultaneous immunoblotting analysis with EMSA has been reported as "shift-western blotting" [7]. In this method, a nitrocellulose filter and an anion-exchange membrane are stacked together for electroblotting of proteins and the radioactive DNA, respectively, following native gel electrophoresis. The protein-DNA complex was detected by autoradiography of the DNA blot and by immunoblotting analysis of the protein blot [7–15]. Simultaneous immunoblotting analysis can also be done with activity gel electrophoresis on the same gel with diffusion blotting. Because the blot and the activity staining are derived from the same gel, the localization of signals in the gel and the replica can be easily aligned for comparison. In this chapter, we describe protocols for combined immunoblotting analysis with proteolytic zymography in identifying nephrosin in carp tissue extracts (Fig. 1) and combined multiple immunoblotting analysis with silver staining of mouse microsomal proteins resolved by 2-dimensional (2-D) gel electrophoresis in identifying endoplasmic reticulum markers such as protein disulfide isomerase (PDI) and glucose-regulated protein 78 (GRP78) (Fig. 2). Potentially, simultaneous immunoblotting analysis with silver staining on a 2-D gel can be applied in large-scale screening and identification of protein markers such as autoantigens and tumor markers

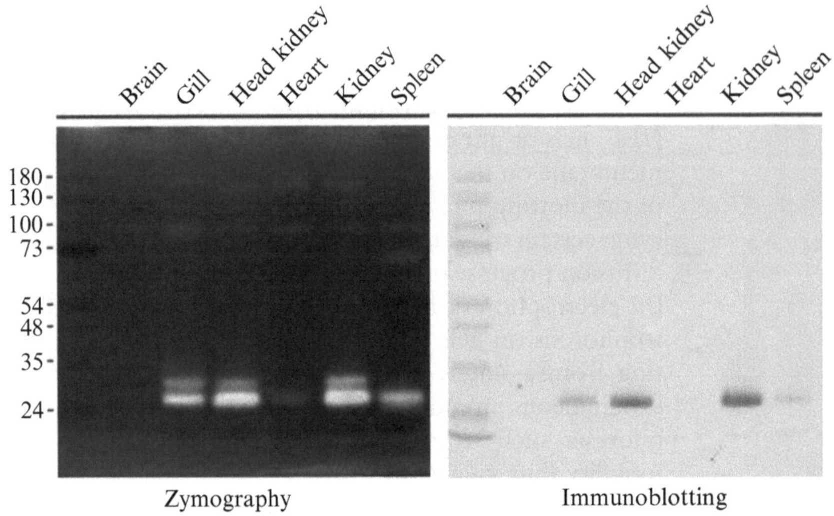

Fig. 1 Simultaneous immunoblotting analysis for nephrosin and proteolytic zymography assay with carp tissue extracts. Four microliter of carp tissue extracts was analyzed by SDS-PAGE in a gel containing 0.2 % gelatin. After electrophoresis, the proteins were blotted by diffusion blotting and immunodetected by an anti-nephrosin antiserum (*right panel*). The remaining gel was treated for the proteolytic zymography assay (*left panel*)

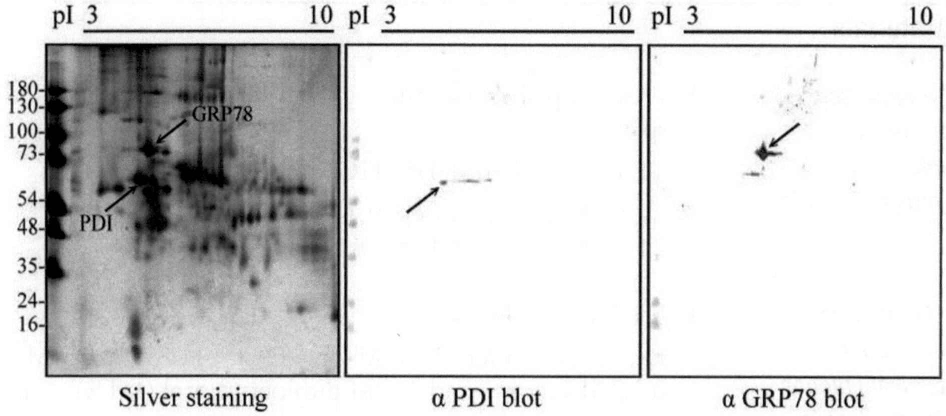

Fig. 2 Simultaneous multiple immunoblotting analysis and silver staining of mouse liver microsomal proteins resolved by 2-DE. Forty micrograms of mouse liver microsomal proteins were resolved by a 7 cm × 8 cm, 2-D polyacrylamide gel. After 2-D gel electrophoresis, the gel was blotted with two pieces of PVDF membrane for 2 h. One membrane was then probed with an anti-PDI and the other with an anti-GRP78 serum. The remaining gel was subjected to silver staining to reveal each protein spot. Theoretical p*I*/Mw of PDI and GRP78 are 4.77/57058.49 and 5.07/72332.96, respectively

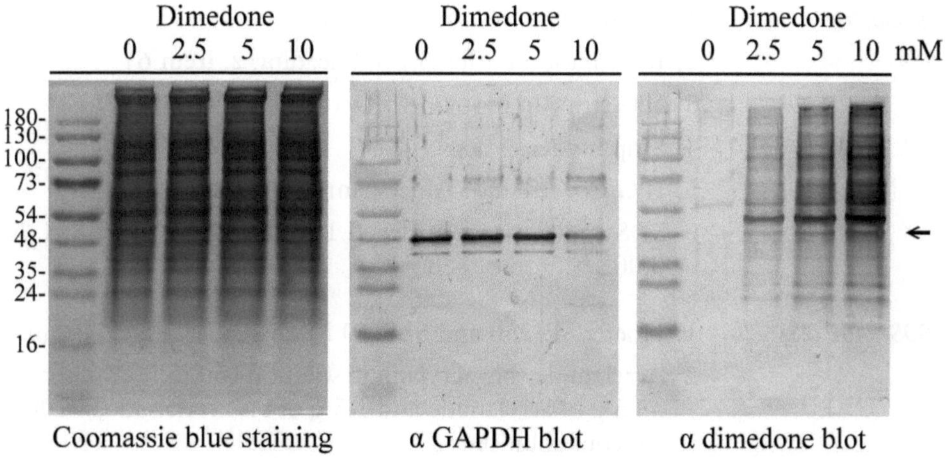

Fig. 3 Simultaneous multiple immunoblotting analysis for protein thiol oxidation detection. Twenty micrograms of MOVAS protein extract treated with dimedone were resolved by SDS-PAGE. After gel electrophoresis, the gel was blotted with two pieces of PVDF membrane for 1 h. One membrane was then probed with an anti-GAPDH and the other with an anti-dimedone serum. The remaining gel was subjected to Coomassie blue R-250 staining to reveal each protein band. The *arrow* indicated the position of GAPDH and dimedone-conjugated GAPDH

[16–18] in the future. For multiple blotting, we describe protocols for simultaneous immunoblotting of glyceraldehyde 3-phosphate dehydrogenase (GAPDH) and dimendone (Fig. 3). Dimedone is used to tag cysteine sulfenic acid of GAPDH on exposure to hydrogen peroxide [19, 20]. Therefore, multiple blotting can be used in simultaneous immunodetection of a given protein and one of the posttranslational modifications associated with the protein.

2 Materials

2.1 Preparation of Carp Tissue Extracts for Zymography [21]

1. Fresh carp tissues: Brains, gills, head kidneys, hearts, kidneys, spleens.
2. TE buffer: 20 mM Tris–HCl, pH 8.0, 5 mM EDTA.
3. 2× SDS sample buffer: 0.1 M Tris–HCl, pH 7.8, 4 % SDS, 24 % glycerol, and 0.002 % bromophenol blue.

2.2 Preparation of Mouse Liver Microsomal Fraction [22] for Isoelectric Focusing

1. Fresh mouse livers (Narl: ICR).
2. Extraction buffer: 20 mM Tris–HCl, pH 8.0, 0.15 M NaCl, 0.25 M sucrose, and 1 mM dithioerythritol (DTE).
3. 10 % Triton X-100 solution.
4. 20 % trichloroacetic acid (TCA) solution.
5. Acetone is kept at −20 °C.
6. Rehydration buffer contains 1.5 % 3–10 IPG buffer and 100 mM DTE (Research Organics) in FOCUS™ Extraction buffer III (*see* **Note 1**).

2.3 Isoelectric Focusing

1. Ettan IPGphor Focusing system.
2. Rehydration buffer (*see* Subheading 2, **item 6**).
3. IPG strip: Immobiline™ DryStrip, pH 3–10, 7 cm.
4. Strip holder, 7 cm.
5. IPG Cover Fluid (Amersham Biosciences Ltd.).
6. SDS equilibration buffer: 0.1 M Tris–HCl, pH 7.8, 1 % SDS, 0.002 % bromophenol blue.

2.4 SDS-PAGE [23]

1. Hoefer SE-250 and SE-260 Mighty small II gel system.
2. Acrylamide/bis-acrylamide solution (50 % T; 3 % C) contains 48.5 g of acrylamide and 1.5 g of bis-acrylamide in 100 mL of aqueous solution.
3. Gel buffer: 1.5 M Tris–HCl, pH 8.45.
4. Stacking gel buffer: 0.1 M Tris–HCl, pH 7.8, 0.4 % SDS.
5. Prepare 10 % ammonium persulfate solution by dissolving 0.1 g of ammonium persulfate into 1 mL of deionized water (*see* **Note 2**).
6. TEMED.
7. Glycerol.
8. Melting 2 % gelatin solution (from porcine skin): Prepare the solution by melting the particles or the gel.
9. TE buffer: 20 mM Tris–HCl, pH 8.0, 5 mM EDTA.
10. Dissolve 3.5, 7, and 14 mg dimedone in 1 mL DMSO to obtain 25, 50, and 100 mM dimedone solution.
11. β-Mercaptoethanol.

12. Prestained protein ladder.
13. Chromatography paper (3MM Chr).
14. Marker strip: Blot one end of a 3-mm-wide strip of 3MM chromatography paper with 5 μL of prestained marker and leave it air-dried for 5 min (*see* **Note 3**).

2.5 Diffusion Blotting [6]

1. Millipore Immobilon-P PVDF membrane: 8.6 cm × 6.5 cm and 8.6 cm × 8.8 cm.
2. Chromatography paper (3MM Chr).
3. Methanol.

2.6 Staining for Gelatin Zymography or SDS-PAGE [6]

1. Tris/Triton solution: 20 mM Tris–HCl, pH 8.0, 2 % Triton X-100.
2. Incubation buffer: 20 mM Tris–HCl, pH 8.0, 0.2 mM $CaCl_2$, 0.1 mM $ZnCl_2$.
3. Coomassie blue R-250 solution: 0.1 % Coomassie brilliant blue R-250, 40 % methanol, and 7 % acetic acid.
4. Destaining solution: 0.1 % Triton X-100 and 7 % acetic acid.

2.7 Staining for 2-DE [24]

1. 20 % TCA solution.
2. Sensitizing solution: Prepare 12.5 mM sodium thiosulfate and 0.8 M sodium acetate in 30 % ethanol as stock solution and add 0.2 mL of glutardialdehyde to 40 mL of the stock solution prior to use (*see* **Note 4**).
3. Silver nitrate solution: Prepare 0.25 % silver nitrate as stock solution and add 20 μL of formaldehyde to 40 mL of the stock solution prior to use (*see* **Note 4**).
4. Developing solution: Prepare 2.5 % sodium carbonate as stock solution and add 10 μL of formaldehyde to 40 mL of the stock solution prior to use (*see* **Note 4**).
5. 4 % acetic acid.

2.8 Immunoblotting [25]

1. Phosphate-buffered saline (PBS).
2. PBST: 0.1 % Tween 20 in PBS.
3. Blocking solution: 3 % skim milk in PBST.
4. Primary antibody solution: Anti-nephrosin, anti-GRP78, anti-PDI, and anti-GAPDH were raised in guinea pigs and anti-dimedone was raised in a rabbit in our laboratory. They are used at a titer of 1:2,000 in PBST containing 0.3 % BSA.
5. Secondary antibody solution: 0.2 μg/mL HRP-conjugated anti-guinea pig or anti-rabbit IgG in 0.3 % BSA in PBST.
6. $NiCl_2$-DAB solution: Prepare 0.5 mg of 3′,3′-diaminobenzidine (DAB) and 0.5 mL of 1 % $NiCl_2$ in 10 mL of PBS prior to use.
7. Hydrogen peroxide.

3 Methods

3.1 Preparation of Carp Tissue Extracts for Zymography [21]

1. Brains, gills, head kidneys, hearts, kidneys, and spleens are freshly obtained from a male carp sacrificed by decapitation.
2. At a ratio of 10 mL of cold TE buffer to 1 g of carp tissues, carp tissues are homogenized in an ice bath and the homogenates are centrifuged at 27,000 × *g* at 4 °C for 30 min. The supernatant fractions are stored at −20 °C.
3. Add the 20 μL of 2× SDS sample buffer to the 20 μL of supernatant fractions, and boil them at 100 °C for 5 min. The samples are ready for SDS-PAGE.

3.2 Preparation of Mouse Liver Microsomal Fraction [22] for Isoelectric Focusing

1. Livers are isolated from ICR mice and immediately frozen on dry ice.
2. Twenty grams of mouse livers are homogenized with the addition of 30 mL of cold extraction buffer in an ice bath and the homogenates are centrifuged at 27,000 × *g* at 4 °C for 30 min. The supernatant fraction contains cytosolic proteins and microsomes except most mitochondria, nuclei, large organelles, cells, and cell debris.
3. The supernatant fractions are centrifuged at 100,000 × *g* for 1 h and the microsomes are pelleted. The pellet is suspended in 60 mL of extraction buffer and centrifuged at 100,000 × *g* for 1 h to remove excess cytosolic proteins. After a second wash with 60 mL of extraction buffer, the pellet is suspended in 5 mL of extraction buffer and stored at −20 °C in 0.6 mL aliquot.
4. Add 100 μL of 10 % Triton X-100 to 100 μL of microsomal fraction (about 2 mg of proteins) to lyse microsomes by vigorous vortex until the solution becomes clear.
5. Add 200 μL of 20 % TCA to the lysate to precipitate proteins by vigorous vortex for few seconds, and centrifuge at 6,000 × *g* for 30 s. Discard the supernatant.
6. Disperse the pellet in 200 μL of deionized water. Add 1 mL of −20 °C precooled acetone, and centrifuge at 6,000 × *g* for 30 s. Discard the supernatant.
7. Repeat **step 6** three times. Finally, disperse the pellet in 20 μL of acetone with repeated pipetting. Dry the pellet in a chemical hood for 2–3 h.
8. Solubilize the pellet in 100 μL of rehydration buffer and centrifuge at 14,000 × *g* for 20 min. Transfer the supernatant into a new microfuge tube and the sample is ready for IEF.

3.3 Isoelectric Focusing

1. IEF is performed with the Ettan IPGphor Focusing system.
2. Ten microliter of sample (about 40 μg of proteins) was diluted with 115 μL of rehydration buffer and centrifuged at 14,000 × *g*

for 20 min. Apply supernatant to the strip holder channel and avoid formation of air bubbles.

3. Remove the protective cover foil from an IPG strip. Position the IPG strip with the gel side down and direct the acidic (anodic) end of the strip toward the anodic end (+) of the strip holder. Hold the basic (cathodic) end of the IPG strip and lower the acidic end of the strip onto the solution. Push the IPG strip toward the anodic end of the strip holder and lower the strip onto the solution at the same time. Slide the IPG strip back and forth along the surface of the solution to coat the entire strip. Be careful not to trap air bubbles under the IPG strip.
4. Add IPG Cover Fluid dropwise into both ends of the strip holder until the entire IPG strip is covered. Place the cover on the strip holder to maintain good contact between the gel and electrodes, and put the holder onto the platform of IPGphor. Position anodic (+) end of the strip holder on anode (+) and cathodic end (−) of the strip holder on cathode (−) and then cover the lid of IPGphor.
5. After rehydration at 30 V for 10 h, start the IEF until the total voltage-hours reach 20,000–30,000.
6. Transfer the IPG strip to a tray and soak the IPG strip in the SDS equilibration buffer for 10 min, twice, with gentle agitation (*see* **Note 5**).

3.4 SDS-PAGE

3.4.1 SDS-PAGE of Sample for Gelatin Zymography [6]

1. Modified tricine SDS-PAGE is used to resolve the sample for gelatin zymography and the electrophoresis is performed with Hoefer SE-250 Mighty small II gel system.
2. Prepare 7.5 % separation gel by mixing 4.5 mL of acrylamide/bis-acrylamide solution, 10 mL of gel buffer, 4 g of glycerol, 8.5 mL of deionized water, 3 mL of 2 % gelatin, 0.1 mL of 10 % ammonium persulfate solution, and 10 μL of TEMED. Cast gel within 8 cm × 10 cm × 0.75 mm gel cassette, leave space for a stacking gel, and overlay with deionized water.
3. Prepare stacking gel by mixing 1 mL of acrylamide/bis solution, 3 mL of stacking gel buffer, 1 g of glycerol, 7 mL of deionized water, 0.1 mL of 10 % ammonium persulfate solution, and 10 μL of TEMED. Then insert a 10-well comb into the stacking gel solution immediately.
4. Apply 4 μL of each sample and 5 μL of prestained marker to each well and complete the electrophoresis at 150 V for 60 min with a water-cooling system.

3.4.2 SDS-PAGE of Sample for 2-DE

1. Modified tricine SDS-PAGE is used to resolve the sample for 2-DE and the electrophoresis is performed with Hoefer SE-260 Mighty small II gel system.

2. Prepare 7.5 % separation gel by mixing 4.5 mL of acrylamide/bis solution, 10 mL of gel buffer, 4 g of glycerol, 11.5 mL of deionized water, 0.1 mL of 10 % ammonium persulfate solution, and 10 μL of TEMED. Cast gel within 10.5 cm × 10 cm × 1.0 mm gel cassette, leave space for a stacking gel, and overlay with deionized water.
3. Prepare stacking gel by mixing 1 mL of acrylamide/bis solution, 3 mL of stacking gel buffer, 1 g of glycerol, 7 mL of deionized water, 0.1 mL of 10 % ammonium persulfate solution, and 10 μL of TEMED. Leave space for positioning an IPG strip and overlay with deionized water.
4. Position an equilibrated IPG strip onto the stacking gel and leave space near the acidic end of the IPG strip for insertion of the marker strip. Insert the marker strip half way into the stacking gel and start SDS gel electrophoresis as soon as possible. Electrophoresis is complete at 150 V for 120 min with water-cooling system.

3.4.3 Protein Thiol Oxidation Determination

1. Modified tricine SDS-PAGE is used to resolve the protein sample and the electrophoresis is performed with Hoefer SE-250 Mighty small II gel system.
2. Prepare 7.5 % separation gel by mixing 4.5 mL of acrylamide/bis solution, 10 mL of gel buffer, 4 g of glycerol, 11.5 mL of deionized water, 0.1 mL of 10 % ammonium persulfate solution, and 10 μL of TEMED. Cast gel within 8 cm × 10 cm × 0.75 mm gel cassette, leave space for a stacking gel, and overlay with deionized water.
3. Prepare stacking gel by mixing 1 mL of acrylamide/bis solution, 3 mL of stacking gel buffer, 1 g of glycerol, 7 mL of deionized water, 0.1 mL of 10 % ammonium persulfate solution, and 10 μL of TEMED. Then insert a 10-well comb into the stacking gel solution immediately.
4. MOVAS cell protein extract is obtained from the supernatant of the cell lysate within TE buffer by centrifugation at 14,000 × *g* for 10 min.
5. Add 4 μL of DMSO or 25, 50, and 100 mM dimedone to 36 μL of protein extract at 1 mg/mL protein concentration and leave the solution to sit at 25 °C for 2 h.
6. Add 40 μL of 2× SDS sample buffer containing 4 % β-mercaptoethanol to each sample and boil the samples at 100 °C for 5 min.
7. Apply 20 μL of each sample and 5 μL of prestained marker to each well and complete the electrophoresis at 150 V for 60 min with a water-cooling system.

3.5 Diffusion Blotting

1. PVDF membranes from different sources are equally efficient to retain proteins blotted by diffusion blotting. However, Millipore Immobilon-P PVDF membrane is used here.
2. Wet PVDF membranes with 100 % methanol and rinse them in deionized water three times before this procedure.
3. After electrophoresis, remove excess cathode buffer from the gel assembly. Lever the glass plate up with one spacer, and remove the glass plate. Most of the time, the gel remains firmly attached on the aluminum plate.
4. Press one piece of moistened PVDF membrane between two pieces of chromatography paper to remove excess water. Cover the gel surface with the PVDF membrane. Remove air bubbles, if any, between the gel and the PVDF membrane with a narrow weighing spatula by gently scratching the PVDF membrane from one end to the other.
5. If a second blot is required, the gel can be overlaid with another PVDF membrane on the other gel surface. With the aluminum plate side up, peel the gel from aluminum plate, set the PVDF membrane down on a piece of chromatography paper, and make the other side of the gel facing up. Cover the gel surface with another piece of PVDF membrane. Remove air bubbles between the gel and the second PVDF membrane.
6. Cover a piece of chromatography paper on the PVDF membrane and place a piece of glass plate of similar size on top to maintain the gel-membrane contact. Diffusion blotting should be satisfactory for 1–2 h (*see* **Note 6**).
7. After diffusion blotting, peel the PVDF membrane and rinse it with deionized water three times. The PVDF membrane can be processed for immunoblotting immediately (*see* **Note 7**). Meanwhile, the gel can be stained to reveal the protein profiles.

3.6 Staining for Gelatin Zymography or SDS-PAGE

1. Activation of proteases within the electrophoresed gelatin gel occurs in conditioning solution after renaturation of proteases by removing bound SDS. However, different proteases should be treated in different incubation conditions.
2. After diffusion blotting, soak the gelatin gel in Tris/Triton solution with constant agitation for 20 min, twice. Rinse the gel with deionized water and develop protease activities by agitating the gel in incubation buffer at 37 °C for 3 h.
3. Rinse the gel with deionized water and agitate the gel in Coomassie blue R-250 solution for 1–3 h until the entire gel becomes deep blue. Then soak the gel in destaining solution with constant agitation for few hours until the appearance of transparent bands.
4. To stain the gel of SDS-PAGE, agitate the gel with deionized water for 30 min and then agitate the gel in Coomassie blue

R-250 solution for 1–3 h until the entire gel becomes deep blue. Then soak the gel in destaining solution with constant agitation for few hours until the appearance of stained bands.

3.7 Staining for 2-DE

1. Modified silver staining protocol for 2-DE gels is based on the description of Heukeshoven and Dernick [24].
2. Wash a 2-DE gel with deionized water for 15 min, twice, and then agitate the gel in 20 % TCA solution for 1 h to fix proteins within the gel after diffusion blotting.
3. Wash the gel with deionized water for 5 min, three times, to remove excess TCA.
4. Soak the gel in sensitizing solution with constant agitation for 30 min.
5. Wash the gel with deionized water for 5 min, twice, to remove excess sensitizing solution.
6. Soak the gel in silver nitrate solution with constant agitation for 20 min.
7. Wash the gel with deionized water for 20 s, twice. After washing the gel, agitate it in developing solution for several seconds to few minutes under observation until staining is satisfactory.
8. Stop development immediately by agitating the gel within refilled 4 % acetic acid for 5 min. After washing the gel in deionized water for 5 min, three times, keep the gel in deionized water.

3.8 Immunoblotting [25]

1. Blotted PVDF membranes can be processed directly for immunoblotting after wash with deionized water. For stored membranes, wet the membranes with 100 % methanol and then briefly rinse the membrane with deionized water, three times.
2. Agitate a blotted PVDF membrane in blocking solution for 1–2 h. When blocking is complete, rinse the membrane with PBS once.
3. Pour the primary antibody solution into the container and agitate the membrane for 1–2 h. Wash the membrane with PBST for 5 min, twice, and PBS for 5 min, once.
4. Pour the secondary antibody solution into the container and agitate the membrane for 1–2 h. Wash the membrane with PBST for 5 min, twice, and PBS for 5 min, once.
5. Submerge and agitate the membrane within 10 mL of $NiCl_2$-DAB solution for about 10 s, and add 10 μL of H_2O_2 to activate HRP activity. Rock the container until the image becomes apparent, pour the waste into a reservoir containing bleach, and rinse the membrane with tap water. Flush the membrane under tap water for few minutes and preserve the membrane after being air-dried.

4 Notes

1. Some proteins are very resistant to reduction, so we have used up to 100 mM DTE or dithiothreitol routinely.
2. The ammonium persulfate solution would expire after 2 weeks.
3. Marker strips should be prepared fresh and avoid prolonged drying because some proteins would absorb onto the paper permanently.
4. Glutardialdehyde and formaldehyde are highly volatile and should not be prepared in stock solution.
5. The IPG strip can be immediately subjected to second dimension of SDS-PAGE or be frozen at −20 °C for later use.
6. The transfer time for diffusion blotting should be adjusted for gels with different thicknesses. We routinely blot gel of 0.75 mm for 1 h and gel of 1 mm for 2 h.
7. After three washes with deionized water, the membrane can be air-dried and stored for later use.

Acknowledgements

This work was supported by grants (to G.D.C.) from the National Science Council, Republic of China.

References

1. Reinhart MP, Malamud D (1982) Protein transfer from isoelectric focusing gels: the native blot. Anal Biochem 123:229–235
2. De Keyser F, Verbruggen G, Veys EM et al (1990) "Microgel diffusion blotting" for sensitive detection of antibodies to extractable nuclear antigens. Clin Chem 36:337–339
3. Heukeshoven J, Dernick R (1995) Effective blotting of ultrathin polyacrylamide gels anchored to a solid matrix. Electrophoresis 16:748–756
4. Olsen I, Wiker HG (1998) Diffusion blotting for rapid production of multiple identical imprints from sodium dodecyl sulfate polyacrylamide gel electrophoresis on a solid support. J Immunol Meth 220:77–84
5. Kurien BT, Scofield RH (1997) Multiple immunoblots after non-electrophoretic bidirectional transfer of a single SDS-PAGE gel with multiple antigens. J Immunol Meth 205:91–94
6. Chen H, Chang GD (2001) Simultaneous immunoblotting analysis with activity gel electrophoresis in a single polyacrylamide gel. Electrophoresis 22:1894–1899
7. Demczuk S, Harbers M, Vennström B (1993) Identification and analysis of all components of a gel retardation assay by combination with immunoblotting. Proc Natl Acad Sci U S A 90: 2574–2578
8. Reddy SV, Alcantara O, Boldt DH (1998) Analysis of DNA binding proteins associated with hemin-induced transcriptional inhibition. The hemin response element binding protein is a heterogeneous complex that includes the Ku protein. Blood 91:1793–1801
9. Iwata T, Sato S, Jimenez J et al (1999) Osmotic response element is required for the induction of aldose reductase by tumor necrosis factor-alpha. J Biol Chem 274:7993–8001

10. Bouet JY, Surtees JA, Funnell BE (2000) Stoichiometry of P1 plasmid partition complexes. J Biol Chem 275:8213–8219
11. Bauvois B, Djavaheri-Mergny M, Rouillard D et al (2000) Regulation of CD26/DPPIV gene expression by interferons and retinoic acid in tumor B cells. Oncogene 19:265–272
12. Kuzmanov U, Jiang N, Smith CR et al (2009) Differential N-glycosylation of kallikrein 6 derived from ovarian cancer cells or the central nervous system. Mol Cell Proteomics 8: 791–798
13. Zhu YS, Pfaff DW (1998) Differential regulation of AP-1 DNA binding activity in rat hypothalamus and pituitary by estrogen. Brain Res Mol Brain Res 55:115–125
14. Ogawa A, Yano M, Tsujinaka T et al (1997) Gene expression of albumin and liver-specific nuclear transcription factors in liver of protein-deprived rats. J Nutr 127:1328–1332
15. Neary JT, Zhu Q, Kang Y et al (1996) Extracellular ATP induces formation of AP-1 complexes in astrocytes via P2 purinoceptors. Neuroreport 7:2893–2896
16. Le Naour F (2007) Identification of tumor antigens by using proteomics. Methods Mol Biol 360:327–334
17. Canelle L, Bousquet J, Pionneau C et al (2005) An efficient proteomics-based approach for the screening of autoantibodies. J Immunol Meth 299:77–89
18. Matsuo K, Xiang Y, Nakamura H et al (2006) Identification of novel citrullinated autoantigens of synovium in rheumatoid arthritis using a proteomic approach. Arthritis Res Ther 8: R175–R187
19. Seo YH, Carroll KS (2009) Profiling protein thiol oxidation in tumor cells using sulfenic acid-specific antibodies. Proc Natl Acad Sci 106:16163–16168
20. Maller C, Schröde E, Eaton P (2011) Glyceraldehyde 3-phosphate dehydrogenase is unlikely to mediate hydrogen peroxide signaling: studies with a novel anti-dimedone sulfenic acid antibody. Antioxid Redox Signal 14:49–60
21. Hung CH, Huang HR, Huang CJ et al (1997) Purification and cloning of carp nephrosin, a secreted zinc endopeptidase of the astacin family. J Biol Chem 272:13772–13778, Correction 272:19632
22. Nigam SK, Goldberg AL, Ho S et al (1994) A set of endoplasmic reticulum proteins possessing properties of molecular chaperones includes Ca^{2+}-binding proteins and members of the thioredoxin superfamily. J Biol Chem 269:1744–1749
23. Schägger H, von Jagow GV (1987) Tricine-sodium dodecyl sulfate-polyacrylamide gel electrophoresis for the separation of proteins in the range from 1 to 100 kDa. Anal Biochem 166:368–379
24. Heukeshoven J, Dernick R (1988) Improved silver staining procedure for fast staining in PhastSystem Development Unit I. Staining of sodium dodecyl sulfate gels. Electrophoresis 9:28–32
25. Harlow E, Lane D (1988) Antibodies. Cold Spring Harbor Laboratory, Cold Spring Harbor, NY, pp 471–510

Chapter 10

Diffusion Blotting: A Rapid and Simple Method for Production of Multiple Blots from a Single Gel

Ingrid Olsen and Harald G. Wiker

Abstract

A very simple and fast method for diffusion blotting of proteins from precast SDS-PAGE gels on a solid plastic support was developed. Diffusion blotting for 3 min gives a quantitative transfer of 10 % compared to 1-h electroblotting. For each subsequent blot from the same gel a doubling of transfer time is necessary to obtain the same amount of protein onto each blot. High- and low-molecular-weight components are transferred equally efficiently when compared to electroblotting. However, both methods do give a higher total transfer of the low-molecular-weight proteins compared to the large proteins. The greatest advantage of diffusion blotting is that several blots can be made from each lane, thus enabling testing of multiple antisera on virtually identical blots. The gel remains on the plastic support, which prevents it from stretching or shrinking. This ensures identical blots and facilitates more reliable molecular weight determination. Furthermore the proteins remaining in the gel can be stained with Coomassie Brilliant Blue or other methods for exact and easy comparison with the developed blots. These advantages make diffusion blotting the method of choice when quantitative protein transfer is not required.

Key words SDS-PAGE, Diffusion blotting, Electrophoresis, Western blotting

1 Introduction

Electrophoretic transfer of protein is the most widely applied method for western blotting after SDS-PAGE. When using precast gels supported on plastic films, removal of the film became necessary to allow electrophoretic transfer. Ultrathin gels (0.1–0.2 mm) are difficult to separate from the supporting material, and a method for diffusion blotting which gave efficient transfer was subsequently developed [1–3]. Furthermore a method for diffusion blotting from 1.5 mm slab gels has been described [4] (*see* Chapter 11). We wished to investigate diffusion blotting as an alternative method for the commercially available precast 0.5 mm SDS-PAGE Excel gels on plastic supports and obtain quantitative data concerning the efficiency of the method [5, 6]. The efficiency of diffusion blotting was compared with 1-h electroblotting, and it was found that the transfer

Biji T. Kurien and R. Hal Scofield (eds.), *Western Blotting: Methods and Protocols*, Methods in Molecular Biology, vol. 1312, DOI 10.1007/978-1-4939-2694-7_10,

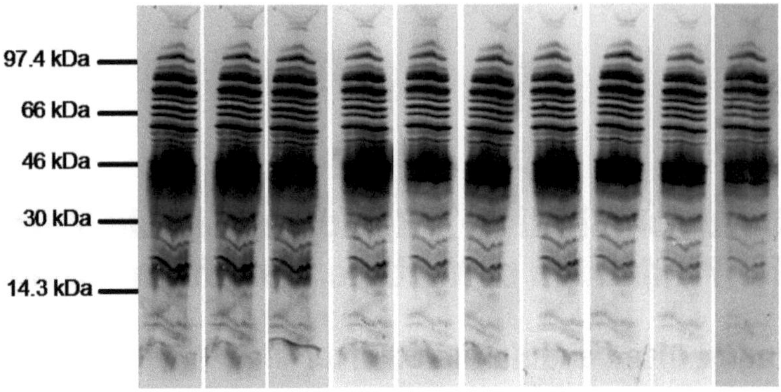

Fig. 1 Multiple blots of the same lane. *Mycobacterium avium* subspecies *avium* proteins (5 μg total protein) were separated on an SDS-PAGE 8–18 % precast gel and transferred to a nitrocellulose membrane by diffusion blotting. The transfer time was kept constant at 3 min for each imprint. The blots were stained with polyclonal anti-*M. avium* antibodies and HRP-labeled anti-rabbit immunoglobulin with diaminobenzidine as substrate. The picture shows the individual blots 1–10 and the position of molecular mass markers at the *left*. Note reproduction of artifacts such as curving of the bands

rate was fast initially and that it declined with time. Compared to electroblotting, 10 % of the proteins was transferred during the first 3 min, 20 % was transferred after 10 min, and after 3 h 45–65 % was transferred. Diffusion blotting also allows for several blots to be made from one gel. A doubling of transfer time from each blot is necessary to obtain similar amount of proteins on each blot [5]. After blotting, the proteins in the gel can be stained by standard methods such as Coomassie Brilliant Blue or silver staining. Diffusion blotting is a simple method which is performed with minimal equipment. The gel can remain on the solid matrix, which maintains the integrity of the gel and permits the generation of multiple blots. Several antisera can thus be tested on identical runs and these can be compared to protein staining of the same gel. The many advantages of diffusion blotting make this an alternative to electroblotting when ultimate sensitivity is not required (Fig. 1).

2 Materials

1. Prestained molecular mass standard from any supplier.
2. Nitrocellulose membrane.
3. 3MM Chr paper.
4. Tris/glycine transblot buffer (pH 8.0): 25 mM Tris, 190 mM glycine, and 20 % (v/v) methanol. Adjustment of pH is not required (*see* **Note 1**).
5. Glass plate.
6. Weight.

3 Methods

1. After SDS-PAGE put the gel on a flat surface leaving the solid support in place (for blotting after 2D gel electrophoresis, *see* **Note 2**).
2. Use gloves when handling membrane. Cut appropriate number of nitrocellulose membranes of the exact same size as the gel. Cut upper right corner corresponding to the cut corner on the precast SDS-PAGE Excel gels to help correct orientation.
3. Soak the nitrocellulose membrane in the transblot buffer until wet.
4. Remove excess buffer by holding two corners of the membrane with your fingers and carefully allowing the edge to touch a sheet of filter paper.
5. Carefully lay the membrane onto the gel. Avoid air bubbles. Do not remove the membrane after it has been in contact with the gel because some protein is transferred immediately and this may result in double bands.
6. Wet three sheets of 3MM paper in transblot buffer. Remove excess buffer by squeezing the sheets between two fingers and place the papers on top of the membrane.
7. Cover the gel, membrane, and filter paper assembly with a glass plate and a weight of approximately 1.5 kg (6 g/cm^2).
8. Leave for 3 min (*see* **Note 3**).
9. Repeat **steps 3–6** for every blot to be made and double the transfer time for each blot to obtain the same amount of protein on each blot.
10. Cut a mark in the nitrocellulose for each band in the prestained standard (*see* **Note 4**).

4 Notes

1. Instead of transblot buffer we have often used PBS pH 7.4 and it appears to work equally well.
2. The method is described for precast 0.5 mm SDS-PAGE gels from GE Healthcare (previously Amersham Biosciences), Piscataway, NJ, USA. These gels are also used in a two-dimensional electrophoresis system where Immobiline DryStrip for isoelectrofocusing is used in the first dimension and the precast gels are used for the second dimension. The method works equally well following two-dimensional gel electrophoresis, where exact pattern reproducibility is difficult to achieve. Creation of multiple identical imprints by diffusion blotting will eliminate problems with gel-to-gel variation and facilitate

correct identification of protein spots as well as easy comparison of different antisera.

3. The time can be changed according to the sensitivity of the detection system. Three minutes will give a transfer of approximately 10 % of what you see for electroblotting while 30 % is transferred after 30 min [5].
4. The prestained standard is visible after blotting, but will weaken considerably after blocking and development. Marking the bands with a knife will help in locating the molecular mass markers. If protein staining of the gel is performed, molecular mass determination can be done on the corresponding band identified on the gel.

References

1. Heukeshoven J, Dernick R (1995) Effective blotting of ultrathin polyacrylamide gels anchored to a solid matrix. Electrophoresis 16:748–756
2. Braun W, Abraham R (1989) Modified diffusion blotting for rapid and efficient protein transfer with PhastSystem. Electrophoresis 10:249–253
3. Jagersten C, Edstrom A, Olsson B, Jacobson G (1988) Blotting from PhastGel media after horizontal sodium dodecyl sulfate-polyacrylamide gel electrophoresis. Electrophoresis 9:662–665
4. Kurien BT, Scofield RH (1997) Multiple immunoblots after non-electrophoretic bidirectional transfer of a single SDS-PAGE gel with multiple antigens. J Immunol Methods 205:91–94
5. Olsen I, Wiker HG (1998) Diffusion blotting for rapid production of multiple imprints from sodium dodecyl sulfate polyacrylamide gel electrophoresis on a solid support. J Immunol Methods 220:77–84
6. Olsen I, Wiker HG (2009) Diffusion blotting for rapid production of multiple identical imprints from sodium dodecyl sulfate polyacrylamide gel electrophoresis on a solid support. Methods Mol Biol 536:35–38

Chapter 11

Multiple Immunoblots by Passive Diffusion of Proteins from a Single SDS-PAGE Gel

Biji T. Kurien and R. Hal Scofield

Abstract

Western blotting enables the detection and characterization of proteins of low abundance. Sodium dodecyl sulfate (SDS) polyacrylamide gel-separated proteins are normally transferred electrophoretically to nitrocellulose or polyvinylidene difluoride membranes. Here we describe the transfer proteins [Ro 60 (or SSA) autoantigen, 220 and 240 kDa spectrin antigens, and prestained molecular weight standards] by diffusion from SDS polyacrylamide gels at 37 °C. Up to 12 immunoblots can be obtained from a single gel by this method.

Key words Non-electrophoretic transfer, Immunoblots, Bidirectional transfer, Nitrocellulose membrane, Autoantibodies, Autoantigens

1 Introduction

Western blotting, involving electrophoretic transfer of proteins to a microporous membrane support with subsequent immunodetection [1], has made a tremendous impact in the field of immunology. Proteins separated on sodium dodecyl sulfate-polyacrylamide gel electrophoresis (SDS-PAGE) are transferred electrophoretically to mainly nitrocellulose or polyvinylidene difluoride [1–4]. Protein transfer from gels to membranes has been achieved in three ways: (a) simple diffusion [3], (b) vacuum-assisted solvent flow [5, 6], and (c) electrophoretic elution [1, 7, 8]. Only electrophoretic elution has been used widely owing to a variety of reasons, including efficiency, simplicity, and length of transfer. Bidirectional transfer procedure was demonstrated in 1980, when Smith and Summers [9] transferred DNA and RNA from polyacrylamide gels to nitrocellulose in 36 h to obtain two blots. Here we demonstrate that proteins can be efficiently transferred non-electrophoretically from SDS-PAGE gels to nitrocellulose membranes. A similar procedure has been used in 1982 [10] to transfer proteins from thin (0.5 mm)

Biji T. Kurien and R. Hal Scofield (eds.), *Western Blotting: Methods and Protocols*, Methods in Molecular Biology, vol. 1312, DOI 10.1007/978-1-4939-2694-7_11, © Springer Science+Business Media New York 2015

native isoelectric focusing gels to nitrocellulose membranes to obtain two blots in 1 h. Diffusion-mediated transfer of immunoglobulins from one side of a native gel after isoelectric focusing to an antigen-coated nitrocellulose sheet has been achieved in a similar fashion [11, 12]. We were able to transfer and immunologically detect a 60,000 molecular weight autoantigen (Ro 60) (Fig. 2) as well as the spectrin antigens (molecular weight >200,000) (Fig. 3) using this method. When prestained molecular weight standards were transferred, only markers up to 118,000 molecular weight could be visualized (these proteins were not visualized immunologically) (Figs. 2 and 3).

Thus, we have obtained up to 12 immunoblots from a single gel using multiple antigens (high-, intermediate-, and low-molecular-weight proteins) and multiple sera. Subsequently several investigators have also obtained similar results and also in a quantitative manner (*see* Chapters 9 and 10).

2 Materials

Prepare all solutions using ultrapure water (prepared by purifying deionized water, to attain a sensitivity of 18 MΩ-cm at 25 °C) and analytical grade reagents. Prepare and store all reagents at room temperature (unless indicated otherwise). Diligently follow all waste disposal regulations when disposing waste materials. We do not add sodium azide to reagents.

2.1 SDS Polyacrylamide Gel

1. Resolving gel buffer: 1.5 M Tris–HCl, pH 8.8. Add about 100 mL water to a 1 L graduated cylinder or a glass beaker (*see* **Note 1**). Weigh 181.7 g Tris–HCl and transfer to the cylinder. Add water to a volume of 900 mL. Mix and adjust pH with HCl (*see* **Note 2**). Make up to 1 L with water. Store at 4 °C.
2. Stacking gel buffer: 0.5 M Tris–HCl, pH 6.8. Weigh 60.6 g Tris–HCl and prepare a 1 L solution as in previous step. Store at 4 °C.
3. Thirty percent acrylamide/Bis solution (29.2:0.8) acrylamide:bis: Weigh 29.2 g of acrylamide monomer and 0.8 g Bis (cross-linker) and transfer to a 100 mL graduated cylinder containing about 40 mL of water. Add a spatula of AG 501-X8 (D) mixed-resin beads and mix for about 30 min. Make up to 100 mL with water and filter through a 0.45 μm Corning filter (*see* **Note 3**). Store at 4 °C, in a bottle wrapped with aluminum foil (*see* **Note 4**).
4. Ammonium persulfate: 10 % solution in water (*see* **Note 5**).
5. *N*,*N*,*N*,*N*′-Tetramethyl-ethylenediamine: Store at 4 °C (*see* **Note 6**).

6. SDS-PAGE running buffer: 0.025 M Tris–HCl, pH 8.3, 0.192 M glycine, 0.1 % SDS (*see* **Note 7**).
7. SDS lysis buffer (5×): 0.3 M Tris–HCl (pH 6.8), 10 % SDS, 25 % β-mercaptoethanol, 0.1 % bromophenol blue, 45 % glycerol. Leave one aliquot at 4 °C for current use and store the remaining aliquots at −20 °C (*see* **Note 8**).
8. Bromophenol blue (BPB) solution: Dissolve 0.1 g BPB in 100 mL water.

2.2 Immunoblotting

1. Nitrocellulose membranes.
2. Western blot transfer buffer: 0.025 M Tris–HCl, 0.192 M glycine, 20 % methanol (*see* **Note 9**).
3. Tris-buffered saline (TBS; 10×): 1.5 M NaCl, 0.1 M Tris–HCl, pH 7.4.
4. TBS containing 0.05 % Tween-20 (TBST).
5. Blocking solution: 5 % milk in TBS (*see* **Note 10**). Store at 4 °C.
6. Diluent solution: 5 % milk in TBST (*see* **Note 10**). Store at 4 °C.
7. Mini PROTEAN® 3 System glass plates.
8. Medium binder clips (1¼ in.).
9. Plastic container.
10. Wypall X-60-reinforced paper.
11. Nitro blue tetrazolium (NBT)/5-bromo-4-chloro-3-indolyl phosphate (BCIP): Dissolve 1 g NBT in 20 mL of 70 % dimethylformamide (DMF). Dissolve 1 g BCIP in 20 mL of 100 % DMF. Add 33 μL of BCIP and 66 μL of NBT to 10 mL of alkaline phosphatase buffer just before adding to membrane. Alternatively, use 1-Step™ NBT/BCIP ready-made mix.

2.3 Antigens and Conjugates

1. Purified red blood cell spectrin, anti-spectrin polyclonal antibody, anti-hemoglobin antibody.
2. BenchMark prestained molecular weight standards.
3. Purified bovine Ro 60: Purify Ro 60 as reported [13, 14] or purchase from Immunovision, Springdale, AK, USA.
4. Prepare human erythrocyte membranes according to Dodge et al. [15].

3 Methods

Carry out all procedures at room temperature unless otherwise specified.

3.1 10 % Sodium Dodecyl Sulfate-Polyacrylamide Gel Electrophoresis

1. Mix 2.5 mL of resolving buffer, 3.33 mL of acrylamide mixture, and 4 mL water in a 50 mL conical flask. Degas with helium for 10 min. Add 100 μL of SDS, 80 μL of ammonium persulfate, and 10 μL of TEMED and cast gel within a 7.25 cm × 10 cm × 1.5 mm gel cassette. Allow space for stacking gel and gently overlay with isobutanol or water (*see* **Note 11**).
2. Prepare the stacking gel by mixing 1.25 mL of resolving buffer, 0.67 mL of acrylamide mixture, and 3 mL water in a 50 mL conical flask. Degas with helium for 10 min. Add 100 μL of SDS, 40 μL of ammonium persulfate, and 5 μL of TEMED. Insert a 10-well gel comb immediately without introducing air bubbles.
3. Heat aliquots of bovine Ro 60, RBC membranes, and human spectrin antigens at 95 °C for 5 min. Do not add lysis buffer to the prestained protein standard or subject it to heat. Centrifuge the heated samples at 3,000 × *g* for 30 s to bring down the condensate. Load increasing amounts of Ro antigen (1–4 μg) on one gel and same amounts of spectrin (3 μg/lane) or RBC membrane antigens on two other gels along with protein standards (10 μL/well-2 μg/marker/lane). Add protein standards in every other lane (alternating with spectrin) in the gel with spectrin. Electrophorese at 15 mA until the sample has entered the gel and then continue at 25 mA till the dye front (from the BPB dye in the samples) reached the bottom of the gel (*see* **Note 12**).
4. Following electrophoresis, pry the gel plates open with the use of a spatula. The gel remains on one of the glass plates. Rinse the gel with water and transfer carefully to a container with western blot transfer buffer.
5. Cut a nitrocellulose membrane to the size of the gel and immerse in methanol. Rinse twice in distilled water and once with transfer buffer.

3.2 Non-electrophoretic Transfer

1. Immediately following SDS-PAGE, when the dye front reaches the end of the gel, turn off the power supply. Separate the gel plates with the help of a spatula or a similar tool. Remove the stacking gel.
2. Rinse the gel (still supported by the bottom glass plate) carefully with deionized water to remove traces of SDS-PAGE running buffer.
3. Excise the gel with spectrin antigen such that there is one lane with the protein markers and one with the spectrin antigen.
4. Leave the gels to air-dry for 5–10 min (*see* **Note 13**).
5. Gently lay one nitrocellulose membrane, cut to the shape of the gel, on top of the gel (*see* **Note 14**).

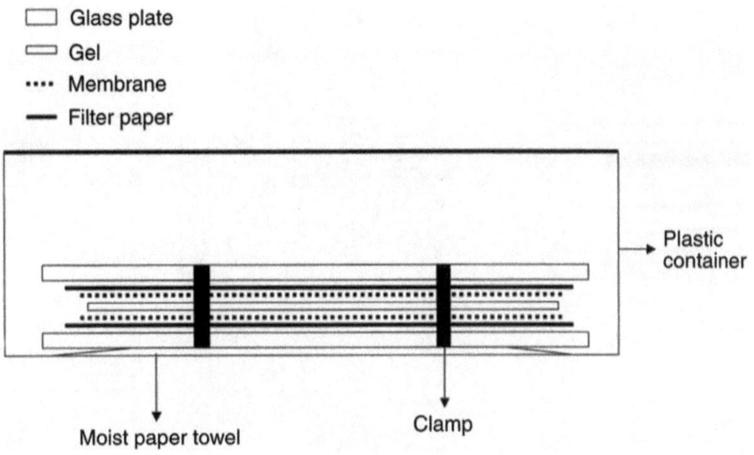

Fig. 1 Gel-membrane assembly for the non-electrophoretic transfer of proteins from SDS-PAGE gels to nitrocellulose membranes to obtain up to 12 blots. The polyacrylamide gel is sandwiched between two membranes, filter paper, and glass plates and incubated at 37 °C for varying periods of time to obtain up to 12 blots (reproduced from ref. 2 with permission from Elsevier)

6. Gently lift the gel-membrane sandwich from the glass plate and place it on a Whatman No. 3 filter (place membrane side directly on the filter paper and the exposed gel side on top) cut to the size of the gel.
7. Place a second nitrocellulose membrane, cut to the shape of the gel, on top of the gel, followed by a Whatman No. 3 filter paper cut similarly (*see* **Note 14**).
8. Place the nitrocellulose-gel-filter paper sandwich between two mini-PROTEAN® 3 System glass plates and secure with clamps.
9. Place this assembly in a pre-warmed humidified plastic container (Fig. 1) and incubate at 37 °C for 30 min (*see* **Note 15**). Remove the membranes for immunoblotting (*see* **Note 16**) (Fig. 2).
10. Repeat this procedure with another set of nitrocellulose membranes and incubate the assembly at 37 °C for 2 h to obtain two more blots from the same gel (*see* **Note 17**).
11. Use the same gel further to obtain blots by repeating this procedure to obtain a total of 12 blots. Incubate gel with the respective membranes for a period of 3 or 4 h to obtain the third and fourth sets of blots. Obtain the fifth and sixth sets of blots (Fig. 3) by incubating with the respective membranes for a period of 9 h or 36 h, respectively (*see* **Note 18**).
12. Cut excess membrane to smoothen edges and also cut the spectrin-containing membrane into individual lanes (*see* **Note 19**).
13. Block the membranes with blocking solution for 1 h.

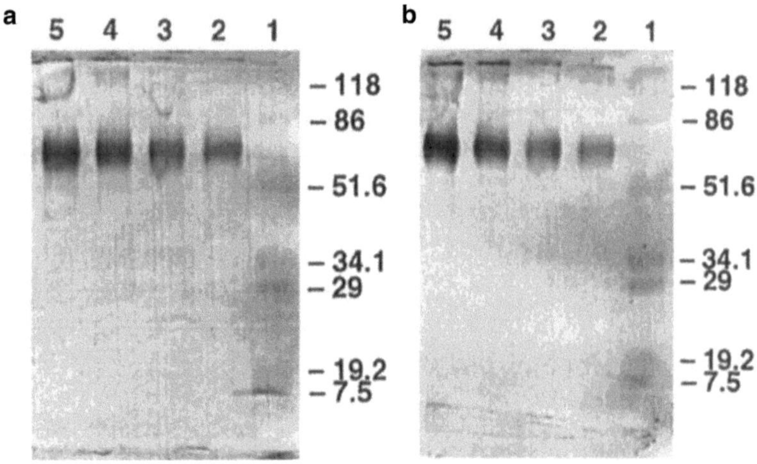

Fig. 2 Ro 60 immunoblots obtained following non-electrophoretic transfer for 30 min or 2 h. (**a**) One of the two Ro 60 blots obtained by the first incubation of nitrocellulose membranes on either side of the gel at 37 °C for 30 min. (**b**) A blot from the second set (blots 3 and 4) obtained from the same gel following incubation at 37 °C for 2 h (reproduced from ref. 6 with permission from Elsevier)

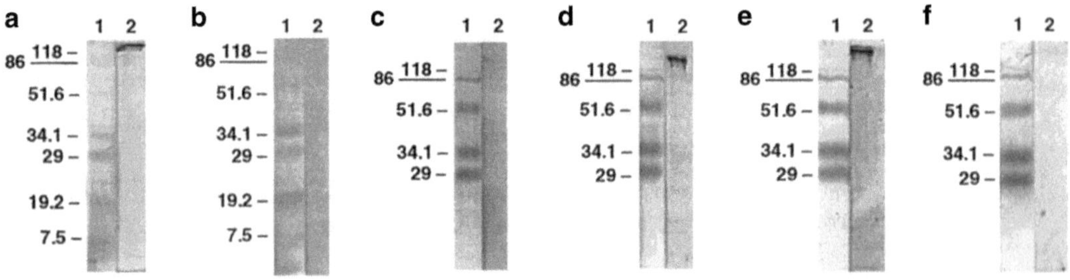

Fig. 3 Immunoblots obtained using spectrin as antigen. *Lane 1* in (**a**)–(**f**) shows prestained SDS-PAGE molecular weight standards. *Lane 2* shows spectrin probed with either preimmune (**b**, **c**, **f**) or anti-spectrin rabbit sera (**a**, **d**, **e**). Panel (**a**) shows one of the first two blots obtained and has been probed with anti-spectrin. Panel (**b**) shows the second of the two blots of the first set obtained from the reverse side of the gel and was probed with preimmune sera. Panels (**c**) and (**d**) show the fifth set (blots 9 and 10) of immunoblots probed with preimmune and anti-spectrin sera, respectively. Panels (**e**) and (**f**) show the sixth set (blots 11 and 12) of blots probed with anti-spectrin and preimmune sera, respectively (reproduced from ref. 6 with permission from Elsevier)

14. Add appropriate anti-sera to the membranes (anti-spectrin, anti-Ro 60, or control sera) and incubate for 2 h.
15. Rinse membrane strips with deionized water 2–3 times (*see* **Note 20**).
16. Wash 5× with TBST, 5 min each time.
17. Add anti-human IgG or anti-rabbit IgG alkaline phosphatase conjugate (1:5,000 dilution, diluted in diluent) and incubate for 1 h.
18. Wash as in **step 14**.

19. Add 500 μL of NBT/5-bromo-4-chloro-3-indolyl phosphate substrate and let bands develop.
20. Rinse 2–3 times with deionized water (*see* **Note 20**).
21. Wash with TBST and arrange strips on paperboard inserts (*see* **Note 21**).

4 Notes

1. Having water at the bottom of the cylinder helps to dissolve the Tris relatively easily, allowing the magnetic stir bar to go to work immediately. If using a glass beaker, the Tris can be dissolved faster if the water is warmed to about 37 °C. However, the downside is that care should be taken to bring the solution to room temperature before adjusting pH.
2. Concentrated HCl (12 N) can be used at first to narrow the gap from the starting pH to the required pH. From then on it would be better to use a series of HCl (e.g., 6 and 1 N) with lower ionic strengths to avoid a sudden drop in pH below the required pH.
3. Wear a mask when weighing acrylamide. To avoid exposing acrylamide to co-workers, cover the weigh boat containing the weighed acrylamide with another weigh boat (similar size to the original weigh boat containing the weighed acrylamide) when transporting it to the fume hood. Transfer the weighed acrylamide to the cylinder inside the fume hood and mix on a stirrer placed inside the hood. Unpolymerized acrylamide is a neurotoxin and care should be exercised to avoid skin contact. Mixed resin AG 501-X8 (D) (anion- and cation-exchange resin) is used when acrylamide solution is made, since it removes charged ions (e.g., free radicals) and allows longer storage. Some investigators store the prepared acrylamide along with this resin in the refrigerator. However, we filter them out before storage. The used mixed resin should be disposed as hazardous waste. Manufacturer's warning states that this resin is explosive when mixed with oxidizing substances. The resin contains a dye that changes from blue-green to gold when the exchange capacity is exhausted.
4. The acrylamide solution can be stored at 4 °C for 1 month. Acrylamide hydrolyzes to acrylic acid and ammonia. The acrylamide mixture, buffer, and water can be prepared in large batches, frozen in aliquots (for greater day-to-day reproducibility), and used indefinitely (*see* ref. 16). Remove the required amount, bring to room temperature, and add the other ingredients for polymerization. However, in our laboratory we make the acrylamide solution fresh about every month when we cast our own gels.

5. We find that it is best to prepare this fresh each time.
6. We find that storing at 4 °C reduces its pungent smell.
7. Simple method of preparing running buffer: Prepare 10× native buffer (0.25 M Tris–HCl, 1.92 M glycine). Weigh 30.3 g Tris–HCl and 144 g glycine, mix, and make it to 1 L with water. Dilute 100 mL of 10× native buffer to 990 mL with water and add 10 mL of 10 % SDS. Care should be taken to add SDS solution last, since it makes bubbles.
8. SDS precipitates at 4 °C. Therefore, the lysis buffer needs to be warmed prior to use.
9. Dilute 100 mL of 10× native buffer to 800 mL with water and add 200 mL of methanol. Avoid adding methanol directly to the 10× buffer, since it precipitates its ingredients. Even in such a scenario the precipitate can be redissolved by the addition of 800 mL water.
10. Add 100 mL of 10× TBS to a 1 L graduated cylinder and make it to about 800 mL with water. Transfer 50 g skim milk powder into the cylinder and stir until dissolved. Make to 1 L with water. Separate 500 mL as the blocking solution. To the remaining 500 mL add 250 μL of Tween-20 (cut end of blue tip to aspirate Tween-20 easily), dissolve, and use it as the diluent.
11. The gel cassette was sealed at the base using 1 % agarose. Overlay the resolving gel with water for gels having acrylamide concentration lower than 8 % and use isobutanol (or isobutanol saturated with water) for gels of 10 % or greater (*see* ref. 17). This overlay prevents contact with atmospheric oxygen (which inhibits acrylamide polymerization) in addition to helping to level the resolving gel solution.
12. Centrifuging the samples prior to the run helps remove insoluble debris, which could produce streaks in the protein lanes (revealed when stained with Coomassie blue). Add a drop of 0.1 % BPB to the upper chamber buffer. This helps to form a much stronger dye front during the electrophoretic run.
13. Membrane contact with the gel is much better when the gel is not moist. Therefore it is important to dry the gel for 5–10 min. The membrane will now stick well to the gel and gel will peel off the bottom glass plate by just lifting the membrane.
14. Hold the two top corners of the membranes with each hand. Lower the bottom part of the membrane first on the lower part of the gel and gently release the membrane little by little to lay the complete membrane on the gel. This will prevent trapping of bubbles in between the gel and the membrane. A 10 mL pipette was used to roll out air bubbles from the

gel-membrane sandwich prior to placing in transfer cassette. In the case of the gel with spectrin, cut the membrane to fit the two lanes of the gel.

15. The humid chamber consisted of a closed plastic container with a moist Terri Wipes paper towel at the bottom. The container must be big enough to contain the nitrocellulose-gel-filter paper assembly encased within the glass plates.
16. The second set of two blots was also obtained following incubation with the gel for 1 h (*see* Fig. 3b, c).
17. While removing the nitrocellulose membranes from the gel for immunoblotting, it would be common to find that the gel comes up stuck to one of the two membranes. To remove this membrane from the gel, place a fresh, dry nitrocellulose membrane on top of the gel and gently lift the gel. The gel becomes stuck to this fresh membrane, thus releasing the other membrane.
18. Gel dries, in spite of placing in humid chamber, when incubated for longer time periods (36 h). Therefore it is best to use the blots obtained after 12-h incubation with the membrane.
19. Cut a tiny wedge from the bottom left side of the marker lane and the main membrane sheet for orientation purposes. Also, in the case of the membranes with spectrin (*see* Subheading 3.1, **step 3**), excise the spectrin lane from the protein marker lane after matching each spectrin lane with its specific protein marker lane with pencil marks.
20. Rinsing the membrane strips with deionized water 2–3 times will help remove a bulk of the nonspecific antibodies and help reduce the amount of TBST used subsequently and also reduce the number of washes. This wash helps to reduce nonspecific binding of NBT/BCIP to the strip. The water, owing to its low ionic strength compared to TBST, will be able to remove contaminants much better than TBST. Water is much cheaper compared to TBST, in terms of money and labor. Other investigators have found no reduction in detection of specific signals due to washing with water [18].
21. We use paperboards placed in between stacks of ELISA plates in packages of ELISA plates (Costar, Cambridge, MA, USA) for this purpose.

Acknowledgement

This work was supported by NIH grant ARO1844 and Oklahoma Center for the Advancement of Science and Technology to R.H.S.

References

1. Towbin H, Staehelin T, Gordon J (1979) Electrophoretic transfer of proteins from polyacrylamide gels to NC sheets: procedure and applications. Proc Natl Acad Sci U S A 76: 4350–4354
2. Kurien BT, Scofield RH (2006) Western blotting. Methods 38:283–293
3. Renart J, Reiser J, Stark GR (1979) Transfer of proteins from gels to diazobenzyloxymethyl paper and detection with anti-sera: a method for studying antibody specificity and antigen structure. Proc Natl Acad Sci U S A 76: 3116–3120
4. Gershoni JM, Palade GE (1983) Protein blotting: principles and applications. Anal Biochem 131:1–15
5. Peferoen M, Huybrechts R, De Loof A (1982) Vacuum-blotting: a new simple and efficient transfer of proteins from sodium dodecyl sulfate-polyacrylamide gels to NC. FEBS Lett 145:369–372
6. Kurien BT, Scofield RH (1997) Multiple immunoblots after non-electrophoretic bidirectional transfer of a single SDS-PAGE gel with multiple antigens. J Immunol Methods 205:91–94
7. Kurien BT, Scofield RH (eds) (2009) Protein blotting and detection. Humana Press, Totowa, NJ
8. MacPhee DJ (2010) Methodological considerations for improving Western blot analysis. J Pharmacol Toxicol Methods 61: 171–177
9. Smith GE, Summers MD (1980) The bidirectional transfer of DNA and RNA to nitrocellulose or diazobenzyloxymethyl-paper. Anal Biochem 109:123–129
10. Reinhart MP, Malamud D (1982) Protein transfer from isoelectric focusing Gels: the native blot. Anal Biochem 123:229–235
11. Knisley KA, Rodkey LS (1986) Affinity immunoblotting. High resolution isoelectric focusing analysis of antibody clonotype distribution. J Immunol Methods 95:79–87
12. Scofield RH, Zhang F, Kurien BT, Anderson CJ, Reichlin M, Harley JB et al (1996) Development of the anti-Ro autoantibody response in a patient with systemic lupus erythematosus. Arthritis Rheum 39:1664–1668
13. Deutscher SL, Harley JB, Keene JD (1988) Molecular analysis of the 60 kD human Ro ribonucleoprotein. Proc Natl Acad Sci U S A 85:9479–9483
14. Ben-Chetrit E, Gandy BJ, Tan EM, Sullivan KF (1989) Isolation and characterization of a cDNA clone encoding the 60 kDa component of the human SS-A/Ro ribonucleoprotein autoantigen. J Clin Invest 83:1284–1292
15. Dodge JT, Mitchell C, Hanahan DJ (1963) The preparation and chemical characteristics of hemoglobin-free ghosts of human erythrocytes. Arch Biochem Biophys 100:119–130
16. Harlow E, Lane D (eds) (1988) Electrophoresis. Appendix I. Antibodies. A laboratory manual. Cold Spring Harbor Laboratory, Cold Spring Harbor, NY, p 638
17. Harlow E, Lane D (eds) (1988) Electrophoresis. Appendix I. Antibodies. A laboratory manual. Cold Spring Harbor Laboratory, Cold Spring Harbor, NY, p 636
18. Wu M, Stockley PG, Martin WJ 2nd (2002) An improved Western blotting technique effectively reduces background. Electrophoresis 23: 2373–2376

Chapter 12

Slice Blotting

Graeme Lowe

Abstract

Slice blotting is a technique for recording the spatial distribution of extracellular signaling molecules released from thin slices of living tissue. Slices are positioned on the surface of a membrane that can trap secreted substances diffusing from the tissue. The pattern of membrane-bound antigens is subsequently visualized by immunoblotting.

Key words Slice, Immunoblot, Secretion

1 Introduction

A variety of functional imaging technologies can be applied to visualize communication between cells in diverse structured systems, ranging from neuronal circuits of the brain to the developing embryo. Many techniques are indirect, inferring intercellular communication by detecting intracellular signals or gene expression. Methods for directly imaging extracellular signals are limited in the kinds of molecules that can be detected. Slice blotting fills a gap in the spectrum of available methods, allowing investigators to map patterns of secretion for a broad range of signaling compounds.

The basic steps of slice blotting are relatively simple and easily implemented in most laboratories without a major investment in specialized equipment. Tissue slices are cut using a vibrating blade slicer and incubated in vitro on a blotting membrane for a set time period, under physiological conditions determined by the experimenter. During the incubation period, secreted molecules escaping the tissue matrix by diffusion are bound and immobilized by the membrane. Afterwards, the tissue slice is removed and the membrane is processed using standard Western blot methods [1–3], using primary antibodies to probe for particular antigens, and secondary antibodies for staining (Fig. 1).

Biji T. Kurien and R. Hal Scofield (eds.), *Western Blotting: Methods and Protocols*, Methods in Molecular Biology, vol. 1312, DOI 10.1007/978-1-4939-2694-7_12, © Springer Science+Business Media New York 2015

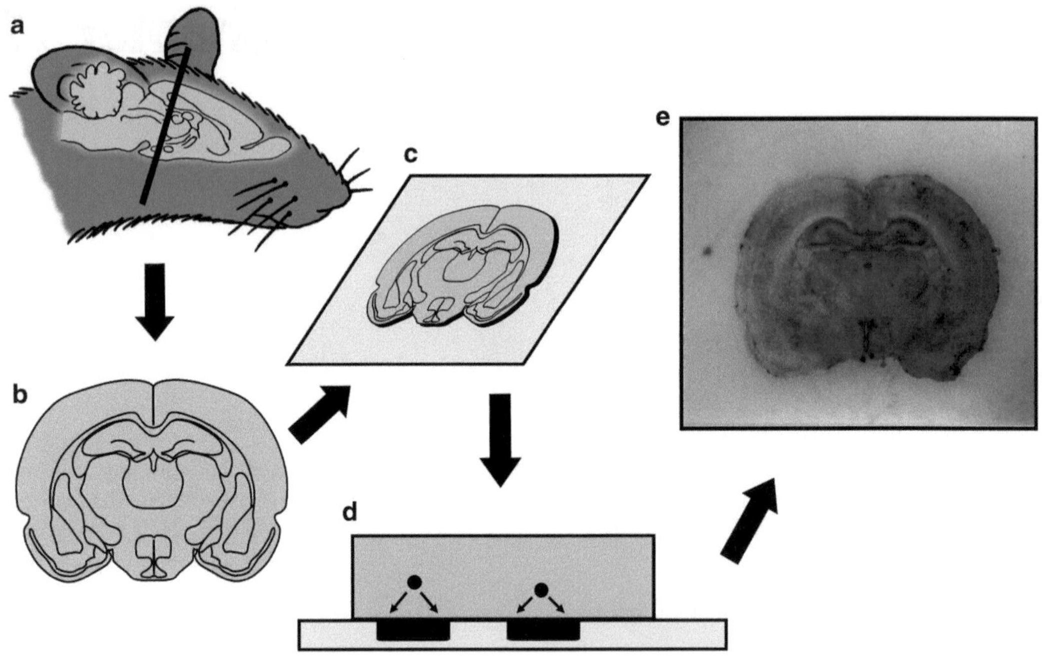

Fig. 1 Steps in slice blotting. A rat brain is extracted (**a**) and in vitro slices are cut (**b**). Slices are placed onto the surface of porous nitrocellulose membranes (**c**) and incubated in saline solution for a fixed time period. Immunoreactive signaling molecules released from slices are bound to the membranes (**d**), which are then developed using western blotting protocols to reveal the distribution of release sites (**e**)

The slice blotting protocols described here were used to detect and visualize secretion patterns of neuropeptide-Y (NPY) from in vitro rat brain slices (midbrain and brain stem) stimulated by high potassium [4]. Potassium treatment caused strong depolarization and action potential firing of NPY-containing nerve terminals, releasing peptide which was collected on nitrocellulose membranes. Similar protocols were applied to detect unstimulated release of a chemotactic guidance molecule, semaphorin III and glial cell line-derived neurotrophic factor (GDNF), from the rat olfactory system [4, 5]. With appropriate modifications, these basic protocols should be adaptable for use in many other signaling systems.

2 Materials

2.1 Brain Slice Preparation

1. Slicing solution, either sucrose based, 240 mM sucrose, 2.5 mM KCl, 10 mM Na-HEPES, 1 mM $CaCl_2$, 4 mM $MgCl_2$, 10 mM glucose, 0.2 mM ascorbic acid, pH 7.3, or glycerol based (same solution with 240 mM glycerol replacing sucrose).
2. High Mg^{2+} artificial cerebrospinal fluid (Mg-ACSF): 122 mM NaCl, 3.75 mM KCl, 26 mM $NaHCO_3$, 1 mM $CaCl_2$, 3.3 mM $MgCl_2$, 0.5 mM NaH_2PO_4, 7.5 mM glucose.

3. Standard ACSF: 122 mM NaCl, 3.75 mM KCl, 26 mM $NaHCO_3$, 2 mM $CaCl_2$, 1 mM $MgCl_2$, 0.5 mM NaH_2PO_4, 10 mM glucose.
4. High K^+ ACSF (same as standard ACSF, but 50 mM Na^+ replaced with 50 mM K^+).
5. Regulated gas supplies: (1) pure O_2, (2) 95 % O_2, (3) 5 % CO_2.
6. Fritted-glass gas dispersion tubes for bubbling solutions.
7. Vibratome™ or other vibrating blade tissue slicer.
8. Slice interface chamber: This is a closed 1 L plastic vessel containing a rack of nylon mesh wells where slices can be incubated prior to blotting. There is an entry port for a gas line (95 % O_2, 5 % CO_2) to enter and bubble the solution, and a small exit port.
9. Heated water bath.

2.2 Blotting

1. Nitrocellulose blotting membrane (e.g., Amersham Hybond™ ECL™, 0.45 μm pore size, 100 μm thickness, GE Healthcare Life Sciences, Piscataway, NJ, USA): Using thin membranes facilitates rapid O_2 diffusion to both sides of the slice.
2. Plastic pipette tips (1 mL capacity), razor blade.
3. Acrylic block 5 cm × 5 cm × 1 cm, with fine slots cut into top surface (3 mm deep, 0.2 mm wide).
4. Fine surgical forceps.
5. Pipettor with 200 μL pipette tips.
6. 500 mL dish or beaker.

2.3 Immunostaining

1. Tris-buffered saline-Tween 20 (TBST): 0.9 % NaCl, 100 mM Tris–HCl, 0.1 % Tween 20, pH 7.5.
2. Goat serum.
3. Primary antibody (e.g., rabbit anti-NPY antibody).
4. Biotinylated secondary antibody (e.g., biotinylated anti-rabbit IgG made in goat).
5. Vectastain™ Elite™ ABC (Avidin Biotinylated enzyme Complex) Kit (Vector Laboratories).
6. Vector™ VIP Peroxidase Substrate Kit (Vector Laboratories).

3 Methods

3.1 Brain Slice Preparation

Techniques for preparing in vitro brain slices have been well documented, and different laboratories may employ slightly different solutions and procedures. The slice blotting procedure yields best results when tissue is optimally healthy, and cell damage or death associated with slicing is kept to a minimum (*see* **Note 1**). The protocol described here is a variant of a widely used method that replaces sodium chloride with equimolar sucrose in the slicing solution, to reduce cell damage caused by chloride influx [1]. A similar strategy using glycerol as the replacement solute to maintain osmolarity has also been applied successfully to improve slice health [6]. Lowered Ca^{2+} and elevated Mg^{2+} are included to suppress depolarization and Ca^{2+} entry associated with NMDA receptor activation by glutamate released when slicing brain tissue. Ascorbic acid is added to scavenge free radicals during slicing.

1. Freeze a volume of slicing solution half-filling the bath vessel used to prepare slices, and layer chilled (4 °C) unfrozen slicing solution on top of the frozen solution (or, use active cooling system of the slice bath if it is available on the slicer). Bubble with O_2 gas.
2. Fill slice interface chamber with Mg-ACSF, bubbled with 95 % O_2 and 5 % CO_2 gas, and warm it to 32 °C in water bath.
3. Decapitate an anesthetized rat, remove the brain, and block off the portion to be sliced. Using a cyanoacrylate adhesive, attach the blocked brain to a slicing stage and submerge in the chilled, bubbled bath of the slicer.
4. Cut 300–400 μm thickness slices.
5. Transfer freshly cut slices to nylon mesh wells of slice interface chamber. Slices should be floated on the air-water interface of each well to maximize their contact with humidified O_2/CO_2 gas. This can be done by momentarily raising the nylon mesh bottom above the air-water interface.
6. Remove the interface chamber from the water bath, and allow it to cool slowly to room temperature (23 °C) over a 30–60-min period.
7. Replace solution in interface chamber with standard ACSF.

3.2 Blotting

1. Cut sheets of nitrocellulose membrane into small rectangles. The size of membrane to use depends on the size of the slices (e.g., 1–1.5 cm × 1.5–2 cm for rat brain).
2. Place a nitrocellulose rectangle onto the cut end of a vertically oriented 1 mL plastic pipette tip, or similar vertical support (e.g., a short acrylic rod). The support is smaller in diameter than the rectangle. This allows a drop of solution to be pipetted onto the membrane without the liquid spreading and spilling off the edges.
3. Pick up a slice from the interface chamber using a truncated 1 mL pipettor tip and deposit it onto the nitrocellulose rectangle. Pipette off excess solution until the slice settles flat onto the membrane surface. Using forceps, pick up the membrane

and lay it onto absorbent tissue or filter paper, slice surface up—this will remove the remaining solution by drawing it through the filter, causing the slice to adhere to the membrane. It is important to make sure that there are no kinks in the slice.

4. Immediately submerge the membrane and attached slice in a bath of standard ACSF bubbled with 95 % O_2 and 5 % CO_2 gas. Many membrane rectangles can be suspended in the same bath by holding them edgewise in thin slots of a machined acrylic block. They might also be submerged in multi-well plates with nylon mesh bottoms, similar to the way individual slices were kept in the interface chamber.
5. To stimulate or modulate release or secretion of signaling molecules, the incubating bath solution can be changed. For example, incubating in high K^+ ACSF will depolarize and drive the release of neurotransmitters or neuromodulators from presynaptic terminals (*see* **Note 2**). Pharmacological agents can be added to modulate secretion. Incubation period is variable, but 10–40-min exposure to high K^+ was sufficient to register immuno-detectable NPY release from rat brain (*see* **Note 3**).
6. After blotting, remove slices from nitrocellulose membranes with a gentle squirt of ACSF from a 200 μL pipette tip. Rinse membranes in water, air-dry, and store at −20 °C until staining. Tissue slices can be retained and further processed using other procedures if necessary.

3.3 Immunostaining

Conventional western blotting procedures can be applied to the nitrocellulose slice blots to visualize patterns of bound antigens.

1. Rinse nitrocellulose membranes for 30 min in TBST.
2. Incubate for 30 min in TBST + 3 % goat serum (blocking serum).
3. Incubate for 30 min in TBST + 1.5 % goat serum + primary antibody (e.g., 1:2,000 dilution of rabbit anti-NPY antibody).
4. Rinse three times, 10 min, in TBST.
5. Incubate for 60 min in TBST + 1.5 % goat serum + biotinylated secondary antibody (e.g., 7.5 μg/mL biotinylated anti-rabbit IgG, made from goat, Vector Laboratories).
6. Rinse three times, 10 min, in TBST.
7. Incubate for 60 min in ABC solution. Prepare the ABC solution 30 min ahead of this step (i.e., two drops solution A + two drops solution B, per 5 mL TBST).
8. Rinse three times, 10 min, in TBST.
9. Incubate for up to 30 min in VIP solution (i.e., three drops Reagent 1 + three drops Reagent 2 + three drops Reagent 3 + three drops hydrogen peroxide). The darkening due to peroxidase reaction can be monitored visually.

10. Rinse three times, 10 min, in TBST, after a strong contrast blot pattern is obtained.
11. Record blot patterns immediately by photographic or digital imaging. Nonspecific staining on membranes (*see* **Note 2**) tends to darken over time, reducing contrast. Darkening can be slowed by storing membranes at low temperature (−20 °C in 80 % glycerol).

4 Notes

1. Slice blotting can be sensitive to nonspecific background staining that may arise when the tissue is not in optimal health. Neurons in some brain regions may be more easily damaged by slicing due to their intrinsic structure, such as having widespread dendritic arbors. In these cases, it is expected that injured cells may more readily lyse during slice handling and blotting, which would contribute to elevated background staining. Some degree of background is expected because direct contact with the nitrocellulose membrane is likely to rupture a thin layer of surviving cells at the surface of the slice, leading to nonspecific release of contents. If slices become sticky (e.g., they cannot be easily lifted off the membrane by a gentle pulse of saline solution), major cell damage is indicated, and slices should be discarded. Cell condition can be checked by examining slices directly under a differential interference contrast microscope.
2. To verify a signal, it is important to set up a series of control blots in which the blotting conditions are varied. For example, in brain slices treated with high K^+, controls would correspond to incubation in a low K^+ ACSF solution. Other controls to determine the level of nonspecific background staining include omitting primary antibody, or preabsorbing primary antibody with antigen before applying to a blot.
3. Slice blotting integrates signal over time, so weak signals can be amplified by incubating slices on a blotting membrane for extended time periods.

References

1. Aghajanian GK, Rasmussen K (1989) Intracellular studies in the facial nucleus illustrating a simple new method for obtaining viable motoneurons in adult rat brain slices. Synapse 3:331–338
2. Kurien BT, Scofield RH (2006) Western blotting. Methods 38:283–293
3. Kurien BT, Scofield RH (2003) Protein blotting: a review. J Immunol Methods 274:1–15
4. Lowe G (1999) Slice blotting: a method for detecting the release of immunoreactive substances from living brain tissue. J Neurosci Methods 90: 117–127
5. Paratcha G, Ibanez CF, Ledda F (2006) GDNF is a chemoattractant factor for neuronal precursor cells in the rostral migratory stream. Mol Cell Neurosci 31:505–514
6. Ye JH, Zhang J, Xiao C, Kong JQ (2006) Patch-clamp studies in the CNS illustrate a simple new method for obtaining viable neurons in rat brain slices: glycerol replacement of NaCl protects CNS neurons. J Neurosci Methods 158:251–259

Chapter 13

Localizing Proteins by Tissue Printing

Rafael F. Pont-Lezica

Abstract

The simple technique of making tissue prints on appropriate substrate material has made possible the easy localization of proteins, nucleic acids, carbohydrates, and small molecules in a tissue-specific mode. Plant tissues can be used to produce prints revealing a remarkable amount of anatomical detail, even without staining, which might be used to record developmental changes over time. In this chapter we will focus on the protocols for the localization of proteins and glycans using antibodies or lectins, probably the most frequently used application, but the localization of other molecules is reported and the sources indicated.

Key words Immunodetection, Tissue blotting, Protein localization, Western blotting

1 Introduction

Fifty years ago, the first series of film printing for the localization of several enzymes (protease, amylase, RNAse, and DNAse) were realized by placing cryostat sections of various organs on gelatin, starch, or gelatin-nucleic acids [1]. Those films were then stained for the substrate, giving a negative image. Since both substrates and enzymes were macromolecules with slow diffusion rates, the images obtained were clear. The availability of several membranes such as nitrocellulose, Nytran, Genescreen, and Immobilon (polyvinylidene di-fluoride: PVDF) designed to bind proteins and nucleic acids opens a new era in tissue blotting, and a book edited by P. D. Reid et al. [2] included numerous protocols for the visualization of enzyme activities, protein and glycan localization, and gene expression, in plants and some animal tissues. Nitrocellulose membrane adsorbs relatively large quantities of proteins that are tightly bound, whereas it usually does not retain salts and hydrophilic small molecules. This type of membrane has been used for transferring proteins separated by sodium dodecyl sulphate-polyacrylamide gel electrophoresis to membranes for immunological detection. The procedure involves the electro elution of the negatively

Biji T. Kurien and R. Hal Scofield (eds.), *Western Blotting: Methods and Protocols*, Methods in Molecular Biology, vol. 1312, DOI 10.1007/978-1-4939-2694-7_13, © Springer Science+Business Media New York 2015

charged proteins from the gel to the membrane, followed by the visualization of the targeted protein with specific antibodies. The basic principle of tissue printing is that much of the contents of the cells at the surface of a freshly cut tissue section can be transferred to an adhesive or absorptive surface with little or no diffusion, by simple contact [3]. The obtained imprints reveal anatomical details that are difficult to see without fixing, embedding, and then sectioning; however, the resolution of anatomical prints is inferior to that of classical fixed and stained sections. The procedure is particularly useful with big samples such as sections of fruits, tubers, or stems, revealing the distribution of particular proteins in the whole organ [4, 5].

A digest of different applications of tissue printing for the detection of particular proteins, glycans, mRNA, enzyme activities, and microorganism in plants is given in Table 1.

The steps to make a print and to show the localization of proteins using specific antibodies, as well as the necessary precautions to avoid nonspecific staining in plant tissues, will be described in this chapter.

Table 1
Different applications of tissue prints for the detection of enzyme activity, specific proteins, virus, gene expression, and carbohydrates in plant tissues

Detected	Tissue	Assay	Film	References
Enzyme activity				
Amylase	Barley aleurone, pole bean hypocotyl, and cotyledons	Negative stain I_2.KI	Starch	[10, 11]
β-Glucosidase	Oil palm	Naphtyl-glucoside	NC	[12]
Catalase	Various	Negative stain. I_2	KI starch-H_2O_2	[13]
Cinnamyl alcohol dehydrogenase	Tomato, poplar	NBT, $NADP^+$, PMS, coniferyl alcohol	NC	[14]
Endo-β-mannanase	Coffea beans	Gel-locust bean gum	Paper	[15]
Laccase	Sycamore cells	4-Methylcatechol, syringaldazine	NC	[16]
Myrosinase	Turnip and horseradish	Sinigrin–glucose oxidase–peroxidase system	PVDF	[17]
Peroxidase	Pea epicotyls	α-Phenylene diamine	NC	[9]
Polyphenol oxidases	Tomato			[18]
Protease	Pea, mung bean	Negative stain	Gelatin-India ink	[19, 20]
Tyrosinase	Mushrooms	Tropolone and L-DOPA	NC	[21]

(continued)

Table 1 (continued)

Detected	Tissue	Assay	Film	References
Proteins				
Allergens	Pollen	Immunodetection	NC	[22]
Antifreeze proteins	Winter rye	Immunodetection	NC	[23]
Cardosin A	Artichoke	Immunodetection	NC	[24]
Cellulase	Abscission zones in various plants	Immunodetection	NC	[25, 26]
Cys-rich proteins	Soybean stem and pod	Fluorescent *S*-carboxymethylation	NC	[27]
Endoxylanase (fungal)	Tobacco	Immunodetection	Nylon	[28]
Expansins	Rice	Immunodetection	NC	[29]
Extensin, HRGPs	Soybean, bean, maize	Immunodetection	NC	[6, 30, 31]
Ferritin	Transgenic rice	Immunodetection	NC	[32]
Glutamine synthase	Potato	Immunodetection	NC	[33]
GRPs	Bean, soybean	Immunodetection	NC	[34–36]
Laccase	Sycamore	Immunodetection	NC	[16]
Lectins	Potato tuber, Canavalia seeds	Immunodetection	NC	[5, 37]
Lipid transfer proteins	Sunflower	Immunodetection	NC	[38]
Lipoxygenase	Tomato fruit	Immunodetection	NC	[39]
Natriuretic peptide	Ivy, potato	Immunodetection	NC	[40]
Polygalacturonase	Tomato fruit	Immunodetection	NC	[4]
Virus				
Bromovirus	Cowpea	Immunodetection	NC	[41]
Pox virus	Plump	Immunodetection	NC	[42]
Several viruses	Plump, Malus	Immunodetection	NC	[43]
Soybean MV	Soybean leaves	Immunodetection	NC	[44]
Tobacco MV	Tobacco leaves	Immunodetection	NC	[45, 46]
Virus and mycoplasma	Soybean leaves	Immunodetection	NC	[47]
Zucchini yellow mosaic potyvirus	Melon leaves	Immunodetection	NC	[48]
Gene expression				
Dirigent proteins	Forsythia	RNA-RNA hybridization	Nylon	[49]
Arabinan	Lemon	Immunodetection	NC	[50]
β-Conglycinin	Soybean seeds	RNA-RNA hybridization	NC	[51]
Extensin, HRGP	Soybean seeds, tomato, petunia, tobacco	RNA-RNA hybridization	NC, nylon	[35, 51]
Pectin Me-esterase	Tomato	RNA-RNA hybridization	NC	[52]
SAUR	Soybean hypocotyls	RNA-RNA hybridization	NC	[53]
Carbohydrates				
Glycoconjugates	Soybean seed, potato	Periodic acid-DNS hydrazine	PVDF	[54]
Man- and Gal-containing proteins	Soybean seed and pod	Fluorescent lectins	NC	[54]
Pectic-galactan	Tomato	Immunodetection		[55]

2 Materials

2.1 Doing a Tissue Print

1. Whatman No. 1 filter paper.
2. Blotting membrane: nitrocellulose, nylon, Immobilon P.
3. Double-edged razor blades, forceps, rubber gloves, paper to protect the membrane, acrylic sheet, marking pen.
4. Hand lens or microscope for viewing the specimen, and the biological material (Fig. 1a).

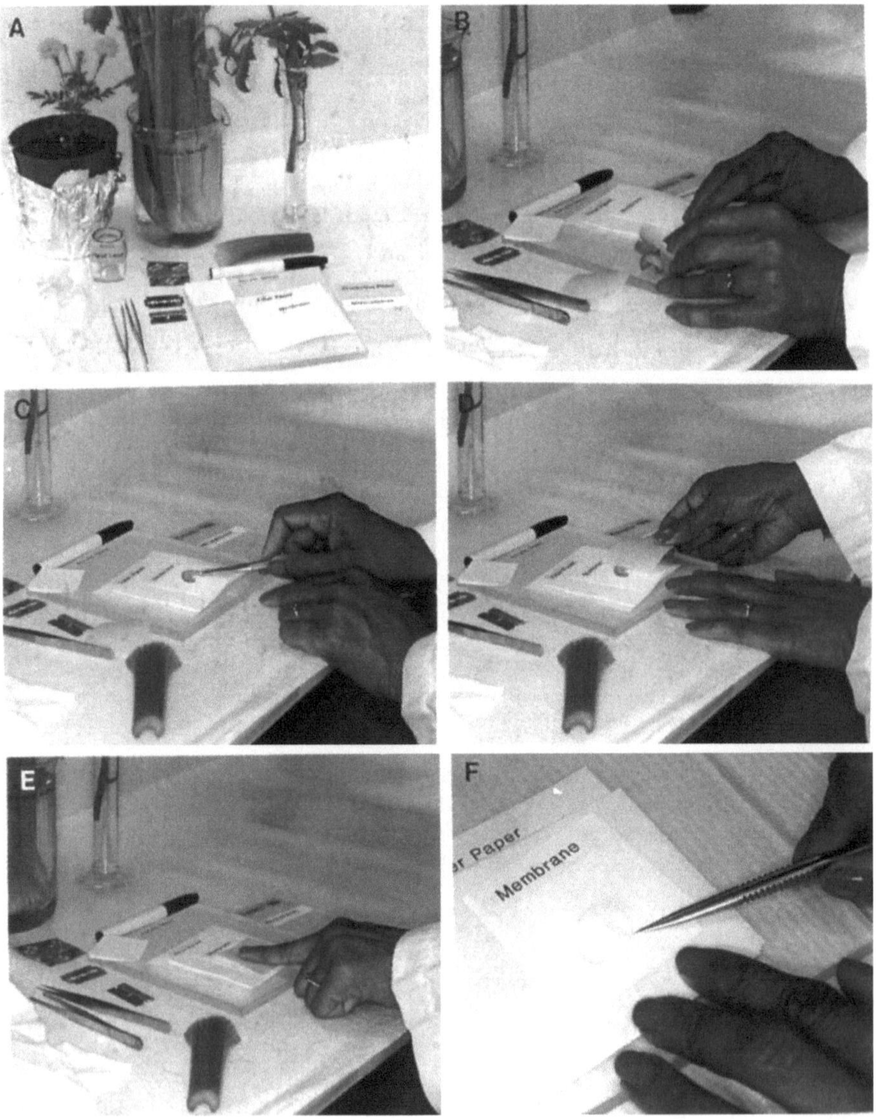

Fig. 1 Detailed description of the steps involved in tissue printing. (**a**) Materials required; (**b**) cutting the tissue; (**c**) placing the section on the membrane; (**d**) protecting the section and the print; (**e**) applying pressure; (**f**) removing the section. (Reprinted from "Tissue Printing," R. Taylor, Tissue printing demonstration, p. 6, 1992 [2] with permission from Elsevier)

2.2 Detection of Total Proteins in a Tissue Blot

1. Nitrocellulose (NC) membrane, 0.45 μm pore size.
2. Tris buffered saline (TBS)-Tween-20 (TBST): 20 mM Tris–HCl (pH 7.5), 0.15 M NaCl, and 0.05 % Tween-20.
3. India ink staining: 1 μL Pelikan India ink/mL of PBST.

2.3 Revealing Proteins with Antibodies

1. Nitrocellulose membrane, 0.45 μm pore size.
2. TBST.
3. Blocking buffer: 0.25 % gelatin and 0.025 % bovine serum albumin (BSA) in TBST.
4. Primary antibody: Antibodies against the targeted protein (potato lectin in this case), in blocking buffer (1:15,000 dilution for the antibody against the deglycosylated lectin).
5. Secondary antibody: Goat anti-rabbit alkaline phosphatase (AP) conjugate $F(ab)^2$ fragment, 1:20,000 dilution in blocking buffer.
6. AP buffer: 0.1 M Tris–HCl (pH 9.5), 0.1 M NaCl, and 5 mM $MgCl_2$.
7. AP substrates: nitro blue tetrazolium (NBT), 50 mg/mL in 70 % methanol diluted 33:10,000 in AP buffer, and 5-bromo-4-chloro-3-indolyl phosphate (BCIP) 50 mg/mL in dimethyl-formamide diluted 66:10,000 in AP buffer.

2.4 Localizing Cell Wall Proteins in Plants

1. NC membrane.
2. 0.2 M $CaCl_2{\sim}2H_2O$.
3. Primary antibody: Specific anti-extensin polyclonal antibodies raised from purified soybean seed coat extensin [6] diluted 1:15,000.
4. Secondary antibody: AP conjugated antibody anti-rabbit immunoglobulin IgG (Fc).
5. Tris buffered saline (TBS): 0.9 % NaCl in 20 mM Tris–HCl (pH 7.4) plus 0.3 % Tween-20 and 0.05 % sodium azide (NaN_3).
6. Blocking buffer: 0.25 % (w/v) BSA, 0.25 % (w/v) gelatin, a 0.3 % (v/v) Tween-20 in TBS.
7. AP buffer: 0.1 M Tris–HCl (pH 9.5), 0.1 M NaCl, and 5 mM $MgCl_2$.
8. AP substrate solution: 66 μL NBT and 33 μL BCIP in 10 mL AP buffer.
9. 3MM Whatman filter paper.

3 Methods

3.1 Doing a Tissue Print

1. Place several layers of filter paper on a smooth, hard surface and place a blotting membrane on top. Use a double-edged razor blade to cut a tissue sample (Fig. 1b) (*see* **Note 1**).
2. Using forceps transfer the tissue section to the membrane (*see* **Note 2**). Several successive sections can be printed on the same piece of membrane (Fig. 1c).
3. Place a small piece of nonabsorbent paper over the section to protect the membrane from fingerprints (Fig. 1d) (*see* **Note 3**).
4. Apply the appropriate amount of pressure to the section for the type of print desired. A chemical print requires only light pressure, but a physical print requires several times as much. The proper pressure also varies with the tissue used (Fig. 1e).
5. Gently remove the protective paper and the section with forceps, and air-dry the print with warm air, and observe (Fig. 1f).
6. Prints may be illuminated from the top or from one side by white or UV light and may also be viewed with transmitted light.

3.2 Detection of Total Proteins in a Tissue Blot

It is useful to have an image stained for total proteins in the blot, which will indicate the pattern of protein distribution in the tissue, as well as some indications on the different tissues present in the section. We have used India ink staining [7] for total protein on nitrocellulose membranes since Coomassie Blue stain gives a strong background. The same tissue slice can be printed several times, showing the tissues with higher amount of proteins (Fig. 2).

1. Wash the dry nitrocellulose filter 2× with TBST for 5 min with constant shaking.
2. Incubate the NC membrane with India ink staining and mix until the image appears clearly (*see* **Note 4**).
3. Wash the print with water for 5 min and dry it.

3.3 Revealing Proteins with Antibodies

Chemical tissue prints result from the molecules that transfer from the freshly cut cells of a tissue section to the surface of a synthetic membrane, where they are retained and immobilized. The principal steps for tissue printing are (1) the release of the protein of interest from the tissue; (2) the contact–diffusion transfer of the protein to the recipient membrane; and (3) the retention and binding of the protein into the synthetic matrix. The protein print is a mirror image of the tissue and can be used to detect and localize specific proteins. Retention and binding of a protein into the recipient matrix depend on the chemistry of the membrane and its ability to interact electrostatically and hydrophobically with the protein of interest. NC membranes have a high binding capacity for proteins and should be tried first for printing a new type of protein molecule.

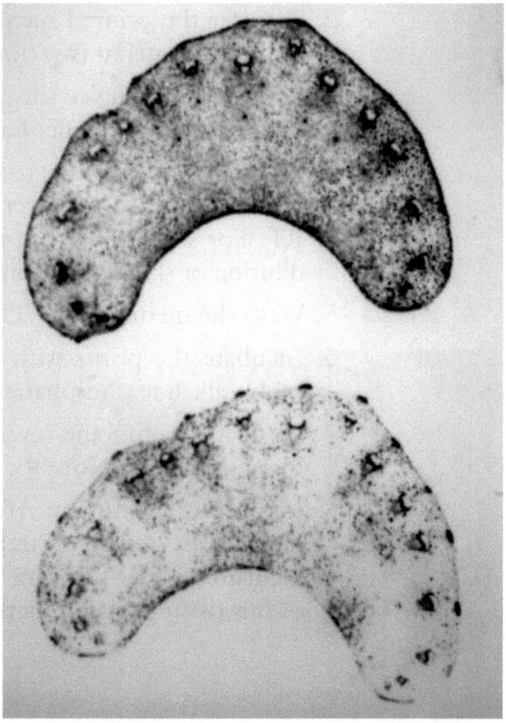

Fig. 2 Tissue print of a cross section from a celery stalk stained for total proteins with India ink. The same section was printed a first time (*upper image*) and a second time (*lower image*). Most of the proteins from the cut cells were transferred on the first print; the second one shows mainly the apoplastic proteins abundant in some tissues

Adding methanol to the transfer buffer increases the capacity and affinity on NC for proteins, presumably by promoting hydrophobic interactions (*see* **Note 5**).

Methods for detecting the protein of interest bound to the imprinted membranes must be specific and sensitive. Best results are achieved by using antibodies raised against the targeted protein. An enzyme-conjugated secondary antibody raised against the primary antibody is commonly used to visualize the binding between the primary antibody and the antigen (*see* **Note 6**). To reduce background from nonspecific cross-reactions, the primary antibody is combined either with 1 % (v/v) normal serum from the species in which the secondary antibody was raised or with a low concentration of sodium dodecyl sulfate or Tween-20.

The following procedure was used to localize potato lectin by means of polyclonal antibodies against the native and deglycosylated potato tuber lectin [5].

1. Cut a section of the tissue about 1 mm thick with a new razor blade and gently wipe the surface with a filter paper to absorb excess liquid. Put the freshly cut surface on the membrane and press for 10–15 s (*see* **Note 7**).

2. Transfer the printed membrane to TBST and wash away the unbound material two times for 5 min each in a shaker.
3. Block the unoccupied site of the membrane by shaking the membrane in blocking buffer for 30 min at room temperature (RT).
4. Transfer the membrane to the first antibody solution (anti-lectin serum) and incubate overnight in a refrigerator or for 2 h at RT. For controls either use preimmune serum at the same dilution or skip the first antibody step (*see* **Note 8**).
5. Wash the membrane in TBST three times for 10 min each.
6. Incubate the prints with the secondary antibody (goat anti-rabbit-alkaline phosphatase in this case) for 2 h at RT.
7. Wash the membrane several times with TBST (five washes of 5 min each) (*see* **Note 9**).
8. Incubate the prints in AP substrates at RT until the reaction product is observed; treated and control prints should be incubated under the same conditions (*see* **Note 10**). The result of this tissue printing method is shown in Fig. 3.

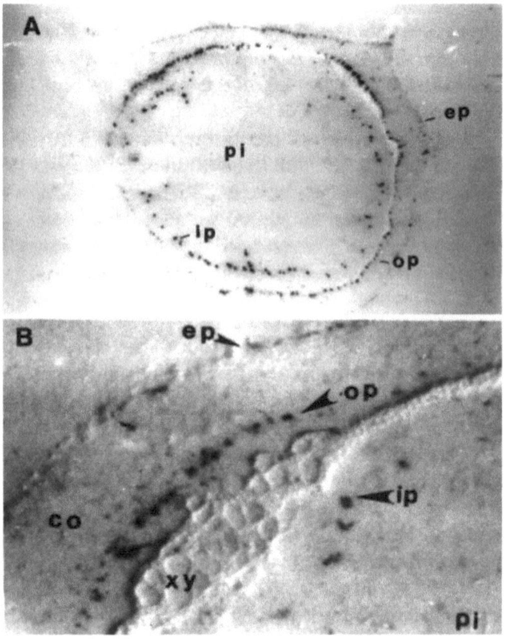

Fig. 3 Tissue prints of potato stem incubated with antibodies raised in rabbit against the deglycosylated lectin, followed by goat anti-rabbit AP. (**a**) Transverse section. (**b**) Magnification of a stem sector with lateral illumination showing greater detail of lectin localization. The lectin is present only in the outer (op) and inner phloem (ip). The stain at the epidermis (ep) is not a positive reaction but a transfer of natural pigments (*xy* xylem, *co* cortex, *pi* pith parenchyma). (Reprinted from "Tissue Printing," R. Pont-Lezica, Lectin and glycans recognition, p. 87, 1992 [54] with permission from Elsevier)

3.4 Localizing Cell Wall Proteins in Plants

The presence of a cell wall is one of the outstanding features distinguishing plant cells from those of animals. The cell wall is not an organelle with relatively constant functions; rather, it is subject to continuous developmental processes that govern cell size, division, shape, and function. In addition to the well-known polysaccharides present in the cell wall, proteins are important modulators of cell wall structure and function [8]. One of the best characterized cell wall proteins is extensin, a basic hydroxyproline-rich glycoprotein important for cell wall structure. It is a difficult protein to isolate, because high proportion of it becomes insolubilized in the wall, but it can be extracted with a solution with high salt concentration. A modification of the tissue print method was developed [6] to transfer cell wall proteins to NC, soaking the membrane previously in 0.2 M $CaCl_2$ and then printing the tissue.

1. Soak the NC membrane in 0.2 M $CaCl_2$ for 30 min and dry on 3-MM Whatman paper.
2. Cut fresh tissue into sections of 0.3–3 mm thick with a new razor blade, previously washed in distilled water for 3 s and dried on Kimwipes. Then transfer each section to the NC membrane as indicated.
3. Block the NC with blocking buffer for 1–3 h at RT with constant shaking.
4. Add the primary antibody to the desired dilution in blocking buffer and incubate the membrane for 1–3 h at RT with shaking.
5. Wash the NC three times for 30 min each in TBS with agitation.
6. Soak the NC membrane in secondary antibody (AP-conjugated anti IgG) diluted 1:20,000 in blocking buffer for 1–3 h with agitation.
7. Wash the NC membrane with AP buffer and add the AP substrate solution. Develop the tissue print until a color signal appears and stop the reaction by washing the membrane in distilled water.

4 Notes

1. Use gloves for all the manipulations to avoid finger prints on the membrane. It may be necessary to gently preblot the section on a separate piece of filter paper before printing to remove excess tissue exudates from cut cells and to ensure an accurate print.
2. To avoid double images be careful when blotting and removing the tissue section from the membrane.
3. When printing a thin section (200–300 μm), place a piece of membrane on top of the section to prevent the nonabsorbent paper from marking the membrane under the section.

4. The incubation time for India ink staining can be very variable, according to the amount of protein present in the tissue. For a seed blot from a legume, 15 min incubation will give a very good image. For other tissues such as stem or petiole, it may be necessary to incubate for several hours or even overnight.
5. In some instances NC membranes do not bind the protein of interest at all or else bind it only weakly. Try next a membrane that can react chemically with the protein, and covalently bind it, such as Immunodyne Immunoaffinity membrane (Pall Corp, Cortland, NY, USA). The type of buffer, salt concentration, and pH of the incubation mixture are important for protein binding and must be determined empirically.
6. Detection of the alkaline phosphatase-conjugated second antibody on tissue prints from plant organs was selected over the peroxidase-conjugated second antibody procedure because the substrates used for detecting the peroxidase, such as *O*-phenylenediamine and hydrogen peroxide, will react with endogenous plant peroxidase activity in the tissue sections [9], making the immunoblotting reaction.
7. The same tissue surface can be reprinted several times; the successive images will be weaker, but a good imprint can be found among these when the proteins are very abundant in a particular tissue.
8. To be sure that the antibody used for tissue printing is specific for the targeted protein, it is important to test it against a total protein extract of the tissue previously by SDS-PAGE, followed by a western blot. Only the targeted protein should be stained, and use the highest dilution to obtain the best results.
9. The last wash is made with TBS without detergent to avoid interference with the alkaline phosphatase reaction.
10. Occasionally some tissues show cross-reaction with the goat serum. To avoid such a reaction, two procedures are available: (1) block the prints with blocking buffer containing 1:3,000 dilution of normal goat serum, or (2) use a secondary antibody raised in a different animal.

Acknowledgment

The author is grateful to the Université Paul Sabatier (Toulouse III, France) and the CNRS for support.

References

1. Daoust R (1957) Localization of deoxyribonuclease in tissue sections. A new approach to the histochemistry of enzymes. Exp Cell Res 12:203–211
2. Reid PD, Pont-Lezica RF, del Campillo E, Taylor R (eds) (1992) Tissue printing. Academic, San Diego, CA

3. Varner J, Ye Z (1994) Tissue printing. FASEB J 8:378–384
4. Tieman DM, Handa A (1989) Immunocytolocalization of polygalacturonase in ripening tomato fruit. Plant Physiol 90:17–20
5. Pont-Lezica RF, Taylor R, Varner J (1991) *Solanum tuberosum* agglutinin accumulation during tuber development. J Plant Physiol 137: 453–458
6. Cassab GI, Varner JE (1987) Immunocytolocalization of extensin in developing soybean seed coats by immunogold-silver staining and by tissue printing on nitrocellulose paper. J Cell Biol 105:2581–2588
7. Hancock K, Tsang VCW (1983) India ink staining for protein on nitrocellulose paper. Anal Biochem 133:157–162
8. Jamet E, Canut H, Boudart G, Pont-Lezica RF (2006) Cell wall proteins: a new insight through proteomics. Trends Plant Sci 11: 33–39
9. Cassab GI, Lin JJ, Lin LS, Varner JE (1988) Ethylene effect on extensin and peroxidase distribution in the subapical region of pea epicotyls. Plant Physiol 88:522–524
10. Jacobsen J, Knox R (1973) Cytochemical localization and antigenicity of α-amylase in barley aleurone tissue. Planta 112:213–224
11. Nolan M, David B (1984) A starch-agar gel method for the localization of starch hydrolyzing enzymes in the cotyledons and hypocotyls of beans. Am J Bot 71:137–141
12. Spruce J, Mayer AM, Osborne DJ (1987) A simple histochemical method for locating enzymes in plant tissue using nitrocellulose blotting. Phytochemistry 26:2901–2903
13. Pearse AGE (1980) Substrate film techniques. In: Pearse AGE (ed) Histochemistry, theoretical and applied. Churchill Livingstone, Edinburgh, pp 253–269
14. Roth R, Boudet A, Pont-Lezica RF (1997) Lignification and cinnamyl alcohol dehydrogenase activity in developing stems of tomato and polar: a spatial and kinetic study through tissue printing. J Exp Bot 48:247–254
15. da Silva E, Toorop P, van Aelst A, Hilhorst H (2004) Abscisic acid controls embryo growth potential and endosperm cap weakening during coffee (*Coffea arabica* cv. Rubi) seed germination. Planta 220:251–261
16. Driouich A, Lainé A, Vian B, Faye L (1992) Characterization and localization of laccase forms in stem and cell cultures of sycamore. Plant J 2:13–24
17. Hara M, Eto H, Kuboi T (2001) Tissue printing for myrosinase activity in roots of turnip and Japanese radish and horseradish: a technique for localizing myrosinases. Plant Sci 160:425–431
18. Maki H, Morohashi Y (2006) Development of polyphenol oxidase activity in the micropylar endosperm of tomato seeds. J Plant Physiol 163:1–10
19. Yomo H, Taylor MP (1973) Histochemical studies on protease formation in the cotyledons of germinating bean seeds. Planta 112:35–43
20. Harris N, Chrispeels MJ (1975) Histochemical and biochemical observations on storage protein metabolism and protein body autolysis in cotyledons of germinating beans. Plant Physiol 56:292–299
21. Moore B, Kang B, Flurkey W (1989) Histochemical localization of mushroom tyrosinase in whole sections on nitrocellulose. Histochemistry 90:379–381
22. Singh MB, Knox R (1985) Grass pollen allergens: antigenic relationships detected using monoclonal antibodies and dot immunobinding assay. Int Arch Allergy Appl Immun 78:300–304
23. Antikainen M, Griffith M, Zhang J, Hon W, Yang D, Pihakaski-Maunsbach K (1996) Immunolocalization of antifreeze proteins in winter rye leaves, crowns, and roots by tissue printing. Plant Physiol 110:845–857
24. Ramalho-Santos M, Pissarra J, Verissimo P, Pereira S, Salema R, Pires E et al (1997) Cardosin A, an abundant aspartic proteinase, accumulates in protein storage vacuoles in the stigmatic papillae of *Cynara cardunculus* L. Planta 203(2):204–212
25. del Campillo E, Reid P, Sexton R, Lewis L (1990) Occurrence and localization of 9.5 cellulase in abscising and nonabscising tissues. Plant Cell 2:245–254
26. Reid PD, del Campillo E, Lewis LN (1990) Anatomical changes and immunolocalization of cellulase during abscission as observed on nitrocellulose tissue prints. Plant Physiol 93: 160–165
27. Pont-Lezica RF, Varner JE (1989) Histochemical localization of cysteine-rich proteins by tissue printing on nitrocellulose. Anal Biochem 182:334–337
28. Bailey BA, Taylor R, Dean JF, Anderson JD (1991) Ethylene biosynthesis-inducing endoxylanase is translocated through the xylem of *Nicotiana tabacum* cv xanthi plants. Plant Physiol 97:1181–1186
29. Cho H, Kende H (1997) Expansins and internodal growth of deepwater rice. Plant Physiol 113:1145–1151
30. Hood K, Baasiri R, Fritz S, Hood E (1991) Biochemical and tissue print analyses of

hydroxyproline-rich glycoproteins in cell walls of sporophytic maize tissues. Plant Physiol 96:1214–1219
31. Ye ZH, Song YR, Marcus A, Varner JE (1991) Comparative localization of three classes of cell wall proteins. Plant J 1:175–183
32. Goto F, Yoshihara T, Shigemoto N, Toki S, Takaiwa F (1999) Iron fortification of rice seed by the soybean ferritin gene. Nat Biotechnol 17:282–286
33. Pereira S, Carvalho H, Sunkel C, Salema R (1992) Immunocytolocalization of glutamine synthase in mesophyll and phloem of leaves of *Solanum tuberosum* L. Protoplasma 167:66–73
34. Keller B, Sauer N, Lamb CJ (1988) Glycine-rich cell wall proteins in bean: gene structure and association of the proteins with the vascular system. EMBO J 7:3625–3633
35. Ye ZH, Varner J (1991) Tissue-specific expression of cell wall proteins in developing soybean tissues. Plant Cell 3:23–37
36. Ryser U, Keller B (1992) Ultrastructural localization of a bean glycine-rich protein in unlignified primary walls of protoxylem cells. Plant Cell 4:773–783
37. Gabius S, Hellmann KP, Hellmann T, Brinck U, Gabius HJ (1989) Neoglycoenzymes: a versatile tool for lectin detection in solid phase assays and glycohistochemistry. Anal Biochem 182:447–451
38. Gonorazky A, Regente M, de la Canal L (2005) Stress induction and antimicrobial properties of a lipid transfer protein in germinating sunflower seeds. J Plant Physiol 162:618–624
39. Kausch K, Handa A (1997) Molecular cloning of a ripening-specific lipoxygenase and its expression during wild-type and mutant tomato fruit development. Plant Physiol 113: 1041–1050
40. Maryani M, Morse M, Bradley G, Irving H, Cahill D, Gehring C (2003) *In situ* localization associates biologically active plant natriuretic peptide immuno-analogues with conductive tissue and stomata. J Exp Bot 54:1553–1564
41. Fujita Y, Fujita M, Mise K, Kobori T, Osaki T, Furusawa I (2000) Bromovirus movement protein conditions for the host specificity of virus movement through the vascular system and affects pathogenicity in cowpea. Mol Plant Microbe Interact 13:1195–1203
42. Dicenta F, Martinez-Gomez P, Bellanger I, Audergon J (2000) Localization of plum pox virus in stem and petiole tissues of apricot cultivars by immuno-tissue printing. Acta Virol 44:323–328
43. Knapp E, da Camara Machado A, Puhringer H, Wang Q, Hanzer V, Weiss H et al (1995) Localization of fruit tree viruses by immuno-tissue printing in infected shoots of *Malus* sp. and *Prunus* sp. J Virol Methods 55:157–173
44. Mansky LM, Andrews RE, Durand DP, Hill JH (1990) Plant virus localization in leaf tissue by press blotting. Plant Mol Biol Rep 8:13–17
45. Holt CA, Beachy RN (1991) *In vivo* complementation of infectious transcripts from mutant tobacco mosaic virus cDNAs in transgenic plants. Virology 181:109–117
46. Jung GL, Hahne G (1992) A simple method to increase resolution in whole leaf blotting. Plant Sci 82:125–132
47. Lin NS, Hsu YH, Hsu HT (1990) Immunological detection of plant viruses and mycoplasma-like organism by direct tissue blotting on nitrocellulose membranes. Phytopathology 80:824–828
48. Polston JE, Bubrick P, Perring TM (1991) Detection of plant virus coat proteins on whole leaf blots. Anal Biochem 196:267–270
49. Burlat V, Kwon M, Davin L, Lewis N (2001) Dirigent proteins and dirigent sites in lignifying tissues. Phytochemistry 57:883–897
50. Willats W, Marcus S, Knox J (1998) Generation of monoclonal antibody specific to (1-5)-α-L-arabinan. Carbohydr Res 308:149–152
51. Varner JE, Song YR, Lin LS, Yuen H (1989) The role of cell wall proteins in the structure and function of cell walls. In: Goldberg R (ed) The molecular basis of plant development. Alan R Lis, New York, pp 161–168
52. Harriman RWF, Tieman DM, Handa AK (1992) Localization of pectin methylesterase mRNA in tomato fruit. In: Reid PD et al (eds) Tissue printing. Academic, San Diego, CA, pp 121–123
53. McClure B, Guilfoyle T (1989) Tissue print hybridization. A simple technique for detecting organ- and tissue-specific gene expression. Plant Mol Biol 12:517–524
54. Pont-Lezica RF (1992) Lectin and glycan recognition. In: Reid PD et al (eds) Tissue printing. Academic, San Diego, CA, pp 71–79
55. Jones L, Seymour G, Knox J (1997) Localization of pectic galactan in tomato cell walls using a monoclonal antibody specific to (1,4)-α-D-galactan. Plant Physiol 113:1405–1412

Chapter 14

Dot-Immunobinding Assay (Dot-Iba)

Sumi Surendran, Annamma Mathai, and Vishnampet Venkataraman Radhakrishnan

Abstract

Dot-immunobinding assay (Dot-Iba) is a simple and highly reproducible immunodiagnostic method. Antibody or antigen is dotted directly onto nitrocellulose membrane (NCM) discs. The diagnostic material to be checked can be incubated on this disc. Presence of antigen-antibody complex in NCM discs can be directly demonstrated with enzyme-conjugated antiglobulins and substrate. Development of a purple-pink colored, insoluble substrate product in the nitrocellulose membrane will be considered a positive result in the assay. This assay allows the processing of multiple specimens at a time and the entire operational procedures required only 4–6 h. Dot-IBA is rapid and the technical steps involved in the assay are much simpler than the other immunoassays such as enzyme-linked immunosorbent assay in detecting circulating antigen and antibody in clinical samples. The Dot-Iba showed an overall sensitivity of 60 % for tuberculous meningitis diagnosis and no false positive results were encountered. Hence this assay is highly specific for the diagnosis of paucibacillary diseases like extrapulmonary tuberculosis. Dot-Iba is best suited to laboratories in developing world where there are constraints in laboratory resources.

Key words Dot-immunobinding assay, Dot-Iba, Nitrocellulose discs, Circulating antibody

1 Introduction

In most of the developing world, infectious diseases continue to remain as one of the major public health diseases. Most of these infectious diseases are potentially curable with appropriate antimicrobials. Any delay in diagnosis and treatment in patients will invariably lead to irreversible sequelae and complications. Development of newer techniques other than bacteriological methods has become not only essential but also relevant for the early and rapid diagnosis of paucibacillary conditions as in tuberculous meningitis. The newly developed assay should be more sensitive and also possess operational advantages over the existing diagnostic methods. The newly designed assay should be specific, cost-effective, and feasible for application in the developing world where there are constraints in laboratory resources [1–5].

Biji T. Kurien and R. Hal Scofield (eds.), *Western Blotting: Methods and Protocols*, Methods in Molecular Biology, vol. 1312, DOI 10.1007/978-1-4939-2694-7_14, © Springer Science+Business Media New York 2015

1.1 Advantages of Dot-Iba

1. NCM discs have superior binding capacity than polystyrene microtiter plates.
2. The results can be read visually and there is no need of any expensive equipment such as microtiter ELISA readers.
3. The technical procedures involved in the assay are user friendly and laboratory personnel can be trained to undertake the assay.
4. Large number of specimens can be handled together.
5. Only limited laboratory space is required and there is no need for extraordinary instrumentation.
6. The NCM discs following the assay can be safely stored at 4 °C and it can be retained for review of previous results in the same patient.
7. The results can be made available within 6 h after the receipt of the specimen in the laboratory. Hence this assay is rapid and will be helpful in making decision at the bedside management of patients.
8. The Dot-Iba is best suited to laboratories in developing world where there is a definite limitation in the laboratory resources and technical expertise.

1.2 Limitations of Dot-Iba

The Dot-Iba will yield negative results if the antigen concentration (as in specimens like CSF) is less than 50 ng/mL. So it is possible that the assay may yield false negative results in patients with tuberculous meningitis who received a course of anti-tuberculosis chemotherapy.

2 Materials

1. Punching machine.
2. 1 cm diameter circular nitrocellulose membranes (NCM).
3. Flat bottom microtiter plates.
4. Phosphate buffered saline (PBS).
5. Bovine serum albumin (BSA).
6. 0.15 M phosphate buffered saline in Tween-20 (PBST).
7. Anti-rabbit IgG-biotin conjugate.
8. Extr-Avidin alkaline phosphatase.
9. *Ortho* dianisidine tetrazotized (0.25 mg/mL).
10. β-Naphthyl acid phosphate (0.25 mg/mL).
11. 0.06 M sodium borate buffer (pH 9.7).
12. Methanol.
13. Acetic acid.
14. Double distilled water.

3 Methods

1. Cut 1 cm diameter circular nitrocellulose membranes (NCM) with a clean "punching" machine.
2. NCM discs should be placed properly in the "flat bottom" microtiter plates with fine forceps (*see* **Note 1**).
3. 5 μL of antigen (*see* **Note 2**) should be carefully spotted in the central portion of the NCM disc and incubated for 12 h at 4 °C (*see* **Note 3**).
4. Another incubation at 37 °C for 1 h (optional).
5. The plates should be washed with PBST.
6. The unbound sites in the NCMs are quenched with 3 % BSA in PBST.
7. Incubate the discs with 1:1,000 diluted specific monoclonal or polyclonal antibody for 1 h at 37 °C followed by repeated washing in PBST.
8. Subsequently the NCM discs are incubated with 1:1,000 diluted anti-rabbit IgG-biotin conjugate and then Extr-Avidin alkaline phosphatase for 1 h respectively and finally washed thoroughly with PBST (*see* **Note 4**).
9. The NCM discs should be completely immersed in substrate containing *o*-dianisidine tetrazotized (0.25 mg/mL), β-naphthyl acid phosphate (0.25 mg/mL) in 0.06 M sodium borate buffer (pH 9.7) for 10 min.
10. Wash in PBST and fix in a solution containing methanol; acetic acid: distilled water in the proportion, 5:1:5.
11. A positive reaction is indicated by the development of purple to purple-pink in the central portion of the NCM discs (Fig. 1; *see* **Note 5**).
12. A positive control and negative controls should be included in the assay.

4 Notes

1. Circular discs should be cleanly cut with the help of a punching device. Discs with irregular edges and with wrinkles should not be used in the assay. NCM strips should be handled with a non-toothed and nonserrated forceps ("do not touch" with fingers).
2. Different concentrations (5–500 ng/mL) should be tried for the standardization procedure. In our experience, the dot-Iba gave positive results in all those NCM discs containing 50 ng/mL and above. In other words, the lowest detection limit in the assay is 50 ng/mL.

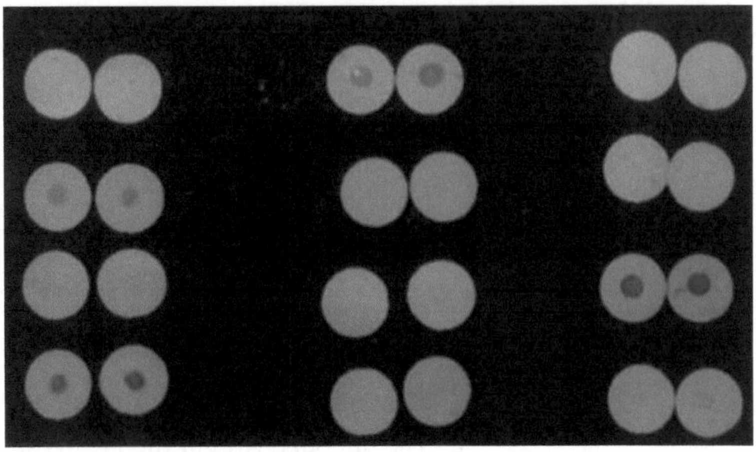

Fig. 1 NCM discs showing positive and negative results in Dot-Iba

3. Do not release more than 5 μL of the antigen or antibody over the NCM disc. Also release the reagent steadily and uniformly over the NCM disc. Use a 5-μL micropipette with disposable tips for this purpose. During the release the reagent should not dribble outside the NCM disc.
4. The immunoreagents used in the assay such as antigen and antibodies must be stored in aliquots in optimal amounts and this should be used during the assay. This will also eliminate repeated effects of freezing and thawing. This will also enhance the shelf life of reagents.
5. As the result of the assay is interpreted by the visual examination of the NCM, the assay should be performed during daylight. However if the assay needs to be undertaken during the night, then a provision should be made to connect a light source to work bench.

References

1. Radhakrishnan VV, Mathai A, Rao SB, Sehgal S (1990) Immunoelectrophoresis for mycobacterial antigens. Indian J Exp Biol 28:812–815
2. Sumi MG, Mathai A, Sarada C, Radhakrishnan VV (1999) Rapid diagnosis of tuberculous meningitis by a dot immunobinding assay to detect mycobacterial antigen in cerebrospinal fluid specimens. J Clin Microbiol 37:3925–3927
3. Sumi MG, Annamma M, Sarada C, Radhakrishnan VV (2000) Rapid diagnosis of tuberculous meningitis by a dot-immunobinding assay. Acta Neurol Scand 101:61–64
4. Sumi MG, Mathai A, Sarada C, Radhakrishnan VV, Indulakshmi R, Sathish M, Ajaykumar R, Manju YK (2002) A comparative evaluation of dot-immunobinding assay (Dot-Iba) and polymerase chain reaction (PCR) for the laboratory diagnosis of tuberculous meningitis. Diagn Microbiol Infect Dis 42:35–38
5. Mathai A, Radhakrishnan VV, Molly M (1991) Rapid diagnosis of tuberculous meningitis with a dot enzyme immunobinding assay to detect antibody in the cerebrospinal fluid. Eur J Clin Microbiol Infect Dis 10:440–443

Chapter 15

Analysis of Antibody Clonotype by Affinity Immunoblotting

Biji T. Kurien and R. Hal Scofield

Abstract

A sensitive and specific method to analyze specific antibody clonotype changes in a lupus patient who developed autoantibodies to the Ro 60 autoantigen under observation is described. Patient sera, collected over several years, were separated by flatbed isoelectric focusing (IEF) and analyzed by affinity immunoblotting utilizing Ro 60-coated nitrocellulose membrane. When the Ro 60-coated nitrocellulose was laid over the surface of the IEF gel, the antibodies present on the surface of the acrylamide gel bound the Ro antigen on the nitrocellulose. Tween-20 was used to prevent nonspecific binding. The bound IgG clonotypes were detected using alkaline phosphatase conjugated anti-IgG. The patient's sera demonstrated an oligoclonal response to the Ro 60 autoantigen that increased in complexity and affinity over time.

Key words Affinity immunoblotting, Clonotype distribution, Systemic lupus erythematosus, Ro 60 autoantigen, Flatbed IEF

1 Introduction

Isoelectric focusing is a very useful method for investigating the heterogeneity of antibody and immunoglobulin (Ig) clonotypes [1]. Antigen-specific antibody clonotype patterns can show whether changes in cell population happen during ongoing immune responses as a response to regulatory influences. It can also tell whether changes in hybridoma cell lines can occur with time [2]. Previously, it was customary to study these changes by immobilizing the separated antibody clonotypes after isoelectric focusing and incubating them with radioactive antigen. In one method, radiolabeled hapten was allowed to diffuse into a gel before precipitation of Ig with sodium sulfate followed by detection of hapten-specific clonotype distribution by autoradiography [3]. In another study Ig was precipitated in the gel with sodium sulfate immediately after completion of the focusing run and was crosslinked with glutaraldehyde followed by the addition of labeled antigen or anti-Ig [4]. Subsequently it was shown that fixation with glutaraldehyde could decrease the antigen-binding

Biji T. Kurien and R. Hal Scofield (eds.), *Western Blotting: Methods and Protocols*, Methods in Molecular Biology, vol. 1312, DOI 10.1007/978-1-4939-2694-7_15, © Springer Science+Business Media New York 2015

ability of certain Ig [5]. Furthermore it was shown that the previous study was unable to define optimal crosslinker (glutaraldehyde or suberimidate) concentration, since certain antibodies could not be fixed at crosslinker concentrations that substantially inactivated others. Another drawback of these methods is the excess time needed to diffuse antigen into the gel and for rinsing the unbound antigen out of the gel, which can take several days especially when using radioactive probes.

One method for immobilizing focused antibodies involved the use of nitrocellulose membranes. Focused antibodies are transferred electrophoretically or nonelectrophoretically to nitrocellulose and labeled antigen was used to detect clonotypes that were antigen specific [6]. Yet another method involved laying the gel with the focused antibodies with agarose containing antigen-coated sheep erythrocytes [7]. In this method, antibodies diffuse into the RBC-containing gel bind the antigen-coated cells and lyse the cells following complement addition.

Here, we describe a method in which a 60,000 molecular weight Ro autoantigen was first passively immobilized on nitrocellulose membrane and placed in contact with an IEF gel that contained autoantibodies (derived from a systemic lupus erythematosus patient who developed antibodies to the Ro 60 autoantigen over time) focused according to its isoelectric point. Following diffusion-mediated transfer to membrane (*see* Chapters 9–11) the antibody clonotypes that are not antigen specific are removed by washing while the antigen-specific antibody clonotypes are detected using alkaline phosphatase conjugated anti-Ig.

Systemic lupus erythematosus (SLE) is a complex, chronic disorder characterized by the production of antibodies to self-antigens, including the Ro (or SS-A) ribonucleoprotein complex. Antibodies to the Ro 60 autoantigen occur in up to 40 % of patients with SLE [8]. The epitopes of the Ro 60 autoantigen bound by SLE patients have been previously characterized [9, 10]. Even though anti-Ro 60 sera were commonly observed to bind to short peptides, it was not found to bind the denatured antigen well. Furthermore, the antibodies that were found to bind to octapeptides were also found to bind the native protein [10].

There have been instances of some SLE autoantibodies appearing and disappearing, at times in association with specific disease manifestations, therapy, or generalized clinical disease activity. For instance, antibodies to native DNA is associated with renal disease, and the detection of this autoantibody may be an indication of disease exacerbation [11]. Antibodies to the P autoantigen (ribosomal P antigens) can appear with an increase of neurologic or renal disease. Autoantibodies such as anti-Ro, on the other hand, occur in some normal subjects as well as in SLE patients before onset of disease [8] and develop only rarely during the course of SLE.

This investigation was carried out following the identification of an SLE patient who developed antibodies to the Ro 60 autoantigen after about 10 years of illness. As shown in Fig. 3 anti-Ro 60 clonality increased in complexity, and affinity to the Ro 60 antigen also increased as the response developed.

2 Materials

Prepare all solutions using ultrapure water (prepared by purifying deionized water, to attain a sensitivity of 18 MΩ-cm at 25 °C) and analytical grade reagents. Prepare and store all reagents at room temperature (unless indicated otherwise). Diligently follow all waste disposal regulations when disposing waste materials. We do not add sodium azide to reagents.

1. 25 % glycerol (v/v): Add 25 mL glycerol to 75 mL of distilled water. Mix well.
2. 5× acrylamide (26.5 % T, 3 % C): Add about 25 mL water to a 100 mL graduated cylinder or a glass beaker. Weigh 12.84 g acrylamide and 0.4098 g bis acrylamide and transfer to the cylinder (*see* **Note 1**). Add a spatula of AG 501-X8 (D) mixed-resin beads and stir using a magnetic stir bar on a magnetic plate for about 30 min. Make up to 50 mL (after removing the stir bar) with water and filter through a 0.45 μm Corning filter (*see* **Note 2**). Store at 4 °C, with bottle wrapped with aluminum foil (*see* **Note 3**).
3. 10 % Tween-20: Add 90 mL of distilled water into a glass beaker. Add 10 mL Tween-20 and mix.
4. 2 % ammonium persulfate: Weigh 0.02 g ammonium persulfate and dissolve in 1 mL of distilled water (*see* **Note 4**).
5. *N*,*N*,*N'*,*N'*-Tetramethyl-ethylenediamine. Store at 4 °C (*see* **Note 5**).
6. Alkaline phosphatase buffer: Weigh 6.1 g of Tris, 2.9 g sodium chloride, and 0.51 g magnesium chloride-6H_2O and make it to 500 mL with water after adjusting pH to 9.3 with HCl (*see* **Note 1**). Store at 4 °C.
7. Nitro blue tetrazolium (NBT)/5-bromo-4-chloro-3-indolyl phosphate (BCIP): Dissolve 1 g NBT in 20 mL of 70 % dimethylformamide (DMF). Dissolve 1 g BCIP in 20 mL of 100 % DMF. Add 33 μL of BCIP and 66 μL of NBT to 10 mL of alkaline phosphatase buffer just before adding to membrane. Alternatively, use 1-Step™ NBT/BCIP readymade mix.
8. Nitrocellulose membrane.
9. 0.5 M sodium bicarbonate solution, pH 9.5.
10. Phosphate buffered saline (PBS), pH 7.4.

11. PBS containing 0.05 % Tween-20 (PBST).
12. Ampholytes: pH 3–10 and pH 8–10.5.
13. Ro 60 autoantigen (Immunovision, Springdale, AK, USA).
14. Glass plates: Two 5″ by 4″ glass plates.
15. Medium binder clips (1¼ in.).
16. Small binder clips (¾ in.).
17. Gasket with three edges, about 3 mm wide, to serve as spacer between the plates.
18. LKB-2117 Multiphor apparatus for IEF.
19. Model 3000/300 power supply.
20. pH 3 and pH 10 solutions.
21. Helium gas.
22. Sample applicator strip.
23. Paper wicks.

3 Methods

All procedures are at room temperature unless otherwise specified.

1. The night before focusing cut a piece of nitrocellulose membrane according to the size that would fit a small pipet box lid (yellow tip box). Add antigen (Ro 60) at 10 μg/mL of sodium bicarbonate, pH 9.5, and incubate this with the membrane overnight with shaking.
2. Pipet 5.6 mL of distilled water into a conical flask. Add 2 mL of 25 % glycerol followed by 2.1 mL of the 5× acrylamide solution. Then add 300 μL of pH 3–10 ampholytes followed by 100 μL of pH 8–10.5 ampholytes.
3. Degas this solution by bubbling helium through it for 15 min. Rinse the metal end of degassing tube first with water and wipe dry with Kimwipes.
4. While the solution is degassing, set up the gel apparatus. Soak the gasket in water for few minutes. Mop dry with Kimwipes.
5. Take one glass plate and lay the gasket on top of the glass plate around the edges so that it will seal the bottom and two sides of the plates. Lay the other glass plate on top of the gasket. Clamp the clips around the edges of the plates (bottom, the left side, and the right side (*see* Fig. 1)). Stand the gel upright on using the base of the clips (*see* Fig. 1) to pour the gel. Prepare a 2 % ammonium persulfate solution fresh.
6. After degassing is complete, the metal end of the degassing tube is cleaned with water. To the degassed solution, add 100 μL of 10 % Tween-20 and mix gently. Then add 100 μL

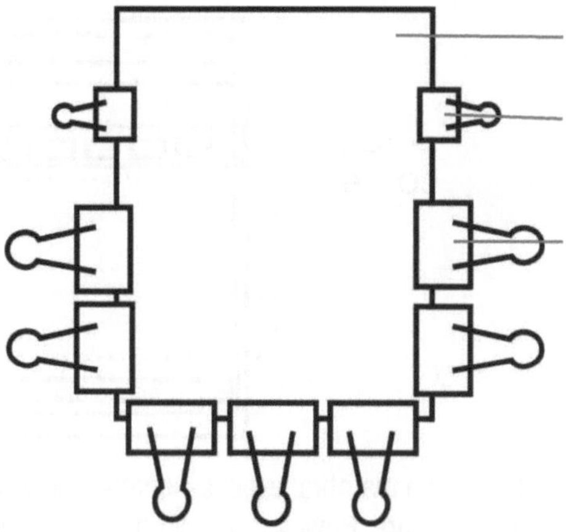

Fig. 1 Gel assembly for isoelectric focusing

of 2 % APS. Have a Pasteur pipette ready for pouring the gel. Add 10 μL of TEMED and mix gently. Pipet the gel mixture into the Pasteur pipette and transfer into the gel apparatus quickly. Attempts should be made to avoid bubbles. Fill up the gel apparatus to the top. Polymerization should begin within minutes. However, let the assembly stand for 2 h without disturbance.

7. Turn cooling unit on and set it on 4 °C in preparation for focusing.
8. After 2 h carefully remove one of the glass plates and gasket. The gel will remain on one of the glass plates.
9. Lay the glass plate on top of the IEF unit, with the gel side facing up (wipe of water on top of the unit beforehand). Place the smaller cover in place and press down slightly so as to make imprints for the wicks. Cut two wicks to the size of the gel (be as close as possible). Soak the top wick in Serva pH 3 solution and the bottom wick in Serva pH 10 solution. Dab off excess solution and place where imprints were made by cover (*see* Fig. 2).
10. Put smaller cover back on, making sure connection is made with both wicks. Connect red and black wires. Put on larger cover and make connections to power supply (red = $+^{ve}$; black = $-^{ve}$).
11. Prefocus by setting constant voltage 200 V for 20 min, then increase voltage to 400 V for another 20 min. Prepare samples for application.
12. Turn of power supply, disconnect wires, and remove covers. Take applicator strip and lay on top of gel 1–2 cm below top wick. Make sure strip is stuck to the gel well. Strip can hang over gel a little (*see* Fig. 2; *see* **Note 6**).

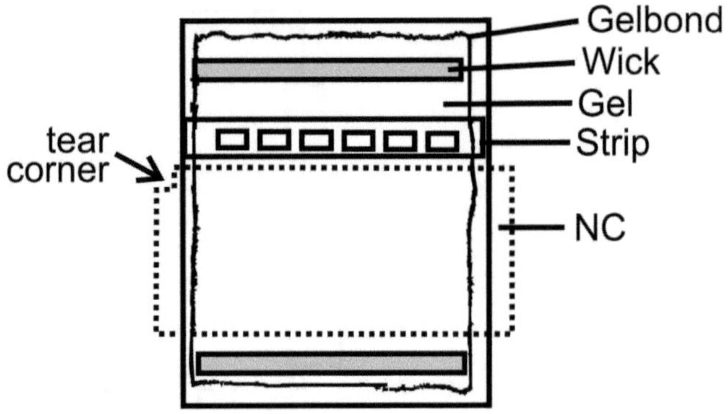

Fig. 2 The membrane-gel assembly following flatbed IEF. The gel bond is shown in this figure. However, we did not use the gel bond to support the gel. The gel was directly in contact with the glass plate

13. Apply the samples, being careful not to spill over into other wells. Replace the covers and make the connections. Turn the power supply on to 12 W constant power. Focus for approximately 1–2 h. When focusing, the voltage will rise, and the current will drop. The rate at which these two parameters change is much faster in the beginning than the end. The run is complete when the voltage is between 1,800 and 2,000 V and the current 3–5 A. When the change appears to be very slow or not at all, turn off the unit (*see* **Note 7**).
14. Thirty minutes prior to end of the run, rinse nitrocellulose membrane three times with PBS, pH 7.4, shake with PBST for 30 min, and rinse three times with PBS.
15. After the run is complete, transfer the focused protein from the gel to membrane. Take gel off the flatbed and remove applicator strip. Place nitrocellulose membrane between two Kimwipes and dab dry. Place the membrane over bottom half of gel above the bottom wick. The membrane may cover the applicator strip area.
16. Place gel in a Tupperware container with a moist towel, cover and place in an oven at 37 °C for 20 min.
17. Take membrane off gel. Rinse with 200 mL of deionized water two to three times (*see* **Note 8**).
18. Rinse three times with PBS. Wash for 20–30 min in PBST and rinse three times with PBS.
19. Add appropriate alkaline phosphatase conjugate (10 mL) to membrane and shake for 1 h.
20. Develop bands with NBT/BCIP (*see* Fig. 3).
21. Scan the membrane and save results [12].

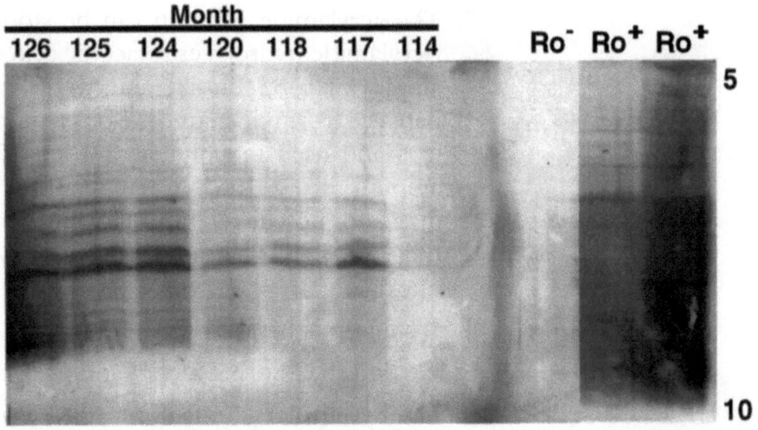

Fig. 3 Affinity immunoblotting of the patient's sera obtained at different months following first observation, showing anti-Ro 60 specific IgG clonotypes. Sera from an anti-Ro 60 negative SLE patient (Ro^-) and from two typical anti-Ro 60 positive patients (Ro^+) are shown for comparison. The pH range of the IEF gel is shown on the *right*

4 Notes

1. Having water at the bottom of the cylinder helps to dissolve the Tris relatively easily, allowing the magnetic stir bar to go to work immediately. If using a glass beaker, the Tris can be dissolved faster if the water is warmed to about 37 °C. However, the downside is that care should be taken to bring the solution to room temperature before adjusting pH.
2. Wear a mask when weighing acrylamide. To avoid exposing acrylamide to coworkers, cover the weigh boat containing the weighed acrylamide with another weigh boat (similar size to the original weigh boat containing the weighed acrylamide) when transporting it to the fume hood. Transfer the weighed acrylamide to the cylinder inside the fume hood and mix on a stirrer placed inside the hood. Unpolymerized acrylamide is a neurotoxin and care should be exercised to avoid skin contact. Mixed resin AG 501-X8 (D) (anion and cation exchange resin) is used when acrylamide solution is made, since it removes charged ions (e.g., free radicals) and allows longer storage. Some investigators store the prepared acrylamide along with this resin in the refrigerator. However, we filter them out before storage. The used mixed resin should be disposed of as hazardous waste. Manufacturer's warning states that this resin is explosive when mixed with oxidizing substances. The resin contains a dye that changes from blue-green to gold when the exchange capacity is exhausted.

3. The acrylamide solution can be stored at 4 °C for 1 month. Acrylamide hydrolyzes to acrylic acid and ammonia. The acrylamide mixture, buffer, and water can be prepared in large batches, frozen in aliquots (for greater day-to-day reproducibility), and used indefinitely (*see* ref. 13). Remove the required amount, bring to room temperature, and add the other ingredients for polymerization. However, in our laboratory we make the acrylamide solution fresh about every month when we cast our own gels.
4. We find it is best to prepare this fresh each time.
5. We find that storing at 4 °C reduces its pungent smell.
6. Large well = 10 all, medium wells = 5 all, small well = 1 all. Only lay one size of wells on to the gel. Cut if necessary. Strips may be used again.
7. During the end of the run, the gel must be watched carefully in case a fire starts. Many times the gel will burn near the applicator strip. If this happens, turn off the unit. The gel can still be used if it had been focused for a long time. The bands are usually below the strip.
8. Rinsing the membrane with deionized water two to three times will help remove a bulk of the nonspecific antibodies and help reduce the amount of TBST used subsequently and also reduce the number of washes. This wash helps to reduce nonspecific binding of NBT/BCIP to the strip. The water, owing to its low ionic strength compared to TBST, will be able to remove contaminants much better than TBST. Water is much cheaper compared to TBST, in terms of money and labor. Other investigators have found no reduction in detection of specific signals due to washing with water [14].

Acknowledgements

This work was supported by NIH grant ARO1844 and Oklahoma Center for the Advancement of Science and Technology to RHS.

References

1. Awed ZL, Williamson AR, Askonas BA (1968) Isoelectric focusing in polyacrylamide gel and its application to immunoglobulins. Nature 219:66–67
2. Knisley KA, Rodkey LS (1986) Affinity immunoblotting. High resolution isoelectric focusing analysis of antibody clonotype distribution. J Immunol Methods 95:79–87
3. Askonas BA, Williamson AR, Wright BE (1970) Selection of a single antibody-forming cell clone and its propagation in syngeneic mice. Proc Natl Acad Sci U S A 67: 1398–1403
4. Keck K, Grossberg AL, Pressmann D (1973) Specific characterization of isoelectrofocused immunoglobulins in polyacrylamide gel by

reaction with 125 I-labeled protein antigens or antibodies. Eur J Immunol 3:99–102

5. Nicolotti RA, Briles DE, Schroer JA, Davie JM (1980) Isoelectric focusing of immunoglobulins: improved methodology. J Immunol Methods 33:101–115
6. Friedenson B, Soong CJ (1984) A simple general method of exactly comparing different binding activities and antigenic properties of antibodies by obtaining duplicate copies from a single isoelectric focusing gel. J Immunol Methods 67:235–242
7. Phillips JM, Dresser DW (1973) Antibody isoelectric spectra visualized by antigen-coated erythrocytes. Eur J Immunol 3:738–740
8. Harley JB, Scofield RH, Reichlin M (1992) Anti-Ro in Sjögren's syndrome and systemic lupus erythematosus. Rheum Dis Clin North Am 18:337–358, Review
9. Scofield RH, Harley JB (1991) Autoantigenicity of Ro/SSA antigen is related to a nucleocapsid protein of vesicular stomatitis virus. Proc Natl Acad Sci U S A 88:3343–3347
10. Huang S-C, Yu H, Scofield RH, Harley JB (1995) Human anti-Ro autoantibodies bind peptides accessible to the surface of the native Ro autoantigen. Scand J Immunol 41: 220–228
11. Lloyd W, Schur PH (1981) Immune complexes, complement, and anti-DNA in exacerbations of systemic lupus erythematosus (SLE). Medicine (Baltimore) 60:208–217
12. Kurien BT, Scofield RH (2009) Affinity immunoblotting for analysis of antibody clonotype distribution in a lupus patient developing anti-Ro 60 over time. Methods Mol Biol 536:45–53
13. Harlow E, Lane D (1988) Electrophoresis. Appendix I. In: Harlow E, Lane D (eds) Antibodies. A laboratory manual. Cold Spring Harbor Laboratory, Cold Spring Harbor, NY, p 638
14. Wu M, Stockley PG, Martin WJ II (2002) An improved western blotting technique effectively reduces background. Electrophoresis 23:2373–2376

Chapter 16

Glycosaminoglycan Blotting and Detection After Electrophoresis Separation

Nicola Volpi and Francesca Maccari

Abstract

Separation of glycosaminoglycans (GAGs) by electrophoresis and their characterization to the microgram level are integral parts of biochemical research. Their blotting on membranes after electrophoresis offers the advantage to perform further analysis on single separated species such as identification with antibodies and/or recovery of single band. A method for the blotting and immobilizing of several nonsulfated and sulfated complex GAGs on membranes made hydrophilic and positively charged by cationic detergent after their separation by conventional agarose-gel electrophoresis is illustrated. This approach to the study of these complex macromolecules utilizes the capacity of agarose-gel electrophoresis to separate single species of polysaccharides from mixtures and the membrane technology for further preparative and analytical uses.

Nitrocellulose membranes are derivatized with the cationic detergent cetylpyridinium chloride (CPC) and mixtures of GAGs are capillary blotted after their separation in agarose-gel electrophoresis. Single purified species of variously sulfated polysaccharides are transferred on derivatized membranes with an efficiency of 100 % and stained with alcian blue (irreversible staining) and toluidine blue (reversible staining). This enables a lower amount limit of detection of 0.1 μg. Nonsulfated polyanions, for example hyaluronic acid (HA), may also be transferred to membranes with a limit of detection of approximately 0.1–0.5 μg after irreversible or reversible staining. The membranes may be stained with reversible staining and the same lanes used for immunological detection or other applications.

Key words Glycosaminoglycans, Electrophoresis, Blotting, Heparin, Chondroitin sulfate

1 Introduction

Blotting of (macro) molecules on membranes after their separation by electrophoresis takes advantage of the possibility to perform further analysis on single separated species by several approaches, e.g., specific binding and identification with antibodies or recovery of single band, and can be used for preparative, quantitative, and qualitative studies.

Glycosaminoglycans (GAGs) are linear, unbranched, complex heteropolysaccharides composed of a variable number of repeating disaccharide units. Each disaccharide consists of one hexosamine,

Biji T. Kurien and R. Hal Scofield (eds.), *Western Blotting: Methods and Protocols*, Methods in Molecular Biology, vol. 1312, DOI 10.1007/978-1-4939-2694-7_16, © Springer Science+Business Media New York 2015

D-galactosamine or D-glucosamine and one uronic acid, D-glucuronic acid or L-iduronic acid (IdoA) or neutral hexose, D-galactose in keratan sulfate. According to the type of the monosaccharide units and the glycosidic bonds between them, GAGs can be divided into four main categories: (1) hyaluronic acid or hyaluronan (HA), (2) chondroitin sulfate (CS) and dermatan sulfate (DS), (3) heparan sulfate (HS) and heparin (Hep), and (4) keratan sulfate (KS) (Fig. 1).

Separation of GAGs by electrophoresis is routine in many laboratories and their characterization to the microgram level forms an integral part of biochemical research, especially with respect to obtaining information from unknown purified polysaccharides.

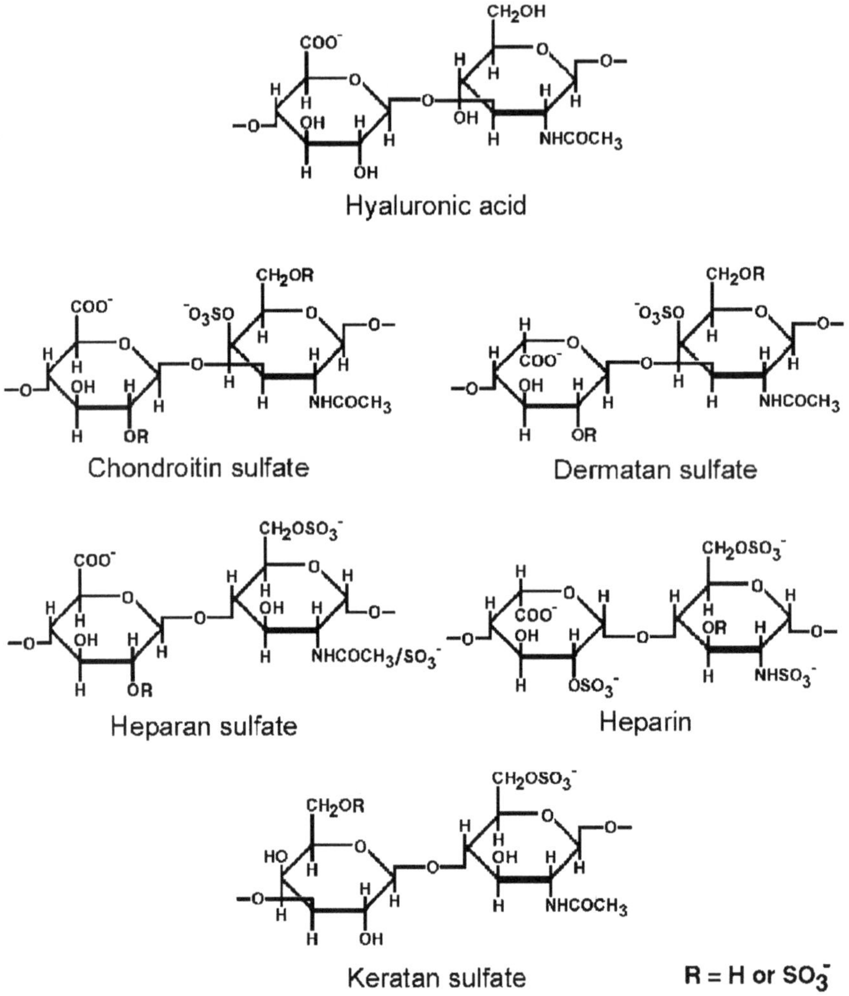

Fig. 1 Structures of disaccharides forming GAGs. Major modifications for each structure are illustrated (R = H or SO_3^-) but minor variations are possible

This is more important for natural (macro) molecules produced by extraction and purification from different animal tissues having several fundamental biological activities, as well as pharmacological properties, making them important drugs for use in clinical and pharmaceutical fields [1–5].

Cellulose acetate [6], nitrocellulose membrane (NC) [7], agarose-gel [8], and polyacrylamide-gel [9, 10] electrophoretic techniques are generally utilized for GAGs isolation and separation and for qualitative and quantitative analyses of their mixtures or single species. However, agarose-gel electrophoresis permits the separation and identification of several GAG species, such as slow (SM heparin) and fast moving heparin (FM heparin) or heparan sulfate, dermatan sulfate, and chondroitin sulfate [8, 11].

GAGs are strongly hydrophilic and negatively charged macromolecules that do not bind well to either polystyrene surfaces or hydrophobic blotting membranes. As a consequence, membranes have been derivatized with cationic detergents to make them hydrophilic and positively charged, like cetylpyridinium chloride (CPC)-treated NC membranes used in this protocol. After their electrophoretic separation, several intact GAGs with high molecular mass, such as HA, CS, highly sulfated CS, DS, HS, Hep, and its two components, FMHep and SMHep species, were transferred on NC membranes treated with a cationic detergent, CPC [12]. Quantitative analysis was performed after visualization of bands by cationic dyes, the recovery of single molecules released from membrane was also examined [12], and the direct and specific recognition of these polysaccharides by antibodies on CPC-treated NC supports has also been described [13].

2 Materials

1. High purity agarose (*see* **Note 1**).
2. Barium acetate and NC membranes 0.45 mm, binding capacity of 80–100 μg/cm^2.
3. 1,2-Diaminopropane (PDA) and cresol red.
4. Cetylpyridinium chloride (CPC).
5. Toluidine blue, alcian blue, and Whatman 3 MM paper.
6. All the other reagents should be of analytical grade.

2.1 Glycosaminoglycans

1. Different GAGs to be used as standard may be purchased from Sigma-Aldrich (http://www.sigmaaldrich.com) or other specialized companies.
2. Extraction and purification protocols for various GAGs are available in specific scientific articles and monographs [3, 5, 14–21].

2.2 Electrophoresis

1. 40 mM barium acetate buffer, pH 5.8 with 1 M acetic acid. Store at 4 °C.
2. 50 mM 1,2-diaminopropane (PDA) buffer: Buffered at pH 9 with glacial acetic acid. Store at 4 °C.
3. Cresol red solution: Dissolve 10 mg of the dye in 100 mL of distilled water (final concentration of 0.1 mg/mL). Store at 4 °C.

2.3 Blotting on Membranes

1. 1 % CPC in 30 % 2-propanol. This solution should be always freshly prepared.
2. 150 mM NaCl.
3. Transfer buffer: 100 mM Tris-acetate buffer pH 7.3. Store at 4 °C.

2.4 Membrane Staining

1. Alcian blue solution: Dissolve 50 mg of the dye in 1 mL 8 M guanidine and 19 mL of 18 mM sulfuric acid-0.25 % Triton X-100. This staining reagent is always freshly prepared.
2. Reversible staining is always freshly prepared with toluidine blue. 20 mg toluidine blue is dissolved in 100 mL of 3 % acetic acid.

3 Methods

3.1 Agarose-Gel Electrophoresis of Glycosaminoglycans

1. Prepare 0.5 % agarose solution in 40 mM barium acetate buffer (*see* **Note 2**). Heat the solution in a microwave oven or on a stirrer mixer, mixing continuously until the agarose completely dissolves and a clear solution is obtained. Do not allow the solution to boil.
2. Thoroughly clean a single glass plate of 7×8 cm (of approx. 2 mm thickness) with alcohol. After drying with paper, place it in a holder of approx. 8×10 cm and pour the warm agarose solution (*see* **Notes 3** and **4**).
3. Prior to electrophoresis, leave the gel at room temperature (RT) for approx. 30 min. Cut the gel by the side of the glass plate and put it with the same glass plate on a grid made of 1×1 cm squares. Make four small wells by using a flat chisel of approx. 5 mm, taking care to leave approx. 5 mm between each well (*see* **Note 5**). Make the wells approx. 2 cm from the edge of the gel.
4. 10 μL of GAGs standard or samples (*see* **Note 6**) may be layered by micropipets into the wells.
5. The electrophoretic run is performed in 50 mM PDA for 150 min at 50 mA by using a Pharmacia Multiphor II electrophoretic cell instrument (*see* **Note 7**).

3.2 Glycosaminoglycans Blotting

1. Cut NC membranes in portions of 7 × 8 cm.
2. Wet the membrane in freshly prepared 1 % CPC in 30 % 2-propanol for 5 min. 6 mL of CPC solution should be used. Shake the membrane manually.
3. After wetting, add 50 mL of 150 mM NaCl and incubate the membrane on a shaker for 15 min.
4. CPC-derivatized membrane is rinsed several times in 150 mM NaCl (*see* **Note 8**) and then equilibrated with continuous shaking in the same NaCl solution until blotting is performed.
5. After agarose-gel electrophoresis, carefully remove the gel from the glass plate.
6. Prepare the blotting sandwich by assembling a Whatman 3 MM paper immersed in the buffer reservoir (*see* **Note 9**). Carefully place the agarose gel on the 3 MM paper with the wells located parallel to the two buffer reservoirs. The detergent-treated NC membrane is then laid on top of the gel (*see* **Note 10**). Three further wetted filter papers, two wetted sponges, and 5 cm of absorbent paper tissue are carefully laid on top of the NC membrane, ensuring that no bubbles are trapped in the resulting sandwich. The blotting sandwich is stabilized by putting a 500 g weight on the top.
7. The capillary blotting is performed overnight at RT.

3.3 Staining Procedures

1. Irreversible staining is performed by means of alcian blue. After 2 h staining, membrane is destained by rinsing in 150 mM NaCl until background staining disappears (*see* **Note 11**). An example of the results produced is shown in Fig. 2a.

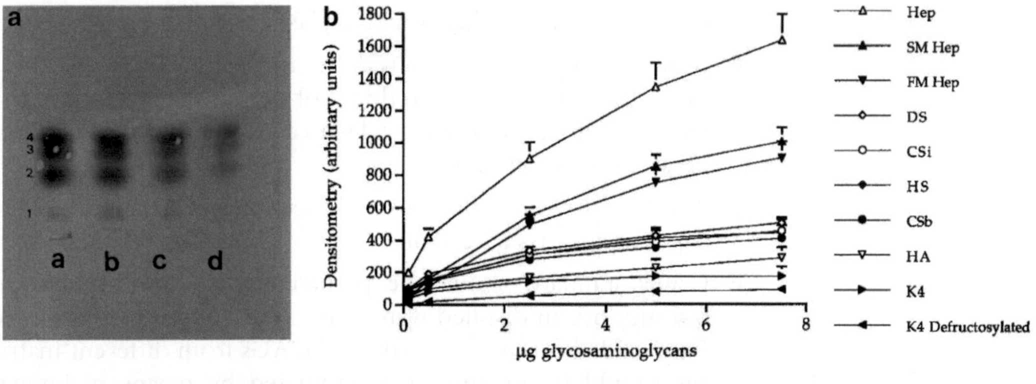

Fig. 2 (**a**) Decreasing amounts (a: 7.5 µg, b: 5.0 µg, c: 2.5 µg, d: 0.1 µg) of mixtures of GAGs composed of (1) slow moving heparin, (2) fast moving heparin, (3) dermatan sulfate, and (4) chondroitin sulfate electrophoretically separated by agarose-gel and blotted on CPC-treated NC membrane. The membrane was stained with alcian blue. (**b**) Quantitation of immobilized sulfated and nonsulfated GAGs on CPC-treated NC membranes after agarose-gel electrophoresis, capillary blotting, staining with alcian blue, destaining, and densitometric analysis. *Hep* heparin, *SM Hep* slow moving heparin, *FM Hep* fast moving heparin, *DS* dermatan sulfate, *CSi* highly sulfated chondroitin sulfate, *HS* heparan sulfate, *CSb* chondroitin sulfate, *HA* hyaluronic acid, *K4* bacterial polysaccharide K4, *K4 Defructosylated* defructosylated bacterial polysaccharide K4. Reprinted with permission

2. Irreversible staining with alcian blue can be used for quantitative studies with a high detection sensitivity (*see* below). Quantitative analysis of GAGs may be performed with a densitometer connected to a computer by using an image processing and analysis software. The wet membranes should be scanned in the RGB mode and saved in gray scale. An example of the results produced is shown in Fig. 2b.
3. GAGs separated by agarose-gel electrophoresis and transferred to a membrane can also be detected by toluidine blue staining (*see* **Note 12**). Reversible staining may be obtained by treating membranes with toluidine blue solution for 5 min. The membrane is rinsed for 30–60 s in 3 % acetic acid in the presence of 0.1 % CPC to remove the excess stain, and further utilized (*see* **Notes 13** and **14**).

4 Notes

1. Agarose should be of very high quality, suitable to run high resolution gels (possibly certified for molecular biology, ideal for the separation of small DNA fragments). Agarose from Sigma-Aldrich or from Bio-Rad is suitable.
2. The volume of the agarose solution is strictly related to the dimension of the gel. For a gel of 7 × 8 cm with a thickness of about 4–5 mm, a volume of 50 mL (250 mg agarose) is advisable.
3. Carefully eliminate possible air bubbles in the warm solution, and allow the agarose solution to convert into a gel at RT for about 30–60 min. The gel may be stored at 4 °C for approx. 4–5 days after covering it with a plastic sheet.
4. It is very important to consider that a gel having a thickness lower than about 4–5 mm does not permit the layering of the samples. On the contrary, a gel with a greater thickness requires greater migration times.
5. Carefully dry the wells by using little pieces of Whatman 3 MM paper of approx. 5 × 20 mm.
6. GAGs standard should be prepared at a concentration of 0.5 mg/mL in distilled water with a final absolute amount of 5 μg loaded on the gel. Extracted GAGs from different matrices should be quantitatively evaluated by means of known assays [1, 13–15, 17] before performing the electrophoretic separation. The optimum concentration range of unknown purified GAGs loaded on the gel should be from 2 to 8 μg.
7. Due to the possible variability of the electrophoretic conditions, 2 μL of cresol red solution (0.1 mg/mL) should be added to each standard or sample solution (10 μL) in order to

make a more accurate evaluation of the electrophoretic migration. A good migration time and electrophoretic separation are obtained at a cresol red migration of approx. 20–25 mm.

8. Derivatized membranes should be rinsed several times in 150 mM NaCl with vigorous shaking until no foaming is observed. This step permits the complete removal of excess CPC.
9. The Whatman 3 MM is immersed in the two buffer reservoirs permitting the migration of the buffer from the tank to the top side of the blotting sandwich. As a consequence, make sure that sufficient buffer volume for a complete GAGs migration (approx. 1 L) is available.
10. Carefully remove possible air bubbles entrapped between the gel and the membrane by using a little glass pestle.
11. It is very important to optimize the irreversible staining period to obtain a good band staining against a clear background. Under the experimental conditions described, an optimum staining time would be 2 h. Furthermore, use several changes of the destaining solution to produce the best results.
12. The sensitivity of staining with toluidine blue is about 10–15 times lower than that with alcian blue (an example is illustrated in Fig. 3), but if the membrane is then destained, the same lanes can be used for immunological detection or other applications.

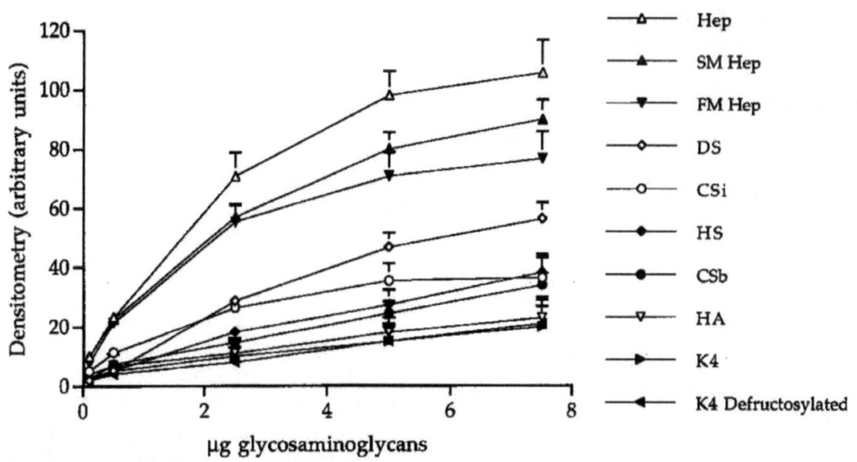

Fig. 3 Quantitation of immobilized sulfated and nonsulfated glycosaminoglycans on CPC-treated NC membranes after agarose-gel electrophoresis, capillary blotting, staining with toluidine blue, destaining, and densitometric analysis. *Hep* heparin, *SM Hep* slow moving heparin, *FM Hep* fast moving heparin, *DS* dermatan sulfate, *CSi* highly sulfated chondroitin sulfate, *HS* heparan sulfate, *CSb* chondroitin sulfate, *HA* hyaluronic acid, *K4* bacterial polysaccharide K4, *K4 Defructosylated* defructosylated bacterial polysaccharide K4. Reprinted with permission

13. After toluidine blue detection, membrane is destained using 3 % acetic acid in the presence of 0.1 % CPC. Under these conditions, the destaining of the bands is completed within 5 min and without any loss of immobilized molecules.
14. The electrophoretically separated GAGs transferred on NC may also be released and recovered from the cationized membranes at the μg level for further analysis, such as disaccharide pattern evaluation, molecular mass determination, and characterization of specifically sulfated sequences inside the polysaccharide chains. The immobilized GAGs are efficiently released from the membrane using a nonionic detergent at high ionic strength (for details *see* ref. 13).

References

1. Mammen EF, Walenga JM, Fareed J (eds) (1991) Development of non-heparin glycosaminoglycans as therapeutic agents, vol 17, Seminars in thrombosis and hemostasis. Thieme Medical Publishers, New York/Stuttgart
2. Ofosu FA, Danishefsky I, Hirsh J (eds) (1989) Heparin and related polysaccharides. Structure and activities, vol 556. Annals of the New York Academy of Sciences, New York
3. Heinegard D, Sommarin Y (1987) Isolation and characterization of proteoglycans. Meth Enzymol 144:319–373
4. Crescenzi V, Dea ICM, Paoletti S, Stivala SS, Sutherland IW (eds) (1989) Biomedical and biotechnological advances in industrial polysaccharides. Gordon and Breach Sc. Pub., New York
5. Volpi N (ed) (2006) Chondroitin sulfate: structure, role and pharmacological activity. Academic, New York
6. Cappelletti R, Del Rosso M, Chiarugi VP (1979) A new electrophoretic method for the complete separation of all known animal glycosaminoglycans in a monodimensional run. Anal Biochem 99:311–315
7. Volpi N (1996) Electrophoresis separation of glycosaminoglycans on nitrocellulose membranes. Anal Biochem 240:114–118
8. Nader HB, Takahashi HK, Guimaraes JA, Dietrich CP, Bianchini P, Osima B (1981) Heterogeneity of heparin: characterization of one hundred components with different anticoagulant activities by a combination of electrophoretic and affinity chromatography methods. Int J Biol Macromol 3:356–360
9. Rice KG, Rottink MK, Linhardt RJ (1987) Fractionation of heparin-derived oligosaccharides by gradient polyacrylamide-gel electrophoresis. Biochem J 244:515–522
10. Lyon M, Gallagher JT (1990) A general method for the detection and mapping of submicrogram quantities of glycosaminoglycan oligosaccharides on polyacrylamide gels by sequential staining with azure A and ammoniacal silver. Anal Biochem 185:63–70
11. Volpi N (1993) "Fast moving" and "slow moving" heparins, dermatan sulfate, and chondroitin sulfate: qualitative and quantitative analysis by agarose-gel electrophoresis. Carbohydr Res 247:263–278
12. Maccari F, Volpi N (2002) Glycosaminoglycan blotting on nitrocellulose membranes treated with cetylpyridinium chloride after agarose-gel electrophoretic separation. Electrophoresis 23:3270–3277
13. Maccari F, Volpi N (2003) Direct and specific recognition of glycosaminoglycans by antibodies after their separation by agarose-gel electrophoresis and blotting on cetylpyridinium chloride-treated nitrocellulose membranes. Electrophoresis 24:1347–1352
14. Volpi N (ed) (2002) Analytical techniques to evaluate the structure and function of natural polysaccharides, glycosaminoglycans. Research Signpost, India
15. Savolainen H (1999) Isolation and separation of proteoglycans. J Chromatogr B Biomed Sci Appl 722:255–262
16. Beaty NB, Mello RJ (1987) Extracellular mammalian polysaccharides: glycosaminoglycans and proteoglycans. J Chromatogr 418:187–222
17. Roden L, Baker JR, Cifonelli JA, Mathews MB (1972) Isolation and characterization of connective tissue polysaccharides. Meth Enzymol XXVIII:73–140
18. Takegawa Y, Araki K, Fujitani N, Furukawa J, Sugiyama H, Sakai H, Shinohara Y (2011)

Simultaneous analysis of heparan sulfate, chondroitin/dermatan sulfates, and hyaluronan disaccharides by glycoblotting-assisted sample preparation followed by single-step zwitter-ionic-hydrophilic interaction chromatography. Anal Chem 83:9443–9449

19. Zhang F, Sun P, Muñoz E, Chi L, Sakai S, Toida T, Zhang H, Mousa S, Linhardt RJ (2006) Microscale isolation and analysis of heparin from plasma using an anion exchange column. Anal Biochem 353:284–286

20. Shao C, Shi X, White M, Huang Y, Hartshorn K, Zaia J (2013) Comparative glycomics of leukocyte glycosaminoglycans. FEBS J 280:2447–2461

21. Mizumoto S, Sugahara K (2012) Glycosaminoglycan chain analysis and characterization (glycosylation/epimerization). Methods Mol Biol 836:99–115

Chapter 17

A Well-Based Reverse-Phase Protein Array of Formalin-Fixed Paraffin-Embedded Tissue

Joon-Yong Chung and Stephen M. Hewitt

Abstract

Biomarkers from tissue-based proteomic studies directly contribute to defining disease states as well as promise to improve early detection or provide for further targeted therapeutics. In the clinical setting, tissue samples are preserved as formalin-fixed paraffin-embedded (FFPE) tissue blocks for histological examination. However, proteomic analysis of FFPE tissue is complicated due to the high level of covalently cross-linked proteins arising from formalin fixation. To address these challenges, we developed well-based reverse-phase protein array (RPPA). This approach is a robust protein isolation methodology (29.44 ± 7.8 µg per 1 mm^3 of FFPE tissue) paired with a novel on electrochemiluminescence detection system. Protein samples derived from FFPE tissue by means of laser capture dissection, with as few as 500 shots, demonstrate measurable signal differences for different proteins. The lysates coated to the array plate, dried up and vacuum-sealed, remain stable up to 2 months at room temperature. This methodology is directly applicable to FFPE tissue and presents the direct opportunity of addressing hypothesis within clinical trials and well-annotated clinical tissue repositories.

Key words Formalin-fixed, Paraffin-embedded, Tissue lysate, Protein extraction, Proteomics, Reverse-phase protein array, Electrochemiluminescence

1 Introduction

The capacity to perform proteomic profiling of formalin fixed, paraffin embedded (FFPE) tissue offers substantial opportunities, leveraging bio-repositories, detailed pathologic diagnosis and clinical annotations. However, the recovery of protein from archival FFPE remains a challenging issue because formalin fixation leads to extensive protein cross-linking through Schiff base formation [1] and results in limited protein extraction, impaired immunoreactivity, and ambiguous identification of protein identification by mass spectrometry. Recent advances in techniques for extracting proteins from FFPE tissue sections have facilitated tissue protein profiling in the clinical proteomics, with varying degree of success [2–8].

Biji T. Kurien and R. Hal Scofield (eds.), *Western Blotting: Methods and Protocols*, Methods in Molecular Biology, vol. 1312, DOI 10.1007/978-1-4939-2694-7_17,

Protein-based array technologies have been used for target identification and characterization [9–11]. Currently there are three types of protein array platforms such as purified recombinant proteins, antibody microarrays, and reverse-phase protein microarrays [9]. Among these platforms, reverse-phase protein array (RPPA) has emerged as a strong candidate proteomic technology for FFPE tissue. Recent studies [7, 8] of FFPE proteome employed RPPA as a validation tool for protein from FFPE tissue specimens. However, limitations are still unsolved and hinder RPPA technology from reaching its full potential. These limitations include the cost of an array printer and complicated study designs where all the specimens to be assayed must be assembled at one time, preventing easy "assay on demand" environments [12]. In addition, the arrays are difficult to store and require extensive antibody and assay validation. In an effort to overcome some of the obstacles of current RPPA, we developed a well-based RPPA platform utilizing an electrochemiluminescence detection system [12]. This platform does not require a printer or arrayer and is applicable to "assay on demand" conditions; however specialized reagents and reader are required. The arrays are stable for over 2 months at room temperature, offering great abilities for antibody affinity validation and alleviating complex study design [12].

As examples of the utility of well-based RPPA, we have demonstrated that a new HER-2 assessment by well-based RPPA significantly correlated with current HER-2 status guideline, suggesting that it could be developed into alternative method for accurately determining HER-2 status in human FFPE breast cancer specimens (*unpublished data*). In addition, we have determined vascular endothelial growth factor expressional levels in archival human FFPE colon cancer tissue specimens by means of this technology [13]. In this study, we showed the new approach could be used for protein profiling analysis in FFPE tissue, with quantification and normalization tools. Especially in secreted proteins, this approach is a sensitive and specific method capable of efficiently unraveling molecular profiles associated with disease status or clinical outcome. Notably, this technology also can be used for an assessment of protein quality in FFPE tissue [14]. Finally, a new proteomic profiling method has the potential to provide better insight into diseases and contribute to the development of clinically applicable biomarkers.

2 Materials

2.1 Protein Extraction from Formalin-Fixed Paraffin-Embedded Tissue

1. Archival formalin-fixed paraffin-embedded (FFPE) prostate tissue specimens.
2. Microtome (Leica, Buffalo Grove, IL, USA).
3. Xylene or Dewaxing reagent (AutoDewaxer, Open Biosystem, Pittsburgh, PA, USA or PROTOCOL™ SafeClear, Fisher Scientific, Pittsburgh, PA, USA) (*see* **Note 1**).

4. Thermomixer.
5. 100, 95, and 70 % ethanol (EtOH).
6. Feather disposable scalpel.
7. SafeSeal microcentrifuge tube.
8. Microcentrifuge tube sealing clips.
9. Disposable Pellet Mixer and Cordless Motor.
10. High pH antigen retrieval buffer (10×, pH 9.9; Dako, Carpentaria, CA, USA). Store at 4 °C.
11. Stock solution of NaN_3 (20×): Prepare 20 % NaN_3 solution in distilled water (DW). Store at room temperature (RT).
12. Stock solution of SDS (20×): 20 % SDS. Store at RT.
13. Stock solution of glycerol (5×): Prepare 50 % glycerol solution in DW. Store at RT.
14. Stock solution of protease inhibitor solution (50×): Prepare 50× protease inhibitor solution (1 tablet/1 mL DW). Store at 4 °C.
15. Protein extraction buffer: 1× high pH antigen retrieval buffer, 1 % NaN3, 1 % SDS, 10 % glycerol, and 1× protease inhibitor solution. Store at RT (*see* **Note 2**).
16. Pascal pressure cooker.
17. Heating block for 1.5 mL microcentrifuge tube.
18. Refrigerated microcentrifuge.
19. Extra-long gel tip.
20. Reynolds Wrap® Aluminum Foil.

2.2 Assessment of Protein Concentration

1. BCA protein assay kit.
2. 96-Well flat bottom plate.
3. Sunrise™ microplate reader.

2.3 Well-Based Reverse-Phase Protein Array

1. Anti-prostate-specific antigen (PSA) and anti-smooth muscle actin antibodies, anti-glyceraldehyde-3-phosphate dehydrogenase (GAPDH) antibodies. Goat anti-mouse or rabbit SULFO-TAG™ antibodies (Meso Scale Discovery, Gaithersburg, MD, USA) (*see* **Note 3**).
2. MesoScale Discovery (MSD) Multi-Spot™ plates (MA2400 96 HB Plate, Meso Scale Discovery) (*see* **Note 4**).
3. 37 °C Incubator.
4. Thermowell™ Sealing Tape.
5. Sealing Film Roller Press (LabSource, Romeoville, IL, USA).
6. Multichannel pipets.
7. Stock solution of phosphate-buffered saline (PBS): 10× PBS.
8. Blocking buffer: 1× PBS, 0.1 % Tween-20 with 5 % w/v nonfat dry milk.

9. Wash buffer PBST: 1× PBS, 0.1 % Tween-20.
10. Microplate shaker.
11. Microplate Strip Washer.

2.4 Signal Detection

1. SECTOR Imager 2400 (Meso Scale Discovery).
2. Molecular grade DW (Quality Biological Inc.).
3. MSD read buffer-T (4×) (Meso Scale Discovery).

3 Methods

The basis of this approach is the determination of the protein concentration of lysates, creation of antigen and antibody titration curve, and direct application of the diluted lysates to a carbon surface of the Meso Scale Discovery plate. Lysates are allowed to dry, and proteins are detected by application of antibody choice (Fig. 1). Initial recovery of protein from the FFPE tissue section involves modified heat-induced antigen retrieval procedures using a high-pressure cooker. This methodology has the merits of a speedy procedure (up to 1 h), high-quality protein extraction from FFPE tissue, and permitting the use of phospho-specific antibodies (Fig. 2). In addition, this method results in excellent protein extraction yield from relatively small amounts of materials, with relatively low concentration of detergent and regardless of the deparaffinization step (*see* **Note 5**). Furthermore, this method was applicable to samples obtained with laser capture microdissection (*see* **Note 6**), as demonstrated with three different antibodies [12]. Subsequently, the proteins extracted from FFPE tissues applied to well-based RPPA which is primarily based on electrochemiluminescence detection system [12]. Once an assay is developed, it is only necessary to apply sufficient lysates for detection within the linear range of the assay. The arrays are stable for over 2 months at room temperature (Fig. 3).

3.1 Protein Extraction from Formalin-Fixed Paraffin-Embedded Tissue

1. Cut two 10 μm sections from FFPE tissue blocks using a microtome.
2. Deparaffinize the FFPE tissue section in xylene (3×5 min) in glass or plastic coplin jars. Transfer to 100 % EtOH (2×5 min), 95 % EtOH (2×5 min), 70 % EtOH (1×5 min), and then to 1× PBS (2×5 min) (*see* **Note 5**).
3. Carefully collect tissue in the SafeSeal microcentrifuge tube using disposable safety scalpel (*see* **Note 7**).
4. Add 1 mL of 1× PBS to the sample and centrifuge the tube for 1 min at 4 °C with maximum speed in a refrigerated microcentrifuge. Remove carefully the PBS solution without disturbing the tissue pellet.

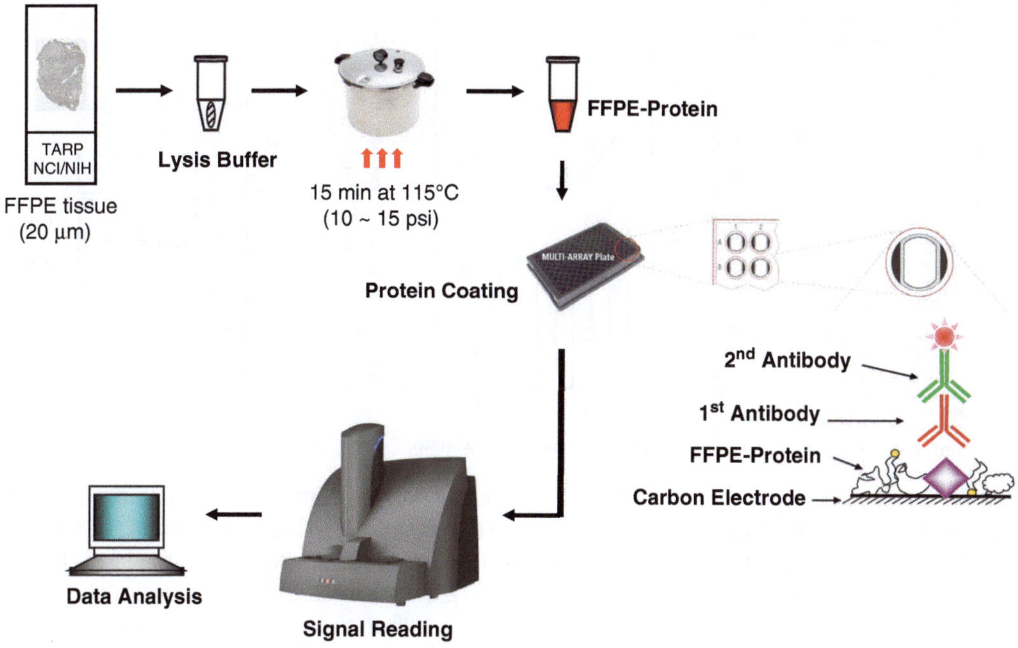

Fig. 1 Schematic diagram of novel protein array from FFPE tissue. Two ten-micrometer-thick formalin-fixed paraffin-embedded (FFPE) tissue sections were trimmed of excess wax and homogenized using a Disposable Pellet Mixer in 200 μL protein extraction solution, followed by incubation for 15 min at 115 °C within a pressure cooker. After incubation, the tissue lysates were centrifuged at 15,000 × *g* for 30 min at 4 °C. The supernatants were collected and used for the protein array. Five microliters of protein extract from FFPE tissue specimen at predetermined protein concentrations was added to Multi-Spot™ plates (MA2400 96 HB Plate, MSD, Gaithersburg, MD, USA), the plate was allowed to dry at room temperature for 90 min, and the plates were subsequently further incubated at 37 °C for 30 min. The antigen-coated plates were preincubated with 5 % BSA in PBST before incubation with specific antibodies at 4 °C for overnight. After washing with PBST, the plates were incubated for 1 h with goat anti-mouse or rabbit SULFO-TAG™ antibodies at a dilution of 1:1,000 (0.5 μg/mL). The plates were then aspirated and washed three times with PBST. Finally, MSD-T read buffer was added to the plates and they were read on the Sector Imager 2400 (MSD, Gaithersburg, MD, USA)

5. Dry the sample for 10 min at 55 °C using Thermomixer (*see* **Note 8**).
6. Add 200 μL of protein extraction buffer to the deparaffinized tissue. Homogenize immediately using the disposable pellet mixers and cordless motor until the sample is uniformly homogeneous.
7. Fix the sealing clips to the microcentrifuge tube and place the sample in a heating block. Subsequently, wrap the heating block with aluminum foil.
8. Place the heat block into a pressure cooker and incubate for 15 min at 115 °C (*see* **Note 9**).
9. Turn off pressure cooker and wait for the pressure to reach 0 psi (*see* **Note 10**).

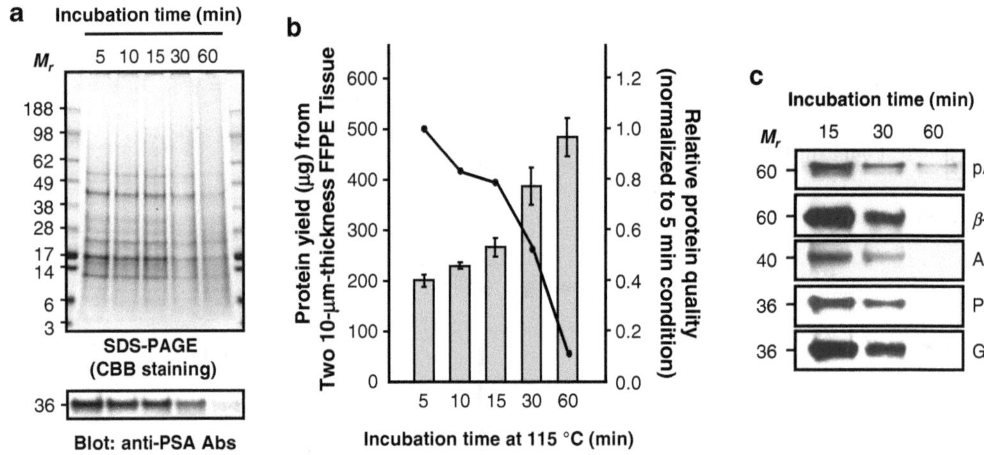

Fig. 2 FFPE-protein quality according to incubation time within a pressure cooker. (**a**) Non-deparaffinized archival human prostate FFPE tissue section was trimmed of excess wax and homogenized using a Disposable Pellet Mixer in 200 µL protein extraction solution, with incubation for 5, 10, 15, 30, and 60 min at 115 °C within a pressure cooker. The FFPE-proteins were separated by 4–12 % reducing SDS-PAGE (CBB staining), electroblotted to nitrocellulose membrane, and probed with anti-PSA antibodies (1:200). (**b**) The amount of protein extracted from each condition was measured using BCA Protein Assay Kit (Pierce). The bar graph shows the relative averages of protein yield; average ± SD. Relative protein quality of each entity is normalized to 5 min incubation condition (1.00). (**c**) Protein integrity of FFPE-derived protein by western blotting. FFPE-proteins were extracted from FFPE tissue specimen with 15, 30, and 60 min incubation within a pressure cooker. 20 µg of FFPE-proteins were subjected to a 4–12 % gradient polyacrylamide gel under reducing condition. After transfer to nitrocellulose membrane, the membrane was probed with anti-pAKT, anti-β-tubulin, anti-AMACR, anti-PSA, and anti-GAPDH antibodies. The signal was detected with a SuperSignal Chemiluminescence kit (Pierce) (*Mr*: protein molecular marker) (reproduced from [12] with permission from Willey)

10. Carefully take out the heating block from pressure cooker and place the tube on ice for 10 min (*see* **Note 11**).
11. Centrifuge the tube for 30 min at 4 °C with maximum speed in a refrigerated microcentrifuge. Carefully transfer supernatant solution to a fresh tube using gel loading tip.

3.2 Assessment of Protein Concentration

1. Centrifuge the tube for 10 min at 4 °C with maximum speed in a refrigerated microcentrifuge (*see* **Note 12**).
2. Take 3 µL of the sample and dilute tenfold with DW.
3. Take 25 µL of the diluted sample and add the sample to flat bottom 96-well plate. Measure protein concentration using BCA protein assay kit.
4. Read absorbance of the sample using a microplate reader and calculate protein concentration based on standard curve.
5. Check the quantity and quality of the protein sample (*see* **Note 13**).
6. Take 10 µg of the protein sample and apply to SDS-PAGE. An example of the results produced is shown in Fig. 2.

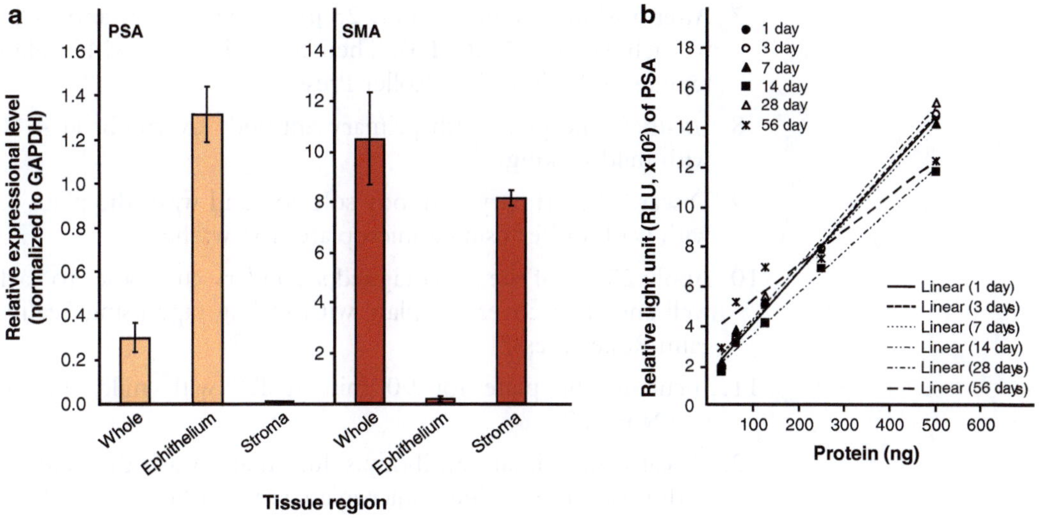

Fig. 3 Sensitivity and stability of the well-based RPPA. (**a**) Five-hundred shots (15-μm diameter laser beam) were microdissected from each of epithelial and stromal region within a 10-μm-thick prostate hematoxylin stained FFPE tissue. Protein extracted from an adjacent whole section of the same prostate FFPE tissue block was also included in this assay. The well-based RPPA using 500 ng extracted protein per well was performed. Primary antibodies were diluted 1:500 (prostate-specific antigen; PSA) or 1:1,000 (smooth muscle actin; SMA & GAPDH) with 3 % BSA. After normalization with GAPDH level, relative expressional signals were represented as ratio. The bar graph shows the average ± SD of three replicated wells. (**b**) Stability of a novel reverse-phase protein array platform. In order to examine the stability of well-based RPPA platform using FFPE-proteins, we stored the vacuum-sealed 96-well plate at RT after protein coating and measured PSA signal over 2 months (1, 3, 7, 14, 28, and 56 days, $R^2 \geq 0.93$). Dynamic ranges in plot are based on the standard curve; results given are the mean of three replicated wells (reproduced from [12] with permission from Willey)

3.3 Well-Based Reverse-Phase Protein Array

1. Pipet 5 μL (1,000–10 ng/well) (*see* **Note 14**) of tissue lysate preparation in wash buffer (PBST) to each well in the Multi-Spot™ plates (MA2400 96 HB Plate, MSD) making sure to touch the bottom of each well with the tip of the pipet and then releasing the solution slowly.
2. After applying the tissue lysate separation to the plate, allow the plate to dry at room temperature for 90 min (*see* **Note 15**).
3. Place the plate in an incubator at 37 °C for 30 min to make sure all fluid has completely evaporated.
4. Remove the plate from the incubator and apply 150 μL of blocking buffer to each well. Then cover the plate with sealing tape using Sealing Film Roller Press.
5. Allow the plate to incubate with the blocking solution for 1 h. Lightly tap the plate to make sure you remove all air bubbles and then place on a plate shaker for 1 h.
6. Discard the blocking buffer and wash the plate 5× with wash buffer using a microplate strip washer.

7. After washing the plate, apply 25 μL of the primary antibodies to each well (*see* **Note 16**). Then cover the plate with sealing tape using Sealing Film Roller Press.
8. Incubate the plate with primary antibody overnight at 4 °C with mild shaking.
9. Discard the primary antibody solution and wash the plate 5× with wash buffer using a microplate strip washer.
10. Apply 25 μL of the sulfo-tagged secondary antibodies to each well and then cover the plate with sealing tape using Sealing Film Roller Press.
11. Incubate the plate for 90 min at RT with mild shaking (*see* **Note 17**).
12. Discard the primary antibody solution and wash the plate 5× with wash buffer using a microplate strip washer.
13. Apply 150 μL of MSD read buffer-T (diluted 1:4 in molecular grade DW) to each well and lightly tap the plate to make sure you remove all air bubbles.
14. Incubate the plate for 2 min and then insert the plate into the SECTOR Imager 2400 instrument for reading (*see* **Note 18**).

4 Notes

1. We found aqueous based *dewaxers* such as AutoDewaxer (Openbiosystems) or PROTOCOL™ SafeClear (Fisher Scientific) could be used as a deparaffinizing agent in place of xylene at high temperature with equal results and greater safety (*see* ref. 15). The temperature of xylene should be controlled under 65 °C.
2. Protein extraction buffer cannot be stored at 4 °C due to precipitation. For this reason, we recommend preparing solution fresh before use.
3. Primary antibody concentration depends on target antibody. Before the assay develops, we recommend a titration test of antigen and antibody, like enzyme-linked immunosorbent assay. The primary antibodies which are suggested for western blotting as well as immunohistochemistry are compatible in this platform.
4. Wells of the plate coated with carbon particles and incorporated in the bottom of each well (for more detailed information *see* www.mesoscale.com).
5. When handling over penny-sized FFPE tissue, the protein extraction yield is consistent regardless of deparaffinization (*see* ref. 12).

6. Protein extraction from laser capture microdissection sample was successful, with average total protein yield 3.38 ± 1.2 μg per 500 shots. We performed the new reverse-phase protein array using 500 ng of the protein extracts against PSA, smooth muscle actin, and GAPDH (Fig. 3a). Expression levels of PSA and smooth muscle actin were in accordance with expected compartments of the tissue (*see* ref. 12).
7. Do *not* dry the deparaffinized tissue slide. Tissue should be collected under wet condition to prevent tissue loss.
8. The pellet should become white.
9. We find that the protein extraction buffer containing 1 % SDS combined with a higher temperature (at 115 °C) and moderate pressure (10–15 psi) resulted in a greater protein yield without change of quality of protein as measured by SDS-PAGE (*see* ref. 12). The protein extraction yield is increased with longer antigen retrieval time; however the protein quality is rapidly dropped after 30 min at 115 °C within a pressure cooker (Fig. 2a, b). In addition, the PSA signal from FFPE-protein extract from prostate tissue after 15 min incubation was approximately 80 % of that probed from 5 min incubation (Fig. 2a, b). To optimize conditions for both protein quality and yield, we utilized a 15 min antigen retrieval time for the protein extraction protocol.
10. When the pressure of pressure cooker reached a low pressure (<5 psi), the user should gently tilt the weight (petcock) located on the lid to one side. This will release residual pressure and enable a safe environment in which to open Pascal press cooker post run.
11. Remove aluminum foil cover from heating block and carefully take out the tube using a pincet. Do *not* touch or handle a heating block with naked hand because heating block is very hot.
12. Precipitation can vary depending on storage times at 4 °C. Recentrifuge the tube and transfer the clear supernatant solution to the fresh tube.
13. Although the extracted protein showed a smearing pattern on SDS-PAGE, the lysate contained relatively broad range of molecular weight proteins ranging from 10 to 180 kDa. The average total protein yield was 29.44 ± 7.8 μg per 1 mm^3 of archival human FFPE tissue. We confirmed that the protein recovery yield from FFPE tissue was approximately 90 % of that recovered from fresh tissue (*see* ref. 12).
14. Optimal amount of coated protein should be determined by research after antigen and antibody titration. The signal in the well-based RPPA correlates with that of western blots ($R^2 = 0.931$) and had greater sensitivity. We have confirmed the

sensitivity that the assay can detect multiple markers using 10 or fewer cells when applied to fresh cell lines (*see* ref. 12).

15. One can tell the plate is dry once the wells no longer appear shiny and all fluid has evaporated. The drying times can be extended depending on tissue lysate dilutions.
16. The typical concentration ranges from 1 to 3 μg/mL. Primary antibody concentration depends on target antibodies. Optimal concentration should be determined by the user for each application.
17. The typical concentration ranges from 0.2 to 2 μg/mL for initial assay development. As a starting point, we recommend diluting secondary antibody to 1 μg/mL. If background signals are higher than desired, reduce the concentration of secondary antibody to 0.5 μg/mL. Optimal concentration should be determined by the researcher for each application.
18. Turn on the SECTOR imager before secondary antibody incubation step. Once the instrument is turned on and software is started, it takes approximately 45–60 min for the CCD chip to stabilize. The SECTOR imager should be operated in a dust-free environment with an ambient temperature between 20 and 26 °C, and humidity levels between 10 and 80 %. Environment or locations with high levels of vibration should be avoided.

Acknowledgments

We thank Kris Ylaya for technical assistance. This research was supported by the Intramural Research Program of the National Institutes of Health, National Cancer Institute, Center for Cancer Research.

References

1. Mason J, O'Leary TJ (1991) Effects of formaldehyde fixation on protein secondary structure: a calorimetric and infrared spectroscopic investigation. J Histochem Cytochem 39:225–229
2. Ikeda K, Monden T, Kanoh T, Tsujie M, Izawa H, Haba A et al (1998) Extraction and analysis of diagnostically useful proteins from formalin-fixed, paraffin-embedded tissue sections. J Histochem Cytochem 46:397–403
3. Shi SR, Liu C, Balgley BM, Lee C, Taylor CR (2006) Protein extraction from formalin-fixed, paraffin-embedded tissue sections: quality evaluation by mass spectrometry. J Histochem Cytochem 54:739–743
4. Chu WS, Liang Q, Liu J, Wei MQ, Winters M, Liotta L et al (2005) A nondestructive molecule extraction method allowing morphological and molecular analyses using a single tissue section. Lab Invest 85:1416–1428
5. Prieto DA, Hood BL, Darfler MM, Guiel TG, Lucas DA, Conrads TP et al (2005) Liquid Tissue™: proteomic profiling of formalin-fixed tissues. Biotechniques 38:S32–S35
6. Yamashita S, Okada Y (2005) Mechanisms of heat-induced antigen retrieval: analyses in vitro employing SDS-PAGE and immunohistochemistry. J Histochem Cytochem 53:13–21
7. Addis MF, Tanca A, Pagnozzi D, Crobu S, Fanciulli G, Cossu-Rocca P et al (2009) Generation of high-quality protein extracts from formalin-fixed, paraffin-embedded tissues. Proteomics 9:3815–3823

8. Becker KF, Schott C, Hipp S, Metzger V, Porschewski P, Beck R et al (2007) Quantitative protein analysis from formalin-fixed tissues: implications for translational clinical research and nanoscale molecular diagnosis. J Pathol 211:370–378
9. Hultschig C, Kreutzberger J, Seitz H, Konthur Z, Büssow K, Lehrach H (2006) Recent advances of protein microarrays. Curr Opin Chem Biol 10:4–10
10. Lueking A, Cahill DJ, Müllner S (2005) Protein biochips: a new and versatile platform technology for molecular medicine. Drug Discov Today 10:789–794
11. Wingren C, Borrebaeck CA (2008) Antibody microarray analysis of directly labeled complex proteomes. Curr Opin Chem Biol 19:55–61
12. Chung JY, Lee SJ, Ylaya K, Braunschweig T, Traicoff JL, Hewitt SM (2008) A well-based reverse-phase protein array applicable to extracts from formalin-fixed paraffin-embedded tissue. Proteomics Clin Appl 2: 1539–1547
13. Chung JY, Braunschweig T, Hong SM, Kwon DS, Eo SH, Cho HJ et al (2014) Assessment of vascular endothelial growth factor in formalin fixed, paraffin embedded colon cancer specimens by means of a well-based reverse phase protein array. Proteome Sci 12:27
14. Xie R, Chung JY, Ylaya K, Williams RL, Guerrero N, Nakatsuka N et al (2011) Factors influencing the degradation of archival formalin-fixed paraffin-embedded tissue sections. J Histochem Cytochem 59:356–365
15. Chung JY, Braunschweig T, Hewitt SM (2006) Optimization of recovery of RNA from formalin-fixed, paraffin-embedded tissue. Diagn Mol Pathol 15:229–236

Chapter 18

Quantitative Computerized Western Blotting in Detail

Dalit Talmi-Frank, Charles L. Jaffe, and Gad Baneth

Abstract

The analysis of antibody reactivity against multiple antigens separated according to their molecular weights is facilitated by western blotting. The distinction between immune dominant and recessive antigens is often difficult and carried out by qualitative or empirical means. Quantitative computerized western blotting (QCWB) analyzes reactivity to specific antigens by providing a statistically measurable value for each band allowing differentiation between immunodominant and immunorecessive determinants. QCWB is useful for both single time point analysis and longitudinal studies where multiple time points are evaluated and the relativities against individual bands compared. This technique can be employed to study humoral responses to complex antigenic mixtures such as allergens and infectious agents, or identify serologic markers for early diagnosis of cancer, autoimmune or infectious diseases, or to monitor patient's clinical status.

Key words Western blot, Quantitative computerized western blot, Immunodominant antigen, SDS-PAGE, Net intensity, Total lane intensity

1 Introduction

Western blotting is a highly sensitive technique for the detection of proteins or monitoring the presence of antibodies that react with discrete antigens in complex mixtures [1]. It can be employed simultaneously against a large number of molecules to test reactivity with a panel of potential antigens [2, 3]. The high sensitivity and specificity of this technique makes western blotting suitable for the detection of low concentrations of particular proteins and allows the identification of specific circulating antibodies to allergens, pathogens, tumor, and auto-antigens in the sera of humans and animals. Early and sensitive disease diagnosis is a crucial factor in the effective and successful treatment and management of various maladies or illness. Early diagnosis prior to the development of clinical symptoms, detection of infection in immunosuppressed patients, and the identification of asymptomatic or subclinical carriers are major diagnostic challenges which often require very sensitive and specific assays [4, 5].

Biji T. Kurien and R. Hal Scofield (eds.), *Western Blotting: Methods and Protocols*, Methods in Molecular Biology, vol. 1312, DOI 10.1007/978-1-4939-2694-7_18, © Springer Science+Business Media New York 2015

Many serologic techniques used for diagnosis are based on reactivity against the whole pathogen, allergen, or tumor. These assays are commonly adjusted to detect high antibody levels, whereas early, subclinical, or latent infections and relapses or diseases are often characterized by low antibody titers against only a few specific antigens in the whole mixture [5]. Western blotting allows complex mixtures to be separated into individual components based on molecular weight and posttranslational modification(s), permitting the detection of even low levels of antibody reactivity with discrete antigens present in the mixture. In general antibody reactions against multiple components are observed when patient serum reactivity is evaluated against complex antigens by western blotting. Often bands that react strongly showing high intensities and in a high percentage of patients are termed immunodominant antigens, whereas other bands that react weakly or moderately in the same or a lower percentage of subjects are termed immunorecessive antigens [6]. The distinction between these band definitions is often difficult and frequently done empirically based on single measurements or an overall qualitative impression. Quantitative computerized western blot (QCWB) analysis addresses this difficulty thus allowing for numerical documentation and statistical analysis of reactivity to specific antigenic bands.

Specific immunodominant antigens identified by this method could be targets for developing rapid serodiagnostic tests as well as effective vaccines. Two parameters can be analyzed using QCWB: first, Net Band Intensity (NBI) that allows the evaluation of antibody reactivity to individual bands; second, Total Lane Intensity (TLI) that measures the total serum antibody reactivity to a whole lane of separated antigens [6]. The computerized quantification of band intensity allows a statistically analyzable distinction between immunodominant and immunorecessive bands.

QCWB can be very useful during a longitudinal follow-up of antibody reaction to infection, exposure to foreign proteins, or changes in clinical status when monitoring cancer or autoimmune diseases. The identification and measurement of reactivity to specific antigenic bands at multiple points over time can facilitate the identification of such bands as sensitive sentinels for early disease detection, as prominent markers of patent infection, or as markers of successful treatment. Sera collected from dogs experimentally infected with *Leishmania infantum*, a fatal protozoan parasite of humans and dogs, will be used to illustrate how QCWB can be used to evaluate and compare antibody reactivity against individual antigens in a whole parasite mixture both at a single time point and over time during a longitudinal study. Other serological methods such as enzyme-linked immunosorbent assay (ELISA) are compared to QCWB using the same sera to demonstrate the advantages of the latter technique [6]. An additional method for reliable quantification of western blot data has been shown to be effective for measuring and comparing data in different experimental settings [7] (*see* Chapter 49).

2 Materials

Sodium dodecylsulfate-polyacrylamide gel electrophoresis (SDS-PAGE) and western blotting are described here essentially as carried out by the authors [6]. Other protocols with slightly different reagents can be analyzed by the quantitative computerized technique detailed in the next paragraphs.

2.1 SDS-PAGE and Western Blotting

1. Protein sample—150 μg/lane/gel.
2. SDS-PAGE equipment.
3. Nitrocellulose paper.
4. Transfer buffer.
5. Blocking solution [Phosphate buffered saline (PBS) containing 0.05 % Tween-20 and 5 % dried milk].
6. Serum from test subjects and controls.
7. Washing buffer—phosphate buffered saline containing 0.01 % Tween 20 (PBS-T).
8. Secondary antibody at 1/1,000 dilution in washing buffer [Horseradish peroxidase (HRP)-conjugated goat anti-canine IgG].
9. DAB (0.0004 % 3-3-diaminobenzidine).
10. Molecular mass protein standard.
11. Digital scanner—the 6200C scanner (Hewlett-Packard Co., Mississauga, Canada).
12. Image analyzing software such as the KODAK 1D program (Eastman Kodak Company, Scientific Imaging Systems, Rochester, NY).

3 Methods

The methods below describe (1) the fundamentals of western blotting procedure in short; (2) application of an image analyzing software on a blotted membrane; (3) lane marking and individual band analysis; (4) determination of net band intensity (NBI); (5) defining immunodominant and immunorecessive bands; and (6) total lane intensity (TLI).

3.1 Western Blotting

While the full western blotting procedure used here was detailed at length [6], any standard procedure for blotting can be employed. Briefly, crude *L. infantum* antigen (CLA, 150 μg/lane/gel) is separated on a 12 % SDS-PAGE gel (*see* Chapter 11) under nonreducing conditions using a mini-electrophoresis apparatus and transferred to nitrocellulose membrane using either a semidry (*see* Chapter 31) or wet transfer apparatus (*see* Chapters 22 and 34). If necessary, stain-free gels can be used for the normalization of total protein electrophoresed on the SDS-PAGE gels for western blot analysis [8, 9]. A molecular mass protein standard, preferably prestained, is also run on the same gel and transferred to the membrane. The membranes are cut into individual strips and undergo four steps including blocking, incubation with dog sera, incubation with an HRP-conjugated goat anti-canine IgG diluted 1:5,000, and development with 0.0004 % 3-3-diaminobenzidine (DAB). Several rinses (3–4 times) with washing buffer are performed between each of these four steps. After a last wash with distilled water, the blots are air-dried and at the end of the western blotting procedure dry nitrocellulose membrane strips with dark bands, corresponding to serum antibody reactions with protein antigens of various molecular weights, are obtained.

3.2 Image Formation

In order to acquire an image of highest quality and resolution, blotted membranes should be scanned using a scanner or other imaging device that permits further digital analysis. Any modification of the scanner or imaging device settings that improves the blot appearance or enhances a particular feature of the blot should be avoided, thus the same settings for brightness, contrast, and gamma values should be strictly maintained during all the analysis. Membranes should be scanned immediately following the blotting procedure in order to produce clearly resolved bands and avoid fading of the bands over time. An optimal scanning procedure is the one that keeps the lanes as straight as possible and avoids color artifacts (Fig. 1a).

3.3 Digital Analysis (See Fig. 1)

3.3.1 Defining Sample Lanes

Sample lanes on an image are automatically and/or manually found by the Kodak image analyzing program using the lane finding algorithm. The algorithm determines the band location and calculates the spacing and width of the different lanes from each other on a selected region of the image. The program identifies the location of lanes according to the pixel peaks for bands that are adjacently localized. Therefore, if bands are not well resolved, homogeneous, or the transfer is poor, the lanes may need to be inserted manually. Lane finding sensitivity can be set to several levels (*see* also Subheading 3.3.5) and corrected to fit straight, slanted or curved lanes. When sensitivity is increased, fainter bands can be identified. However, it's advisable after adjusting the sensitivity to a desired level, to keep it constant throughout the project. It is critical to ensure that the lines defining each lane pass directly through the center of all the bands in each lane (Fig. 1b).

3.3.2 Lane Marker and Iso-molecular Weight Lines

In addition to the sample lane lines, marker lines should be drawn parallel to the bands at the upper and lower edges, or top and bottom of the blotted gel (Fig. 1c, *see* **Note 1**). It is important that

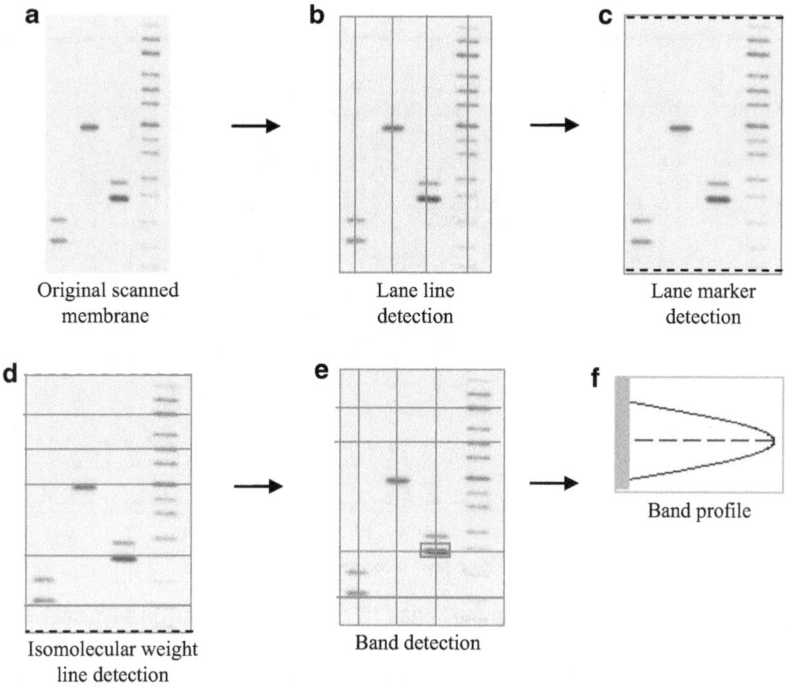

Fig. 1 Flow chart describing the different steps in image analysis of quantitative computerized western blot. Panels (**a**) Scanning of the blot and automatic or manual lane line detection based on pixel peaks of bands. (**b**) Establish the lane markers at the *top* and *bottom* of the scanned membrane and designation of the first background reading. (**c**) Marking iso-molecular weight lines enabling adjustment of individual and multiple blot membranes and determination of the molecular weight for individual bands. (**d**) Automatic software analysis and definition of *rectangles* around the detected bands. (**e**) Manual adjustment of the *rectangle borders* if needed. The software uses an optimized band sensitivity chosen according to the bands of interest. The sensitivity chosen should distinguish the bands of interest from the background while ignoring the extraneous ones. (**f**) Remodeling of profile data if needed

the marker lines accurately reflect the skew of the gel. The first lane marker at the top of the gel blot designates the first background reading, so it's important not to draw lines through bands or wells. The top and bottom lane markers set the region of original image selection and should be moved minimal or not at all. For a better adjustment of skewed gels, iso-molecular weight lines should be added (Fig. 1d). The intersecting points between lane lines and the iso-molecular weight lines serve as control points that enable adjustment for mobility variations (*see* **Note 2**).

3.3.3 Labeling Lanes

The next step is to label each lane as standard or experimental. It's possible to use either the formats given by the software or create a new standard lane according to the molecular weights given by the ladder. A standard lane will define the band size in daltons for proteins, base pairs for DNA, or bases for RNA depending on the type of gel viewed (Fig. 1b).

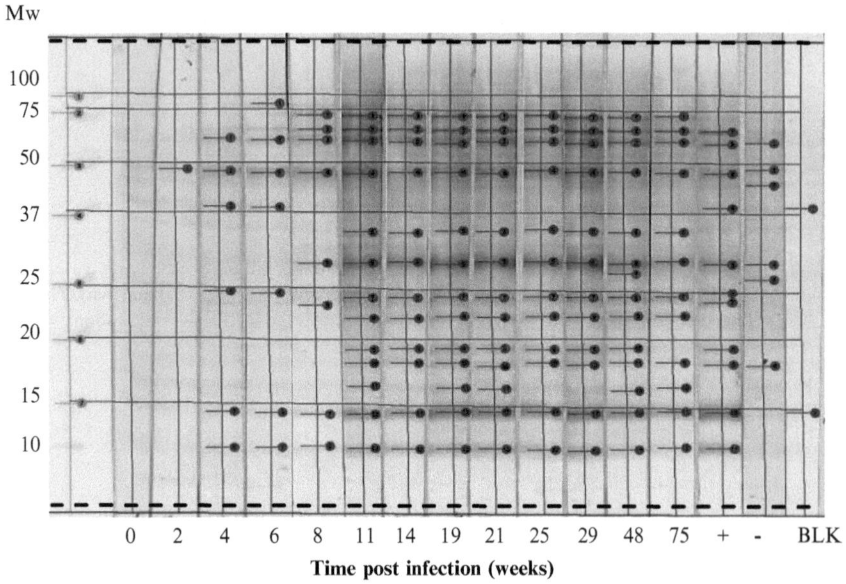

Fig. 2 An example of an image created using the Kodak 1D analyzing software on a typical western blot membrane. The image shows the reaction of sera taken from one dog experimentally infected with *Leishmania infantum* parasites on homologous parasite crude antigen at successive time points during a longitudinal follow-up study. The dog was treated from weeks 32–45 with allopurinol. The *upper* and *lower* long *horizontal dashed lines* indicate the lane markers and the *short horizontal lines* ending with a *dot mark* the iso-molecular weight lines. Positive and negative control sera are depicted in lanes P and N, respectively, and a blank lane depicting reactivity with diluting reagents is marked as BLK

3.3.4 Receiving and Analyzing Band Data

Once the lanes have been marked, bands can be automatically located. The band finding algorithm depends on two parameters that define the band sensitivity and profile width. These two parameters should be set at the beginning of the project and kept constant. A completely marked image can be seen in Fig. 2.

3.3.5 Band Sensitivity

In order to choose the correct sensitivity for a given project, resolution of the individual bands must be considered. If the bands of interest do not stand out from the background, a higher sensitivity should be chosen. Optimally the correct sensitivity should allow the program to find all the bands of interest while minimizing those bands that are irrelevant.

3.3.6 Profile Width and Band Numbering

Profile width defines the width of the band used for the analysis in those cases where the band is narrower than the total strip width. It sets the area of the rectangles that surround each band. Optimal profile width encompasses the whole band without getting false readings from the background that surrounds it.

Each band is then labeled and numbered, and a horizontal marking line is stretched across it. It is important to manually verify that the horizontal line is positioned exactly at the center of the band because the molecular weight is calculated based on the position of this line. A rectangle is marked around the band (Fig. 1e). As soon as the central horizontal line is drawn, it can be used to compare the band positions between the standard and experimental lanes, and it is possible to also adjust the position and boundaries of any band while examining the peak intensity viewed in the profile exhibited (Fig. 1f).

3.3.7 Data Modeling

After the bands are found and the band profile is created, it is important to determine whether the data needs to be modeled. This is especially important for determining the values of unresolved bands, oversaturated bands, or bands with uneven or high background.

When to Remodel Data?

1. When bands are unresolved (closely spaced) (*see* **Note 3**).
2. If bands are saturated (flat peaked) (*see* **Note 4**).
3. When faint bands are located adjacent to strong bands (*see* **Note 5**).
4. If the background is uneven or variable.
5. If false bands are present.

In order to calculate the net intensity, the program uses the pixel values in the band rectangle. Quantitative analysis of overlapping or saturated bands is best approximated by fitting the data to a Gaussian curve [10]. For saturated bands the Gaussian technique better estimates bands that are beyond the dynamic range of the capture device.

Information received from the band analysis includes:

1. Molecular weight.
2. Mobility—measured location of the band from the top lane marker.
3. Net band intensity—the band pixel value minus the background rectangle pixel values.
4. Relative intensity—the percent intensity contribution of a band within a lane.

3.4 Applications of Quantitative Computerized Western Blot Analysis

3.4.1 Net Band Intensity (NBI)

NBI is defined as the sum of pixels within the rectangle area formed around the band minus the background, reflecting the reaction at the band. NBI enables the quantitative comparison of antibody reactivity between specific antigen bands on the same or different blots. It facilitates comparison of antibody reactivity with antigens between individuals or by the same individual at different times, thus allowing one to follow and analyze longitudinally experimental or natural infections. The ability to quantify antibody reactions in a standardized way allows the statistical assessment of response to treatment confirming whether remission or elimination of infection and decrease of antibody responses have occurred [11]. This is especially applicable in the case of diseases where infection does not correlate with clinical signs.

3.5 Immunodominant Bands

NBI can be determined individually or in comparison to other bands allowing one to decide whether specific antigens generate an immunodominant or immunorecessive serological response based on comparative intensity and prevalence (number of individuals reacting with each band). In the experimental model for canine visceral leishmaniasis longitudinal quantification of serologic reactivity (NBI) to *L. infantum* antigens was recorded in six dogs over 75 weeks [6]. Computerized quantification was done and after taking the prevalence of the bands into consideration, statistical analysis identified the immunodominant and immunorecessive antigen bands. QCWB could be applied at one time point (Table 1) or used to follow changes in seroreactivity against specific antigenic bands over time.

Table 1
Quantitative comparative Western blot analysis of antibodies against *Leishmania* antigens following experimental infection with *L. infantum*

Mean intensity[b]	Band (kDa)	Band category[a]
12 ± 4(5)	12	ID
35.7 ± 22.3(6)	14	ID
15.2 ± 5.2(6)	24	ID
23.9 ± 9.7(6)	29	ID
5.8 ± 3.0(3)	48	ID
14.0 ± 0.1(4)	68	ID
12.0 ± 8.4(3)	18	IR
1.2(1)	19	IR
2.5 ± 1.1(3)	34–35	IR
13.1 ± 2.4(2)	71	IR
7.3(1)	102	IR
------	136	IR

Comparison of the mean net band intensities ($n = 6$ dogs) at 53 weeks postinfection for each *L. infantum* antigen and categorization into immunodominant or immunorecessive bands. Bands were categorized according to both the mean net band intensity and the number of animals reacting with each band (prevalence)

[a]ID—immunodominant; IR—immunorecessive

[b]Mean intensity ± SEM was determined at 10^6 pixels. Values in parentheses indicate numbers of dogs expressing bands

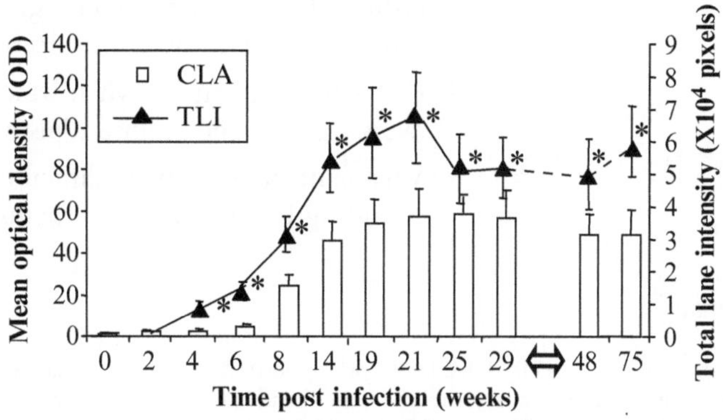

Fig. 3 Comparison of anti-leishmanial antibody reaction with crude leishmanial antigen (CLA) by Quantitative Computerized Western Blotting (QCWB) and Enzyme-linked immunosorbent assay (ELISA). Dogs ($n=6$) were infected intravenously with the parasite *Leishmania infantum* and the increase in total lane intensity (TLI), determined by QCWB, and ELISA reactivity were measured over a 75-week period. The *arrow* on the *X*-axis indicates the period when the dogs were treated with (32–45 weeks postinfection). *Triangular blocks* and *columns* represent mean values standardized to time zero, with *bars* indicating the standard error. A significant difference ($p<0.01$) is indicated by *

3.6 Total Lane Intensity (TLI)

This parameter is defined as the sum of all valid band intensities for each individual lane. TLI is equated to total antibody reactivity against all the components found in a complex crude antigen and roughly parallels serologic reactivity against whole antigen by other serological methods such as ELISA. The serological response against crude *L. infantum* antigen (CLA), following infection of beagles in an experimental canine visceral leishmaniasis model, was measured by TLI and ELISA at different times postinfection and compared. Antibody response measured by TLI showed significantly higher folds increase. At 21 weeks postinfection, TLI was 100 times greater than preinfection, whereas CLA found by ELISA was only 60 times the preinfection value ($p<0.01$) (Fig. 3).

4 Notes

1. Adding lane markers is only possible at this stage and will not be permitted in later steps when using the KODAK 1D software.
2. It is important to locate these lines correctly for the accurate identification of bands with identical molecular weights that migrate differently on the same or different gels.
3. When the program identifies unresolved bands, it might be difficult for it to determine the boundaries of the rectangle, thus not allowing manual insertion of a band close to the one that was found. The way to address this problem is to manually

reduce the size of the rectangle confining the band and then insert the desired neighboring band.

4. Oversaturated bands—when bands are flat peaked, they indicate oversaturation and require adjustment using the Gaussian model.
5. Very faint bands can be difficult to differentiate from the surrounding background and manual insertion may be needed.

References

1. Laemmli UK (1970) Cleavage and structural proteins during the assembly of the head of bacteriophage T4. Nature 227:680–684
2. Zollner TM, Spengler K, Podda M, Ergezinger K, Kaufmann R, Boehncke H (2001) The western blot is a highly sensitive and efficient technique in diagnosing allergy to wasp venom. Clin Exp Allergy 11:1754
3. Belov L, Meher HV, Putaswamy V, Miller R (1999) Western blot analysis of bile or intestinal fluid from patients with septic shock or systemic inflammatory response syndrome, using antibodies to TNF-α, IL-1α and IL-1β. Immunol Cell Biol 77:34–40
4. Riera C, Fisa R, Udina M, Gallego M, Portus M (2004) Detection of *Leishmania infantum* cryptic infection in asymptomatic blood donors living in an endemic area (Eivissa, Balearic Islands, Spain) by different diagnostic methods. Trans R Soc Trop Med Hyg 98:102–110
5. Iniesta L, Fernandez-Barredo S, Bulle B, Gomez MT, Piarroux R, Gallego M, Alunda JM, Portus M (2002) Diagnostic techniques to detect cryptic leishmaniasis in dogs. Clin Diagn Lab Immunol 9:1137–1141
6. Talmi FD, Strauss AD, Jaffe CL, Baneth G (2006) Kinetics and diagnostic and prognostic potential of quantitative Western blot analysis and antigen-specific enzyme-linked immunosorbent assay in experimental canine leishmaniasis. Clin Vaccine Immunol 13:271–276
7. Taylor SC, Berkelman T, Yadav G, Hammond M (2013) A defined methodology for reliable quantification of western blot data. Mol Biotechnol 55:217–226
8. Colella AD, Chegenii N, Tea MN, Gibbins IL, Williams KA, Chataway TK (2012) Comparison of stain-free gels with traditional immunoblot loading control methodology. Anal Biochem 430:108–110
9. Gürtler A, Kunz N, Gomolka M, Hornhardt S, Friedl AA, McDonald K, Kohn JE, Posch A (2013) Stain-free technology as a normalization tool in Western blot analysis. Anal Biochem 433:105–111
10. Vohradsky J, Panek J (1993) Quantitative analysis of gel electrophoretograms by image analysis and least squares modeling. Electrophoresis 14:601–612
11. Hassler D, Schnaufer M, Ehrfeld H, Muller E (2004) Disappearance of specific immune response after successful therapy of chronic Lyme borreliosis. Int J Med Microbiol 37:161–164

Chapter 19

Cationic Electrophoresis and Eastern Blotting

Engelbert Buxbaum

Abstract

Denaturing, discontinuous electrophoresis in the presence of SDS has become a standard method for the protein scientist. However, there are situations where this method produces suboptimal results. In these cases electrophoresis in the presence of positively charged detergents like cetyltrimethylammonium bromide (CTAB) may work considerably better. Methods for electrophoresis, staining, and blotting of such gels are presented.

Key words Disk electrophoresis, Cationic electrophoresis, CTAB electrophoresis, Blotting, Eastern blotting, Detergent, CTAB

1 Introduction

Sodium dodecyl sulfate gel electrophoresis (SDS-PAGE) [1] and western blotting [2, 3] have become indispensable in the protein laboratory to separate and detect proteins with high resolution. Separation is usually performed on an analytical scale, but preparative equipment is available commercially.

With most proteins the R_f value in SDS-PAGE is proportional to size because proteins bind about 1.4 g SDS per g of protein, equivalent to one molecule of SDS per three amino acids [4]. The number of negative charges introduced by the detergent far outweighs the charges on the protein itself, thus the charge/mass ratio, and hence the acceleration in an electrical field, is identical for all proteins. The restriction by the gel matrix however increases with the size of the protein.

There are however some situations where SDS-PAGE performs less well:

1. Very hydrophobic proteins (i.e., transmembrane proteins) bind more than the usual 1.4 g SDS per g of protein, increasing the charge/mass ratio of the protein. Thus these proteins run faster in SDS-PAGE than expected for their molecular mass.

Biji T. Kurien and R. Hal Scofield (eds.), *Western Blotting: Methods and Protocols*, Methods in Molecular Biology, vol. 1312, DOI 10.1007/978-1-4939-2694-7_19, © Springer Science+Business Media New York 2015

2. Proteins with a large number of positively charged amino acids (e.g., histones [5]) also run faster than expected for their molecular mass.
3. Glycoproteins contain a highly variable number of negative charges in their sugar side chains. Thus they do not run as crisp bands in SDS-PAGE, but as broad smears, reducing the achievable resolution of the method.

In the latter two cases, it should be possible to circumvent the problems by using a positively charged detergent in lieu of the negatively charged SDS. Of course the proteins then run from the positive to the negative electrode. Such attempts have been reported several times in the literature [6–12], but resolution of the gels was usually low, caused by relatively broad protein bands.

The high resolution of SDS-PAGE is the result of band stacking in a discontinuous (multiphasic) buffer system, an effect first described by Ornstein [13] and later theoretically elaborated by Jovin and others [14–18]. While the protein moves through the stacking gel, it is electrophoretically concentrated from below 1 mg/mL (as present in usual samples) to several 100 mg/mL. Since the amount of protein cannot change due to mass conservation, the only way in which this can happen is by reducing the volume of the protein band, that is, its height. Changing from an anionic to a cationic detergent requires a change in buffer composition so that stacking is still possible. With selected buffer systems high resolution electrophoresis of proteins in cationic detergents is possible [19]. As with SDS-PAGE separation in CTAB (cetyltrimethylammonium bromide)-PAGE is based on protein size, as noted also by others [6, 7, 10, 11].

An additional advantage of CTAB compared to SDS is that it efficiently solubilizes membrane proteins, often without damaging their structure [8, 19–22]. Thus one can use the same detergent for electrophoresis that is also used for solubilisation and purification, an advantage since extraneous detergent can interfere in PAGE by competing with SDS for binding to the protein.

2 Materials

All chemicals were of the highest purity available and were obtained mostly from Fluka (Buchs, Switzerland). Antibodies were from Accurate (Westbury, NY). Water came from a MilliQ-system (Millipore, Billerica, MA), who also supplied the Immobilon P blotting membranes. Immobiline DryStrip IEF-strips, IPG-buffer, and electrophoresis units were from Pharmacia (now part of GE Healthcare Bioscience, Piscataway, NJ, USA), detergents from Anatrace (Maumee, OH, USA). The blotting tank was from BioRad (Hercules, CA, USA).

2.1 Casting of CTAB Gels

1. 40 % acrylamide/Bis (37:1): 1.08 g bisacrylamide and 38.9 g acrylamide made to 100 mL with water (*see* **Note 1**). Stable for months at 4 °C, especially when stored over an anion exchanger.
2. 40 % Acrylamide/Bis (19:1): 2.11 g bisacrylamide and 37.9 g acrylamide made to 100 mL with water. Stable for months at 4 °C, especially when stored over an anion exchanger.
3. Potassium hydroxide (KOH; 1 M): 5.611 g KOH made to 100 mL with water. Stable at room temperature (RT) if protected from air.
4. 16.6 M acetic acid (commercial 99.5 % glacial acetic acid): Stable at RT. Exact molarity is determined once by titration and noted onto the bottle.
5. 10 % CTAB: 10 g CTAB made to 100 mL with water. Store at 37 °C to increase solubility.
6. Malachite green (1 %): 10 mg/mL malachite green in water, stable at 4 °C.
7. Water-saturated butanol: *n*-butanol shaken with some water, after phase separation the upper, organic phase is used. Stable at RT.

2.2 For Photopolymerization

1. 100 mM methylene blue: 780 mg methylene blue made to 20 mL with water. Stable for months at 4 °C.
2. 100 mM sodium toluene 4-sulfinate (T4S): 356 mg T4S (anhydrous) made to 20 mL with water. Stable for months at 4 °C.
3. 1 mM diphenyl iodonium chloride (DPIC): 6.3 mg DPIC made to 20 mL with water. Stable for 1 week at 4 °C.

2.3 For Fenton-System

1. 10 mM ferrous sulfate ($FeSO_4$): 27.8 mg $FeSO_4 \times 7H_2O$ made to 10 mL with water, make fresh daily.
2. 40 mM ascorbic acid: 70.5 mg ascorbic acid made to 10 mL with water, make fresh daily.
3. 30 % Hydrogen peroxide: commercially available, store at 4 °C (*see* **Note 2**).

The recipes for both photopolymerization [5] and a Fenton-system [8] are given, obviously only one needs to be prepared.

2.4 Running of CTAB Gels

1. Upper tank buffer: 40 mM (3.56 g/L) β-alanine, 70 mM (2.29 mL/L) acetic acid, 0.1 % CTAB. Make fresh each time.
2. Lower tank buffer: 50 mM KOH, 187 mM (3.18 mL/L) acetic acid, 0.1 % CTAB. Make fresh each time.
3. Sample buffer (2×): 1.27 mL 1 M KOH (127 mM final), 107 μL acetic acid (187 mM final), 2 mL 10 % CTAB (2 % final), 100 μL β-mercaptoethanol (βME) (1 % final), 1 mL glycerol (10 % final), 7.21 g urea (12 M final), and 50 μL 1 % basic fuchsine (0.005 % final), make to 10 mL with water. Stable at RT for a week.

2.5 Staining of CTAB Gels with Ponceau S

1. Fixative: 100 mL glacial acetic acid and 400 mL methanol made to 1 L with water. Stable at RT.
2. Ponceau S solution: 0.1 g Ponceau S and 10 mL glacial acetic acid made to 1 L with water. The solution is stable at RT and may be reused several times.

2.6 Eastern Blotting

1. Blotting buffer: 10 mL 1 M KOH (10 mM final) and 636 μL acetic acid (11 mM final), water to 1 L. Make fresh.
2. Washing solution: 50 mg SDS (0.05% final) in 100 mL methanol. Stable at RT.
3. Phosphate buffered saline (PBS; 10×stock): 2.0 g KH_2PO_4 (14.7 mM final), 2.0 g KCl (26.9 mM final), 80.0 g NaCl (1369 mM final), and 21.6 g $Na_2HPO_4 \times 7H_2O$ (or 11.44 g of the anhydride, 80.6 mM final) made to 1 L with water. Store at RT.
4. Blotto: 1 g low fat milk powder (0.1 % final), 0.5 mL Tween-20 (0.05 % final), and 0.1 g thimerosal (0.01 % final) made to 1 L with PBS. Make fresh each day.
5. Blocking solution: 0.5 g low fat milk powder in 10 mL Blotto. Make fresh each time.
6. Horseradish peroxidase (HRP) substrate (20×): 6.01 g Tris (1 M final), 14.6 g NaCl (3.2 M final), 950 mg NiCl2 × $6H_2O$ (80 mM final) dissolved in 40 mL water. Adjust to pH 7.6 with HCl and add 250 mg DAB (13 mM final) (*see* **Note 3**). Dilute to 50 mL with water, filter, and freeze in 1 mL aliquots. Stable for years at −20 °C.
7. Chemiluminescence reagent: 400 mg luminol (2.2 mM final) and 9.5 mg p-iodophenol (432 μM final) in 19 mL dimethylsulfoxide diluted to 100 mL with water. Immediately before use take 10 mL and add 4.8 μL 30 % hydrogen peroxide (420 nM final).
8. Primary antibody: 1 μg monoclonal antibody in 10 mL Blotto. Stable for many months at 4 °C if the milk powder in Blotto is replaced by BSA (fraction V). For antisera the dilution has to be worked out on a case-by-case basis.

2.7 2D-Electrophoresis

1. IEF sample buffer (2×): 100 μL β-ME (1 % final), 100 μL IPG-buffer (same pH-range as IEF strip, 1 % final), 7.21 g urea (12 M final), 0.4 g CHAPS (4 % final), and a trace of bromophenol blue, made to 10 mL with water. Store in aliquots at −20 °C.
2. Equilibration buffer: 3.6 g urea (6 M final), 3 mL glycerol (3 % final), 0.2 g CTAB (2 % final), 640 μL 1 M KOH (64 mM final), 54 μL glacial acetic acid (94 mM final), and 25 μL 1 % basic fuchsin (0.0025 % final), make to 10 mL with water.
3. Mounting medium: 50 mg agarose in 10 mL stacking gel buffer (0.5 % final), heated to boiling in a microwave oven (*see* **Note 4**). Best prepared fresh for each run.

3 Methods

3.1 Casting of CTAB Gels

1. Pour the mixture, containing either the reagents for photo-cross-linking or the Fenton-system, into an Erlenmeyer flask with magnetic stirrer (*see* Table 1). The stirrer is adjusted for vigorous movement without foam production. Then apply a vacuum for at least 10 min to remove dissolved oxygen. This is essential as oxygen inhibits the polymerization reaction. Do not apply the vacuum before starting the stirrer!
2. Add the catalyst: 18 μL 10 mM methylene blue or 2 μL hydrogen peroxide depending on whether the photopolymerization or the Fenton-system is used. Mix by gentle inversion, do not shake oxygen into the solution again. At this stage, do not add the catalyst to the stacking gel (*see* **Note 5**).
3. Pour the separating gel into the casting sandwich and overlay with water-saturated *n*-butanol. The Fenton-system will polymerize by itself; the photopolymerization system requires exposure to a strong source of white light. A sun-exposed window or a halogen lamp may be used.
4. Polymerization is finished when the interface between gel and butanol becomes prominent. Methylene blue will become colorless during polymerization. Normally polymerization should be complete within 30 min.
5. Pour the butanol from the top of the gel, add catalyst to the stacking gel mixture, and cast the stacking gel. Insert combs immediately and allow the gel to polymerize (*see* **Note 6**).

Table 1
Mixing table for CTAB-PAGE gels (all volumes in mL)

Solution	5 %	7.5 %	10 %	12.5 %	15 %	17.5 %	20 %	Stack
Acrylamide 19:1	–	–	–	–	–	–	–	3.0
Acrylamide 37:1	7.5	11.3	15.0	18.8	22.5	26.3	30.0	–
1 M KOH	2.6	2.6	2.6	2.6	2.6	2.6	2.6	1.91
Glacial acetic acid	0.962	0.962	0.962	0.962	0.962	0.962	0.962	0.161
Urea	10.8 g	10.8 g	10.8 g	10.8 g	10.8 g	10.8 g	10.8 g	5.4 g
10 % CTAB	0.6	0.6	0.6	0.6	0.6	0.6	0.6	0.3
100 mM T4S	0.3	0.3	0.3	0.3	0.3	0.3	0.3	0.15
1 mM DPIC	1.5	1.5	1.5	1.5	1.5	1.5	1.5	0.75
10 mM $FeSO_4$	0.048	0.048	0.048	0.048	0.048	0.048	0.048	0.024
40 mM Ascorbate	6.0	6.0	6.0	6.0	6.0	6.0	6.0	3.0
Water to	60	60	60	60	60	60	60	30

Note that either the reagents for photopolymerization or those for the Fenton-system should be used

3.2 Running of CTAB Gels

1. Mount the gel in the running chamber according to manufacturer's instruction, add upper and lower tank buffer. Sometimes air bubbles get trapped in the wells; these can be rinsed out using a tuberculin syringe with 27G needle. This process and the loading of samples are aided by malachite green in the stacking gel, which makes the wells easier to see (for SDS-PAGE use phenol red for the same purpose [23]).
2. Load the sample with a 25 μL Hamilton syringe, the needle should have a flat point. Between samples rinse the syringe with upper tank buffer, finally with water.
3. For a standard minigel, electrophoresis is performed at 20 mA per gel (10 mA during stacking), with a maximum voltage of 200 V. Do not forget to reverse the electrode polarity compared to SDS-PAGE (*see* **Notes 5–12**).

3.3 Staining of CTAB Gels with Ponceau S

1. Fix the gel for 2 × 10–15 min on an orbital shaker, then replace the fixative with dye solution for 5 min. The gel can be differentiated by incubation with several changes of fixative (in the same way as is commonly done with CBB-R250), but this procedure is time consuming (*see* **Notes 13–18**).

3.4 Eastern Blotting

1. Soak the PVDF blotting membrane first in methanol, then in water, and finally in blotting buffer (nitrocellulose membranes are soaked directly in buffer).
2. Place one of the sponges and three sheets of filter paper (e.g., Whatman No. 3), all soaked in blotting buffer, onto the red part of the blotting cassette. Place the gel on top and flood it with a small amount of blotting buffer (plastic Pasteur pipettes are ideal for this). Add the blotting membrane carefully on top without trapping air bubbles between gel and membrane, then place three sheets of filter paper and the second sponge, all soaked in blotting buffer, on top. Close the cassette.
3. Place the cassette (in the correct orientation, red to red and black to black) into the blotting tank already filled with blotting buffer. In this way the gel is on the positive and the membrane on the negative side. This orientation is opposite to the one used in western blotting, hence the name "eastern blotting." The entire procedure is best performed on a plastic tea tray or similar, so that the unavoidable spillage of buffer is contained.
4. Electrotransfer is achieved at 40 V constant voltage (resulting in a current of ~200 mA); no reversal of current is required if the sandwich is assembled as described above. The BioRad blotting tank comes with a plastic vessel for ice to keep the buffer cold during blotting. In my experience 2 h blotting time is sufficient to transfer even high molecular weight proteins at the interface between stacking and separating gel. It is good

practice to stain the gel afterwards to check for residual protein, at least occasionally (*see* **Note 19**).

5. Remove bound CTAB by incubating the membrane twice for 1 h in washing solution (*see* **Note 20**).
6. To check whether proteins have been transferred to the membrane, incubate the membrane with staining solution for a few minutes and then rinse with running tap water. It is convenient to mark the position of the molecular mass standards at this time with a soft pencil (*see* **Notes 21–23**).
7. Block unspecific protein binding sites on the membrane with 5 % low fat milk powder in Blotto for about 15 min. Longer times are often suggested in the literature, but are not required and may reduce sensitivity by covering the bands to be detected with milk proteins. This step also removes any Ponceau that may still be on the membrane.
8. Wash the membrane briefly with Blotto and incubate with the primary antibody. Incubation can be performed for 1 h at 37 °C if one is in a hurry. However, if the incubation is performed overnight at 4 °C in the cold room on an orbital shaker, sensitivity is higher and the antibody solution can be reused many times. In fact it is quite possible to do ones entire PhD work with a single 10 mL batch of primary antibody solution.
9. Wash the membrane three times for 15 min with blotto, incubate with the secondary antibody (HRP-conjugated, diluted as by manufacturer's recommendation) for 1 h and wash again three times with blotto.
10. For chemical staining of the blot [24], defrost a 1 mL aliquot of substrate solution, dilute to 20 mL, and add 1 μL of 30 % H_2O_2 (*see* **Note 2**). Incubate the membrane with that solution on an orbital shaker until you see well-developed bands. Rinse briefly with water and air-dry. During the reaction a black precipitate of elementary Ni is formed that photographs well and is stable for years in a notebook (*see* **Notes 24** and **25**).

3.5 2D-Electrophoresis

1. To use an IPGphor horizontal electrophoresis unit, mix the sample 1:1 with IEF sample buffer and place this solution into the ceramic strip holder.
2. Remove the protective plastic cover from the gel and place it gel side down into the sample. The sample volume depends on the length of the strips, for the 7 cm strips compatible with minigels use 125 μL. Add covering solution (obtained from GE) and place the strip holder onto the IPGphor unit.
3. Allow the sample to soak into the strip at a temperature of 18 °C and an applied voltage of 50 V for 8 h. Then increase the voltage to 500, 1,000, and 4,000 V for 30 min each, finally 8,000 V for a total of 25,000 Vh. 8,000 V is the maximum

voltage available on the IPGphor unit; depending on the conductivity of the sample this value may not be reached. This is no problem however, since runs can be normalized by the applied Vh. It is convenient to run the IEF overnight, as it does not need supervision.

4. Remove the strip from the sample holder and place it into a test tube with equilibration buffer. Seal the tube with Parafilm and place it on an end-over-end mixer for 10 min at 37 °C. This allows sufficient time for CTAB binding to the protein bands without diffusional band spreading.
5. Mount the equilibrated strip carefully onto a CTAB-gel without introducing air bubbles. A thin layer of stacking gel buffer on top of the gel can aid this process and is wicked away once the strip is in place. Use gels with one small lane for the molecular mass standard and a wide lane for the IEF strip. Strip and gel should be hand-warm to allow mounting the strip with agarose solution (300 μL). Once the agarose has set, the gel is ready to be run.

4 Notes

1. Acrylamide is neurotoxic, handle with great care. Ready-made acrylamide/bisacrylamide solutions are commercially available which avoid the development of dust during weighing. Store the solution over an anion exchanger to remove any acrylic acid which may form to prevent electroendosmosis during runs.
2. Store hydrogen peroxide solutions in the bottles supplied by the manufacturer, which have release valves to prevent the build-up of pressure from decomposition.
3. DAB is a suspected carcinogen. It is safest to add the buffer directly to the bottle and assume that the manufacturer has sold the correct weight. When opening a bottle containing DAB, do not wear gloves as the electrostatic charge on the gloves would blow the fine powder directly into your face. Wear eye protection. DAB is also available as tablets for safer handling.
4. Some experimentation is required with the settings of the microwave oven so that the agarose dissolves in the mounting medium without boiling over.
5. Fenton-polymerization results in gels which are somewhat more brittle than those produced by photopolymerization. However, if the Pharmacia multicasting stand is used, the non-transparent alumina back-plates prevent the use of a photopolymerization system; in that case the Fenton-system must be used. The separation achievable does not depend on the polymerization reaction used, however.

Table 2
Volumes of 5 and 15 % gel solutions to be mixed for step gradients

% Desired	5 % (μL)	15 % (μL)
15.0	0	392
14.0	36	356
13.1	71	320
12.3	107	285
11.3	142	249
10.5	178	214
9.5	214	178
8.6	249	142
7.7	285	107
6.8	320	71
5.9	356	36
5.0	392	0

The total volume of the gradient is 2.35 mL, enough for 1 minigel

6. Both systems are even more sensitive to the presence of oxygen than the TEMED/APS system used for Laemmli gels. Proper degassing of the gel mixture is essential. In addition it is important that the combs prevent access of air to the polymerizing stacking gel. Neither those produced by Pharmacia nor those from BioRad do that however, resulting in ill-formed wells, sometimes so badly that the gel must be discarded. I have solved that problem by having better combs manufactured in a workshop [19].
7. Gel concentration depends on the molecular mass range of the proteins of interest [26, 27]. I have found 5–15 % gradient gels most convenient for proteins of 10–200 kDa.
8. If the 10 gel multicaster is used, the gradient can be easily formed with a gradient maker, but for single gels the volume required is too small. In that case cast a step gradients by mixing 5 and 15 % solutions according to Table 2.

   These solutions are carefully layered on top of each other; this is made easier when the heavy solution contains 10 % glycerol (which does not otherwise interfere with electrophoresis). With the photopolymerization chemistry, you can start polymerization once the gel is cast.
9. Proteins with very high molecular mass (>200 kDa) require low acrylamide concentration that result in very soft, difficult to handle gels. Add 0.5 % agarose to stabilize them without any effect on separation.

10. The buffer system used here was originally described in [17] as suitable for cationic electrophoresis. More recently, another system has been suggested by Kramer [28].
11. CTAB has a relatively high Krafft point and precipitates if the temperature drops below 18 °C. If lower temperatures are desired during electrophoresis, consider replacing CTAB with a detergent with lower cmc and/or Krafft point like 16-BAC. Note that the stacking properties of buffer systems are temperature dependent.
12. Prestained molecular mass markers, available for SDS-PAGE from several manufacturers, are unsuitable for CTAB-PAGE even after detergent exchange. The bound dye influences the R_f value and separation is no longer according to molecular mass. It can only be hoped that such standards will also become available for CTAB-PAGE.
13. The staining of gels with Ponceau S is sensitive and fast, but staining with CBB-R250 is also possible. Phenol red can be used in the same way as Ponceau S; the same is probably true for other acidic dyes. Phenol red is fluorescent under acidic conditions, making very sensitive detection of proteins possible.
14. Silver staining of CTAB gels is possible; the method of Heukeshoven and Dernick [29] achieves a higher sensitivity than that of Merril [30]. You can execute the former method at a constant temperature of 37 °C for all steps; replace the glutardialdehyde with formaldehyde for maximum sensitivity.
15. PAS staining of CTAB gels [31] works, but considerable savings in time and chemicals are possible if the staining is performed on eastern blots rather than on gels [32].
16. CTAB is a very mild detergent, which can retain the enzymatic activity of enzymes solubilized with it. Try zymograms [11], proteins in gels specifically stained by their enzymatic activity, at least with monomeric enzymes. Another paper in this volume discusses zymograms in detail. Fluorescent staining of gels with rubidium (II) tris(bathophenanthroline disulfonate) [36] or with Sypro stains may also be used [37].
17. Gels can be dried after incubation with 1 % glycerol between two sheets of cellophane. Clamp the assembly between two plastic frames and leave on the bench to dry overnight. This is a more convenient approach than drying gels onto filter paper; rehydration is easier should that become necessary. For autoradiography with ^{14}C or ^{35}S, replace one of the sheets by saran wrap and remove after drying.
18. Alternatively destain by putting the gel between filter paper (3 sheets of Whatman No. 3 on both sides of the gel) and place it in a tank blotter with blotting buffer. Thirty min at 40 V (~200 mA) removes the background stain, while the protein/stain complex is immobile. If need be, proteins

become mobile again after incubating the gel in 1 m*M* KOH for 30 min.

19. Large membrane proteins like Na/K-ATPase and Mdr1 blot well in a tank blotter but there is no obvious reason why you couldn't use other blotting methods (as described in this volume) if desired.
20. Before eastern blots can be immuno-stained, the CTAB needs to be washed off, as the protein/CTAB complexes would unspecifically bind antibodies. Wash PVDF membranes with SDS in methanol. For nitrocellulose, a mixture of 50 % methanol and 10 % acetic acid is suitable, but results in weaker bands.
21. After removal of CTAB, most staining procedures used for western blots can also be used for easterns, for example India ink [permanent, [33]] or Ponceau S [34, 35]. The latter dye is washed off during the blocking step and does not interfere with immune detection.
22. When you use horseradish peroxidase conjugated antibodies for detection, do not use sodium azide as preservative, since it inactivates the enzyme. Thimerosal® (sodium ethylmercurithiosalicylate) is a suitable alternative.
23. In some cases phosphate or milk powder present in the Blotto may interfere (e.g., anti-phosphotyrosine antibodies). In those cases use Tris-buffered saline (20 mM Tris–HCl, 137 mM NaCl, 4 mM KCl, pH 7.4). Replace milk powder by fish skin gelatin (mammalian gelatin or bovine serum albumin works less well).
24. Detection of HRP-conjugated secondary antibodies can be done either by forming a colored precipitate directly on the blot [24] or by chemiluminescence [25]. The direct method is easier to perform and the stained blot can be glued into the notebook for a permanent record. Chemiluminescence, however, is more sensitive by at least 1 order of magnitude. Note that for chemiluminescence detection the secondary antibody must be more dilute to reduce background. Refer to manufacturer's instruction. Very pure luminol and p-iodophenol (should be colorless, not brown) are required; commercially available reagent mixes may also be used.
25. Phosphatase-labeled secondary antibodies with the appropriate detection reagents should also be usable in Eastern blotting; however I have not tested this.

Acknowledgement

This work was supported in part by Kuwait University grant MPB029.

References

1. Laemmli U (1970) Cleavage of structural proteins during the assembly of the head of bacteriophage T4. Nature 227:680–685
2. Towbin H, Staehelin T, Gordon J (1979) Electrophoretic transfer of proteins from polyacrylamide gels to nitrocellulose sheets: procedure and some applications. Proc Natl Acad Sci U S A 76:4350–4354
3. Dunn S (1986) Effects of the modification of transfer buffer composition and the renaturation of proteins in gels on the recognition of proteins on western blots by monoclonal antibodies. Anal Biochem 157:144–153
4. Shirahama K, Tsujii K, Takagi T (1974) Free boundary electrophoresis of sodium dodecyl sulphate-protein polypeptide complexes with special reference to SDS polyacrylamide gel electrophoresis. J Biochem 75:309–319
5. Rabilloud T, Girardot V, Lawrence J-J (1996) One-and two dimensional histone separations in acidic gels: usefulness of methylene blue-driven photo-polymerisation. Electrophoresis 17:67–73
6. Panyim S, Thitipongpanich R, Supatimusro D (1977) A simplified gel electrophoretic system and its validity for molecular weight determinations of protein cetyltrimethylammonium complexes. Anal Biochem 81:320–327
7. Eley M, Burns P, Kannapell C, Campbell P (1979) Cetyltrimethylammonium bromide polyacrylamide gel electrophoresis: estimation of protein subunit molecular weights using cationic detergents. Anal Biochem 92:411–419
8. MacFarlane D (1983) Use of benzyldimethyl-n-hexadecylammonium chloride ("16-BAC"), a cationic detergent, in an acidic polyacrylamide gel electrophoresis system to detect base labile protein methylation in intact cells. Anal Biochem 132:231–235
9. Mócz G, Bálint M (1984) Use of cationic detergents for polyacrylamide gel electrophoresis in multiphasic buffer systems. Anal Biochem 143:283–292
10. Akin D, Shapira R, Kinkade J Jr (1985) The determination of molecular weights of biologically active proteins by cetyltrimethylammonium bromide polyacrylamide gel electrophoresis. Anal Biochem 145:170–176
11. Akins R, Levin P, Tuan R (1992) Cetyltrimethylammonium bromide discontinuous electrophoresis: *Mr*-based separation of proteins with retention of enzymatic activity. Anal Biochem 202:172–178
12. Hartinger J, Stenius K, Högemann D, Jahn R (1996) 16-BAC/SDS-PAGE: a two-dimensional gel electrophoresis system suitable for the separation of integral membrane proteins. Anal Biochem 240:126–133
13. Ornstein L (1962) Disk electrophoresis: I. Background and theory. Ann N Y Acad Sci 121:321–351
14. Jovin T (1973) Multiphasic zone electrophoresis. I. Steady-state moving-boundary systems formed by different electrolyte combinations. Biochemistry 12:871–879
15. Jovin T (1973) Multiphasic zone electrophoresis. II. Design of integrated discontinuous buffer systems for analytical and preparative fractionation. Biochemistry 12:879–890
16. Jovin T (1973) Multiphasic zone electrophoresis. III. Further analysis and new forms of discontinuous buffer systems. Biochemistry 12:890–898
17. Jovin T (1973) Multiphasic zone electrophoresis. IV. Design and analysis of discontinuous buffer systems with a digital computer. Ann N Y Acad Sci 209:477–496
18. Chrambach A (1985) The practice of quantitative gel electrophoresis. VCH, Weinheim
19. Buxbaum E (2003) Cationic electrophoresis and electrotransfer of membrane glycoproteins. Anal Biochem 314:70–76
20. Pritchard D, Crawford C, Duce I, Behnke J (1985) Antigen stripping from the nematode epicuticle using the cationic detergent cetyltrimethylammonium bromide (CTAB). Parasite Immunol 7:575–585
21. Freedman D, Nutman T, Ottesen E (1988) Enhanced solubilization of immunoreactive proteins from *Brugia malayi* adult parasites using cetyltrimethylammonium bromide. Experim Parasitol 65:244–250
22. Maki K, Sagara J, Kawai A (1991) A cationic detergent, cetyltrimethylammonium bromide (CTAB), selectively dissociates the intermediate filament of the fibroblast. Biochem Biophys Res Commun 175:768–774
23. Smith I, Cromie R, Stainsby K (1988) Seeing gel wells well. Anal Biochem 169:370–371
24. Lazar J, Taub F (1992) A highly sensitive method for detecting peroxidase in situ hybridization or immunohistochemical assays. In: Kessler C (ed) Nonradioactive labeling and detection of biomolecules. Springer, Berlin, pp 135–142
25. Leong M, Fox G (1990) Luminescent detection of immunodot and Western blot. Meth Enzymol 184:442–451
26. Rodbard D, Chrambach A (1971) Estimation of molecular radius, free mobility, and valence using polyacrylamide gel electrophoresis. Anal Biochem 40:95–134

27. Stellwagen N (1998) Apparent pore size of polyacrylamide gels. Electrophoresis 19: 1542–1547
28. Kramer ML (2006) A new multiphasic buffer system for benzyldimethyl-n-hexadecylammonium chloride polyacrylamide gel electrophoresis of proteins providing efficient stacking. Electrophoresis 27(2):347–356 doi:10.1002/elps.200500563
29. Heukeshoven J, Dernick R (1988) Improved silver staining procedure for fast staining in phastsystem development unit I. Staining of sodium dodecyl sulfate gels. Electrophoresis 9:28–32
30. Merril C, Goldman D, Sedman S, Ebert M (1981) Ultrasensitive stain for proteins in polyacrylamide gels shows regional variations in cerebrospinal fluid proteins. Science 211: 1437–1438
31. Segrest J, Jackson R (1972) Molecular weight determination of glycoproteins by polyacrylamide gel electrophoresis in sodium dodecyl sulfate. Meth Enzymol 28:54–63
32. Thornton D, Holmes D, Sheehan J, Carlstedt I (1989) Quantitation of mucus glycoproteins blotted onto nitrocellulose membranes. Anal Biochem 182:160–164
33. Hancock K, Tsang V (1983) India ink staining of proteins on nitrocellulose paper. Anal Biochem 133:157–162
34. Hassan J, Feighery C, Whalan A (1987) Staining molecular weight markers on nitrocellulose using Ponceau S. J Clin Lab Immunol 24:104
35. Moore M, Viselli S (2000) Staining and quantification of proteins transferred to polyvinylidene difluoride membranes. Anal Biochem 279:241–242
36. Rabilloud T, Strub J-M, Luche S, Dorsselaer AV, Lunardi J (2001) A comparison between Sypro ruby and rubidium (II) tris (bathophenanthroline disulfonate) as fluorescent stains for protein detection in gels. Proteomics 1:699–704
37. Buxbaum E (2012) Fluorescent staining of gels. Meth Mol Biol 869:5

Chapter 20

A Miniaturized Blotting System for Simultaneous Detection of Different Autoantibodies

Ulrich Canzler, Holger Bartsch, Kai Großmann, Werner Lehmann, Karsten Conrad, Biji T. Kurien, Yaser Dorri, R. Hal Scofield, and Michael P. Bachmann

Abstract

Sera of tumor patients frequently contain autoantibodies to tumor associated antigens. Here we describe a miniaturized immunoblot platform allowing us to screen sera of patients for the presence of autoantibodies to ten autoantigens in parallel.

Key words Multiparametric assay, Autoantibodies, Tumor-associated antigens

1 Introduction

Autoantibodies to cellular components are not only found in sera of patients with autoimmune diseases: using a variety of different techniques including for example cDNA library screening (SEREX technique), enzyme linked immunosorbent assay, immunoblotting, immunocoprecipitation, and epifluorescence microscopy, autoantibodies were also detected in tumor patients (e.g., [1–6]). The presence of such antibodies are of interest for several reasons: (1) these antibodies indicate that tumor cells are indeed recognized by both, the cellular and humoral arm of the immune system, and (2) the antibodies against tumor associated antigens (TAAs) can perhaps be used as early indicators of a developing or already existing tumor or a relapse of a cancer. Many investigators have, therefore, been interested in the use of autoantibodies to TAAs as serological markers for cancer diagnosis (reviewed in ref. 7). Enthusiasm for this approach has been tempered by the low sensitivity when individual antigen–antibody reactions were studied: According to these studies, antibodies to any individual antigen such as for example p53, c-myc, or p62 do not reach levels of sensitivity which could become routinely useful in diagnosis.

Biji T. Kurien and R. Hal Scofield (eds.), *Western Blotting: Methods and Protocols*, Methods in Molecular Biology, vol. 1312, DOI 10.1007/978-1-4939-2694-7_20, © Springer Science+Business Media New York 2015

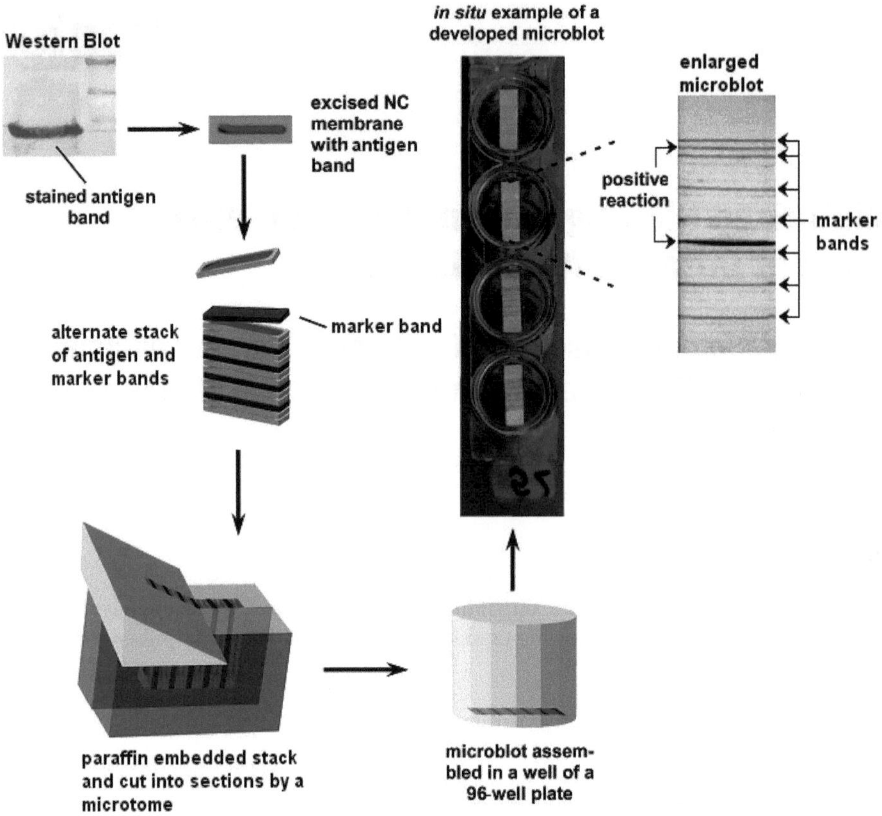

Fig. 1 Schematic view of the microblot manufacturing steps

However, it was proposed that this drawback can be overcome by using a panel of selected TAAs. Indeed, some published data support the idea that a multiparametric assay improves the sensitivity and specificity for a specific tumor entity when several selected TAAs are analyzed in parallel (e.g., [8]). Here we describe a novel platform, a miniaturized immunoblot system (Fig. 1), allowing us to screen sera for the presence of different autoantibodies in parallel.

2 Materials

2.1 Bacterial Expression and Purification of Antigens

1. Luria-Bertani (LB) medium (10 g/L NaCl, 5 g/L yeast extract, 10 g/L tryptone, pH 7.0), store at 4 °C.
2. Isopropyl β-D-1-thiogalactopyranoside (IPTG), 1 M stock solution in water, store in aliquots at –20 °C (*see* **Note 1**).
3. Bacterial expression clones (pET28 or equivalent) in BL21 (DE3)pLysS bacteria (*see* **Note 2**).
4. Ni-NTA agarose.

5. Lysis buffer (8 M urea, 100 mM NaH_2PO_4, 10 mM Tris–HCl, pH 8).
6. Wash buffer I (10 mM imidazole, 8 M urea, 100 mM NaH_2PO_4, 10 mM Tris–HCl, pH 8.0) and wash buffer II (20 mM imidazole, 8 M urea, 100 mM NaH_2PO_4, 10 mM Tris–HCl, pH 8).
7. Elution buffer (350 mM imidazole, 8 M urea, 100 mM NaH_2PO_4, 10 mM Tris–HCl, pH 8)
8. Suitable antibiotic stock, dependent on the used expression system.

2.2 Western Blotting for Preparing of the Microblot

1. Transfer buffer: Roti-Blot A and Roti-Blot K (Carl-Roth, Karlsruhe, Germany).
2. Nitrocellulose membrane: Porablot NCP, not enforced (Macherey Nagel, Düren, Germany) (*see* **Note 3**).
3. 3MM Chr chromatography paper.
4. Tris-buffered saline with Tween 20 (TBST) (10×): 1.5 M NaCl, 0.5 M Tris–HCl, pH 8, 1 % Tween 20. Dilute 100 mL with 900 mL water for use.
5. Blocking buffer: 5 % (w/v) Blocking reagent in TBST.
6. Ponceau S staining solution: 0.2 % [w/v] Ponceau S in 0.3 % [v/v] trichloroacetic acid.

2.3 Development of Microblot

1. The manufacturing principle of the microblot [9, 10] is shown in Fig. 1. Briefly, protein antigen bearing lines excised from stained western blots as well as marker bands for software based data analysis are stacked and embedded in paraffin. Sections of 10 μm are cut by a microtome, paraffin is removed and nitrocellulose slices are mounted to a solvent resistant support membrane by organic solvents. The resulting microblots (6×2 mm) are fixed to a plastic holder and integrated into modules of a standard 96-well microtitre plate. Each microblot contains ten autoantigen bands, seven marker bands and a conjugate reaction control line.
2. The solutions requested for development of the microblots are provided by Attomol GmbH (Lipten, Germany) upon delivery of your microblot test system. These ready-to-use solutions include a sample diluent and a precipitating 3,3′,5,5′-tetramethylbenzidine (TMB) substrate solution for horseradish peroxidase (HRP), concentrated wash buffer (10×), and sheep anti-human IgG-HRP-conjugate (27×). The wells containing the microblots are stored at room temperature (RT). The 10× washing buffer has to be stored at RT. The anti-IgG-HRP-conjugate as well as the TMB substrate has to be stored at 4 °C. The sample diluent solution should be stored at −20 °C.

3. For data collection, a scanner with a suitable depth of focus (e.g., Plustek Optic Pro ST48) and a PC for documentation and analysis is required. The results can be automatically evaluated with the Attomol® Microblot-Analyzer software.
4. For washing of the 96-well microtitre plates, an ELISA washer, adjusted to 400 μL wash volume set to overflow can be used. Adjust the washer needle maximal eccentrically in order to provide the microblot from being touched (*see* **Note 4**). An automatic plate washer from Tecan GmbH (Germany) can be employed.

3 Methods

As patient sera are often limited, it is desirable to evaluate as much parameters with as little sera probe as possible. Therefore, multiparameter assays have been developed. Other problems often confronted with when screening patient sera for specific antigen recognition are limitations concerning the used antigens. Often, these are difficult to purify or, if purchased, very expensive.

One technique, which is able to address both problems at one time (limited availability of serum sample and antigen) is the use of miniaturized multiparametric assays as for example the microblot technique used herein. Here, one can use very little amount of antigen, which does not need to be absolutely pure, because there is an additional SDS-PAGE step for separation of the protein of interest from undesired contaminants. In addition, in one well of a 96 well plate the simultaneous screening of up to ten different antigen reactivities can be accomplished. Also, the procedure is related to a normal ELISA assay, which makes it easy to perform. Hence, the microblot is very suitable for a multiparametric evaluation of precious patient sera.

3.1 Preparation of Samples; Bacterial Expression and Affinity Purification of His_6 Tagged Proteins

1. If not already available, start with cloning of the desired antigens into a bacterial expression system. We have good experience using the pET system provided by Merck KGaA (Darmstadt, Germany), but other systems might work as well.
2. The day before the bacterial expression pick an isolated colony of the desired clone and incubate in 10 mL LB medium supplemented with the appropriate antibiotic in a rotating incubator at 37 °C.
3. Transfer 8 mL of the overnight culture to 800 mL of freshly prepared LB medium and continue to incubate at 37 °C in a rotating incubator (*see* **Note 5**).
4. Monitor the growth of the bacterial culture by measuring the optical density$_{600}$ (OD_{600}) and continue the culture until the OD_{600} reaches 0.5.

5. Induce the culture by adding 1 mM IPTG.
6. Grow the culture for additional 3 h and keep monitoring OD_{600} at least every hour. If the OD_{600} is not increasing any further, you can stop the culture.
7. Pellet the bacterial suspension at 5,000 × *g* for 20 min at 4 °C. Discard the supernatant.
8. Store the cell pellet at −20 °C until use or proceed immediately with the cleaning procedure.
9. Resuspend the pellet in 10 mL lysis buffer, add 1 mM PMSF (*see* **Note 6**).
10. Disintegrate the bacterial suspension by incubating for 15 min at room temperature and periodical mixing. The suspension will become very viscous due to the released nucleic acids.
11. Degrade the nucleic acids by applying ultrasound in short pulses. Keep the protein suspension on ice during the whole procedure (*see* **Note** 7).
12. Centrifuge the lysate for 15 min at 10,000 × *g* (4 °C) and save the supernatant for purification.
13. Prepare a Ni-NTA column as follows: fill the bottom of a 5 mL syringe with some sterile glass wool and wet the glass wool with water to remove any air that is trapped. Add 500 μL Ni-NTA agarose beads (approximately 1 mL Ni-NTA slurry) on top of the glass wool and equilibrate the prepared column with 10 mL lysis buffer.
14. Apply the supernatant saved after the centrifugation step onto the column. Keep the flow-through and reapply once more. If desired keep an aliquot for control purpose.
15. Wash the column once with 10 mL lysis buffer.
16. Wash four times with 1 mL wash buffer I each, followed by four additional wash steps with 1 mL wash buffer II each. If desired keep an aliquot for control purpose.
17. Elute the protein with elution buffer. Apply six times 500 μL each and save in separate sample tubes.
18. Continue with an SDS-PAGE (*see* Subheading 3.2) in order to analyze the efficacy of your protein purification. Load samples of each elution fraction and, if desired samples of your washing steps, flow through and crude extract.

3.2 SDS-PAGE

1. Carry out SDS-PAGE (1 mm thick mini gels) essentially according to Laemmli [11] (*see* Chapters 11 and 34).
2. Load 30 μL (approximately 3 μg protein) of each sample in a well. Include one well for prestained molecular weight markers Page Ruler TM. Load either the samples for the control or for the western blot.

3. Stain the control gel in Coomassie Blue staining solution for 1 h and destain until protein bands are clearly visible with a nearly clear background, or continue with western blotting.

3.3 Western Blotting

1. After separation by SDS-PAGE, the samples are transferred to nitrocellulose membranes electrophoretically. These directions assume the use of a semidry system provided by Bio-Rad (*see* Chapter 31). Two trays with transfer buffer Roti-Blot A and transfer buffer Roti-Blot K respectively are prepared with a size slightly bigger than the dimension of the gel.
2. Cut the nitrocellulose membrane and eight pieces of Whatman 3MM paper to the size of the separating gel. Cut one edge of the membrane for later orientation.
3. Four sheets of 3MM are moistened in transfer buffer Roti-Blot A and transferred to the anode plate of the blotting device. On top of this stack, the nitrocellulose membrane is added.
4. The gel unit is disconnected from the power supply and disassembled. The stacking gel is removed and discarded and one corner (corresponding to the membrane) is cut from the separating gel to allow its orientation to be tracked. The separating gel is then laid on top of the nitrocellulose membrane.
5. Another four sheets of 3MM paper are wetted in transfer buffer K and carefully laid on top of the gel, ensuring that no bubbles are trapped in the resulting sandwich. You can remove air bubbles by carefully rolling a glass pipette on top of the stack.
6. The lid is put on top of the stack and the power supply activated. Transfers can be accomplished at 0.8 mA/cm^2 blot size for 1 h.
7. Once the transfer is complete the stack is carefully disassembled. The 3MM paper and the gel can then be discarded. If you want, you can stain the gel in Coomassie blue staining reagent for 1 h and destain thereafter in order to check for complete transfer of the separated proteins. On the nitrocellulose membrane the colored molecular weight markers should be clearly visible on the membrane.
8. Stain the membrane with Ponceau S staining solution for 5 min at room temperature (RT). Destain the blot with water until the protein band is clearly visible.
9. This protein band together with similarly prepared nine additional protein bands can then be used for the assembly of a stack of protein blots as schematically summarized in Fig. 1. This sophisticated assembly of the microblot is made commercially available from the company Attomol GmbH (Lipten, Germany). At present, up to ten different proteins have been used for the assembly of one microblot.

3.4 Detection of Autoantibodies Using Microblots

1. Equilibrate assay reagents and microtitre modules to room temperature (RT), vortex reagents and insert the desired modules into a microplate frame.
2. Prepare 1× washing buffer by adding 900 mL aqua dest. to 100 mL concentrated washing buffer. The ready-to-use buffer is stable for at least 2 weeks if stored at 4 °C.
3. Warm sample buffer to 37 °C for 15 min in order to solubilize all components. Mix thoroughly. After usage, store the remaining buffer at −20 °C.
4. Thaw the needed sera and warm to RT, mix well. Dilute the sera 1:100 with sample buffer and mix well.
5. Fill each well with 100 μL diluted sera sample and cover the microtitre modules. Incubate for 60 min at RT on a rotary shaker.
6. Carefully remove the diluted sera and wash each well six times with 300 μL diluted wash buffer for 200 s at RT (*see* **Note 8**). Remove any remaining liquid by tapping the plate on a pile of filter paper.
7. During the washing steps prepare the HRP-conjugate always freshly by diluting 1–27 in ready-to-use wash buffer.
8. Fill each well with 100 μL diluted HRP-conjugate. Incubate the covered microtitre modules for 60 min at RT on the rotary shaker.
9. Carefully remove the HRP-conjugate and wash each well six times with 300 μL ready-to-use wash buffer for 200 s at RT. Remove any remaining liquid by tapping the plate on a pile of filter paper.
10. Fill each well with 100 μL substrate solution. Incubate the covered microtitre modules for 20 min at RT on the rotary shaker. *Protect from light!*
11. Carefully remove the substrate solution and wash each well by applying 300 μL ready-to-use wash buffer. Incubate for 5 min at RT on the rotary shaker. Wash finally with 300 μL aqua dest. per well for 5 min on the rotary shaker. Remove any remaining liquid by tapping the plate on a pile of filter paper.
12. Dry microtitre modules for 2 h at RT. Clean the lower surface of all wells free of lint. Scan the microblots from the bottom of modules (within the microplate frame) by using the scanner (resolution 1,200 dpi). Analyze the immunoreactions on the enlarged scan image. An immunostained microblot is shown in Fig. 1. The control band displays a blue staining if the detection reagent does function and if the test was correctly carried out. Differentially stained antigen bands reflect the autoantibody content of the serum sample. The use of the automatic image processing algorithms of the Microblot-Analyzer software

facilitates the evaluation as it detects marker bands and measures signals in all antigen areas densitometrically. The data are displayed as negative, borderline or positive results in relation to negative control sera defined cut-off values.

4 Notes

1. Unless stated otherwise, all solutions should be prepared in water that has a conductivity of 0.056 μS/cm and total organic content of less than five parts per billion. This standard is referred to as "water" in this text.
2. Different expression systems are commercially available, some of them for special purpose (e.g., for expression of sequences including rare used codons in bacteria). You should evaluate the system which works best for your purpose.
3. The usage of the nitrocellulose membrane *porablot NCP*, not enforced from Macherey Nagel (Düren, Germany) is mandatory, as the preparation of the microblots is adjusted to that membrane. Other membranes might not work with this system. *Especially the use of reinforced membrane must be avoided!*
4. If there is no automated washer available, use an eight channel pipette for washing instead.
5. Use a flask which has ten times the size of the used expression volume or e.g., Fernbach-Flasks as alternative.
6. There is no need for native purification, as the proteins are run on denaturing SDS-PAGE later on. Normally, the denaturing cleanup procedure is more effective. If purification under native condition is superior, one can use this procedure as well.
7. The settings one will need for this ultrasound step are extremely dependent on your specific device and have to be determined individually.
8. If possible use a commercial automated washer for all washing steps.

Acknowledgement

This work was supported in part by grants from the Medical Faculty of the TU-Dresden and the BMBF (Germany). We thank Martina Franke for her excellent technical support.

References

1. Tan EM (1989) Antinuclear antibodies: diagnostic markers for autoimmune diseases and probes for cell biology. Adv Immunol 44: 93–151
2. Cheever MA, Disis ML, Bernhard H, Gralow JR, Hand SL, Huseby ES et al (1995) Immunity to oncogenic proteins. Immunol Rev 145: 33–59
3. Tan EM (2001) Autoantibodies as reporters identifying aberrant cellular mechanisms in tumorigenesis. J Clin Invest 108:1411–1415
4. Stockert E, Jager E, Chen YT, Scanlan MJ, Gout I, Karbach J et al (1998) A survey of the humoral immune response of cancer patients to a panel of human tumor antigens. J Exp Med 187:1349–1354
5. Soussi T (2000) p53 Antibodies in the sera of patients with various types of cancer: a review. Cancer Res 60:1777–1788
6. Forti S, Scanlan MJ, Invernizzi A, Castiglioni F, Pupa S, Agresti R et al (2002) Identification of breast cancer-restricted antigens by antibody screening of SKBR3 cDNA library using a preselected patient's serum. Breast Cancer Res Treat 73:245–256
7. Conrad K, Bachmann M (2006) Tumor-associated autoantibodies. In: Shoenfeld Y, Gershwin ME, Meroni PL (eds) Autoantibodies, 2nd edn. Elsevier, Amsterdam, pp 423–436
8. Zhang JY, Casiano CA, Peng XX, Koziol JA, Chan EK, Tan EM (2003) Enhancement of antibody detection in cancer using panel of recombinant tumor-associated antigens. Cancer Epidemiol Biomarkers Prev 12:136–143
9. Lehmann W (2001) Procedure for production of an array for detection of components of a biological sample. Patent DE 101 10 511
10. Canzler U, Bartsch H, Grossmann K et al (2009) A miniaturized blotting system for simultaneous detecting of different autoantibodies. Methods Mol Biol 536:129–137
11. Laemmli UK (1970) Cleavage of structural proteins during the assembly of the head of bacteriophage T4. Nature 227:680–685

Chapter 21

Proteomic Expressional Profiling of a Paraffin-Embedded Tissue by Multiplex Tissue Immunoblotting

Joon-Yong Chung and Stephen M. Hewitt

Abstract

In the functional proteome era, the proteomic profiling of clinicopathologic annotated tissues is an essential step for mining and evaluations of candidate biomarkers for disease. Previously, application of routine proteomic methodologies to clinical tissue specimens has provided unsatisfactory results. Multiplex tissue immunoblotting is a method of transferring proteins from a formalin-fixed, paraffin-embedded tissue section to a stack of membranes which can be applied to a conventional immunoblotting method. A single tissue section can be transferred to up to ten membranes, each of which is probed with antibodies and detected with fluorescent tags. By this approach, total protein and target signals can be simultaneously determined on each membrane; hence each antibody is internally normalized. Phosphorylation specific antibodies as well as antibodies that do not readily work well with paraffin-embedded tissue are applicable to the membranes, expanding the menu of antibodies that can be utilized with formalin-fixed tissue. This novel platform can provide quantitative detection retaining histomorphologic detail in clinical samples and has great potential to facilitate discovery and development of new diagnostic assays and therapeutic agents.

Key words Formalin-fixed, Paraffin-embedded, Tissue, Immunodetection, Histomorphology, Proteomics, Expressional profiling

1 Introduction

Proteomic profiling of tissue specimens, having pathologic and histologic relevance, promises to the development of biomarkers to guide diagnosis and therapy in biomedicine [1, 2]. Many of the traditional approaches such as western blots are based on a "grind and bind" means of isolating proteins from tissue. This "grind and bind" approach fails to provide a histomorphologic perspective of protein expression. The only means of gaining a histomorphologic understanding of protein expression have been immunohistochemistry and laser capture microdissection based collection of samples for traditional analysis. Laser capture microdissection does provide the capacity to perform a directed western blot on tissue [3–5]; however, it is time-consuming and does not provide a global

Biji T. Kurien and R. Hal Scofield (eds.), *Western Blotting: Methods and Protocols*, Methods in Molecular Biology, vol. 1312, DOI 10.1007/978-1-4939-2694-7_21, © Springer Science+Business Media New York 2015

expression view of a target protein. Immunohistochemistry, yet providing excellent localization, lacks quantification without sophisticated instrumentation [6] and lacks normalization. However, formalin-fixed and paraffin-embedded tissue, the gold standard of diagnostic histopathology, is not routinely applicable to the "grind and bind" approach due to the high level of covalently cross-linked proteins arising from formalin fixation. In translational research, there is the great desire to utilize the vast archive of formalin-fixed and paraffin-embedded tissue that has been collected [7]. Most research antibodies do not perform well in paraffin-embedded tissues. This failure is thought to be related to protein cross-linking, inadequate deparaffinization and issues of epitope presentation. This problem has become a bottleneck in translational research [2].

To address these challenges, a number of protein-based arrays have been developed and evaluated in clinical research fields. Although these techniques are generally superior in expression profiling and quantitation of protein changes associated with disease states, each still has significant limitations [8–10]. Investigators continue to seek a convenient and reliable proteomic tool, which can detect protein in tissue, while having retention of both quantitative and histomorphologic features. Multiplex tissue immunoblotting meets the criteria. This method provides a level of histomorphologic correlation, but at the same time, presents the proteins on a membrane platform that widens the number of antibodies that can be utilized. Additionally the total amount of protein present on each membrane can be determined and used for normalization [11–15]. As a research tool, this method expands the capacity of a tissue microarray to a protein array with quantitative data that can be normalized and directly composed for different antigens detected on a single stack of membranes [13–15]. When applied to a whole section of tissue [11, 12], it allows the ability to profile a tumor for multiple antigens with the use of single paraffin-embedded slide.

As an example of the utility of multiplex tissue immunoblotting, we have quantified the change in seven proteins in the transition from normal epithelium to invasive tumor [12]. This approach allows the quantification of changes in the expression of potential biomarkers in normal, in situ and invasive disease. This approach provided insight into the timing and magnitude of protein changes seen in this transition of from benign to invasive cancer as is useful in the development of novel biomarkers for prevention and screening of cancer. Additionally, the application of this method can be expanded to studying the proteomic profile of a signaling pathway, and is a promising means of assessment of molecular-targeting therapies as well. We previously demonstrated the feasibility of this method in studying PI3K/AKT pathway in extrahepatic cholangiocarcinoma and premalignant lesions. Finally, this approach can be applied to tissue microarrays, creating a form of protein array [15].

2 Materials

2.1 Deparaffinization and Enzyme Treatment

1. Xylene or dewaxing reagents.
2. 100, 95, and 70 % ethanol (EtOH) (molecular biology grade).
3. 50 mM ammonium bicarbonate (NH_4HCO_3) buffer (pH 8.2). Store room temperature (RT).
4. Stock solution of trypsin (200×): 0.2 % trypsin solution in 50 mM ammonium bicarbonate solution (pH 8.2). Immediately freeze in single use (200 μL) aliquots at −20 °C.
5. Stock solution of Proteinase-K (400×): ready-to-use Proteinase-K, freeze in single use (50 μL) aliquots at −20 °C (*see* **Note 1**).
6. Enzyme cocktail solution (prepare freshly before use): 0.001 % trypsin plus 1:400-fold diluted the ready-to-use Proteinase-K in 50 mM ammonium bicarbonate solution, pH 8.2 (*see* **Note 2**).
7. ProBuffer: One tablet of complete protease inhibitor, 0.5 mL phosphatase inhibitor I, 0.5 mL phosphatase inhibitor II, in 50 mL PBS (pH 7.2). Store at 4 °C.
8. Phosphate-buffered saline (PBS, pH 7.2).
9. Plastic or glass coplin jars for slide processing.
10. Incubation chamber for enzyme reaction (five-slides mailer box).

2.2 Transferring from the Tissue Slide to Membrane

1. Transfer buffer (5×): 250 mM Tris base (do not adjust pH), 1,900 mM glycine. Store at RT. Dilute 200 mL with 800 mL distilled water for use.
2. A stack of membranes (five or ten sheets) (P-Film, 20/20 GeneSystems, Rockville, MD, USA) (*see* **Note 3**).
3. Spacer membrane (GE polycarbonate PVPF membrane, GE Osmonics Labstore, Minnetonka, MN, USA) (*see* **Note 4**).
4. Nitrocellulose membrane.
5. Absorbent pads (Blot Absorbent Filter paper).
6. Slide glass (Opticlear® Microscope slide).
7. Kapak SealPAK pouches.
8. Impulse sealer.
9. Heat block (Dry bath).

2.3 Immunoblotting

1. Biotinylation solution (1 μg/mL, EZ-link Sulfo-NHS-Biotin, Pierce, Rockford, IL, USA): Prepare freshly solution before use in PBS.
2. Blot FastStain kit (Chemicon International, Temecula, CA, USA).
3. Tris-buffered saline (TBS, 10×): 1.5 M NaCl, 0.5 M Tris–HCl, pH 8. Store at RT.

4. Tris-buffered saline with Tween-20 (TBST): TBS plus 0.05 % (w/v) Tween-20. Store at RT.
5. Primary and secondary antibody dilution buffer: TBST supplemented with 0.5 % (w/v) fraction V bovine serum albumin (BSA).
6. Primary antibodies (*see* **Note 5**).
7. Streptavidin linked Cy5.
8. FITC conjugated anti-rabbit IgG or anti-mouse IgG.

2.4 Image Scanning and Quantitation

1. GE Typhoon 9410 Imager (GE Healthcare Life Sciences, Pittsburgh, PA, USA).
2. ImageQuant 5.2 software (GE Healthcare Life Sciences, Pittsburgh, PA, USA).

3 Methods

This method begins with routinely processed formalin-fixed, paraffin-embedded tissue section on a regular glass slide and utilizing routine laboratory procedures for microtomy. Initial retrieval of protein from the FFPE tissue section involves enzymatic digestion using trypsin and proteinase K. Subsequently, the tissue section is placed in an assembled heat-facilitated capillary transfer unit (Fig. 1). The unit is sealed in a pouch and applied multiple-serial heating system so that protein is transferred from the slide to a stack of membrane. Typically, we can obtain five replicate membranes from a 5 μM-thick formalin-fixed, paraffin-embedded tissue section. After transfer, the membranes stack is disassembled and each membrane can be probed using conventional Western blotting or immunoblotting method. This method allows quantification of the specific target signal based on total protein normalization, which is expressed as an intensity ratio between the protein of interest and total protein. In addition, post-transfer tissue on the original slide can be stained with routine Hematoxylin and Eosin (H and E) for direct correlation of histopathology with the immunoblot results. The residual tissue on the section after transfer should be limited, and may be difficult to interpret. If complex diagnostic features are to be examined, an adjacent H and E section should be utilized. Spatial resolution is determined by the scanning methodology, as well as transfer conditions and abundance of the protein. "Acinar" resolution on the order of 100–200 μm is obtainable with optimized conditions.

3.1 Deparaffinization and Enzyme Treatment

1. Deparaffinize the formalin-fixed, paraffin-embedded tissue section in xylene (3 × 5 min) in glass or plastic coplin jars. Transfer to 100 % EtOH (2 × 5 min), 95 % EtOH (2 × 5 min), 70 % EtOH (1 × 5 min), and then to PBS (2 × 5 min) (*see* **Note 6**).

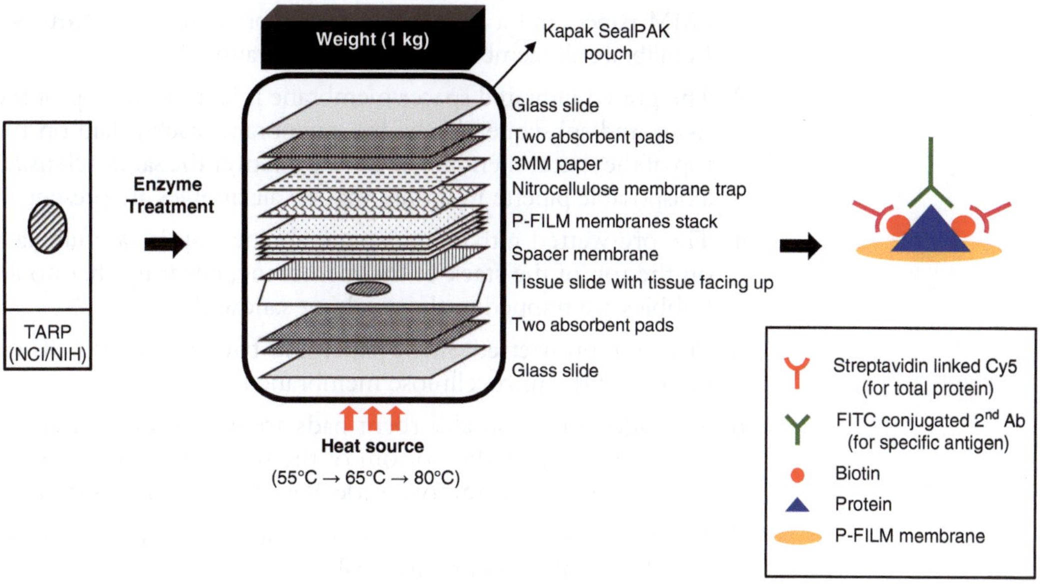

Fig. 1 A schematic diagram of the multiplex tissue immunoblotting for tissue protein transfer. The tissue section is treated with enzymes and then assembled for heat-facilitated capillary transfer. Assemble the transfer unit carefully and then seal the assembled transfer unit using a Kapak SealPAK pouch. After the sealing, the transfer unit is moved to the heat block and then applied to a multiple-serial heating system (1 h for 55 °C, 0.5 h for 65 °C, and 2 h for 80 °C). The protein leaves the tissue slide and is deposited onto the membrane stack. A weight applied to the top of the transfer assembly unit helps to ensure a tight connection between the layers of material used in the transfer system. Total protein is detected by biotinylation and Cy5 fluorescence (red emission). Individual target proteins are detected with specific antibodies and FITC fluorescence (green emission)

2. Equilibrate the deparaffinized tissue slide for 5 min with 50 mM ammonium bicarbonate buffer (pH 8.2).
3. Prepare the enzyme cocktail solution by mixing 100 μL of trypsin stock solution, 50 μL of Proteinase-K stock solution, and 20 mL of 50 mM ammonium bicarbonate buffer (pH 8.2).
4. Place the enzyme cocktail solution in an incubation chamber. After the equilibration is completed, incubate the slide for 30 min at 37 °C with enzyme cocktail solution (*see* **Note 7**).
5. Pour off the enzyme cocktail solution and rinse the surface of the tissue twice with PBS.
6. Place the slide on a flat surface and then apply immediately 2 mL of proBuffer to the slide for 15 min at RT.
7. Briefly wash the slide with transfer buffer. The slide is ready for transfer.

3.2 Transferring from the Tissue Slide to Membrane

1. After deparaffinization and enzyme treatment are completed, equilibrate the tissue slide in 1 mL of transfer buffer for 15 min.
2. During the equilibration of the tissue slide, prepare a spacer membrane, a stack of membranes (*see* **Note 8**), four absorbent pads,

3MM paper, and a nitrocellulose membrane trap (*see* **Note 9**). Equilibrate all membranes for 5 min in transfer buffer.

3. The pre-equilibrated spacer membrane is laid on the top of the tissue and subsequently the five-membrane stack is laid on the top of the spacer membrane. Gently roll over the sandwich using a disposable pipette to ensure that no air bubbles are present.
4. The pre-wetted nitrocellulose membrane trap is carefully laid on the top of the stack of the membrane, ensuring that no air bubbles are trapped in the resulting sandwich.
5. Place the pre-wetted 3MM paper and two absorbent pads on the top of the nitrocellulose membrane trap.
6. Two additional two absorbent pads are wetted in the transfer buffer and laid on the bottom of the tissue slide. Support the transfer assembly using two slide glasses as shown in Fig. 1.
7. Place the transfer assembly unit in a Kapak SealPAK pouch and then heat seal by an impulse sealer.
8. After the assembly of the transfer unit is completed, the transfer unit is transferred under serial conditions for 1 h at 55 °C, for 0.5 h at 65 °C, and for 2 h at 80 °C using heat block (*see* **Note 10**).

3.3 Immunoblotting

1. Once the transfer is completed, the SealPAK pouch is opened and the membrane stack carefully disassembled, with the top absorbent pads and 3MM paper removed.
2. Remove excess transfer buffer by washing the membranes (3 × 5 min) in PBS.
3. Stain the nitrocellulose and the spacer membranes using the Blot FastStain kit (*see* **Note 11**).
   (a) Prepare 1:7-fold diluted working solutions of reagent A and B with distilled water.
   (b) Incubate the nitrocellulose and spacer membranes in 7 mL of reagent A for 10 min, followed by incubation of the both membranes in reagent B for 10 min or until spot visualized.
   (c) Move the staining container to 4 °C and then let stand for 10–30 min.
4. The five-membrane stack is then incubated in 20 mL biotinylation solution for 10 min at RT on a rocking platform (*see* **Note 12**).
5. The biotinylation solution is discarded and then the membrane washed three times for 5 min each with TBST.
6. Incubate each membrane by adding 0.5 mL of an appropriately diluted primary antibody in the antibody dilution buffer for overnight (16–18 h) at 4 °C on a Kapak SealPAK pouch. Rocking platform is recommended (*see* **Note 13**).

7. The primary antibody is then removed and the membrane washed three times for 5 min each with TBST.
8. Prepare the mixture of secondary antibodies by mixing 5 μL of streptavidin linked Cy5 and 5 μL of FITC conjugated anti-rabbit IgG or anti-mouse IgG, and 5 mL of the antibody dilution buffer.
9. Add the mixture of secondary antibodies to the membrane and then incubate for 30 min at RT on a rocking platform. The membrane should be protected form light until scan is acquired.
10. The secondary antibodies are discarded and the membrane washed five times for 10 min each with PBST.

3.4 Image Scanning and Quantitation

1. Dry the membranes after the final wash between two sheets of 3MM paper.
2. Scan each membrane at appropriate wavelength with GE Typhoon 9410 Imager. Examples of the signals for total protein and pan-cytokeratin are shown in Fig. 2a (*see* **Note 14**).
3. Analyze the scanned images using ImageQuant 5.2 software. Example of quantitation of scanned image is shown in Fig. 2b.

4 Notes

1. We have found that Dako Proteinase-K is best for this method. Substitution with other Proteinase-K can diminish reproducibility of results.
2. This protocol can be adapted for ethanol-fixed, paraffin-embedded tissue sections. In that case, the enzyme solution should be changed to 0.001 % trypsin only for 15 min at 37 °C. Overall the final condition of enzyme digestion should be optimized depending on tissue type with a minor change of enzyme concentration and time.
3. When handling membranes, always wear gloves to prevent contamination. The P-Film membrane is very thin and flexible and requires care in handling to avoid bubbles and folds.
4. The spacer is an uncoated polycarbonate PVPF membrane (pore size: 0.4 μm) and is used for filtrations of inappropriately digested proteins and debris during heat-facilitated capillary transfer procedure.
5. The primary antibodies which are suggested for western blotting as well as immunohistochemistry are compatible in this platform. We recommend 1:200-fold starting antibody dilution in this protocol. The cytoplasmic markers are in general better targets than nuclear and membrane-bound molecules;

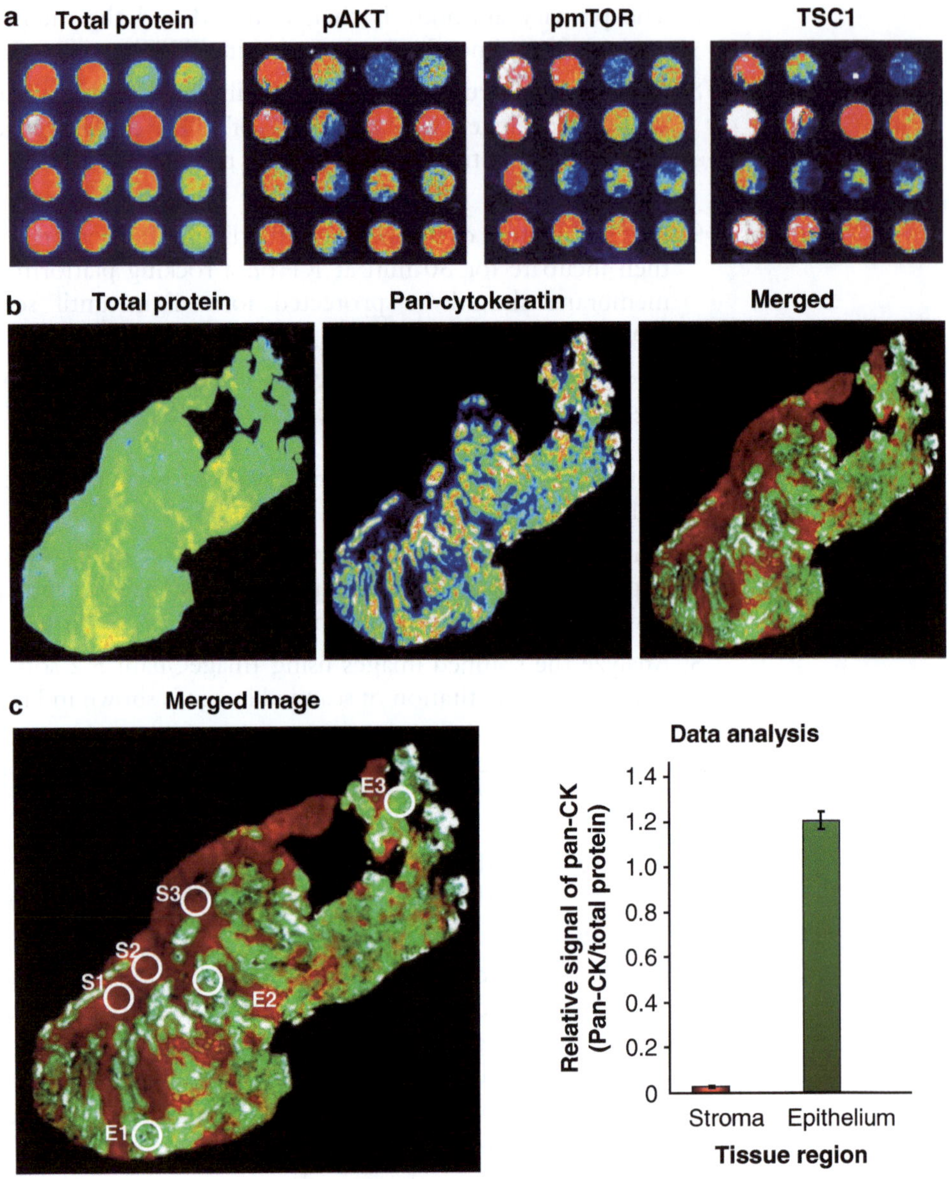

Fig. 2 Representative images and quantitation of whole tissue section by multiplex tissue immunoblotting. (**a**) Prior to immunoblotting, lung cancer TMA tissue section (5 μm thickness) was transferred to the five-membrane stack by heat-facilitated capillary transfer system. The transferred proteins are detected on the membranes by conventional immunoblotting methodologies. (**b**) A formalin-fixed and paraffin-embedded tissue section of a gastrointestinal tumor (5 μm thickness) was transferred to the five-membrane stack by heat-facilitated capillary transfer system. The third membrane was incubated in biotinylation solution and then was reacted with anti-pan-cytokeratin antibodies (1:200-fold diluted). After primary antibody incubation is completed, total protein and specific target signals were detected using streptavidin linked Cy5 and FITC conjugated anti-mouse IgG (1:1,000-fold diluted). Membranes were imaged with a microarray scanner. Fluorescent scans are represented in pseudo-color, where signal intensity is *white-red-yellow-green-blue-black* from maximum to minimum signal. (**c**) We selected three different representative stroma and epithelium tissue regions based on the H and E slide. We subsequently quantified those areas using ImageQuant 5.2 software. After normalization with total protein level, relative expressional signals were represented as a ratio. The bar graph shows the average ± SD of three circle areas (*S* stroma, *E* epithelium)

however, we have detected proteins in all these locations. For membrane-bound targets, thicker tissue sections (10 μm) may produce better results.

6. We found aqueous based *dewaxers* such as AutoDewaxer (Openbiosystems, Huntsville, AL, USA) could be used as a deparaffinizing agent in place of xylene at high temperature with equal results and greater safety (*see* ref. 16). The temperature of xylene should be controlled under 65 °C. Inappropriate deparaffinization can result in poor protein transfer to the membrane.
7. The dynamic range of the enzyme condition is relatively narrow. For this reason we recommend avoiding repeated freezing and thawing of the enzyme stock solution. The stock solution can be stored up to 6 months in the freezer (−20 °C).
8. There are two different features (glossy vs. non-glossy sides) on the membrane. Before pre-soaking the membrane, we marked membrane and case numbers in the margin of non-glossy side for further information such as number of membrane and case using a regular ballpoint pen. The glossy side of the membrane stack should face the tissue on the slide (Fig. 1).
9. This protocol is used for a regular tissue slide, and can be adapted for irregular slide size with appropriate membrane size. Cut all membrane and pads with size of approximately 2.2×4.5 cm. The cover of cover slip box can be used a container for incubation chamber, to prevent excess buffer use.
10. The use of a multi-serial temperature condition produce an even distribution of proteins across membranes compared to a single temperature, which resulted in uneven transfer or bubble spots. This procedure generated a linear decrease in protein concentration through the membrane stack, with a great correlation coefficient ($R2 = 0.985$) (*see* ref. 11).
11. This step is just a confirmation stage for protein transferred to the membrane stack. The staining kit is very sensitive but is a transient staining. If you want to keep original image you should scan the semi-wet membrane between two transparent films and e save the image.
12. Do *not* use a plastic petri dish coated for cell culture. A 20 mL of biotinylation solution can be covered up to 20 membranes (2.2×4.5 cm).
13. It is not necessary to block the membranes. The carrier protein (BSA) in the antibody dilution is sufficient to inhibit nonspecific binding. Do *not* use dry milk as a carrier protein.
14. Excitation at 633 nm induces the Cy5 fluorescence (red emission) for total protein, while excitation at 488 nm induces FITC fluorescence (blue emission). This fluorescence labeling system can be adapted for user purpose.

Acknowledgment

The authors wish to thank Michael R. Emmert-Buck for his advice and insights into applications for this platform.

References

1. Chuaqui RF, Bonner RF, Best CJ, Gillespie JW, Flaig MJ, Hewitt SM et al (2002) Post-analysis follow-up and validation of microarray experiments. Nat Genet 32:509–514
2. Hewitt SM, Dear J, Star RA (2004) Discovery of protein biomarkers for renal diseases. J Am Soc Nephrol 15:1677–1689
3. Simone NL, Paweletz CP, Charboneau L, Petricoin EF, Liotta LA (2000) Laser capture microdissection: beyond functional genomics to proteomics. Mol Diagn 5:301–307
4. Ornstein DK, Englert C, Gillespie JW, Paweletz CP, Linehan WM, Emmert-Buck MR et al (2000) Characterization of intracellular prostate-specific antigen from laser capture microdissected benign and malignant prostatic epithelium. Clin Cancer Res 6:353–356
5. Craven RA, Totty N, Harnden P, Selby PJ, Banks RE (2002) Laser capture microdissection and two-dimensional polyacrylamide gel electrophoresis: evaluation of tissue preparation and sample limitations. Am J Pathol 160: 815–822
6. Camp RL, Chung GG, Rimm DL (2002) Automated subcellular localization and quantification of protein expression in tissue microarrays. Nat Med 11:1323–1327
7. Liotta LA, Petricoin E (2000) Molecular profiling of human cancer. Nat Rev Genet 1: 48–56
8. Miyaji T, Hewitt SM, Liotta LA, Star RA (2002) Frozen protein arrays: a new method for arraying and detecting recombinant native tissue proteins. Proteomics 2:1489–1493
9. Lopez MF (2000) Better approaches to finding the needle in a haystack: optimizing proteome analysis through automation. Electrophoresis 21:1082–1093
10. Molloy MP, Phadke ND, Maddock JR, Andrews PC (2001) Two-dimensional electrophoresis and peptide mass fingerprinting of bacterial outer membrane proteins. Electrophoresis 22:1686–1696
11. Chung JY, Braunschweig T, Baibakov G, Galperin M, Ramesh A, Skacel M et al (2006) Transfer and multiplex immunoblotting of a paraffin embedded tissue. Proteomics 6:767–774
12. Chung JY, Braunschweig T, Hu N, Roth M, Traicoff JL, Wang QH et al (2006) Profiling of biomarkers in the normal to tumor transition zone of esophageal squamous cell carcinoma by multiplex tissue immunoblotting. Cancer Epidemiol Biomarkers Prev 15:1403–1408
13. Chung JY, Braunschweig T, Tuttle K, Hewitt SM (2007) Tissue microarrays as a platform for proteomic investigation. J Mol Histol 38: 123–128
14. Traicoff JL, Chung JY, Braunschweig T, Mazo I, Shu Y, Ramesh A et al (2007) Expression of EIF3-p48/INT6, TID1 and Patched in cancer, a profiling of multiple tumor types and correlation of expression. J Biomed Sci 14:395–405
15. Chung JY, Hong SM, Choi BY, Cho HJ, Yu ES, Hewitt SM (2009) The expression of phospho-AKT, phospho-mTOR, and PTEN in extrahepatic cholangiocarcinoma. Clin Cancer Res 15:660–667
16. Chung JY, Braunschweig T, Hewitt SM (2006) Optimization of recovery of RNA from formalin-fixed, paraffin-embedded tissue. Diagn Mol Pathol 15:229–236

Chapter 22

Post-Staining Electroblotting for Efficient and Reliable Peptide Blotting

Der-Yen Lee and Geen-Dong Chang

Abstract

Post-staining electroblotting has been previously described to transfer Coomassie blue-stained proteins from polyacrylamide gel onto polyvinylidene difluoride (PVDF) membranes. Actually, stained peptides can also be efficiently and reliably transferred. Because of selective staining procedures for peptides and increased retention of stained peptides on the membrane, even peptides with molecular masses less than 2 kDa such as bacitracin and granuliberin R are transferred with satisfactory results. For comparison, post-staining electroblotting is about 16-fold more sensitive than the conventional electroblotting for visualization of insulin on the membrane. Therefore, the peptide blots become practicable and more accessible to further applications, e.g., blot overlay detection or immunoblotting analysis. In addition, the efficiency of peptide transfer is favorable for N-terminal sequence analysis. With this method, peptide blotting can be normalized for further analysis such as blot overlay assay, immunoblotting, and N-terminal sequencing for identification of peptide in crude or partially purified samples.

Key words Blotting, Peptides, Electroblotting, Coomassie brilliant blue

1 Introduction

Electrophoresis has been improved to resolve peptides by various modifications. For achieving optimal resolution, efforts have been made, such as inclusion of high concentrations of urea in the resolving gel buffer [1–5], using high concentrations of Tris-buffer in the resolving gel [6], and running against high concentrations of polyacrylamide [2, 4, 6, 7]. Most critical improvement is the development of a Tris-tricine buffer system which has provided the superior resolution of peptides without the use of urea or high concentrations of polyacrylamide in gels [8].

After electrophoresis, if peptides require further assay on a solid matrix, e.g., PVDF membrane, the efficiency of transferring peptides becomes critical. Electrophoretic blotting (electroblotting), either by wet transfer or by semidry transfer, is most widely used in immunoblotting analysis. Electroblotting is rapid and

Biji T. Kurien and R. Hal Scofield (eds.), *Western Blotting: Methods and Protocols*, Methods in Molecular Biology, vol. 1312, DOI 10.1007/978-1-4939-2694-7_22, © Springer Science+Business Media New York 2015

complete for protein elution from polyacrylamide gels. However, the retention of proteins on the membrane during transfer determines the transfer efficiency. Transferring time and membrane pore sizes can affect the retention of proteins on the membranes, especially for low-molecular-weight proteins. Therefore, peptide transfer requires controlling these factors to prevent the loss of peptides from the blotted membrane. However, flow-through of proteins occurs in transfer even using 0.2-μm (Bio-Rad Transblot) in an appropriate transferring time [9].

To improve the peptide transfer, we introduce an alternative blotting method, appropriately named post-staining electroblotting. In brief, stained proteins and peptides can be transferred onto membranes after routine fixation, dye staining and destaining. The method has been used in protein immunoblotting [10–16], for gel-drying, convenient localization of immuno-reactive protein in relation to other major proteins on the membrane and for providing protein staining and immunoblot from a single gel. Coomassie blue-stained peptides in post-staining electroblotting are highly adsorptive to PVDF membrane (Fig. 1a).This novel method provides blots with about 16-fold higher sensitivity than the conventional electroblotting in detecting insulin (Fig. 1b). Moreover, three compatible staining protocols described in this chapter are

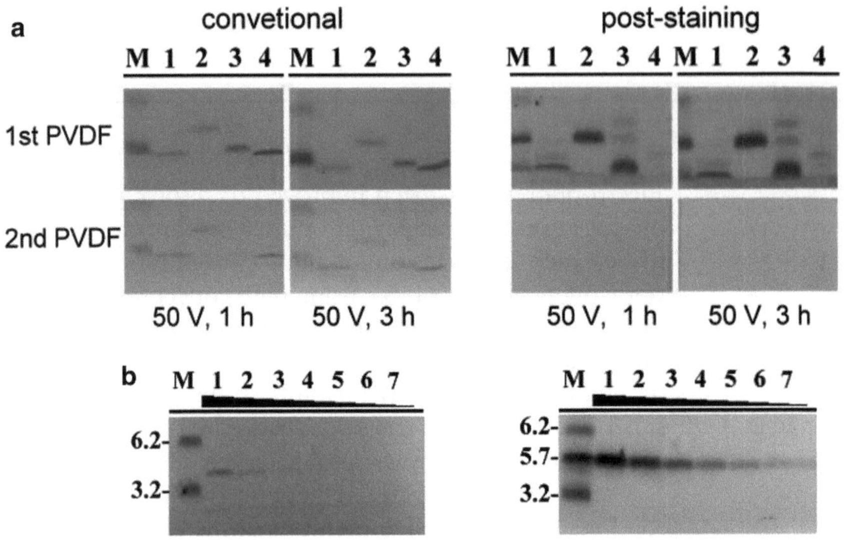

Fig. 1 Retention of peptides on PVDF membrane by conventional and post-staining electroblotting. (**a**) Peptides, 4 μg of bacitracin (*lane 1*), 2 μg of insulin (*lane 2*), 2 μg of reduced insulin (*lane 3*), and 2 μg of granuliberin (*lane 4*) were subjected to Tricine SDS-PAGE. Two sheets of PVDF membrane were set to capture peptide and each membrane was stained with Amido black except the first PVDF of post-staining electroblotting. (**b**) Twofold serial diluents of 2 μg of insulin was resolved by Tricine SDS-PAGE, and duplicates of the gel were applied to conventional (*left*) and post-staining electroblotting (*right*), respectively

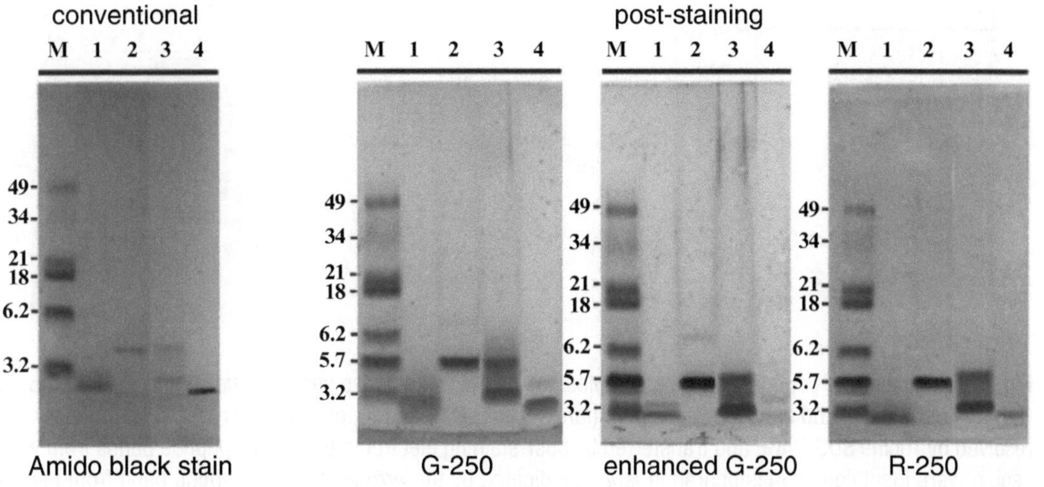

Fig. 2 Options of peptide staining for post-staining electroblotting. Peptide samples as Fig. 1a were resolved by Tricine SDS-PAGE and transferred at 50 V for 1 h by conventional or post-staining electroblotting. For post-staining electroblotting, other similar gels were stained with GelCode G-250, enhanced G-250, or Coomassie blue R-250, respectively and then transferred by electroblotting onto the PVDF membrane at 50 V for 1 h. *Lane 1*: 4 μg of bacitracin, *lane 2*: 2 μg of insulin, *lane 3*: 2 μg of reduced insulin, and *lane 4*: 2 μg of granuliberin

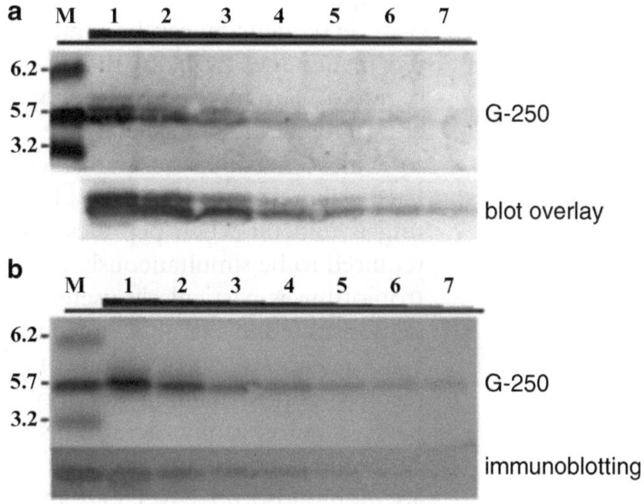

Fig. 3 Blot overlay assay and immunoblotting with post-staining electroblotting. Twofold serial diluents of 2 μg of (**a**) biotinylated insulin or (**b**) insulin were applied to Tricine SDS-PAGE and transferred at 50 V for 1 h by post-staining electroblotting. (**a**) Biotinylated insulin was probed with peroxidase-conjugated streptavidin, and (**b**) insulin was probed with polyclonal anti-insulin antibodies against bovine insulin

optional for targets of interest (Fig. 2). Except for providing the staining image on the membrane, the blotted peptides are detectable by affinity probing, such as streptavidin overlay blotting or immunoblotting (Fig. 3).

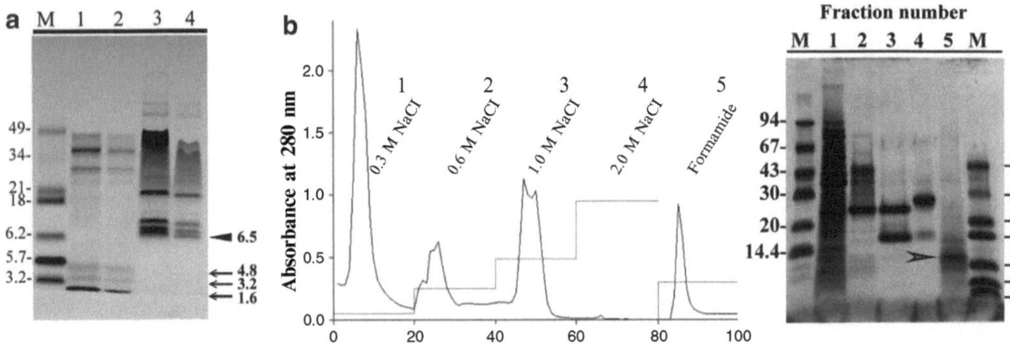

Fig. 4 Identification of peptides in crude samples. (**a**) Yellow jacket wasp venom (*lanes 1* and *2*: 10 and 5 μg) (*Vespula germanica*) and banded krait venom (*lanes 3* and *4*: 10 and 5 μg) (*Bungarus fasciatus*) proteins were resolved by Tricine SDS-PAGE and transferred by post-staining electroblotting. Three peptide bands from wasp venom were identified as mastoparan in *lane 1*, indicated by the *arrows*, and one peptide band from banded krait venom was identified as chymotrypsin inhibitor in *lane 3*, indicated by the *arrowhead*. (**b**) Carp kidney extract was applied to the heparin affinity column. Elution fraction 1–4: 0.3, 0.6, 1.0 and 2.0 M NaCl in TE buffer, and fraction 5: 1 M NaCl in 50 % formamide. Fractions 1–5 were applied to SDS-PAGE in the order (*lanes 1–5*). After post-staining electroblotting, the Coomassie blue G-250 stained peptide band was identified as granulin-3 in *lane 5*, indicated by the *arrowhead*

Applications such as N-terminal sequencing and internal peptide sequencing can be achieved for further analysis from blotted membranes [17]. In these cases, higher transfer efficiency is required. The increase in binding efficiency on the membrane is critical for peptide transfer. These peptides require special attention to achieve maximal transfer efficiency, however very few articles have elaborated on this issue [18]. Besides, optimization of blotting is difficult when peptides and proteins of extreme sizes are required to be simultaneously transferred [19]. Post-staining electroblotting is particularly useful for handling this requirement as demonstrated in Fig. 4. Crude yellow jacket venom, banded krait venom (Fig. 4a), and fractions of carp kidney extract (Fig. 4b) are resolved as post-staining electroblotted bands and each individual band is successfully determined as the indicated identity by N-terminal sequencing. Because Coomassie blue-stained peptides are highly adsorptive to PVDF membranes, post-staining electroblotting for peptides is preferable in the applications of N-terminal sequencing, immunoblotting, or blot overlay assay.

2 Materials

2.1 Tricine SDS-PAGE

1. Hoefer SE-260 Mighty small II gel system.
2. Acrylamide/bis-acrylamide solution (50 % T; 3 % C) contains 48.5 g of acrylamide and 1.5 g of bis-acrylamide in 100 mL of aqueous solution.

3. Gel buffer: 1.5 M Tris–HCl, pH 8.45.
4. Stacking gel buffer: 0.1 M Tris–HCl, pH 7.8 containing 0.4 % SDS.
5. 10 % ammonium persulfate solution: Dissolve 0.1 g of ammonium persulfate into 1 mL of deionized water (*see* **Note 1**).
6. TEMED.
7. Cathode buffer: 0.1 M Tris base with 0.1 % SDS adjusted with tricine to pH 8.25.
8. Anode buffer: 0.2 M Tris–HCl, pH 8.9.
9. 2× SDS sample buffer: 0.1 M Tris–HCl, pH 7.8 containing 8 % SDS and 24 % glycerol and 100 ppm Coomassie brilliant blue G-250.
10. 1 mg/mL insulin (Calbiochem), insulin with 2 % β-mercaptoethanol, bacitracin, granuliberin R, wasp venom from *Dolichovespula arenaria* (Yellow Hornet), or snake venom from *Bungarus fasciatus* (Banded Krait) (all from Sigma-Aldrich) in 1× SDS sample buffer.

2.2 Conventional Electroblotting

1. Hoefer TE22 transfer tank.
2. The transfer buffer: 50 mM Tris base, 40 mM glycine, 0.04 % SDS in 10 % methanol.
3. Chromatography paper (3MM Chr).
4. PVDF membrane: Immobilon P or Immobilon P^{SQ}.
5. Amido black staining solution: 0.2 % (w/v) amido black in 7 % acetic acid.

2.3 Gel Staining

2.3.1 Coomassie Brilliant Blue R-250 Staining

1. Gel fixer: 1.3 % H_3PO_4 (w/v) within 20 % methanol.
2. Coomassie brilliant blue R-250 staining solution: 0.1 % Coomassie brilliant blue R-250 (w/v) in 40 % methanol and 7 % acetic acid. 0.1 % Triton X-100 and 7 % acetic acid.

2.3.2 Coomassie Brilliant blue G-250 Staining

1. Gel fixer: 1.3 % H_3PO_4 (w/v) in 20 % methanol.
2. Coomassie brilliant blue G-250 staining solution (GelCode blue).

2.4 Enhanced Coomassie Brilliant Blue G-250 Staining [20]

1. Gel fixer: 1.3 % H_3PO_4 (w/v) in 20 % methanol.
2. Enhanced Coomassie brilliant blue G-250 solution: 0.1 % Coomassie brilliant blue G-250 (w/v) in 2 % w/v H_3PO_4, 10 % ammonium sulfate and 20 % methanol solution.
3. 0.1 M Tris–H_3PO_4, pH 6.5.
4. 25 % methanol.

2.5 Post-staining Electroblotting

1. Hoefer TE22 transfer tank.
2. The transfer buffer: 50 mM Tris base, 40 mM glycine, 0.04 % SDS in 10 % methanol.

3. Chromatography paper (3MM Chr).
4. PVDF membrane: Immobilon P or Immobilon P^{SQ}.

2.6 Streptavidin-Peroxidase Blot Overlay

1. NHS-LC-biotin.
2. *N*,*N*-dimethylformamide.
3. Insulin.
4. Prepare 50 mM sodium bicarbonate.
5. Prepare 1 M ethanolamine.
6. Methanol.
7. Phosphate buffered saline (PBS).
8. PBST: 0.1 % Tween-20 in PBS.
9. Blocking solution: 3 % skim milk in PBST.
10. Horseradish peroxidase-conjugated streptavidin solution: 0.1 mg/10 mL horseradish peroxidase-conjugated streptavidin in PBST containing 3 mg/mL of BSA.
11. $NiCl_2$-DAB solution: Prepare 0.5 mg of 3′,3′-diaminobenzidine (DAB) and 0.5 mL of 1 % $NiCl_2$ in 10 mL of PBS prior to use.
12. Hydrogen peroxide (35 %).

2.7 Immunoblotting of Insulin

1. Phosphate buffered saline (PBS).
2. PBST: 0.1 % Tween-20 in PBS.
3. Blocking solution: 3 % skim milk in PBST.
4. Primary antibody solution: Anti-insulin serum is raised in guinea pigs in our laboratory. They are used at a titer of 1:2,000 in PBST containing 0.3 % BSA.
5. Secondary antibody solution: 0.2 μg/mL horseradish peroxidase (HRP)-conjugated anti-guinea pig IgG (Jackson ImmunoResearch Inc.) in 0.3 % BSA in PBST.
6. $NiCl_2$-DAB solution: Prepare 0.5 mg of 3′,3′-diaminobenzidine (DAB) and 0.5 mL of 1 % $NiCl_2$ in 10 mL of PBS prior to use.
7. Hydrogen peroxide (35 %).

2.8 Fractionation of Carp Kidney Extract

1. Carp kidneys.
2. TE buffer: 20 mM Tris–HCl containing 5 mM EDTA, pH 8.0.
3. TE-2 M NaCl buffer: 20 mM Tris–HCl containing 5 mM EDTA and 2 M NaCl, pH 8.0.
4. FPLC system (Amersham Biosciences Ltd.).
5. Heparin-HyperD ceramics (Biosepra).
6. Prepare 1.0 M NaCl in 50 % formamide.
7. GelCode blue.

3 Methods

3.1 Tricine SDS-PAGE

1. Modified Tricine SDS-PAGE is used to resolve the peptides and electrophoresis is performed with Hoefer SE-260 Mighty small II gel system.
2. Prepare 10 % separation gel by mixing 6 mL of acrylamide/bis-acrylamide solution, 10 mL of gel buffer, 10 mL of deionized water, 4 g glycerol, 0.1 mL of 10 % ammonium persulfate solution, and 10 μL of TEMED. Cast gel within 10.5 cm × 10 cm × 0.75 mm gel cassette, leave space for a stacking gel, and overlay with deionized water.
3. Prepare stacking gel by mixing 1 mL of acrylamide/bis solution, 3 mL of stacking gel buffer, 8 mL of deionized water, 0.1 mL of 10 % ammonium persulfate solution, and 10 μL of TEMED. Then insert a 10-well comb into the stacking gel solution immediately.
4. Apply samples within 15 μL to each well and complete the electrophoresis at 150 V for 60 min with water cooling system.

3.2 Conventional Electroblotting

3.2.1 Electrophoretic Transfer

1. After finishing the electrophoresis, remove one glass plate and leave the gel on the other plate.
2. Cover the gel with a size-matched chromatography paper and let the gel attached to the chromatography paper.
3. Cover the other side of the gel with a moisturized PVDF membrane, soak the stack into deionized water perfectly attached to each other and then cast the assembly in a cassette.
4. Soak the cassette in transfer buffer in the tank and tap the cassette to get rid of trapped bubbles (*see* **Note 2**).
5. Regular transferring is achieved in 3 h at 50 V.
6. Rinse the blotted membrane with deionized water five times (*see* **Note 3**).
7. Drain the residual water and immerse the membrane into amido black solution.
8. Drain the amido black solution after constant agitation for 1 min and then reveal the stained bands by destaining the background with agitation in deionized water three times, each for 2 min.

3.3 Gel Staining

3.3.1 Coomassie Brilliant Blue R-250

1. Soak the polyacrylamide gel in gel fixer for 30 min to fix the peptides and wash with deionized water three times, each for 5 min.
2. Stain the gel in Coomassie brilliant blue R-250 staining solution with constant agitation for 10 min.
3. Destain the gel in 0.1 % Triton X-100 and 7 % acetic acid for 30 min and then in deionized water three times, each for 10 min.

3.3.2 Coomassie Blue G-250 Staining

1. Soak the polyacrylamide gel in gel fixer for 30 min to fix the peptides and wash with deionized water three times, each for 5 min.
2. Stain gel in Coomassie brilliant blue G-250 staining solution (GelCode blue) with constant agitation for 60 min.
3. Destain the gel in deionized water three times, each for 20 min.

3.3.3 Enhanced Coomassie Blue G-250 Staining [20]

1. Soak the polyacrylamide gel in gel fixer for 30 min to fix the peptides and wash with deionized water three times, each for 5 min.
2. Stained the gel overnight in enhanced Coomassie brilliant blue G-250 staining solution with constant agitation.
3. Wash the gel in 0.1 M Tris–H_3PO_4, pH 6.5 for 1 min and then rinse the gel in 25 % methanol three times, each for 1 min.
4. Wash the gel with deionized water three times, each for 5 min and then destain the gel with deionized water to appropriate background.

3.4 Post-staining Electroblotting

1. Remove background of the stained gels as much as possible before electrophoretic transfer by following the destaining protocol (*see* **Note 4**).
2. Cover the gel with moisturized PVDF membrane.
3. Cover the other side of the gel with another moisturized PVDF membrane and a wet chromatography paper on top and then cast the assembly in a cassette (*see* **Note 5**).
4. Soak the cassette in transfer buffer in the tank and tap the cassette to get rid of trapped bubbles (*see* **Note 2**).
5. Regular transferring is achieved in 3 h at 50 V.
6. Rinse the blotted membrane with deionized water five times (*see* **Note 3**).

3.5 Streptavidin-Peroxidase Blot Overlay

3.5.1 Preparation of Biotin Labeled Insulin

1. Dissolve 0.1 mg of NHS-LC-biotin in 25 μL of dimethylformamide.
2. Prepare 2.0 mg of insulin in 0.25 mL of 50 mM $NaHCO_3$.
3. Mix NHS-LC-biotin reagent with the insulin solution prepared in **steps 1** and **2**.
4. After incubating 2 h on ice, add 0.1 mL of 1 M ethanolamine to the reactant and incubate for 30 min at 25 °C.
5. Add 125 μL of deionized water and 500 μL of 2× SDS sample buffer to the reactant as the 2 mg/mL biotinylated insulin sample.

3.5.2 Streptavidin-Peroxidase Blot Overlay

1. Wash the blotted membrane of biotinylated insulin with methanol until peptide band destained and then with deionized water twice, each for 5 min, if using post-staining electroblotting.

2. Agitate the membrane in blocking solution for 1 h and then rinse the membrane with deionized water five times.
3. Wash the membrane with PBST three times, each for 5 min, and probe with horseradish peroxidase-conjugated streptavidin solution (0.1 mg/10 mL) for 30 min.
4. After washing the membrane with PBST three times, each for 5 min, submerge and agitate the membrane within 10 mL of $NiCl_2$-DAB solution for 10 s, and add 10 μL of hydrogen peroxide to activate HRP activity [21].
5. Rock the container until the image becomes apparent, pour the waste into a reservoir containing bleach and rinse the membrane with tap water. Flush the membrane under tap water for few minutes and preserve the membrane after being air-dried.

3.6 Immunoblotting of Insulin

1. Blotted PVDF membranes can proceed directly for immunoblotting after washing with deionized water. For stored membrane, wet the membrane with 100 % methanol and then briefly rinse the membrane with deionized water, three times.
2. Agitate the blotted PVDF membrane in blocking solution for 1 h. When blocking is complete, rinse the membrane with PBS once.
3. Pour the primary antibody solution into the container and agitate the membrane for 1 h. Wash the membrane with PBST twice, each for 5 min, and PBS once for 5 min.
4. Pour the secondary antibody solution into the container and agitate the membrane for 1 h. Wash the membrane with PBST twice, each for 5 min, and PBS once for 5 min.
5. Submerge and agitate the membrane within 10 mL of $NiCl_2$-DAB solution for about 10 s, and add 10 μL of hydrogen peroxide to activate HRP activity [21]. Rock the container until the image becomes apparent, pour the waste into a reservoir containing bleach and rinse the membrane with tap water. Flush the membrane under tap water for few minutes and preserve the membrane after being air-dried.

3.7 Fractionation of Carp Kidney Extract

1. Homogenize carp kidneys (20 g) in 300 mL of TE buffer in a glass homogenizer.
2. Collect the supernatant by centrifugation at 20,000 × *g* for 30 min.
3. Apply the supernatant to a heparin-HyperD column equilibrated with TE buffer and elute stepwise by 0.3, 0.6, 1.0 and 2.0 M NaCl in TE buffer and by 1.0 M NaCl in 50 % formamide (*see* **Note 6**).

4. Collect all fractions and resolve by SDS-PAGE. The gel is stained with GelCode blue and subjected to post-staining electroblotting.
5. After rinsing the membrane with deionized water, stained band of interest is cut and subjected to Edman degradation reactions.

4 Notes

1. The ammonium persulfate solution would expire after 2 weeks.
2. Make sure that trapped bubbles within the gel stack or cassette are removed.
3. The blotted membranes rinsed with deionized water can be applied for further applications or air dried for storage till next usage.
4. Use each destaining condition to decrease the background of gel to obtain a clear image because excessive background would be enhanced after post-staining electroblotting.
5. Additional PVDF membrane on the cathode side of the gel prevents dye molecules from condensing on the blotted membrane to form patterns of artifacts. Actually, the assembly of additional PVDF membrane improves the fidelity of patterns from a gel to a membrane in any type of electroblotting.
6. The concentration of NaCl can be created by mixing TE and TE-2 M NaCl buffer with various ratios.

Acknowledgement

This work was supported by grants (to G.D.C.) from the National Science Council, Republic of China.

References

1. Swank RT, Munkres KD (1971) Molecular weight analysis of oligopeptides by electrophoresis in polyacrylamide gel with sodium dodecyl sulfate. Anal Biochem 39:462–477
2. Kyte J, Rodriguez H (1983) A discontinuous electrophoretic system for separating peptides on polyacrylamide gels. Anal Biochem 133: 515–522
3. Anderson BL, Berry RW, Telser A (1983) A sodium docecyl sulfate-polyacrylamide gel electrophoresis system that separate peptides and proteins in the molecular weight range of 2500 to 90,000. Anal Biochem 132:365–375
4. Hashimoto F, Horigome T, Kanbayashi M et al (1983) An improved method for separation of low-molecular-weight polypeptides by electrophoresis in sodium dodecyl sulfate polyacrylamide gel. Anal Biochem 129:192–199
5. Sheng HZ, Martenson RE, Carnegie PR et al (1988) Electroimmunoblotting of small peptides separated on urea-dodecyl sulphate (SUDS) gels. Application to myelin basic protein. J Immunol Meth 107:13–22
6. Fling SP, Gregerson DS (1986) Peptide and protein molecular weight determination by electrophoresis using a high-molarity tris buffer system without urea. Anal Biochem 155:83–88
7. Bothe D, Simonis M, von Döhren H (1985) A sodium dodecyl sulfate-gradient gel electrophoresis system that separates polypeptides in

the molecular weight range of 1500 to 100,000. Anal Biochem 151:49–54

8. Schägger H, von Jagow G (1987) Tricine-sodium dodecyl sulfate-polyacrylamide gel electrophoresis for the separation of proteins in the range from 1 to 100 kDa. Anal Biochem 166:368–379
9. Wang MS, Pang JS, Selsted ME (1997) Semidry electroblotting of peptides and proteins from acid-urea polyacrylamide gels. Anal Biochem 253:225–230
10. Phelps DS (1984) Electrophoretic transfer of proteins from fixed and stained gels. Anal Biochem 141:409–412
11. Pryor JL, Xu W, Hamilton DW (1992) Immunodetection after complete destaining of coomassie blue-stained proteins on immobilon-PVDF. Anal Biochem 202: 100–104
12. Thompson D, Larson G (1992) Western blots using stained protein gels. Biotechniques 12:656–658
13. Ranganathan V, De PK (1996) Western blot of proteins from Coomassie-stained polyacrylamide gels. Anal Biochem 234:102–104
14. Dionisi HM, Checa SK, Viale AM (1995) Protein immunoblotting of stained gels. Biotechniques 19:348–350
15. Wang D, Dzandu JK, Hussain M et al (1989) Western blots from sodium dodecyl sulfate-polyacrylamide gels stained with metal salts. Anal Biochem 180:311–313
16. Dibas AI, Yorio T (1996) A novel use for Coomassie Brilliant blue (R-250) in protein-drying procedure and assessing the electro-transferring efficiency. Biochem Biophys Res Comm 220:929–932
17. Cook RG (1994) Amino acid sequence analysis of blotted proteins. In: Dunbar BS (ed) Protein blotting: a practical approach. Oxford University, Oxford, pp 207–220
18. Kurien BT, Scofield RH (2003) Protein blotting: a review. J Immunol Meth 274:1–15
19. Bolt MW, Mahoney PA (1997) High-efficiency blotting of proteins of diverse sizes following sodium dodecyl sulfate-polyacrylamide gel electrophoresis. Anal Biochem 247:185–192
20. Neuhoff V, Arold N, Taube D et al (1988) Improved staining of proteins in polyacrylamide gels including isoelectric focusing gels with clear background at nanogram sensitivity using Coomassie Brilliant Blue G-250 and R-250. Electrophoresis 9:252–262
21. Harlow E, Lane D (1988) Antibodies. Cold Spring Harbor laboratory, New York, pp 471–510

Chapter 23

Multistrip Western Blotting: A Tool for Comparative Quantitative Analysis of Multiple Proteins

Edita Aksamitiene, Jan B. Hoek, and Anatoly Kiyatkin

Abstract

The qualitative and quantitative measurements of protein abundance and modification states are essential in understanding their functions in diverse cellular processes. Typical Western blotting, though sensitive, is prone to produce substantial errors and is not readily adapted to high-throughput technologies. Multistrip Western blotting is a modified immunoblotting procedure based on simultaneous electrophoretic transfer of proteins from multiple strips of polyacrylamide gels to a single membrane sheet. In comparison with the conventional technique, Multistrip Western blotting increases data output per single blotting cycle up to tenfold; allows concurrent measurement of up to nine different total and/or posttranslationally modified protein expression obtained from the same loading of the sample; and substantially improves the data accuracy by reducing immunoblotting-derived signal errors. This approach enables statistically reliable comparison of different or repeated sets of data and therefore is advantageous to apply in biomedical diagnostics, systems biology, and cell signaling research.

Key words Western blotting, Electrophoretic transfer, Gel cutting, Quantitative protein analysis, High-throughput, Blotting errors, Systems biology, Cell signaling

1 Introduction

Qualitative measurement of protein abundance is one of the common tasks in biomedical diagnostics in the search for therapeutic targets and biomarkers of various diseases and disorders [1–8]. The quantitative analysis of protein expression, posttranslational modification states (e.g., phosphorylation), recruitment to specific subcellular compartments, and interaction with other proteins is a paramount goal in systems biology, which explores, predicts, and explains how signaling networks govern cellular behavior by exploiting experimental data-driven mathematical models. To achieve this goal, the cellular response to external stimuli in vivo is often compared to the response obtained under one or more perturbations (e.g., pharmacological inhibitors, drugs, exposure to physiochemical stresses, or the down- or upregulation of

Biji T. Kurien and R. Hal Scofield (eds.), *Western Blotting: Methods and Protocols*, Methods in Molecular Biology, vol. 1312, DOI 10.1007/978-1-4939-2694-7_23, © Springer Science+Business Media New York 2015

protein expression). In addition, variations in the concentration of a ligand and/or the time course of stimulation provide deeper insight into the spatiotemporal functioning of a specific cell signaling pathway [9]. Obviously, these tasks require the generation of large amounts of high-quality data points.

Quantitative Western blotting used for the immunodetection and densitometric analysis of relative expression levels of electrophoretically resolved proteins is a sensitive and widely used technique, which, however, has several drawbacks [1, 10]. It is a low-throughput, time-consuming and expensive multistep procedure that often results in images that make comparison of paired experimental samples difficult. Each step of Western blotting (sampling, gel loading, electrophoretic separation and transfer of proteins, immunoblotting, and signal detection) is performed under slightly differing conditions in sequential blotting cycles. This eventually increases data variability, which makes it difficult to compare the signals obtained from different series of biological samples [11]. Therefore the improvements of typical Western blotting procedure are in high demand [12–18].

Multistrip Western blotting (MSWB) is a modified immunoblotting procedure based on simultaneous electrophoretic transfer of proteins from multiple strips of polyacrylamide gels to a single membrane sheet. The proposed modification has several advantages over a classical Western blotting procedure [19].

First, the transfer and the sequential procedures with the blot such as membrane washing, incubation with antibodies, and protein detection are performed under similar conditions. This significantly improves the data accuracy by reducing immunoblotting-derived signal errors.

Second, instead of detection of a single target protein *per* each blotting cycle, many additional proteins of interest that sufficiently differ in their molecular weight (MW) can be simultaneously obtained from a single gel and using same sample load. These proteins can be further visualized by immunoblotting (IB) or by dye staining [20–22]. The analytical power of Western blotting is increased, because a one-step analysis of numerous signaling proteins is more productive, saving time as well as costly materials.

Third, the MSWB approach eliminates the need to reuse single blots by stripping and reprobing, which is known to cause inconsistent and undetermined protein loss from a membrane and therefore remains a subject for improvement [23–25]. Blot reprobing is commonly used for the sequential detection of house-keeping proteins (e.g., actin, tubulin, GAPDH, COX-IV, PBGD, mATPsy6) that are believed to have a stable level of expression across all tissue/cell line samples under various experimental conditions. Therefore, despite the susceptibility to error, these are widely used to serve as internal reference controls for loading normalization, which, in turn, poses distinct challenges [26–35].

Fourth, in MSWB, only specific target protein bands transferred from a narrow area of the gel are synchronously detected on the membrane that is incubated in a single-primary antibody solution. This prevents the antibody cross-reactivity and nonspecific binding problems associated with multiplex detection of different-sized proteins following blot incubation in a mixture of two or more primary antibodies.

Fifth, the gel strip(s) containing serial dilutions of any recombinant protein could be included and used as a calibration curve to quantify an absolute amounts of protein of interest in a sample.

Finally, when the number of samples to be analyzed exceeds the number of wells in a gel, the concurrent quantitative protein analysis can be readily achieved by the MSWB technique, which increases the data output *per* single blotting cycle up to tenfold. As a consequence, a large number of data points can be measured, integrated, and compared on the same graph.

Herein we provide some practical examples of routine signal transduction biological experiments that streamline the strategic planning of how to use the MSWB approach to obtain, visualize, analyze, and represent data graphically in an easy-to-read format. Although the hands-on protocol presented here is developed and optimized for denaturing polyacrylamide gel electrophoresis (PAGE) run in MOPS-SDS buffer under reducing conditions using NuPAGE Novex 4–12 % gradient Bis-Tris mini-gels and XCell™ devices (Life Technologies, Carlsbad, CA), the MSWB is an adaptable technique, which can be used with other gel types and various protein transfer systems and is compatible with any protein electrophoresis methods using discontinuous denaturing buffer systems.

2 Materials

2.1 Preparation of Samples

1. Cell lysis buffer (*see* Table 1 and **Note 1**).
2. 100× protease inhibitor cocktail (*see* Table 2).
3. 100× phosphatase inhibitor cocktail (*see* Table 3).
4. 4× LDS sample buffer, pH 8.5 (*see* Table 4).
5. 10× sample reducing agent (*see* Table 5 and **Note 2**).
6. Dry heated bath.
7. Refrigerated high-speed centrifuge.
8. 1.5–2 mL Eppendorf tubes.
9. Cell scrapers.

2.2 Lithium-Dodecyl Sulfate-Polyacrylamide Gel Electrophoresis (LDS-PAGE)

1. Electrophoresis unit: XCell SureLock Mini-Cell units (Life Technologies) (*see* **Note 3**).
2. Gels: Precast 10- or 12-well NuPAGE Novex 4–12 % gradient Bis-Tris Mini-gels, 10×10 cm, 1.0 mm thickness (available from Life Technologies) or 4–12 % gradient ExpressPlus™

Table 1
Suggested composition of self-made 1× lysis buffer for total protein whole-cell lysis (WCL), plasma membrane permeabilization (MP) or protein immunoprecipitation (IP), final concentration (FC) of the compounds in 1× working solution, directions for buffer preparation and storage

	WCL		MP[a]		IP[a]	
Compound	Amount	FC	Amount	FC	Amount	FC
1 M HEPES[b], pH 7.4	2.5 mL	50 mM	2.5 mL	50 mM	1 mL	20 mM
1 M NaCl	7.5 mL	150 mM	7.5 mL	150 mM	7.5 mL	150 mM
50 mM EGTA	1 mL	1 mM	1 mL	1 mM	2.5 mL	2.5 mM
10 % Triton X-100	5 mL	1 %	–		5 mL	1 %
Glycerol[c]	5 mL	10 %	5 mL	10 %	2.5 mL	5 %
Digitonin	–		7.5 mg	150 μg/mL	–	
10 % SDC[d]	2.5 mL	0.5 %	–		–	
10 % SDS	0.5 mL	0.1 %	–		–	
dH_2O	Up to 50 mL		Up to 50 mL		Up to 50 mL	

Store at +4 °C. Prior to lysis, supplement the buffer with phosphatase and protease inhibitors. Keep on ice

[a]Details of cell fractionation into cytosolic and particulate fractions and immunoprecipitation are described elsewhere [36]

[b]HEPES interferes with the Lowry, but not the Bradford protein assay and is not suitable for redox studies, because it can form radicals. It may be substituted with Tris–HCl, pH 8.0 or MES, pH 6.8. Be aware that Tris possesses a potentially reactive amine, has high temperature sensitivity, and participates in some enzymatic reactions (e.g., alkaline phosphatase)

[c]Similar to glycerol, 250 mM sucrose also stabilizes integrity of lysosomal membranes, reduces protease release, helps prevent aggregation, and acts as protein cryoprotectant

[d]Sodium deoxycholate may be substituted or supplemented with 70 mM *n*-octyl-β-D-glucoside (add 1 g per 50 mL) to extract low-density Triton X-100-insoluble caveolin-rich membrane domains

Table 2
Suggested composition of self-made 100× protease inhibitor cocktail, final concentration (FC) of enlisted compounds in 1× working solution, directions for preparation, and storage

Compound	Mr (g/mol)	Amount	FC (1×)
AEBSF-HCl	239.69	25 g	104 mM
Aprotinin from bovine lung	6511.44	0.52 g	80 μM
Bestatin-HCl	344.8	1.5 g	4.5 mM
E-64	357.4	0.54 g	1.5 mM
Leupeptin hemisulfate salt	475.6	1 g	2.1 mM
Pepstatin A	685.91	1 g	1.5 mM
DMSO		Up to 10 mL	

Store at −20 °C in 1 mL aliquots. Dilute to 1× in ice-cold lysis buffer immediately prior to lysis

Table 3
Suggested composition of self-made 100× phosphatase inhibitor cocktail, final concentration (FC) of the compounds in 1× solution, target, directions for preparation, and storage

Compound	Mr (g/mol)	Amount	FC (1×)	Target
Sodium fluoride	42	84 mg	2 mM	Acid phosphatases
Imidazole	68.1	136.2 mg	2 mM	Alkaline phosphatases
Sodium molybdate	205.9	236.8 mg	1.15 mM	Acid phosphatases
Sodium tartrate, dihydrate	230.1	920.4 mg	4 mM	Acid phosphatases
Sodium pyrophosphate, decahydrate	416.1	416.1 mg	1 mM	PP1 and PP2A
b-Glycerophosphate	306.1	306.1 mg	1 mM	Ser/Thr phosphatases
200 mM activated sodium orthovanadate solution[a]	183.9	10 mL	2 mM	Tyrosine and alkaline phosphatases

Dissolve the reagents in 200 mM activated sodium orthovanadate solution. Aliquot 1.5 mL of stock cocktail in 2 mL Eppendorf tubes and store at −20 °C (stable at least 6 months). Dilute to 1× in ice-cold lysis buffer immediately before use

[a]Dissolve 3.68 g of Na_3VO_4 in 90 mL dH_2O and adjust to 100 mL. Adjust pH to 10 while stirring. Adding HCl will make the solution yellow. Boil the solution by heating in a microwave for 5–15 s or place it on a heated platform until it becomes clear and colorless. Cool on ice until the solution reaches RT. Add a small amount of 1 M HCl while stirring to adjust pH back to 10. Repeat these steps a total of 3–5 times. After several cycles of boiling, cooling, and pH adjustment, the solution should reach a point of a stable pH at ~10, where adding HCl should result in little, if any, appearance of yellow color. Aliquot and store activated Na_3VO_4 at −20 °C. Stable for at least 6 months

Table 4
The composition of self-made 4× NuPAGE LDS Sample Buffer, final concentration (FC) of the compounds in 4× solution, directions for preparation, and storage

Compound	Mr (g/mol)	Amount	FC (1×)
Tris–HCl[a]	157.56	6.68 g	106 mM
Tris-Base[a]	121.14	6.83 g	141 mM
EDTA disodium salt	372.24	76 mg	0.51 mM
Coomassie G250	854.02	75.15 mg	0.22 mM
Phenol Red	354.38	25.51 mg	0.18 mM
LDS	272.33	4–8 g	1–2 %
Glycerol (≥99 %)	92.09	40 mL	10 %
Ultrapure H_2O		Up to 100 mL	

Mix the reagents well in 30 mL of H_2O, add glycerol, and adjust the volume to 100 mL. pH should be ~8.5. Store at +4–25 °C room temperature (RT) (stable for 6 months)

[a]Tris can be substituted with 0.8 M triethanolamine-Cl, pH 7.6 (FC for 1× is 0.2 M)

Table 5
The composition of self-made 10× Reducing reagent, its final concentration (FC) in 1× solution, directions for preparation and storage

Compound	Mr (g/mol)	Amount	FC (1×)
DL-Dithiothreitol (DTT), anhydrous	154.25	3.85 g	50 mM
dH_2O		Up to 50 mL	

Dissolve in 40 mL of dH_2O. Adjust volume to 50 mL. Store at +4 °C in foil-wrapped or dark centrifuge tubes

Table 6
Recipes for making Bis-Tris resolving gels from 6.0 to 20.0 % and stacking gels of 4 % T

	Resolving gel monomer %						
Compound	6 %	8 %	10 %	12 %	15 %	20 %	Stacking gel (4 %)
4× Bis-Tris–HCl buffer[a], mL	5	5	5	5	5	5	5
30 % T/2.67 % C[b], mL	4	5.3	6.67	8	10	13.3	2.66
dH_2O, mL[c]	10.94	9.64	8.27	6.94	4.94	1.64	12.33
N,N,N′,N′-tetra-methyl-ethylene-diamine (TEMED)[d], μL	5	5	5	5	5	5	10
10 % (w/v) ammonium persulfate (APS)[e, f], μL	50	50	50	50	50	50	100

Total Monomer VOLUME: 20 mL

[a]Dissolve 149.4 g of Bis-Tris in 400 mL of dH_2O and titrate with 37 % HCl until pH reaches 6.5–6.8. Bring volume to 500 mL

[b]Dissolve 0.8 g N′N′-Methylene-Bis-acrylamide and 29.2 g Acrylamide in 70 mL of dH_2O. Bring volume to 100 mL. Filter through a 0.45 μm filter and store at +4 °C in amber glass bottle (30 days maximum). Alternatively, the researcher may use pre-weighted commercially available mixture of 37.5:1 ratio

[c]Adjust water volume if using optional Rhinohide Polyacrylamide Gel Strengthener Concentrate

[d]Use undiluted TEMED

[e]Dissolve 1 g APS in 10 mL of dH_2O. Sterilize by passing through a 0.22 μm filter using 20 mL Luer-Lok syringe. Aliquot to 0.5 mL tubes and store at −20 °C. APS slowly decays in solution, so replace the stock every 2–3 weeks

[f]Degas monomer solution prior to adding catalysts for copolymerization of gels

PAGE Gels with adapters for XCell [available from GenScript (Piscataway, NJ)] (*see* **Note 4**). Alternatively, the researcher may hand-pour the homogenous or gradient Bis-Tris gels using 1.0 mm Gel Cassettes and a gradient former of choice [37]. For instance, homogenous Bis-Tris separating and stacking gels can be prepared from a 30 % T/2.67 % C acrylamide/Bis (37.5:1 ratio) stock solution and a 4× Bis-Tris–HCl stock buffer (Table 6). Be aware that acrylamide and bisacrylamide are carcinogens and neurotoxins when in solution.

3. Prestained molecular weight marker: Precision Plus Protein™ All Blue standards containing a mixture of ten blue-stained recombinant proteins (10–250 kD), including three reference bands (25, 50, and 75 kDa). Store at −20 °C (*see* **Note 5**).

Table 7
The composition of self-made 20× MOPS-SDS Running buffer, final concentration (FC) of the compounds in 1× solution, directions for preparation, and storage

Compound	Mr (g/mol)	Amount	FC (1×)
3-(*N*-morpholino)-propanesulfonic acid (MOPS) (free acid)	209.26	209.26 g	50 mM
Tris (free base)	121.14	121.14 g	50 mM
EDTA, pH 7.7	292.24	5.84 g	1 mM
Sodium-dodecyl sulfate (SDS)	288.372	20 g	0.1 % (w/v)
dH_2O		Up to 1 L	

Mix the reagents well in 800 mL of dH_2O and prior to adding SDS, adjust the pH to 7.7. Adjust the volume to 1 L. Store at RT or at +4 °C (stable for 6 months). For PAGE, dilute this buffer to 1× with dH_2O
Check the pH of this buffer if it is obtained from commercial source other than Life Technologies (#NP0001) or Boston BioProducts (Worcester, MA, #BP-178)

Table 8
The composition of self-made NuPAGE Antioxidant, final concentration (FC) of the compounds in 1× solution, directions for preparation, storage, and commercial source of pre-made reagent (CS)

Compound	Mr (g/mol)	Amount	FC (1×)
Sodium Bisulfite	104.061	7.5 g	15 % (w/w)
N,N-Dimethylformamide	73.09	5 mL	10 % (w/w)
dH_2O		Up to 50 mL	

Dissolve sodium bisulfite in 40 mL dH_2O, then under the fume hood VERY SLOWLY add *N,N*-Dimethylformamide, and adjust volume to 50 mL with dH_2O. DO NOT SHAKE! Store in sealed dark centrifuge tube at +4 °C. Formation of crystals after some time is acceptable
CS Life Technologies (#NP0005)

4. 20× NuPAGE MOPS-SDS Running buffer (*see* Table 7 and **Note 6**).
5. 1× NuPAGE antioxidant (*see* Table 8).
6. Gel loading tips.

2.3 Western Blotting: Protein Transfer

1. Blotting unit: XCell II™ Blot wet transfer module (*see* **Note 3**).
2. Nitrocellulose membrane, one roll, pore size 0.22 μm (*see* **Note 7**).
3. Filter paper (FP) sheets.
   (a) Extra-thick narrow: 7 × 10 × 0.248 cm (W × L × H), 320 grade. For mini-gels cut to 7 × 9 cm.
   (b) Extra-thick medium: 7.5 × 10 × 0.248 cm (W × L × H), 320 grade. For mini-gels cut to 7.5 × 9 cm.

Table 9
The composition of self-made 1× NuPAGE Transfer buffer, final concentration (FC) of the compounds in 1× solution, directions for preparation, storage, and commercial source of pre-made reagent (CS)

Compound	Mr (g/mol)	Amount	FC (1×)
Bicine	163.17	10.2 g	25 mM
Bis-Tris (free base)	209.24	13.1 g	25 mM
EDTA, pH 7.2	292.24	0.75 g	1 mM
Chlorobutanol (optional preservative)	177.46	0.177 g	0.05 mM
dH_2O		Up to 1 L	

Store at RT. For transfer, dilute 50 mL of 20× buffer with 849 mL dH_2O, add 100 mL Methanol (FC 10 % w/w), and supplement with 1 mL of NuPAGE Antioxidant. The pH of the 1× solution is 7.2. Store at +4 °C
CS Life Technologies (#NP0006-1)

Table 10
The composition of self-made 1× Setup buffer, final concentration (FC) of the compounds in 1× solution, directions for preparation and storage

Compound	Mr (g/mol)	Amount	FC (1×)
Tris (free base)	121.14	12.114 g	25 mM
Glycine	75.066	57.65 g	192 mM
SDS	288.372	4 g	0.1 %
dH_2O	18.01	3,200 mL	
Methanol (add last)	32.04	800 mL	20 % (v/v)

Total VOLUME: 4 L
Buffer pH should be 8.1–8.6, but no adjustment is required. Store at +4 °C

(c) Extra-thick wide: 10×15×0.248 cm (W×L×H), 320 grade. For mini-gels cut to 8.5×9 cm.

(d) Thin: 7.5×10×0.083 cm (W×L×H), grade 222; For mini-gels cut to 7.5×9 cm.

4. A set of firm sponge pads (four pads for one blot). Alternatively, manually cut the 15.2×22.8 cm (6″×9″) light-duty scour pad into four pieces that fit well in the XCell II Blot module.
5. Gel cutting knife.
6. Flat and upward bent tip tweezers.
7. Gel/Blot assembly trays.
8. Blotting roller, 8.6 cm wide.
9. 20× NuPAGE Transfer buffer (*see* Table 9) and refrigerated dH_2O.
10. 1× setup buffer (*see* Table 10).

Table 11
The composition of self-made 10× TBS-T buffer, final concentration (FC) of the compounds in 1× solution, directions for preparation and storage

Compound	Mr (g/mol)	Amount	FC (1×)
1 M Tris, pH 8.0	121.14	200 mL	10 mM
NaCl	58.44	175.32	150 mM
100 % Triton X-100	647	10 mL	0.5 % (v/v)
dH_2O		Up to 2 L	

Store at RT. For solutions, dilute this buffer to 1× with dH_2O and store at +4 °C

2.4 Western Blotting: Protein Immunodetection

1. Blot incubation dishes: Square Petri Dishes with Grid (*see* **Note 8**).
2. 10× TBS-T buffer: Tris-buffered saline (TBS) with Triton X-100 (*see* Table 11).
3. Blocking buffer: 3 % (w/v) bovine serum albumin (BSA) in 1× TBS-T buffer.
4. Antibody solutions:
   (a) Unconjugated or HRP-linked primary antibodies (1°Ab) of choice diluted in 1× TBS-T.
   (b) Secondary HRP-linked antibodies (2°Ab) of choice diluted in 1× TBS-T.
5. SNAP i.d.® Protein Detection system (EMD Millipore, Upstate, NY) (optional). If used, this system significantly shortens the time of membrane blocking and/or incubation with primary/secondary antibodies. Requires heavily concentrated primary antibody solution.
6. Western blotting detection ECL reagents: SuperSignal West Dura and/or West Pico Extended Duration Chemiluminescent substrates.
7. Titanium Bonded Scissors: 1 for membrane cutting, 1 for filter papers, 1 for opening gel envelopes, 1 for cutting sheet protectors. We strongly advise against using the same scissors for these tasks.
8. 3 M Scotch Magic Transparent Tape and desktop tape dispenser.
9. Imaging system with CCD sensor and zoom for image visualization and densitometric analysis: Image Station 440CF (Eastman Kodak Scientific Imaging Systems, New Haven, CT).
10. Avery® Diamond Clear Heavyweight Quick-Load Sheet Protectors. Each protector can be cut horizontally into three or four pieces, depending on the width of the blot to be protected during its visualization and subsequent storage.

3 Methods

3.1 Preparation of Samples for Multistrip Western Blotting

1. Scrape the stimulated or nonstimulated cells into ice-cold 1× lysis buffer. Suggested volume of buffer used for total protein whole-cell lysis (WCL) of 70–90 % confluent cells grown in 150 × 15 cm cell tissue culture dishes is 1.6–1.8 mL, in 100 × 20 cm dishes—1–1.2 mL, for 60 × 15 cm dishes—0.5–0.6 mL. To homogenize 1 mg of tissue, use 1.5× fold volume of lysis buffer (in mL) (e.g., for 100 mg tissue, add 1.5 mL of lysis buffer). Keep tubes on ice throughout all experiment.
2. Spin the samples at least at 10,000 × *g* for 10 min at 4 °C.
3. Mix the supernatant of each cell lysate with 4× NuPAGE LDS sample buffer and 10× NuPAGE sample reducing agent in a ratio of 65:25:10 in prelabeled Eppendorf tubes.
4. Heat samples at 75 °C for 5 min. Cool samples to RT.
5. Run samples within 2 week period. For extended storage, store the samples at 4 °C (reheat prior to loading). Unused amount of cell lysates can be stored for further use at −20 °C/−80 °C.

3.2 Sample Loading and LDS-PAGE

Sample loading strategy highly depends on the design of an optimal experiment to test a hypothesis or answer a biological question. The number of gels to be loaded depends on the number of data series and the number of samples within each series to be analyzed. For example, when the time-course expression of protein of interest in control cells (A) is compared to that under perturbed conditions (e.g., in the presence of inhibitor of protein X (B) and the suppression of protein Y by siRNA (C)) at 0, 1, 3, 5, 7, 10, 20, 30, and 60 min, one will have to load three data series (A, B, and C), consisting of nine time-points each (A1, A2... A9, B1–B9, C1–C9) into three 10-well gels. There are two alternative ways of loading such number of samples (Fig. 1). Further, the loaded samples onto one gel (excluding protein MW marker(s)) will be referred to as a "*set of sample*," sometimes, indicating how many samples are in the set (e.g., one set of nine samples; three sets of eight samples).

1. Carefully remove a comb from each precast gel and rinse its wells and the whole gel under a running stream of dH_2O.
2. Place the first gel in front of the buffer core (FRONT), and the second gel—behind (BACK). In both cases, the shorter (notched) side of the cassette should face in toward the core. Lock the tension wedge.
3. Fill the upper chamber of each XCell SureLock Mini-Cell unit with 200 mL of properly cooled 1× MOPS-SDS Running

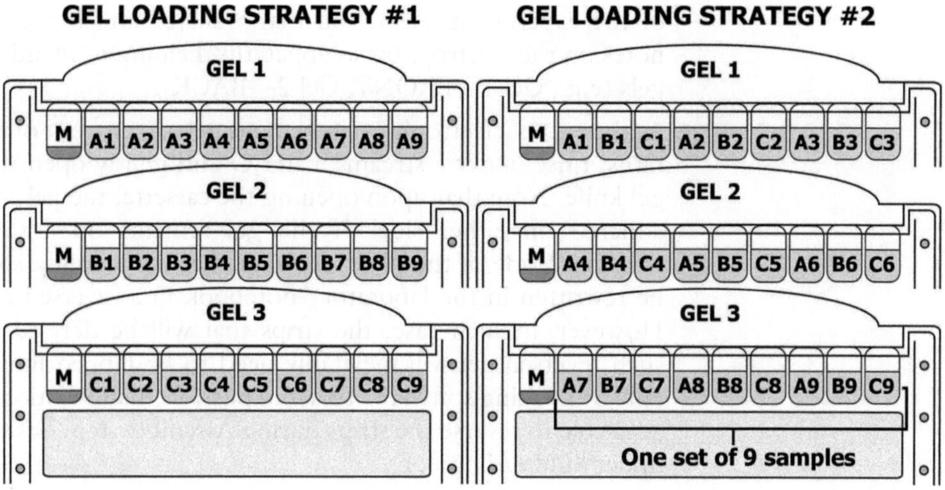

Fig. 1 Gel loading strategies when the number of samples exceeds the number of wells in one gel. M, prestained protein MW marker

buffer to completely cover the sample wells of a gel. Pour ~600 mL of the same buffer into the lower chamber.

4. Prior to sample loading and electrophoresis, supplement the 1× Running buffer in the upper (cathode) chamber of unit with 0.5 mL NuPAGE Antioxidant.
5. Use a pipette equipped with prolonged gel loading tip to underlay 7 μL of prestained protein MW marker into the first and/or the last well of the gel. Loading both wells with a marker can substantially ease and speed up the subsequent step of Gel Cutting (described under Subheading 3.3).
6. Load an equal volume of each sample into the rest of gel wells (*see* **Note 9**). Maximum sample loading volume per well of 1.0 mm 10-well gel is 25–30 μL, 12-well—20–25 μL, 15-well—15 μL. Do not overload! If there are empty wells without loaded sample left, fill them with similar amount of sample buffer diluted in phosphate-buffered saline (PBS buffer, pH 7.4). Overall, load the amount of protein according to the sensitivity of your detection method.
7. Add the lid on the buffer core and connect the apparatus to a power supply. Separate proteins according to their electrophoretic mobility at constant voltage of 125–150 V until the blue dye front (BDF) reaches the bottom of a gel.
8. If running more than two gels, make an interval of at least 5–10 min before loading the next tandem of gels and powering on the electrophoresis unit. This will reserve enough time for follow-up steps. During this pause, mark a sequence of loaded samples in a laboratory notebook along with the details

of the experiment. Also, you may want to attach the sticky notes on the electrophoresis apparatus, helping to identify the gels (e.g., Gel 1—FRONT, Gel 2—BACK).

9. At the end of electrophoresis, remove gel cassette out of apparatus, rinse under a stream of dH_2O and gently open with a gel knife. Note that upon opening the cassette, the gel can be adhered on either side. If the gel remains on a shorter (notched) side of the plate, the sequence of sampling should be rewritten in the laboratory notebook in a reversed order. However, in such case, the strips that will be derived from such reversed gel will inevitably need to be flipped horizontally by imaging software. To avoid possible manipulation, it is advisable to reverse the strips during Assembly step (described under Subheading 3.4).
10. Discard the plate of gel cassette without the gel. Rinse the side with an adhered gel with dH_2O.
11. Proceed to Gel Cutting step as instructed under Subheading 3.3. Preferably, each gel should be cut immediately after opening the cassette as soon as electrophoresis is complete. If you decided to wait until electrophoresis is over and BDF reaches the bottom of ALL gels, then cover adhered gel with extra-thick 7.5 × 8 cm filter paper (further referred as CFP, i.e., covering filter paper), which has been submerged once in cold 1× Setup buffer. Thereafter, put the cassette with covered whole gel aside and stick the label, helping to identify a gel.
12. Perform **steps 9–11** with gel cassettes from other electrophoresis units.

3.3 Gel Cutting

Figure 2 illustrates the plate with attached gel after protein separation according to their molecular weight by LDS-PAGE. The prestained protein MW marker is visibly separated into the bands corresponding to protein molecular weights of 250, 150, 100, 75, 50, 37, 25, 20, 15, and 10 kDa (Fig. 2, *M*). Though this is an optional step, the researcher may firmly position a millimeter-scaled transparent ruler near the edge lane with separated marker so that zero (0 cm) aligns with the middle of the *BDF* (Fig. 2). The distance from the *BDF* to the center of each marker band (Fig. 2, *H*) is measured in millimeters and can be registered in a statistical table (*see* **Note 10** and Table 12).

The distance between two electrophoretically separated marker bands corresponds to the migration range of certain molecular weight proteins. For instance, the distance between *H250* and *H150* defines a migration range of electrophoretically separated proteins with molecular sizes between 150 and 250 kDa. This area is termed zone ①. Each sample provides up to nine protein-containing zones that may be simultaneously cut out from a single gel (Fig. 2).

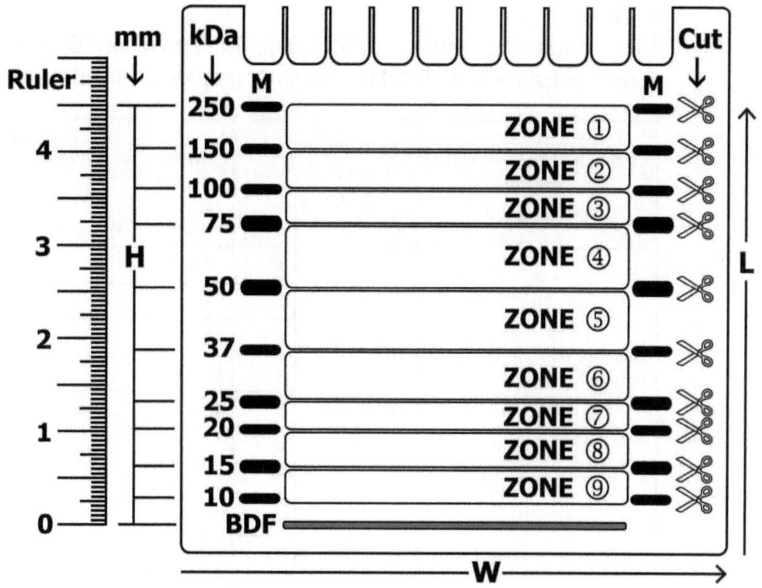

Fig. 2 Identification of protein migration zones in Multistrip Western blotting procedure. First and last lanes show separated prestained protein MW marker (*M*). *BDF* indicates blue dye front; *H* is the distance from *BDF* to the center of particular marker band. *Scissors symbols* indicate the cutting lines, which would separate the entire gel into nine strips. Protein migration zones are enumerated by *numbers in circles*. *L*—gel plate length, *W*—gel cassette width

The number of strips to be cut out from the gel depends on the number of distinct proteins to be analyzed. Most frequently studied signal transduction proteins migrating in various zones are listed in Table 13, which also indicates the appropriate areas that can be cut out of the gel for subsequent detection of these proteins. These areas may be narrow (containing only one Zone) or wide (spanning two or more Zones).

For example, Fig. 3 depicts the strategy of gel cutting into strips by the researcher who seeks to examine changes in the levels of multiple biochemical markers of programmed cell death (apoptosis). Only four blotting cycles (or eight if two gels are used) allow the researcher to separate and detect a group of distinct apoptosis- and cell survival-related proteins whose migration in a gel often overlap. Importantly, if the researcher wants to detect only those proteins enlisted in the *left lower panel of* Fig. 3 and has loaded only one set of eight samples for concurrent measurement and comparison of signals, there is no need to cut a gel into strips. He/she should proceed to the Assembly of Western blotting sandwich step (described under Subheading 3.5), transfer the proteins onto the membrane from the entire uncut gel, and thereafter (using clean Titanium scissors) cut the membrane into four strips containing Zone ①②, Zone ③④, Zone ④⑤, and Zone ⑤⑥⑦⑧⑨, respectively.

Table 12
The example of a template of Statistical Table to record the migration patterns of various prestained molecular standards (markers) in various types of gels

Protein MW standards (M) title, commercial source	Running Buffer	Zone	MW bands (kDa)	Cut area (H) (distance in mm from BDF) in various gels		
				10% Bis-Tris	4-12% Bis-Tris	4-20% Tris-Glycine
BlueRay Prestained Protein Marker (10-180 kDa) Jena Bioscience #PS-103	MES-SDS	1	125-165			
		2	93-125			
		3	72-93			
		4	57-72			
		5	42-57			
		6	31-42			
		7	24-31			
		8	15-24			
		9	8-24			
Precision Plus Protein Dual Xtra Standards (2-250 kDa) Bio-Rad #161-0377	MOPS-SDS	1	150-250		49 to 55	
		2	100-150		44 to 49	
		3	75-100		38 to 44	
		4	50-75		29 to 38	
		5	37-50		22 to 29	
		6	25-37		12 to 22	
		7	20-25		9 to 12	
		8	15-20		5 to 9	
		9	10-15		2 to 5	
		10	5-10		0 to 2	
		11	2-5		BDF	

If you need to compare several sets of samples that were resolved in *two* or *more* gels, then proceed to Subheading 3.3, **step 1**.

1. Starting from the bottom and moving toward the top of the adhered gel, use a regular gel knife (or blade) to cut out the strip, which covers an area with a protein of interest located in the middle, from the gel across its entire width, including the lane(s) with a separated protein MW marker (Figs. 2 and 3, *scissors symbol*).
2. Proceed to cutting the next strip above.
3. Discard the gel pieces outside the strips.
4. Cover the first plate with prepared multiple gel strips with a sheet of moistened *CFP* and place on the bench top. Similarly, cut the second gel, cover it with another sheet of CFP, and place it next to the previously laid plate. Repeat above procedure with the rest of the gels.

Table 13
Migration range of signal transduction proteins in NuPAGE 4–12 % gradient mini-gel and their cutting areas, based on the migration pattern of prestained Precision Plus All Blue Protein marker

Zone	Protein MW range (kDa)	Cut area, H (mm from BDF)	Suggested cutting guidelines of gradient 4-12% gel strips for subsequent detection of indicated signal transduction proteins				
1	150 – 250	from 49 ± 1 to 54 ± 1	EGFR, ErbBs, MRP2, PDGFR, CSFR, Rictor	c-Met, PLCs, FGFR, c-Kit, SHIP, Tyk2, MDR1, Raptor, Eps15, ROCK1, ASK1, Collagen I, eNOS	SOS, IRS1/2/3/4, ZEB1, Filamin, Alk	VEGFR, c-Ret, BRCA1, Fibronectin	Bcr-Abl, RhoGAP
2	100 – 150	from 44 ± 1 to 49 ± 1	RasGAP, FAK, c-Cbl, SirT-1, PLD1, PTPα, Pecam-1, Vinculin, JAK1/2/3		PI3K-p110, Gab1/2, Pyk2, Vav, MLK3, β-Catenin, STATs, Fer, Rb, p105 NF-κB1, VE-Cadherin, FoxO3a, PARP (FL and CL)	E/N-Cadherins, PKD, cPLA2, c-Abl, gp130, ERK5, HIF-1α, MCM2, p130Cas	MMP-9, PLD2, Exportin, TRIF, Integrin β1
3	75 – 100	from 38 ± 1 to 44 ± 1	PI3K-p85, FRS-2, GRK, APS, HIF-1β, Wee1, Eps8	IGF-1R, InsR, PKCs, p90RSK, IKKα, Hsp90, FoxO1		Raf family, MXR, Grb10, FKHR, p70 S6K, Calpain, Gab3, Egr1, MMP-2, Lamins A/B/C, PAK 4/5/6, cdc25A/B/C	
4	50 – 75	from 29 ± 1 to 38 ± 1	SHP1/2, c-Src family, PTEN, Csk, Akt 1/2/3, AFP, Grb14, Grb7, Myc, PAK1/2/3, CDT1, RIP3, MT-MMP-1	α-Tubulin, Shc, p53, JNK1/2, SGK, ILK1, GATA-2/3, PP2A, PAI-1, CaMKII, GSK-3 α/β, PTP1B, β-Arrestin, IκBβ,	Dok-R, Paxillin, PDK1, Sam68, SRF, Chk1/2, AMPKα/γ, STAM1, Hsp70, Myt1, Cyclin A, Cyclin B, RPA1		FoxO4, WASP, AIF, ZAP-70, Syk, GATA-6, LIMK-1/2, p65 NF-κB1, Merlin
5	37 – 50	from 22 ± 1 to 29 ± 1	Crk, ERK1/2, MEK, Nck, CREB, AMPKβ, FRA1, GATA-1, PAR-4	Caspase-9 (FL), c-Fos, Cyclin E, Vimentin, MMP-1/3, p50 NF-κB1, Keratin 8/17/18	β-Actin, PKA, MKKs, c-Jun, Flotillin-1/2, SMA MAPKAPK2	Sprouty, PCNA, CDK6, LAT, GAPDH, siRT-2, p38 MAPK, Caspase-3/7 (FL), Caspase-9 (CL), Cyclin D, Pim-1/2/3	TGF-β1/2/3, VEGF isoforms, TNF-α, TGF-α IL-6, IL-10, IL-1 α/β, PDGF-A/B, HB-EGF (all mature and precursor) MMP-7
6	25 - 37	from 12 ± 1 to 22 ± 1	14-3-3, Bic, VDAC, Slug, Snail, Twist	Bcl-2, PP1, S6RP, HO-1, cdc2, CDK2, CDK4	Bak, GRB2, Bcl-xL, Hsp27, p27 Kip1, Lamin A (CL small subunit)		
7	20 - 25	from 9 ± 1 to 12 ± 1	Bad, Rac1/cdc42, Puma	Bax, Caveolin-1, p21, Ras, DAP1, RKIP, Claudin-1, Cofilin		BID, MCP1, IGF-1, Caspase-3/7/9 (CL)	
8	15 - 20	from 5 to 9 ± 1	Survivin, Bmf, p18, COX IV		TCL1, Cytochrome c, Angiogenin 1, Histone H3/H2A.X		
9	10 - 15	from 2 to 5					

3.4 Assembly of Gel Strips

During this step, the gel strips that are derived from different gels are assembled onto a single sheet of filter paper (*AFP*, for assembling filter paper) for the subsequent electrophoretic protein transfer onto the same piece of nitrocellulose membrane (*see* **Note 7**). The strategy of assembly depends on the quantity of gels used for PAGE as well on the number of strips containing the appropriate proteins of interest (*see* **Note 11**).

Here we provide two exemplar cases of gel strip assembly:

(i) If one loaded five sets of samples in five gels (GEL1–5) and subsequently cut five strips to detect five proteins of interest (PARP, p-Chk1, β-Actin, Bcl-2, and Cytochrome c) as shown in *left upper panel of* Fig. 3.

(ii) If one loaded three sets of samples in three gels (GEL1, GEL2, and GEL3) and subsequently cut five strips to detect five proteins of interest (p-BRCA1, E-Cadherin, c-Myc, p-c-Jun, and Bax) as shown in *right lower panel of* Fig. 3.

1. In case (i), flip and gently lift the plate containing GEL1 so that all strips would stick to the moistened CFP. Use gel knife if the strips do not independently detach from the plate.

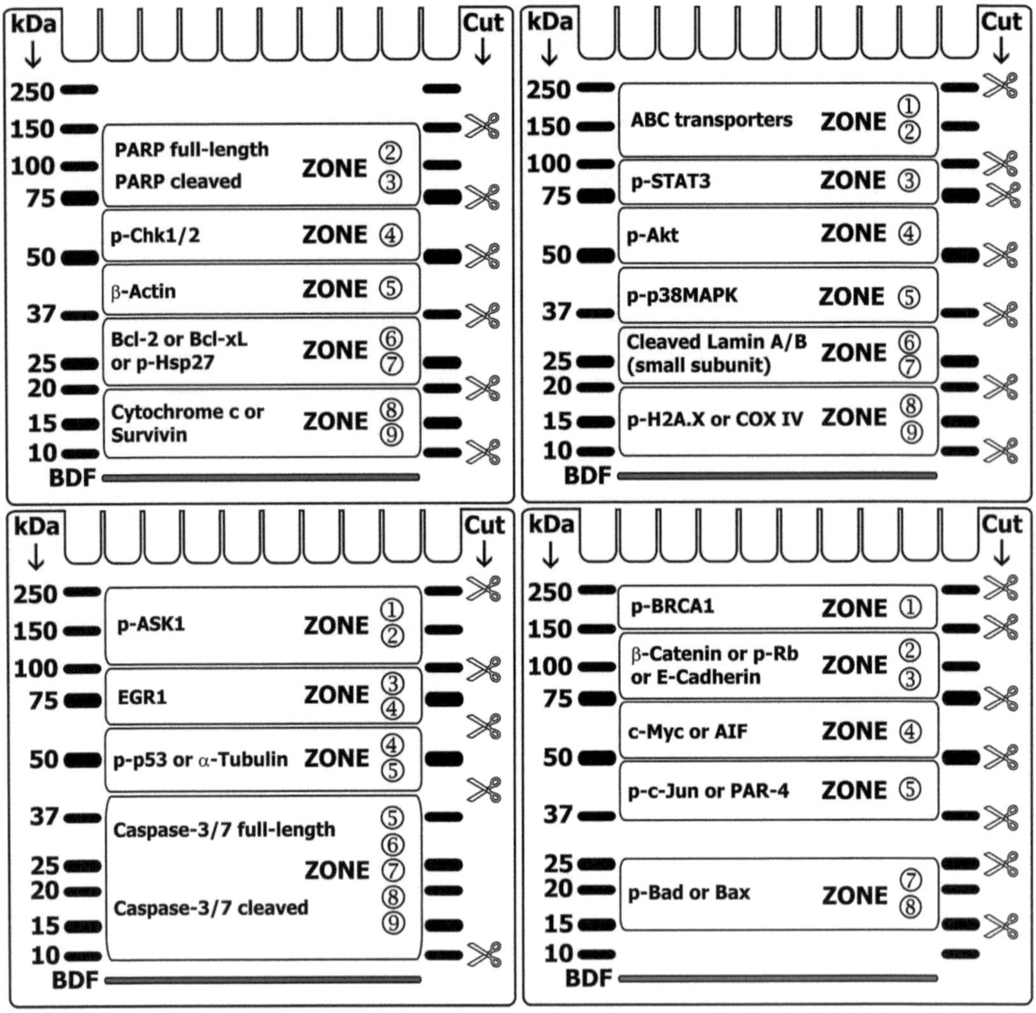

Fig. 3 Suggested gel cutting for analysis of proteins implicated in apoptosis and cell survival

2. Lift the first gel strip from the top (possessing Zone ②③ proteins) with a gloved hand and carefully transfer onto AFP #1 that must be briefly presoaked in 1× Setup buffer. Return the CFP with remaining gel strips onto the plate by flipping it back.
3. Similarly, top gel strips derived from GEL2–GEL4 and GEL5 are sequentially transferred onto the AFP #1 so that the strips would lay side by side and parallel to each other. AFP #1 is now ready for immediate protein transfer (*see* **Note 12**).
4. **Steps 1–3** are repeated with the strips derived from GEL1 to GEL5 that contain the proteins migrating in Zone ④, Zone ⑤,

then Zone ⑥⑦, and finally in Zone ⑧⑨. This procedure will yield five AFPs (AFP #1–#5) with collected five gel strips on each (Fig. 4). Now they are ready for assembly of Western Blotting sandwich and subsequent electrophoretic protein transfer onto the same nitrocellulose membrane.

5. In case (ii), perform **steps 1–2** with GEL1, followed by a sequential transfer of gel strips derived from GEL2 and GEL3 onto the AFP #1. Then, place gel strips with Zone ②③ onto the AFP #1 below previously laid triplet of strips. Leave a small gap between the triplets (Fig. 5, *upper panel*).
6. AFP #2 is processed in the same manner so that it would contain triplet of strips with Zone ④ and triplet of strips with Zone ⑤ (Fig. 5, *middle panel*). After protein transfer, the resulting membrane is cut into two pieces across the gap between the triplets (Fig. 5, *scissors symbol*). The pieces are then treated in separate dishes (*see* **Notes 8** and **13**).
7. The remaining three strips containing Zone ⑦⑧ should be assembled onto AFP #3 (Fig. 5, *lower panel*).

3.5 Assembly of Western Blotting Sandwich and Protein Transfer

Instructions provided below assume the use of XCell II Blot module, which is used for transfer of protein from one AFP.

1. Fill one side of gel/blot assembly tray with 500 mL of refrigerated 1× Setup buffer, while another side—with 400 mL of 1× Transfer buffer.
2. Presoak four sponge pads in 1× Setup buffer. Remove air bubbles by squeezing the pads while they are submerged in buffer. Cut a sheet of nitrocellulose membrane to the dimensions of AFP and presoak it in Transfer buffer for 5 min before using. Briefly moisten three additional extra-thick and one thin filter papers in Setup buffer immediately before using.
3. Place two wet sponge pads into the cathode (−) core of the blot module. Place the AFP with collected gel strips on the top. Subsequently, cover the surface of gel strips with a sheet of membrane. Remove any trapped air bubbles by rolling a blotting roller over the membrane surface. Place three moistened extra-thick filters onto the surface of the membrane followed by tandem of wet sponge pads (*see* **Note 14**).
4. Place the anode (+) core on the top of the pads. Slide the blot module into the rails on the lower chamber. Lock the gel tension lever.
5. Fill the blot module with 1× Transfer buffer until the blotting sandwich is completely submerged. Fill the outer chamber with chilled dH_2O.

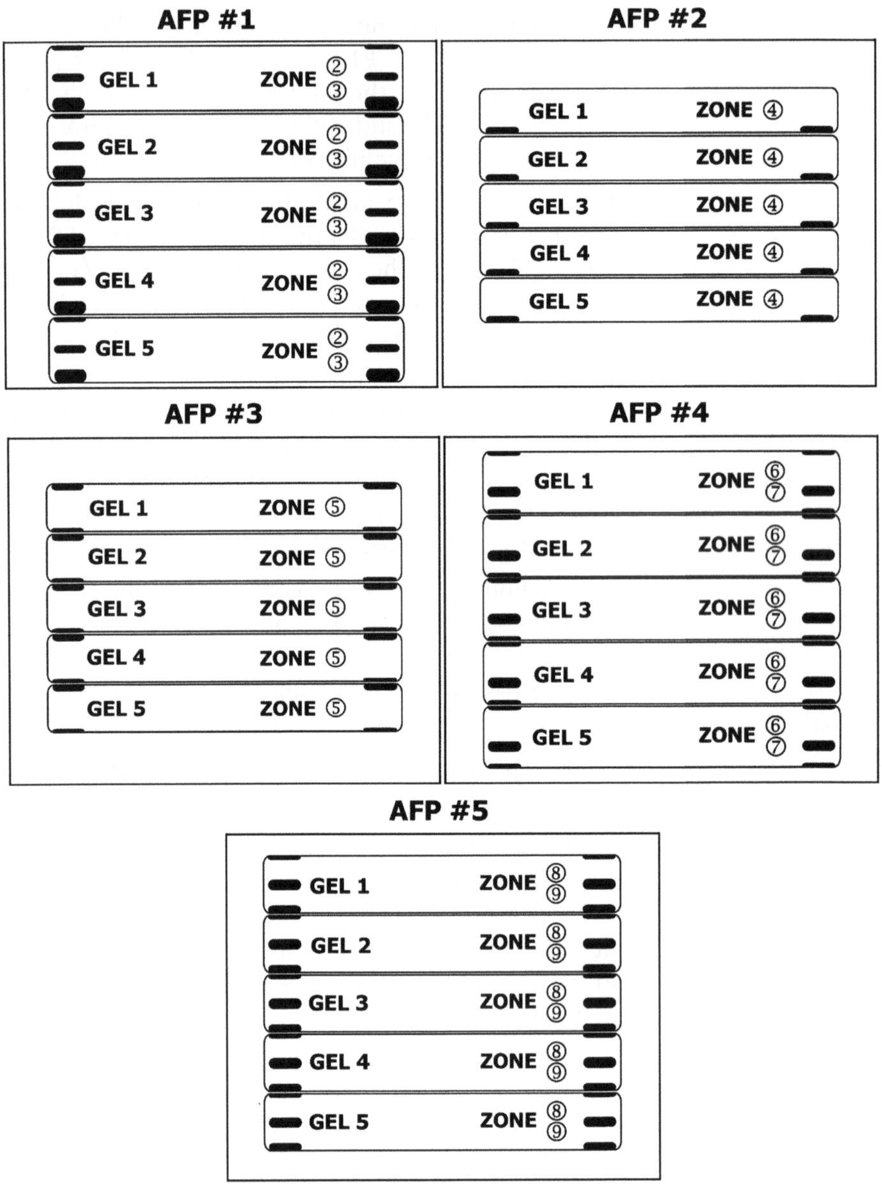

Fig. 4 A strategy for assembly of five gel strips following LDS-PAGE of five sets of samples. LDS-PAGE of five gels was performed. Five strips of the identical length, but of varying width, were cut out of each gel and combined onto appropriate assembling filter papers (*AFP*, *#1* through *#5*)

6. The unit is completely assembled by adding the lid on the buffer core, and connected to a power supply. Transfer the proteins at 30 V constant for 90 min.

3.6 Immunoblotting and Chemiluminescent Detection

1. After transfer is stopped, remove the membrane out of the blot module and attach to the middle of a square Petri dish by stick-

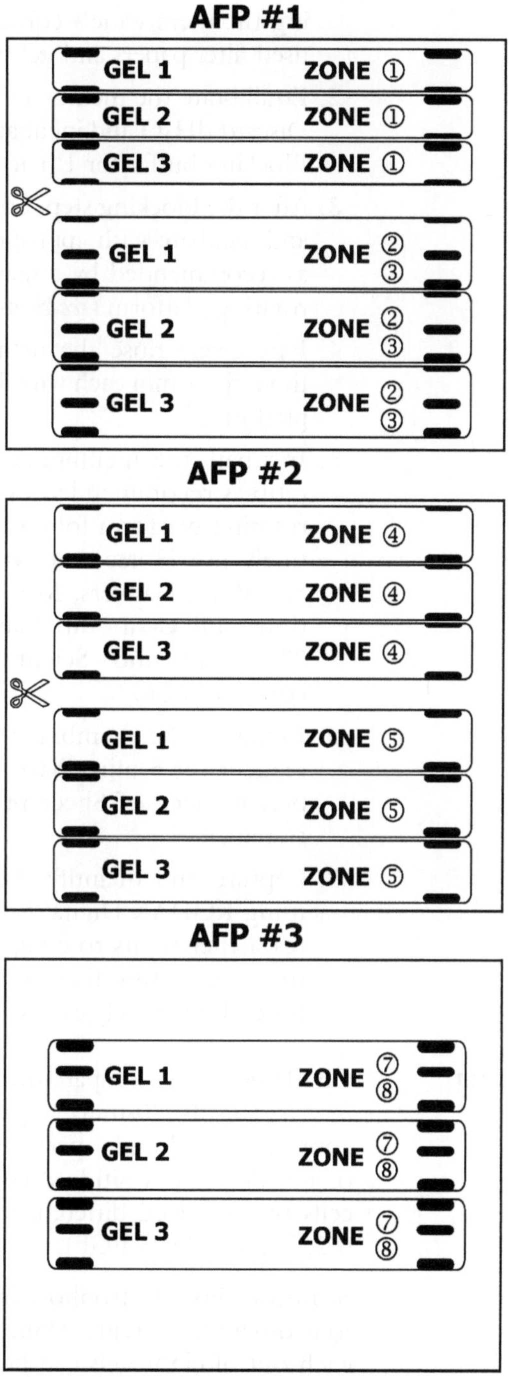

Fig. 5 A strategy for assembly of five gel strips following LDS-PAGE of three sets of samples. LDS-PAGE of three gels was performed. Five strips of the identical length, but of varying width, were cut out of each gel and combined onto appropriate assembling filter papers (*AFP #1* through *#3*)

ing the membrane's corners with a transparent tape. Discard used filter papers and gel strips.

2. Equilibrate the membrane by rinsing with dH_2O for 3–5 min. Discard dH_2O and incubate the membrane with 25–30 mL of Blocking buffer for 1 h at RT on a rotating platform.
3. After the blocking step, briefly rinse the membrane with dH_2O and incubate with appropriate 1°Ab solution at dilution ratio as recommended by a manufacturer overnight at +4 °C on a rotating platform (*see* **Note 15**).
4. Extensively rinse the membrane with dH_2O and wash four times for 7 min each with 1× TBS-T buffer at RT on a rotating platform.
5. Incubate the membrane with appropriate 2°Ab at dilution ratio as recommended by a manufacturer for 1 h at RT on a rotating platform followed by **step 10** once again. We routinely use Horse Anti-Mouse HRP-linked IgG 2°Ab (Cell Signaling, Danvers, MA, #7076) at 0.1 μL/mL concentrations and Goat Anti-Rabbit HRP-conjugated IgG (H + L) 2°Ab (Thermo Scientific, #31460) at 0.025 μL/mL concentration.
6. Incubate the membrane with a working solution of ECL reagent for 5 min. Place the membrane upside down in the precut piece of Sheet protector and place in the Imaging system.
7. Capture and quantify the signal intensity of protein bands using KODAK Digital Science software (*see* **Note 16**). If the researcher wants to compare the signals from different blots, then the capture time and number of frames should be equal for each separately exposed membrane.

3.7 Application Examples

Challenge 1. To compare the activation kinetics (e.g., at 0, 1.5, 3, 5, 7.5, 10, 20, 30 min) of protein of interest (e.g., ERK1/2) in control (K) and under perturbed conditions (e.g., in the presence of the Phosphoinositide 3-kinase inhibitor wortmannin, WT) in cells that received different strengths of stimulation (e.g., 0.02, 0.2, 2, and 20 nM EGF).

Solution. For electrophoresis, the samples may be loaded using four different strategies (Fig. 6a, b). After separation of proteins, each out of eight gels can be cut into desired number of strips. The given experimental task requires analyzing the activating phosphorylation of ERK1/2 (Zone ⑤, p-ERK1/2 (T202/Y204), 42/44 kDa) by the upstream kinase. In addition, it may be desired to check the activity of ERK1/2 by measuring the phosphorylation status of its immediate downstream target p90 ribosomal S6 kinase (Zone ③, p-p90RSK (S380), 90 kDa). Since the cells were treated with phosphoinositide 3-kinase (PI3K) enzyme inhibitor WT, it is important to verify the efficiency of PI3K inhibition. There are many pleckstrin homology (PH) domain-containing proteins whose recruitment to the plasma membrane and subsequent activation/phosphorylation may be altered due to depletion of PIP_3, which is generated only by active PI3K. One of such proteins is a serine/threonine kinase Akt (Zone ④—p-Akt (S473), 60 kDa). S6 ribosomal protein (Zone ⑦⑧, p-S6RP (S240/244), 32 kDa) is a common substrate for p70 S6 kinase (Zone ③④,

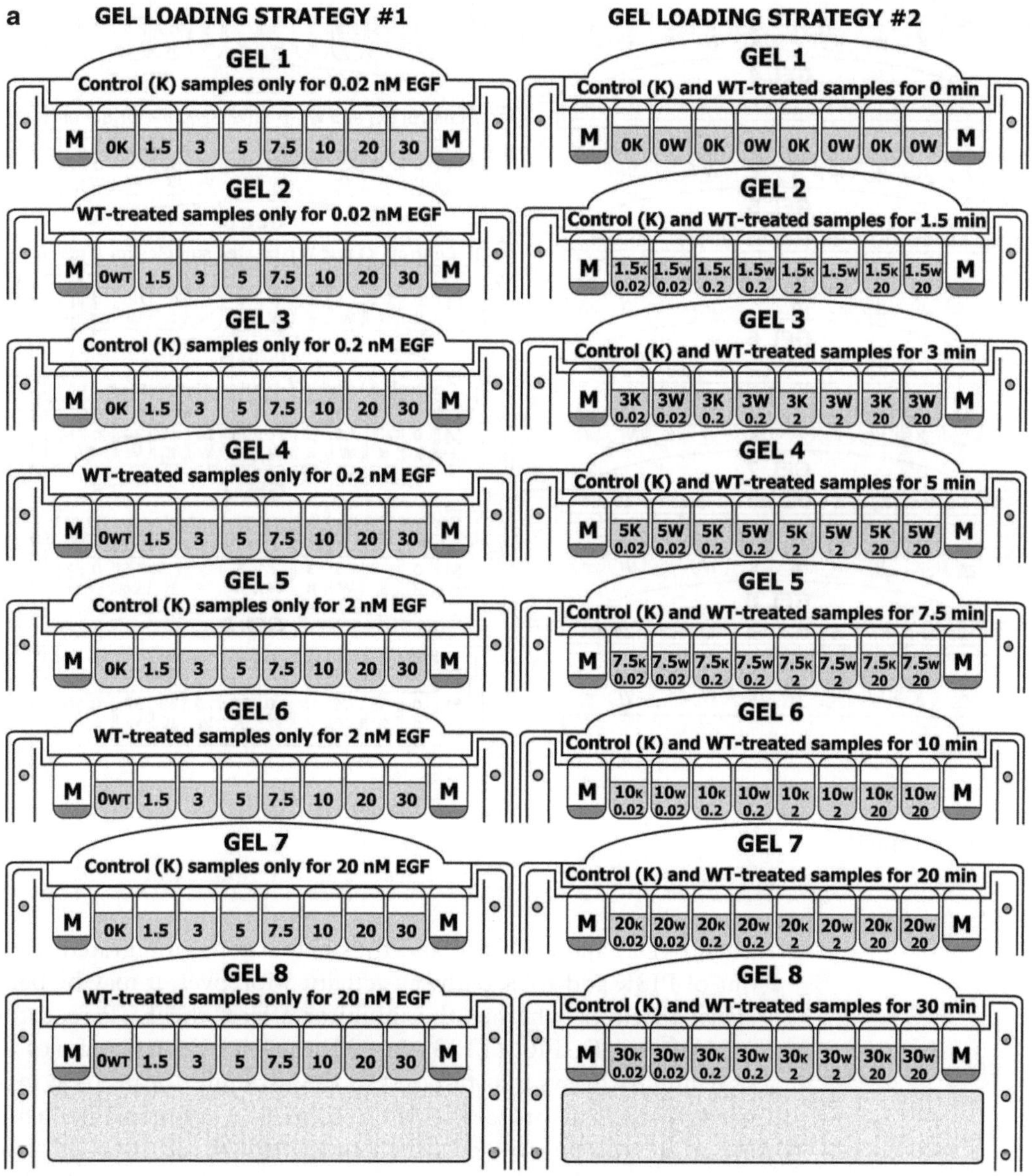

Fig. 6 Various Gel loading strategies (A—#1 and #2; B—#3 and #4) for electrophoresis of eight sets of samples

b

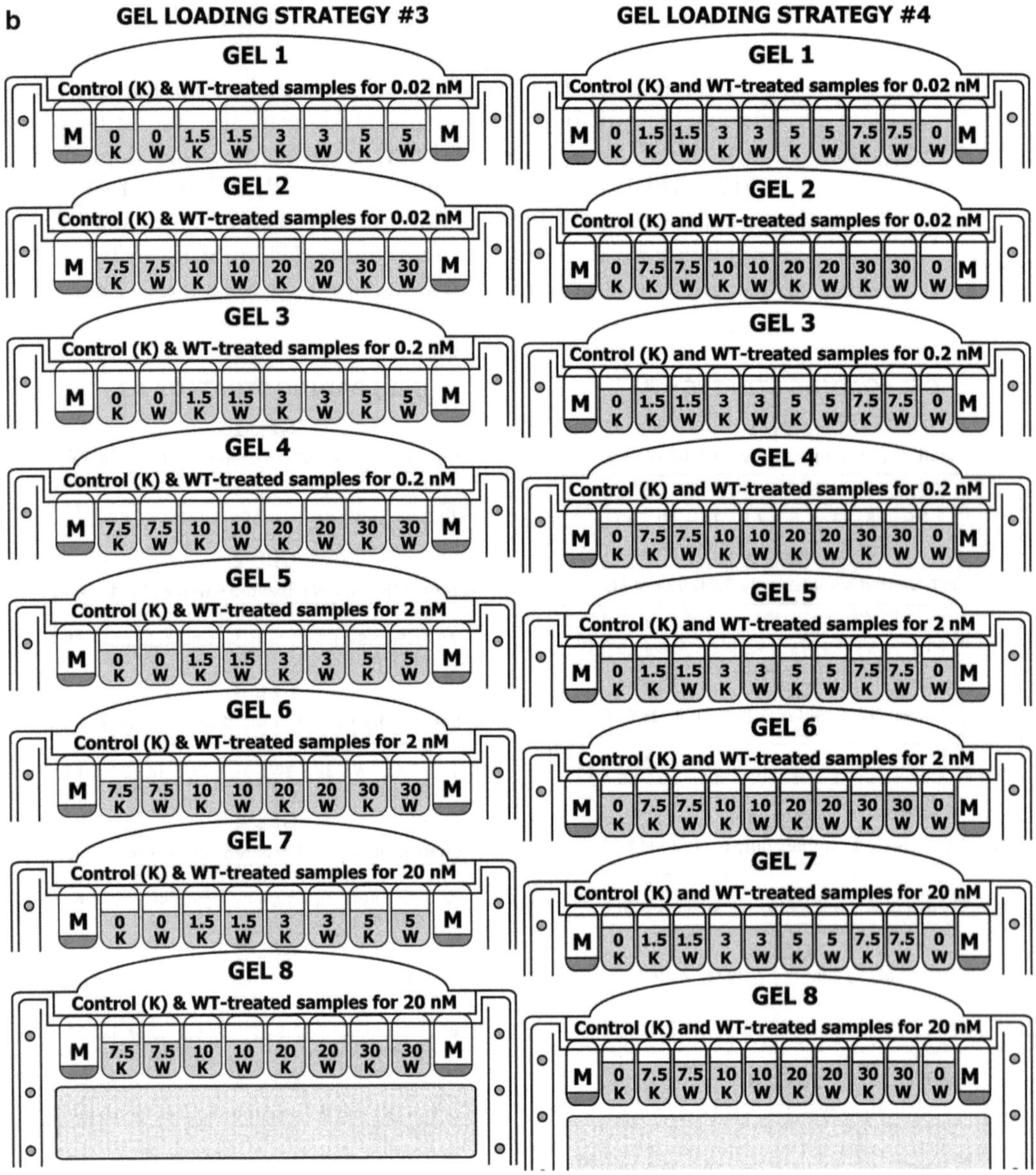

Fig. 6 (continued)

70–85 kDa) and p90RSK, and thus may serve as integrated readout of PI3K and ERK pathway activity. Moreover, it may be beneficial to check whether the inhibitor affects the EGF receptor (Zone ①—p-EGFR (Y1173), 160 kDa), which lies upstream of all signaling pathways that respond to EGF stimuli. Ideally, one may need to have a protein loading control. Unfortunately, actin (45 kDa), tubulin (50–55 kDa), and GAPDH (37 kDa) migrate in the Zones that, if cut out of the gel, would prevent further detection of p-Akt and p-ERK. However, there are lower-molecular weight proteins that are expressed at steady-state levels (at least within a 30 min time frame of experiment) and thus may serve as good protein loading control not only for total cell lysates (e.g., Grb2 (Zone ⑥⑦, 25 kDa)) but also for separated cellular fractions: mitochondrial (COX IV, Zone ⑧⑨, 17 kDa; VDAC, Zone ⑥, 32 kDa), nuclear (Histone H3, Zone ⑧⑨, 17 kDa), plasma membrane (Ras, Zone ⑦⑧, 21 kDa), or caveolin-rich plasma membrane domains (caveolin-1/2, Zone ⑥⑦⑧, 21–24 kDa) (*see* Table 13). Gel cutting into six strips for analysis of p-EGFR, p-p90RSK, p-Akt, p-ERK, p-S6RP, and COX IV is illustrated in the *left upper panel of* Fig. 7.

Please note that if the strip for Zone ⑤ is cut out along with the strips from the upper and lower neighboring Zones (i.e., Zone ④ and Zone ⑥), then the gel cutting knife should pass through the bottom or the center of 50 and 37 kDa MW marker band (Fig. 7, *left middle panel, A*), depending how far away from the MW band migrates the next protein of interest. If the strip for Zone ⑤ is cut out only with the upper neighboring Zone ④, then the cutting may be done slightly below 37 kDa MW marker band (Fig. 7, *left middle panel, B*). If the strip for Zone ⑤ is cut out solely, then the cutting should encompass whole bands of MW marker to ensure risk-free detection of your protein of interest (Fig. 7, *left middle panel, C*). However, in this case the strip width will significantly increase, which would require aligning the strips onto a larger (e.g., extra-thick wide) AFP (Fig. 7, *right upper panel*).

Left lower panel of Fig. 7 shows the actual blot of p-ERK1/2 following the MSWB procedure performed with the protein samples that were loaded onto nine 10-well gels using Gel loading strategy #2 (Fig. 6), while the example of a blot derived from a more traditional Gel loading strategy #1 using four 10-well gels is shown in the *right middle panel of* Fig. 7. All four Gel loading strategies are suitable for densitometric analysis of the intensities of signals *upon equal conditions* including the incubation time of blot in ECL substrate, exposure time, and distance of the blot from the CCD camera. Obtained signals further can be normalized for protein loading and plotted onto a scatter chart for a comparative analysis of two or more groups of biological samples (Fig. 7, *right lower panel, AU—arbitrary units*). The signals may be expressed as fold over basal level, a percentage of the maximal signal intensity values for the respective phosphorylated proteins, a percentage of phosphorylated protein signal intensity to total protein, or as arbitrary units.

Gel loading strategy #4 (Fig. 6b) may be more convenient in case when protein activation is expressed as fold over the basal level, since each strip contains a separated control and inhibitor-

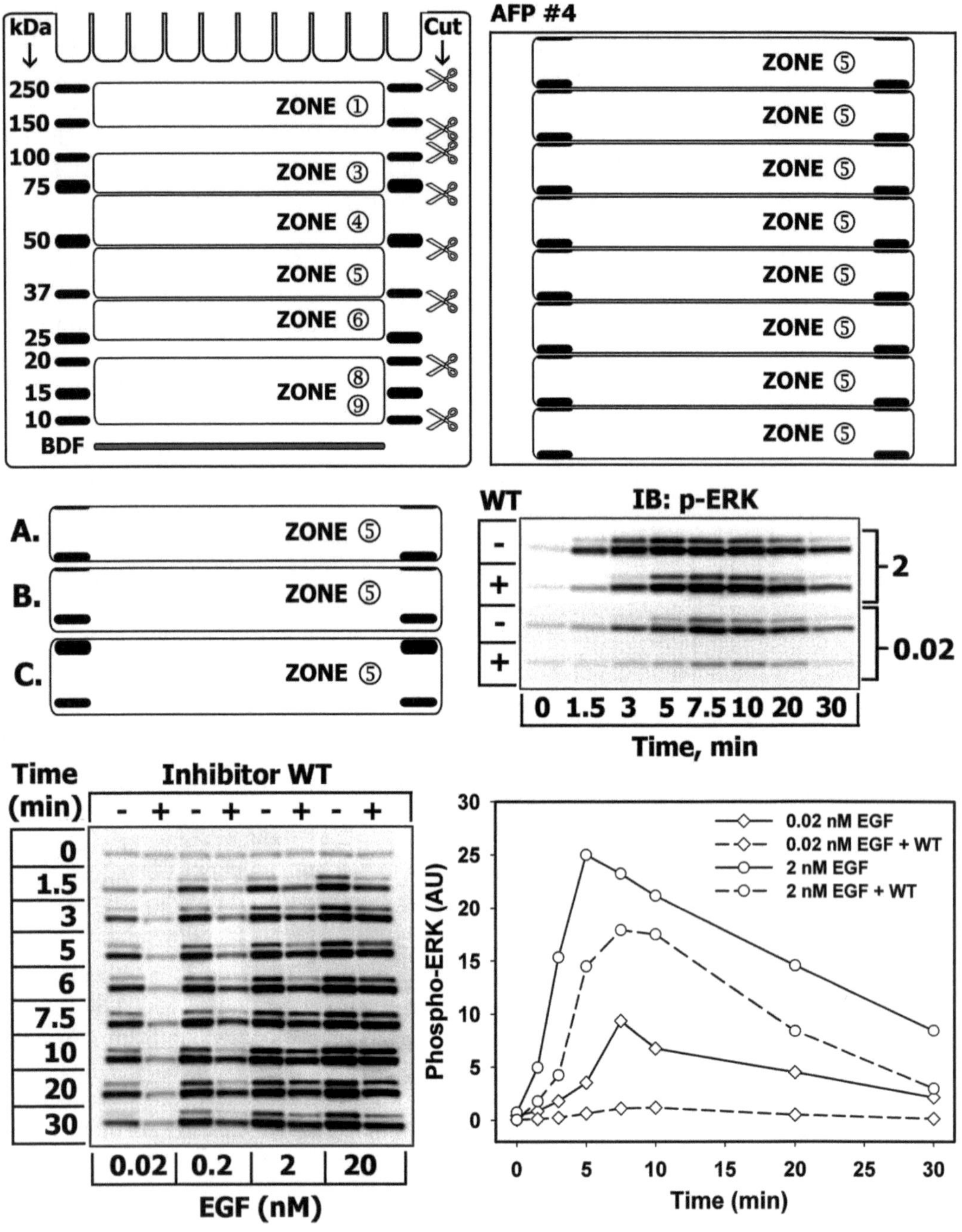

Fig. 7 MSWB strategy for concurrent measurement of EGF-induced ERK1/2 activation kinetics in response to inhibition of PI3/Akt cell survival signaling pathway

treated sample of nonstimulated cells. 0 K and 0 W may be the same sample loaded over and over again throughout all eight gels; biological duplicates (e.g., 0 K_1, 0 K_2 and 0 W_1, 0 W_2) loaded as technical quadruplicates; or biological quadruplicates loaded in technical duplicates (0 K_1, 0 K_2, 0 K_3, 0 K_4 and 0 W_1, 0 W_2, 0 W_3, 0 W_4). Upon Gel loading strategies #1–3 (Fig. 6a, b), 0 K and 0 W can represent the same sample loaded in technical quadruplicate, a biological duplicate loaded in technical duplicate or biological quadruplicate. Their signals later should be averaged, respectively. A repetitive sample loading in 12-well gel (*see 7.5 min in all gels of* Fig. 7, *right panel*) can be avoided by including an extra time-point during the experimental procedure, (e.g., 6 min). However, for visual representation of the blot showing full continuous kinetics, the repetitive region with bands should be merged using Photoshop software. No merging of blot image is needed if the samples are visualized following Gel loading strategies #1 and #2 (Fig. 6a).

Challenge 2. To compare the activation kinetics (e.g., at 0, 5, 15, and 30 min) of EGFR, FAK, Akt, ERK1/2, and Histone 3 (H3) in untreated cells (control, *K*) and in cells that were pretreated with six different agents (*A–F*) prior to stimulation with 5 nM EGF.

Solution. For electrophoresis, the samples were loaded onto six 10-well Novex 4–12 % Bis-Tris gradient gels as shown in left upper panel of Fig. 8. Each out of six gels with separated proteins was cut into six strips containing indicated Zones for proteins of interest as shown in the *right upper panel*. Gel strips were then transferred onto appropriate AFP sheets and horizontally aligned together with identical protein Zone strips derived from the other five gels (*right middle panel*). After protein transfer and blocking steps, the membrane bearing protein Zone ① was probed with 1°Ab against p-EGFR (Y1173) (160 kDa; Cell Signaling #4407 at 1:1,000), membrane with Zone ②③—p-FAK (Y397) (125 kDa; Cell Signaling #8556 at 1:1,000), Zone ④—p-Akt (S473) (60 kDa; Cell Signaling #4051 at 1:1,000), Zone ⑤—p-MEK (S217/221) (Cell Signaling #9154 at 1:1,000), Zone ⑥⑦—Grb2 (25 kDa; Santa Cruz Biotechnology #sc-255 at 1: 500), and Zone ⑧⑨—p-H3 (S10) (17 kDa; Cell Signaling #9701). The resulting blots of Zone ④ (*left lower panel*) and Zone ⑥⑦ (*right lower panel*) after respective protein detection by chemiluminescence are shown to demonstrate the increased throughput, expected final signal/noise ratio, and sample-to-sample reproducibility for comparative study of cell signaling events under perturbed and unperturbed conditions. Please note that half of the first and the last lanes containing resolved MW marker are obscured from the visual field (*left and right lower panels*) by zooming in the CCD camera, because some antibodies tend to cross-react with marker proteins.

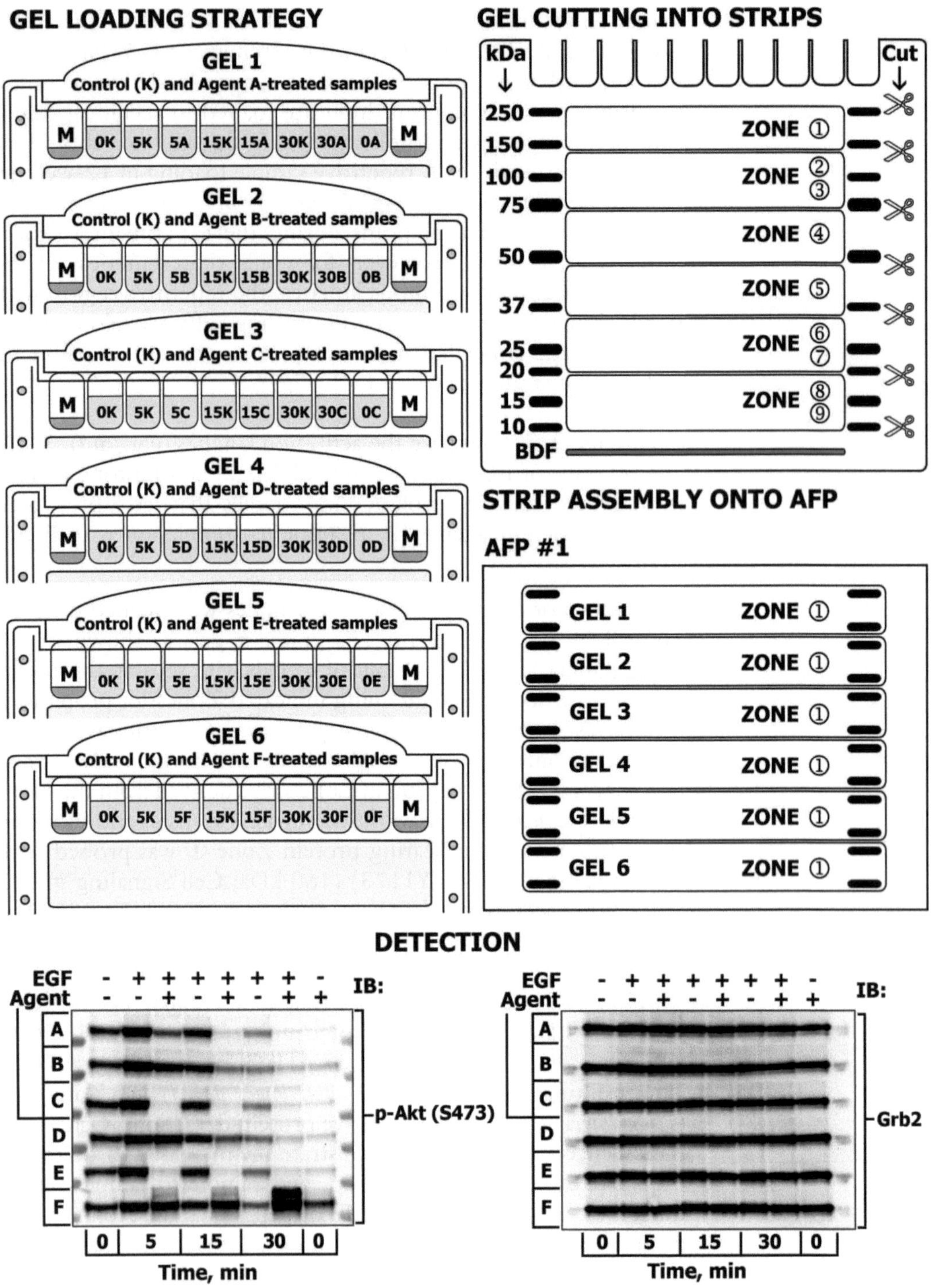

Fig. 8 Major MSWB steps for simultaneous analysis of temporal expression of different proteins engaged in EGFR signaling network

4 Notes

1. Pre-made lysis buffers, containing different detergents and additives that vary in their ability to facilitate cell/tissue solubilization as well as to extract and denature proteins or preserve protein–protein interactions, are commercially available.
2. Laemmli instead of LDS sample buffer can be used with appropriate running and transfer buffers. Electrophoresis can be performed under reducing (add DTT or β-mercaptoethanol) as well as nonreducing (omit adding reducing agent) conditions. Please note that under nonreducing conditions proteins may migrate differently than your MW marker.
3. Choose the type of apparatus for electrophoresis and protein transfer suitable for the size of your gels (e.g., mini-, midi-, or maxi-gels). If iBlot® Dry Blotting System is used, the transfer time can be reduced from 1.30 h to ≤7 min, but it requires further optimization of actual performance by user.
4. This protocol can be adapted for gels of any percentage, composition, size, and number of wells. For instance, lower percentage Bis-Tris or Tris-Acetate gels may be selected for separation of large MW proteins of interest. We routinely use gradient Bis-Tris gels that, when compared to Laemmli system's Tris-Glycine SDS-PAGE gels, have longer shelf-life; their neutral pH environment minimizes protein modifications and significantly delays acrylamide hydrolysis resulting in high-resolution separation of small to mid-size MW protein bands that are sharper and rarely overlap.
5. The researcher may adapt this protocol for various protein MW markers, but it will require further optimization by user. The reference bands of protein MW marker other than that used in this protocol will have different migration patterns, generating a different number of designated protein migration Zones that may also vary in width.
6. The researcher may adapt this protocol for other NuPAGE® Running buffers (MES-SDS and Tris-Acetate-SDS). Be aware that it will change the width and/or quantity of designated protein migration Zones depicted in this procedure and will require further optimization by user.
7. If desired, PVDF or nylon membranes can be also used. Most proteins can be successfully blotted using a 0.45 μm pore size membrane; however, for proteins of low molecular weight or peptides, a 0.2 μm pore size membrane is recommended. We have successfully used 0.2 μm for proteins with MW ranging from 14 to 220 kDa without compromising the efficiency of their transfer onto the membrane.

8. For more flexibility in handling, higher working volumes and higher rotating speeds, we prefer using individual over multi-well square dishes.
9. The samples to be loaded can be different or repetitive.
10. The table is designed to track the statistics of certain protein MW marker migration patterns in the gel of selected percentage under certain buffered conditions. The statistics is required for successive Multistrip Western blotting procedures if one needs to cut out the gel strip(s) containing protein(s) of interest of known MW, but no prestained marker has been loaded onto a gel.
11. The maximal number of gel strips that can be combined onto a single AFP depends on the overall dimension of the transfer unit, hence on the size (length and width) of AFP. Routinely we use *extra-thick narrow*, *medium*, *or wide* filter sheets that provide space for maximum of 10–11 strips of 0.7 cm width each. However, regularly we place fewer amounts of gel strips (e.g., six), especially when they are wider and/or the membrane should be cut into two or more pieces after electrophoretic protein transfer.
12. If some pauses occur, regularly wet the surface of gel strips by dropping dH_2O.
13. Alternatively, the whole piece of nitrocellulose membrane can be treated with blocking reagent and then incubated with the mixture of primary antibodies (be sure that they do not cross-react) in a single dish.
14. The pads should rise at least 0.5 cm over the rim of the cathode core. If not, then insert one or more thin filter paper sheets on the top of a thick filter paper in the tank.
15. 1°Ab can be collected into the 50 mL tube and reused several times if supplemented with 0.05–0.1 % (w/v) sodium azide. If precipitation occurs, filter the solution through 0.22 μm filter.
16. The chemiluminescent signal can be visualized by another imaging instrument and quantified using appropriate software. Alternatively, the signal can be captured on the film followed by densitometric quantification.

Acknowledgments

The authors gratefully acknowledge Dr. Boris N Kholodenko for his support. This work was supported by National Institutes of Health Grants GM59570, AA018873, AA017261, AA007463, and AA022417.

References

1. Soundy P, Harvey B (2005) Western blotting as a diagnostic method. In: Walker JM, Rapley R (eds) Medical biomethods handbook. Humana, Totowa, NJ, pp 43–62
2. Omenn GS (2006) Strategies for plasma proteomic profiling of cancers. Proteomics 6: 5662–5673
3. Hueber W, Robinson WH (2006) Proteomic biomarkers for autoimmune disease. Proteomics 6:4100–4105
4. Fardilha M, Wu W, Sa R, Fidalgo S, Sousa C, Mota C, da Cruz e Silva OA, da Cruz e Silva EF (2004) Alternatively spliced protein variants as potential therapeutic targets for male infertility and contraception. Ann N Y Acad Sci 1030:468–478
5. Ducruet AP, Vogt A, Wipf P, Lazo JS (2005) Dual specificity protein phosphatases: therapeutic targets for cancer and Alzheimer's disease. Annu Rev Pharmacol Toxicol 45: 725–750
6. Gaiger A, Heintel D, Jager U (2004) Novel molecular diagnostic and therapeutic targets in chronic lymphocytic leukaemia. Eur J Clin Invest 34(Suppl 2):25–30
7. Sawyer TK (2004) Cancer metastasis therapeutic targets and drug discovery: emerging small-molecule protein kinase inhibitors. Expert Opin Investig Drugs 13:1–19
8. Krause DS, Van Etten RA (2005) Tyrosine kinases as targets for cancer therapy. N Engl J Med 353:172–187
9. Kholodenko BN (2006) Cell-signalling dynamics in time and space. Nat Rev Mol Cell Biol 7:165–176
10. Kurien BT, Scofield RH (2006) Western blotting. Methods 38:283–293
11. Schilling M, Maiwald T, Bohl S, Kollmann M, Kreutz C, Timmer J, Klingmuller U (2005) Quantitative data generation for systems biology: the impact of randomisation, calibrators and normalisers. Syst Biol (Stevenage) 152:193–200
12. Bergendahl V, Glaser BT, Burgess RR (2003) A fast Western blot procedure improved for quantitative analysis by direct fluorescence labeling of primary antibodies. J Immunol Methods 277:117–125
13. Bolt MW, Mahoney PA (1997) High-efficiency blotting of proteins of diverse sizes following sodium dodecyl sulfate-polyacrylamide gel electrophoresis. Anal Biochem 247:185–192
14. Kashino Y, Koike H, Satoh K (2001) An improved sodium dodecyl sulfate-polyacrylamide gel electrophoresis system for the analysis of membrane protein complexes. Electrophoresis 22:1004–1007
15. Kurien BT, Scofield RH (2002) Heat-mediated, ultra-rapid electrophoretic transfer of high and low molecular weight proteins to nitrocellulose membranes. J Immunol Methods 266:127–133
16. Swank MW, Kumar V, Zhao J, Wu GY (2006) A novel method of loading samples onto minigels for SDS-PAGE: Increased sensitivity and Western blots using sub-microgram quantities of protein. J Neurosci Methods 158:224–33
17. Wu M, Stockley PG, Martin WJ 2nd (2002) An improved western blotting technique effectively reduces background. Electrophoresis 23:2373–2376
18. Schilling M, Maiwald T, Bohl S, Kollmann M, Kreutz C, Timmer J, Klingmuller U (2005) Computational processing and error reduction strategies for standardized quantitative data in biological networks. Febs J 272:6400–6411
19. Aksamitiene E, Hoek JB, Kholodenko B, Kiyatkin A (2007) Multistrip Western blotting to increase quantitative data output. Electrophoresis 28:3163–3173
20. Welinder C, Ekblad L (2011) Coomassie staining as loading control in Western blot analysis. J Proteome Res 10:1416–1419
21. Romero-Calvo I, Ocon B, Martinez-Moya P, Suarez MD, Zarzuelo A, Martinez-Augustin O, de Medina FS (2010) Reversible Ponceau staining as a loading control alternative to actin in Western blots. Anal Biochem 401:318–320
22. Aldridge GM, Podrebarac DM, Greenough WT, Weiler IJ (2008) The use of total protein stains as loading controls: an alternative to high-abundance single-protein controls in

semi-quantitative immunoblotting. J Neurosci Methods 172:250–254

23. Kar P, Agnihotri SK, Sharma A, Sachan R, Lal Bhatt M, Sachdev M (2012) A protocol for stripping and reprobing of Western blots originally developed with colorimetric substrate TMB. Electrophoresis 33:3062–3065
24. Sennepin AD, Charpentier S, Normand T, Sarre C, Legrand A, Mollet LM (2009) Multiple reprobing of Western blots after inactivation of peroxidase activity by its substrate, hydrogen peroxide. Anal Biochem 393:129–131
25. Upadhaya R, Mizunoya W, Anderson JE (2011) Detecting multiple proteins by Western blotting using same-species primary antibodies, precomplexed serum, and hydrogen peroxide. Anal Biochem 419:342–344
26. Suzuki O, Koura M, Noguchi Y, Uchio-Yamada K, Matsuda J (2011) Use of sample mixtures for standard curve creation in quantitative western blots. Exp Anim 60:193–196
27. Dittmer A, Dittmer J (2006) Beta-actin is not a reliable loading control in Western blot analysis. Electrophoresis 27:2844–2845
28. Khimani AH, Mhashilkar AM, Mikulskis A, O'Malley M, Liao J, Golenko EE, Mayer P, Chada S, Killian JB, Lott ST (2005) Housekeeping genes in cancer: normalization of array data. Biotechniques 38:739–745
29. Janssens N, Janicot M, Perera T, Bakker A (2004) Housekeeping genes as internal standards in cancer research. Mol Diagn 8:107–113
30. Caradec J, Sirab N, Revaud D, Keumeugni C, Loric S (2010) Is GAPDH a relevant housekeeping gene for normalisation in colorectal cancer experiments? Br J Cancer 103:1475–1476
31. Waxman S, Wurmbach E (2007) De-regulation of common housekeeping genes in hepatocellular carcinoma. BMC Genomics 8:243
32. Rubie C, Kempf K, Hans J, Su T, Tilton B, Georg T, Brittner B, Ludwig B, Schilling M (2005) Housekeeping gene variability in normal and cancerous colorectal, pancreatic, esophageal, gastric and hepatic tissues. Mol Cell Probes 19:101–109
33. Sikand K, Singh J, Ebron JS, Shukla GC (2012) Housekeeping gene selection advisory: glyceraldehyde-3-phosphate dehydrogenase (GAPDH) and beta-actin are targets of miR-644a. PLoS One 7:e47510
34. Ferguson RE, Carroll HP, Harris A, Maher ER, Selby PJ, Banks RE (2005) Housekeeping proteins: a preliminary study illustrating some limitations as useful references in protein expression studies. Proteomics 5:566–571
35. Taylor SC, Berkelman T, Yadav G, Hammond M (2013) A defined methodology for reliable quantification of Western blot data. Mol Biotechnol 55:217–226
36. Aksamitiene E, Achanta S, Kolch W, Kholodenko BN, Hoek JB, Kiyatkin A (2011) Prolactin-stimulated activation of ERK1/2 mitogen-activated protein kinases is controlled by PI3-kinase/Rac/PAK signaling pathway in breast cancer cells. Cell Signal 23:1794–1805
37. Simpson RJ (2010) Pouring linear gradient gels with a gradient former. Cold Spring Harbor Protoc. 2010: pdb prot5411

Chapter 24

Western Blotting Using PVDF Membranes and Its Downstream Applications

Setsuko Komatsu

Abstract

Western blotting using polyvinylidene difluoride (PVDF) membranes is one of the most popular techniques for detection and characterization of proteins. If this technique is combined with immunodetection, the behavior of a particular protein can be clarified. On the other hand, if it is combined with Edman sequencing, the primary structure of the protein can be determined. A protein sample is transferred from an SDS-polyacrylamide gel electrophoresis (PAGE) gel onto a PVDF membrane by electroblotting. The membrane carrying the protein is either used for immunodetection or protein sequencing. SDS-PAGE followed by Western blotting combined with immunodetection using antibodies can easily detect protein behavior in crude protein mixtures. Furthermore, two-dimensional PAGE followed by Western blotting and Edman sequencing allows effective sequence determination of crude protein mixtures that may not be easily purified by conventional column chromatography.

Key words Western blotting, Edman sequencing, PVDF membrane, Immunodetection, Deblocking, Cleveland method

1 Introduction

Western blotting is a widely accepted analytical technique used to detect and identify specific proteins in a crude protein extract. Gel electrophoresis is first used to separate structural or denatured proteins. The proteins are transferred to a nitrocellulose or polyvinylidene difluoride (PVDF) membrane, where they are allowed to react with an antibody specific to the target protein, which often involves staining for identification. Although methods without an electrophoresis step have been improved, Western blotting based on an immunoreaction is widely used in the fields of molecular biology, biochemistry, immunogenetics and related disciplines.

Edman sequencing of proteins separated on two-dimensional polyacrylamide gel electrophoresis (2D-PAGE) gels became possible with the introduction of protein electroblotting methods that allow efficient transfer of sample from the gel matrix onto

Biji T. Kurien and R. Hal Scofield (eds.), *Western Blotting: Methods and Protocols*, Methods in Molecular Biology, vol. 1312, DOI 10.1007/978-1-4939-2694-7_24, © Springer Science+Business Media New York 2015

supports suitable for gas-phase sequencing or related techniques [1]. Picomole amounts of protein are first separated by 2D-PAGE [2], and then electroblotted from 2D-PAGE gels onto a PVDF membrane. The amino acid sequence of the electroblotted protein is determined by Edman sequencing. Direct N-terminal sequencing is the most sensitive method (1–5 μg of protein), but when gaps or ambiguous assignments are seen, verification of the sequence by other means often demands much more material. Proteins are often posttranslationally modified, and N-terminal blockage is one of the more common posttranslational modifications. Proteins can become N-terminally blocked not only in vivo but also in vitro. However, it is possible to prevent in vitro blocking, which is caused during protein extraction, 2D-PAGE, and blotting. The use of very pure reagents during these procedures, the addition of thioglycolic acid as a free radical scavenger to the extraction buffer, electrophoresis, and electroblotting buffers, and preelectrophoresis to remove free radicals from the gel may all be effective in preventing in vitro blocking [3]. However, if proteins are blocked in vivo, a chemical or enzymatic deblocking procedure or peptide mapping procedure is required to determine the N-terminal or internal sequence.

2 Materials

1. Sodium dodecyl sulfate (SDS) sample buffer: 60 mM Tris–HCl (pH 6.8), 2 % SDS, 10 % glycerol, and 5 % β-mercaptoethanol [4].
2. Acrylamide for the separating gel (acrylamide/bis-acrylamide = 30:0.135): 30.00 g acrylamide, 0.135 g bis-acrylamide. Make volume to 100 mL with Milli-Q water, and keep in the dark (brown bottle).
3. Separating gel buffer (pH 8.8): 12.11 g Tris–HCl to a 1 M final concentration, 0.27 g SDS to a 0.27 % final concentration. Dissolve in 80 mL Milli-Q water, adjust pH to 8.8, and make the volume to 100 mL.
4. Acrylamide for stacking gel (acrylamide/bis-acrylamide = 29.2:0.8): 29.2 g acrylamide, 0.8 g bis-acrylamide. Make volume to 100 mL with Milli-Q water, and keep in the dark (brown bottle).
5. Stacking gel buffer (pH 6.8): 3.03 g Tris–HCl to a 0.25 M final concentration, 0.20 g SDS to a 0.2 % final concentration. Dissolve in 80 mL Milli-Q water, adjust the pH to 6.8, and make the volume to 100 mL.
6. SDS-PAGE running buffer: 9 g Tris–HCl, 43.2 g glycine, and 3 g SDS. Dissolve in 3 L Milli-Q water.
7. Bromophenol blue (BPB) solution: Dissolve 0.1 g BPB and 10 g glycerol in 100 mL Milli-Q water.

8. Blotting buffer A: 36.33 g Tris–HCl to a 0.3 M final concentration, 200 mL methanol to a 20 % final concentration, and 0.20 g SDS to a 0.02 % final concentration. Make the volume to 1 L with Milli-Q water, and keep at 4 °C.
9. Blotting buffer B: 3.03 g Tris–HCl to a 25 mM final concentration, 200 mL methanol to a 20 % final concentration, and 0.20 g SDS to a 0.02 % final concentration. Make the volume to 1 L with Milli-Q water, and keep at 4 °C.
10. Blotting buffer C: 3.03 g Tris–HCl to a 25 mM final concentration, 5.20 g ε-aminocaproic acid to a 40 mM final concentration, 200 mL methanol to a 20 % final concentration, and 0.20 g SDS to a 0.02 % final concentration. Make the volume to 1 L with Milli-Q water, and keep at 4 °C.
11. Separating gel solution (amounts are for one gel [18 %]): 10 mL acrylamide for separating gel, 6.3 mL separating gel buffer (pH 8.8), 120 μL 10 % ammonium persulfate (APS), and 20 μL TEMED.
12. Stacking gel solution (amounts are for one gel [5 %]): 1 mL acrylamide for stacking gel, 3 mL stacking gel buffer (pH 6.8), 2 mL Milli-Q water, 30 μL 10 % APS, and 20 μL TEMED.
13. Hydration buffer: Make this buffer with 8 M urea, 2 % w/v CHAPS, 50 mM DTT, 0.2 % ampholyte (pH 3.5–10) and a trace of 0.001 % BPB.

3 Methods

3.1 Two-Dimensional Polyacrylamide Gel Electrophoresis

For 2D-PAGE, there are two options for the separation in the first dimension: immobilized pH gradient (IPG) strips or isoelectric focusing tubes. The first option is described here.

3.1.1 Separation in Immobilized pH Gradient Strips as the First Dimension

1. Use a nonlinear IPG strip (p*I* 3.5–10.0, 18 cm) for separation in the first dimension. It offers high resolution, great reproducibility and allows high protein loads.
2. Hydrate the strips overnight in the preswelling cassette with 25 mL hydration buffer.
3. When the rehydration cassette is thoroughly emptied and opened, transfer the strips to the strip tray. After placing IPG strips, humid electrode wicks, electrodes, and sample cups in position, cover the strips and cups with low-viscosity paraffin oil. Apply samples slowly at the cathode end of the IPG strips and continue this manner without touching the strip gels.
4. Increase the voltage linearly from 300 to 3,500 V over 3 h, followed by three additional hours at 3,500 V, and then increase it to 5,000 V. Total volt-h will be 8–10,000 V h.
5. After separation using IPG strips, perform SDS-PAGE in the second dimension.

3.1.2 SDS-PAGE as the Second Dimension

1. Clip together two glass plates (100 × 140 × 1 mm) with a clip, keeping a 1-mm space between the plates.
2. Prepare separating gel solution in a 100-mL beaker. Mix the solutions and fill the plates to about 2 cm from the top. (Caution: Pour the solutions into the plates immediately after adding 10 % APS and TEMED.)
3. Overlay the separating gel solution with 1 mL Milli-Q water.
4. Leave the gel for 40–60 min at room temperature for polymerization.
5. Remove the overlaid water.
6. Prepare the stacking gel solution in a 100-mL beaker. Mix well and pour on the separating gel.
7. Leave the gel for 20 min at room temperature for polymerization.
8. Apply the first-dimension IPG strip directly on the top of the stacking gel. Overlay the first dimension strip with 1 % agarose.
9. Assemble the slab gel for electrophoresis. Add a few drops of BPB in the SDS-PAGE running buffer.
10. Run the sample at 35 mA (constant current) until the tracking dye reaches the bottom of the separating gel.
11. Separate the stacking gel using cutter, and take out the separation gel for the next step.

3.2 Cleveland Peptide Mapping [5]

First separate the samples by 2D-PAGE. Then stain the gels with Coomassie brilliant blue (CBB), and remove the gel pieces (5–20 pieces) containing protein spots and soak them for 1 h in Milli-Q water in a 2 mL microcentrifuge tube. Remove the Milli-Q water and add 750 μL electroelution buffer. Electroelute the protein from the gel pieces using an electrophoretic concentrator run at 2 W constant power for 2 h. After electroelution, dialyze the protein solution against Milli-Q water for 48 h and lyophilize them.

3.2.1 Peptide Mapping Protocol

1. Cut out stained protein spots from 2D gels and soak them in Milli-Q water for 1 h.
2. Fill the 2-mL Eppendorf tube containing the protein spots (5–20 gel pieces) with 750 μL electroelution buffer. Shake for 30 min.
3. Cut seamless cellulose tubing (small size, no. 24, Wako, Osaka, Japan) into pieces 12- to 15-cm long, as space is needed for clipping. Fill a 300 mL beaker with 250 mL Milli-Q water, boil it for 5 min, and keep the tubing membrane in it after boiling. Wet small piece of cellophane film in a small beaker with Milli-Q water.

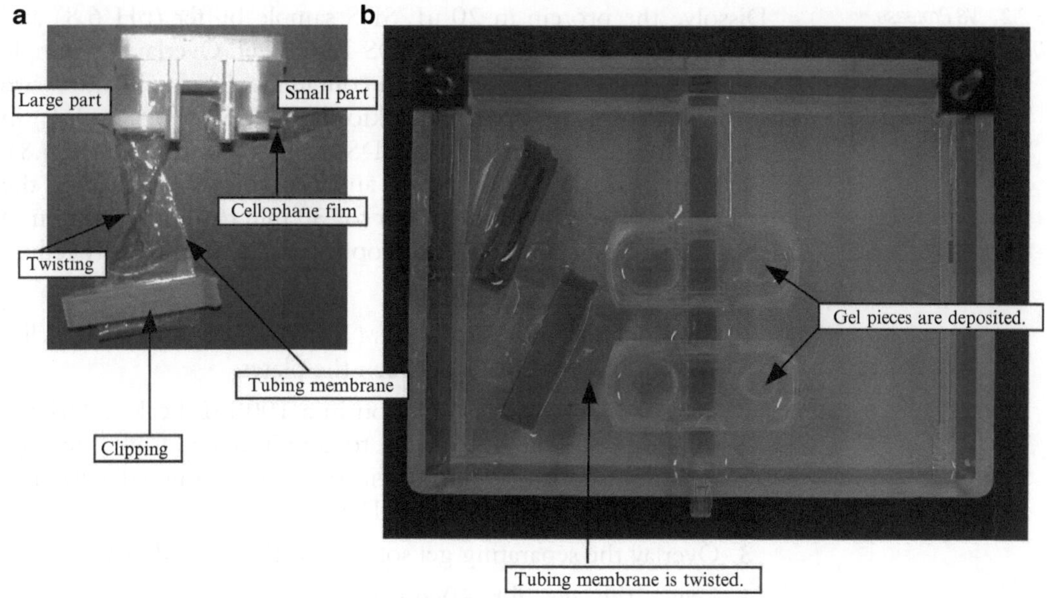

Fig. 1 Electrophoretic concentrator. (**a**) Close-up of the cup with long seamless cellulose tubing. (**b**) The cups fixed to the electrophoretic concentrator. (Reproduced from Komatsu [7] with permission from Springer)

4. Close the bottom of the small part of the cup of electrophoretic concentrator with cellophane film, open the bottom of the large part of the cup, connect and twist the tubing membrane, and close the distal end by clipping (Fig. 1a).
5. Fix the cup in the electrophoretic concentrator (Nippon-Eido, Tokyo, Japan) (Fig. 1b). Deposit gel pieces containing proteins on the cellophane film (the small part of the cup), and add 750 μL of electroelution buffer from the Eppendorf tube. Fill the small part of the cup with electroelution buffer, and then fill the large part of the cup with electroelution buffer in such a way that a layer of buffer joins both parts of the cup, allowing movement of protein from the small part of the cup to the tubing membrane. Fill the apparatus with electroelution buffer. The small part of the cup containing the protein spots should be toward the positive side.
6. Run at 2 W constant power for 2 h.
7. Remove the tubing membrane and clip to close the end. Dialyze in a cold room (4 °C). Change the Milli-Q water three times the first day. The next day, change the Milli-Q water two times.
8. Transfer the protein solution to two to six 2 mL microcentrifuge tubes. Freeze-dry overnight.
9. Dissolve the protein in 20 μL SDS sample buffer.

3.2.2 V8 Protease Digestion

Dissolve the protein in 20 μL SDS sample buffer (pH 6.8) and apply it to a sample well of an SDS-PAGE gel. Overlay the sample solution with 20 μL of a solution containing 10 μL *Staphylococcus aureus* V8 protease (Pierce, Rockford, IL, USA) (0.1 μg/μL) in Milli-Q water and 10 μL SDS sample buffer (pH 6.8). Electrophorese until the sample and protease are stacked in the stacking gel. Switch off the power for 30 min to allow digestion of the protein, and continue electrophoresis. A detailed procedure follows.

1. Clip together two glass plates (100 × 140 × 1 mm) with a clip, keeping a 1-mm space between the plates.
2. Prepare separating gel solution in a 100-mL beaker. Mix the solutions and fill the plates to about 3 cm from the top. (Caution: Pour the solutions into the plates immediately after adding 10 % APS and TEMED.)
3. Overlay the separating gel solution with 1 mL Milli-Q water.
4. Leave the gel for 40–60 min at room temperature for polymerization.
5. Remove the overlaid water.
6. Prepare the stacking gel solution in a 100-mL beaker. Mix well, pour on the separating gel, and insert comb.
7. Leave the gel for 20 min at room temperature for polymerization.
8. Take out the comb, clips, and silicon tubes as spacer.
9. Clean the wells using Milli-Q water with a syringe.
10. Insert the gel plates into the apparatus. Pour SDS-PAGE running buffer into the apparatus.
11. Dissolve the protein in 20 μL SDS sample buffer (pH 6.8), and apply to a sample well of the SDS-PAGE gel. Overlay the sample with 20 μL of a solution containing 10 μL *Staphylococcus aureus* V8 protease at 1 μg/μL in Milli-Q water and 10 μL SDS sample buffer (pH 6.8). Add 30 μL BPB solution.
12. Electrophorese until the sample and protease are stacked in the upper gel and then interrupt the run for 30 min to digest the protein.
13. Run the gel at 35 mA until the BPB line reaches about 5 mm from the bottom.
14. Disconnect the electrical leads, and take out the plates.
15. Separate the two plates with a spatula.
16. Separate the stacking gel, and take out the separating gel for the next step.

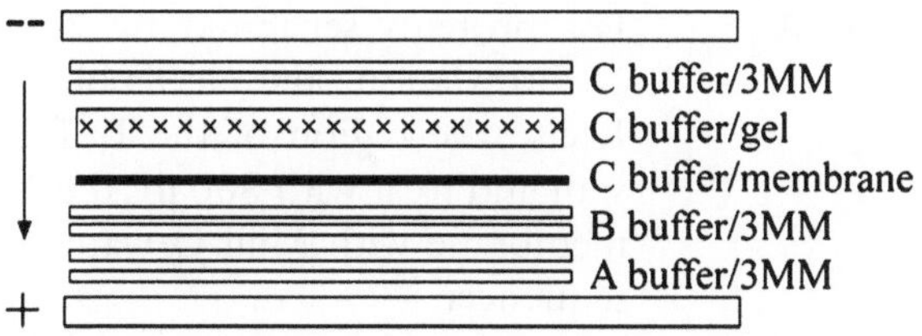

Fig. 2 Western blotting. Following separation by 2D-PAGE or by the Cleveland method, the proteins in the gel are electroblotted onto a PVDF membrane using a semidry transfer blotter. Layers of Whatman 3MM filter paper are wetted in A, B and C blotting buffer, and gel and PVDF membrane are wetted in C blotting buffer. (Reproduced from Komatsu [7] with permission from Springer)

3.3 Western Blotting

Following separation by 2D-PAGE or by the Cleveland method, electroblot the proteins onto a PVDF membrane (Fig. 2) using a semidry transfer blotter, and detect the proteins by CBB staining.

1. Cut the PVDF membrane to a size equal to that of the gel.
2. Cut Whatman 3MM filter paper to a size equal to that of the gel.
3. Wash the PVDF membrane in methanol for a few seconds, and transfer the membrane to 100 mL blotting buffer C, and shake for 5 min.
4. Wet two sheets of Whatman 3MM filter paper in blotting buffer A, B or C (blotting paper A, B or C).
5. Place the separating gel in 100 mL blotting buffer C and shake for 5 min.
6. Wet the semidry transfer blotter with Milli-Q water. Place blotting paper A on the blotting plate followed by blotting paper B. Remove air bubbles, if any. Place the PVDF membrane on the plate followed by the gel and blotting paper C.
7. Connect the power supply. Run the blot at 1 mA/cm^2 for 90 min.
8. Wash the PVDF membrane in 100 mL Milli-Q water.
9. Stain the PVDF membrane for 2–3 min in CBB.
10. Destain the PVDF membrane in 60 % methanol for 3 min twice.
11. Wash with Milli-Q water and air-dry at room temperature.

3.4 Deblocking of Blotted Proteins [6] (See Note 1)

3.4.1 Acetylserine and Acetylthreonine

For proteins separated by 2D-PAGE that have an acetylserine or acetylthreonine block at their N-termini, first electroblot the gel onto a PVDF membrane. Excise the region of the PVDF membrane carrying the protein spot and treat with trifluoroacetic acid at 60 °C for 30 min. These samples can then be sequenced directly (*see* **Note 2**).

3.4.2 Formyl Group

For proteins separated by 2D-PAGE that are blocked due to N-formylation, first electroblot the gel onto a PVDF membrane. Excise the region of the PVDF membrane carrying the protein spot and treat with 300 μL 0.6 M HCl at 25 °C for 24 h. Wash the membrane with Milli-Q water, dry it, and apply it to the protein sequencer.

3.4.3 Pyroglutamic Acid

The following procedure will remove pyroglutamic acid from the N-termini of proteins.

1. After separation by 2D-PAGE, electroblot proteins with pyroglutamic acid at their N-termini onto a PVDF membrane.
2. Excise the region of the PVDF membrane carrying the protein spot and treat with 200 μL 0.5 % (w/v) polyvinylpyrrolidone-40 in 100 mM acetic acid at 37 °C for 30 min (*see* **Note 3**).
3. Wash the PVDF membrane at least ten times with 1 mL Milli-Q water.
4. Soak the PVDF membrane in 100 μL 0.1 M phosphate buffer (pH 8) containing 5 mM dithiothreitol and 10 mM EDTA.
5. Add pyroglutamyl peptidase (5 μg), and incubate the reaction solution at 30 °C for 24 h.
6. Wash the PVDF membrane with Milli-Q water, dry it, and apply it to the protein sequencer [6].

3.5 N-Terminal and Internal Amino Acid Sequence Analysis and Homology Search of Amino Acid Sequence

1. Excise the stained protein spots or bands from the PVDF membrane and apply to the upper glass block of the reaction chamber of a gas-phase protein sequencer, such as Procise 494 or cLC (Applied Biosystems, Foster City, CA, USA) or PPSQ (Shimazu, Kyoto, Japan). Perform Edman degradation according to the standard program supplied by Applied Biosystems or Shimazu. Separate the released phenylthiohydantoin amino acids by an online high-performance liquid chromatography system and identify them by retention time.
2. Compare the amino acid sequences obtained with those of known proteins in the Swiss-Prot, PIR, GenPept, and PDB databases with the web-accessible search program FastA.

3.6 Immune Reaction with Antibody

3.6.1 Blocking and Incubating

1. Incubate the membrane with blocking buffer on a shaker for 1–2 h at 37 °C or overnight at 4 °C.
2. Dilute primary antibody with primary antibody dilution buffer and incubate the membrane with the diluted primary antibody on a shaker for 1 h at 37 °C or overnight at 4 °C.
3. Wash the membrane four times with washing buffer on the shaker for 10 min each time.
4. Dilute secondary antibody with blocking buffer and incubate the membrane with the diluted secondary antibody conjugated

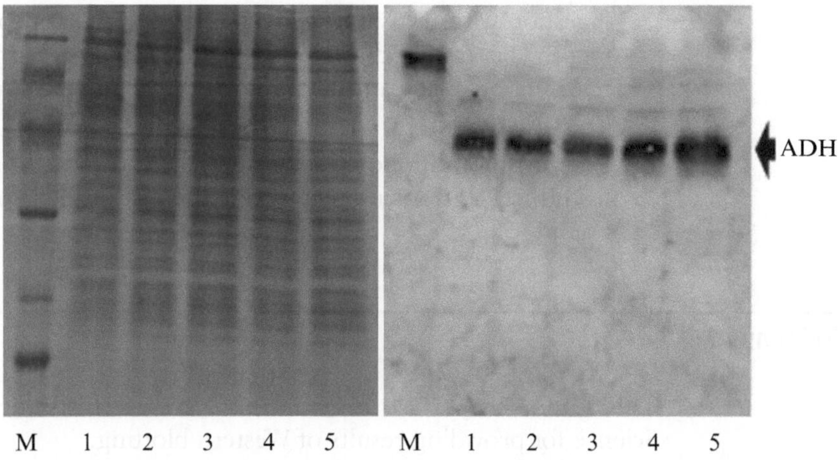

Fig. 3 Example of immune blotting. Proteins were extracted from soybean, separated by SDS-PAGE, Western blotted, incubated with anti-alcohol dehydrogenase antibody, and detected with the ECL method. CBB staining was used to control for loading

with horseradish peroxidase (HRP) on a shaker for 1 h at 37 °C or overnight at 4 °C.

5. Wash the membrane four times with washing buffer on the shaker for 10 min each time.

3.6.2 Detection of HRP-Conjugated Secondary Antibody

1. Detect protein with an ECL kit. In a separate tube, mix black and white ECL solutions in a 1:1 ratio.
2. Aliquot solution onto membranes and wait for 1 min. Drain the ECL, wrap the membrane in plastic and expose it to film. Expose the blots for 10 s, 1 min, 5 min, and more to visualize the chemiluminescence signal that corresponds to the specific antibody-antigen reaction (Fig. 3).

4 Notes

1. These deblocking techniques may be combined to allow the sequential deblocking and sequencing of unknown proteins that have been immobilized onto PVDF membranes. A protein on the PVDF membrane can be directly used for gas-phase sequencing. If sequencing fails at this step, remove the PVDF membrane from the sequencer, remove the acetyl group, the formyl group, and then the pyroglutamic group.
2. The advantage of this method is that deblocking is easy and rapid, although overall sequencing yields obtained by this procedure are low compared with acylamino acid-releasing enzyme digestion. N-acetylated proteins can be enzymatically deblocked with acylamino acid-releasing enzyme after on-membrane

digestion with trypsin to generate an N-terminal peptide fragment. This tryptic digestion is required since acylamino acid-releasing enzyme can only remove the acylamino acid from a short *peptide* [6].

3. Polyvinylpyrrolidone-40 is used to unbind pyroglutamic acid from the PVDF membrane, while the rest of the protein stays bound to it.

Acknowledgement

The author thanks Mr. Yin at the National Institute of Crop Science for providing results of Western blotting.

References

1. Eckerskorn C, Mewes W, Goretzki H, Lottspeich F (1988) A new siliconized-gas fiber as support for protein-chemical analysis of electroblotted proteins. Eur J Biochem 176:509–512
2. O'Farrell PF (1975) High resolution two-dimensional electrophoresis of proteins. J Biol Chem 250:4007–4021
3. Hirano H, Komatsu S, Kajiwara H, Takagi Y, Tsunasawa S (1993) Microsequence analysis of the N-terminally blocked proteins immobilized on polyvinylidene difluoride membrane by Western blotting. Electrophoresis 14:839–846
4. Laemmli UK (1970) Cleavage of structural proteins during the assembly of the head of bacteriophage T4. Nature 227:680–685
5. Cleveland DW, Fischer SG, Krischer MW, Laemmli UK (1977) Peptide mapping by limited proteolysis in sodium dodecyl sulphate and analysis by gel electrophoresis. J Biol Chem 252:1102–1106
6. Hirano H, Komatsu S, Nakamura A et al (1991) Structural homology between semidwarfism-related proteins and glutelin seed protein in rice (Oryza sativa L.). Theor Appl Genet 83: 153–158
7. Komatsu S (2009) Western blotting/Edman sequencing using PVDF membrane. In: Kurien BT, Scofield RH (eds) Protein blotting and detection, vol 536, Methods in molecular biology. Springer, New York, pp 163–171

Chapter 25

Blotting from PhastGel to Membranes by Ultrasound

Joseph Kost and Aharon Azagury

Abstract

Ultrasound based approach for enhanced protein blotting is proposed. Three minutes of ultrasound exposure (1 MHz, 2.5 W/cm^2) was sufficient for a clear transfer of proteins from a polyacrylamide gel (PhastGel) to nitrocellulose or Nylon 66 Biotrans membrane. The proteins evaluated were prestained sodium dodecyl sulfate-polyacrylamide standards (18,500–106,000 Da) and 14C-labeled Rainbow protein molecular weight markers (14,300–200,000 Da).

Key words Ultrasound, Protein blotting, Mass transfer, PhastGel, Polyacrylamide gel, Nitrocellulose membranes, Nylon 66 Biotrans membranes

1 Introduction

1.1 Ultrasound

Ultrasound, or sound of frequency higher than 20 kHz, is inaudible to the human ear. Irradiation with ultrasound is widely used in medical applications (imaging, drug delivery systems, and gene therapy) [1–3], sonochemical processing [4], ultrasonic cleaning of surfaces [5], ultrasonic scalpels and osteotome for surgery and bone removal [6, 7], and in food-processing applications [8].

In the last decade, two new approaches for utilizing ultrasound have been developed. The first approach termed high intensity focused ultrasound (HIFU) is a system that uses multiple small piezoelectric crystals. Piezoelectric crystals can generate an electrical charge resulting from an applied mechanical force and vice versa, they can generate a mechanical strain resulting from an applied electrical field. In HIFU system, these multiple crystals are arranged and used in a manner that enables the focusing of ultrasound beams to a desired specific focal zone without harming the tissues on the way. This enables the use of ultrasound on internal organs.

Enhanced membrane permeation (a phonophoretic effect) of ultrasound on cells has been widely reported [9]. Ultrasound has been used successfully to induce transfer of genetic material into live animal [10–12] and plant cells [13]. Ultrasound has been shown to

Biji T. Kurien and R. Hal Scofield (eds.), *Western Blotting: Methods and Protocols*, Methods in Molecular Biology, vol. 1312, DOI 10.1007/978-1-4939-2694-7_25,

facilitate the delivery of drugs from liposomes [14, 15] polymers [16], across the skin [17] as well as enhance the transdermal noninvasive extraction for continuous detection of glucose [18]. In addition, ultrasound has been shown to enhance the mass transport across other biological membranes such as the tympanic membrane, buccal mucosa, nasal membrane, and even the blood retinal barrier [19–23]. At sufficiently high acoustic power inputs, ultrasound is known to rupture cells and ultrasonication is a well-established laboratory technique of cell disruption [24].

1.2 Mass Transport Enhancement

Ultrasound has the potential for enhancing mass transfer within a cell and across membranes. At certain ultrasound intensities, intracellular microstreaming has been reported inside animal and plant cells [25]. Similarly, rotation of organelles and induced circulation within vacuoles of plant cells have been associated with ultrasound [26].

Ultrasound-enhanced diffusion of nutrients through gels has been used to explain improved dehydrogenation of hydrocortisone by gel-entrapped cells of *Arthrobacter simplex* [27]. Ultrasound has also been shown to increase the concentration of traces of metal ions in anodic stripping voltammetry (ASV) [28].

Ultrasound can induce several effects that in turn may enhance mass transport. These effects include: cavitation (the main effect attributed to enhancement effect in mass transport), thermal effect, acoustic streaming, shock waves and mechanical strains [an effect termed bilayer sonophoresis [29]].

Many reviews have been published summarizing ultrasound enhancing effect on mass transport of various biological agents across membranes (predominantly, the skin) both in vitro, in vivo and in clinical studies [30–33]. For example, ultrasound was found to increase drug release from non-erodible polymers due to the contribution of a convective term, generated by cavitation, without any destructive effect on morphology of the polymer [34]. In another research, the effect of ultrasound on drug release from hydrogels was evaluated. It was found that ultrasound (20 kHz, 4 W/cm^2 and 25 % duty cycle) increases the release rate of a drug at different loadings from ethylene-vinyl acetate copolymer (EVAc) matrices by a factor of 50–300 [35]. Since mass transfer can be a significant limitation in many bioprocessing situations [36], mass transfer enhancing effect of ultrasound has many potential applications.

1.3 Protein Blotting

We previously proposed the application of ultrasound in order to accelerate protein blotting [37, 38]. The very pronounced effect of ultrasound on blotting performance is shown in Fig. 1. Three minutes of ultrasound exposure (1 MHz, 2.5 W/cm^2) was sufficient for a very clear transfer of prestained SDS-PAGE standards from

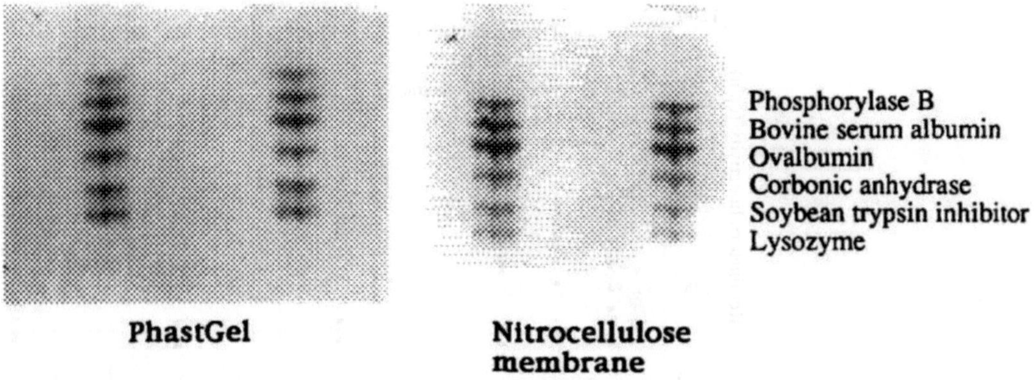

Fig. 1 Low-range SDS-PAGE standards were electrophoretically loaded on polyacrylamide gels (PhastGel) and ultrasonically (3 min in TB-1) transferred to nitrocellulose (Trans-Blot) membrane [19]

a polyacrylamide gel (PhastGel) to nitrocellulose (NC) or nylon 66 (BR) membranes. In control experiments, which were performed following similar procedures without turning the ultrasonic generator on, no protein blotting could be seen. For comparable blotting results, 30 min for electroelution or 240 min for the elution by convection blotting procedure was required.

The kinetics of protein transfer from gel to BR or NC membranes, enhanced by (a) ultrasound, (b) electroelution or (c) convection are shown in Fig. 2. As seen, the highest rates of protein transfer from gels to membranes were for systems exposed to ultrasound. The proteins blotted covered a span of molecular weights from 14,300 to 200,000.

The fraction of protein retained in the gel and the fraction transferred from the gel to NC or BR membranes for the different blotting procedures are displayed in Figs. 3 and 4. At least 85 % of the initial protein amount loaded on the gel was transferred to the membranes when blotted by ultrasound. We could not detect differential efficiencies between transfers of high and low molecular weight proteins. For ultrasonic transport, the highest fraction of proteins was eluted from PhastGels and retained on the membranes, while the fractions of protein lost during blotting were much smaller than those lost during the electrophoretic or convectional methods (Fig. 5). The lower fraction of protein lost during ultrasonic transfer might be due to the smaller volumes of transfer buffer required for this procedure relative to those required for electroelution or convectional blotting.

The mechanism by which ultrasound enhances the transfer of BSA from Phastgel to membranes is unclear but few assumptions may be deduced relying on the ultrasound conditions used in this application. First, denaturation of BSA is not optional due to the low ultrasound intensity used (2.5 W/cm^2) and since

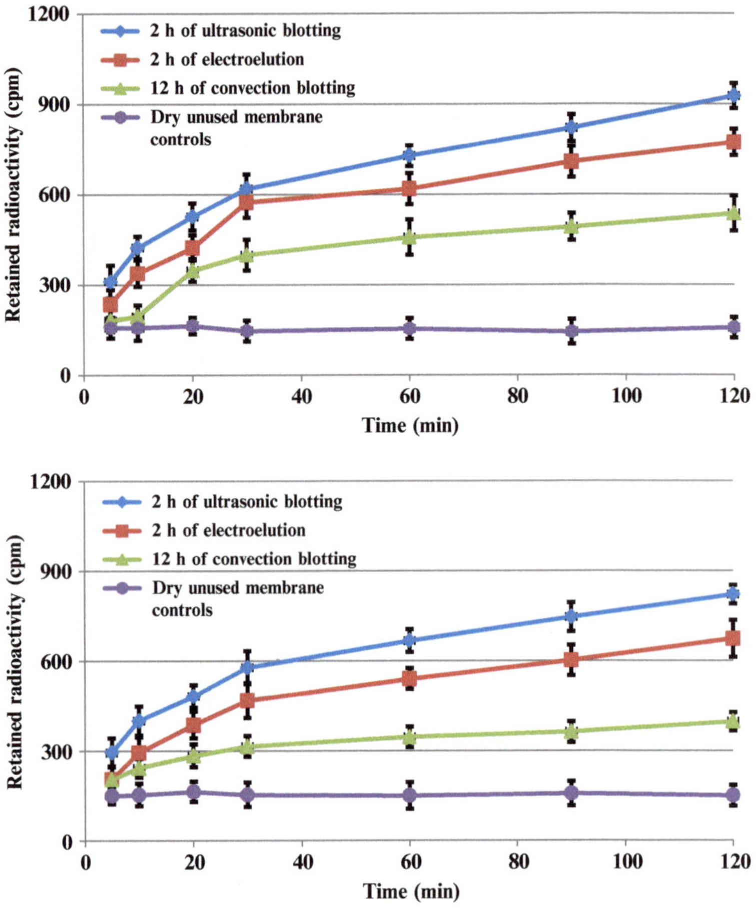

Fig. 2 Retained radioactivity of 14C-methylated protein markers vs. blotting time for 2 h of ultrasonic blotting (*open circles*), 2 h of electroelution (*triangles*), 12 h of convection blotting (*rectangular*) and dry unused membrane controls (*filled circles*) in (**a**) nylon 66 (BR) membranes and (**b**) nitrocellulose (NC) membranes. The transfer solution in all experiments was TB-1 [19]

BSA denaturation begins in temperatures above 55 °C [39]. Although cavitation effect is mainly considered to be responsible for enhanced mass transport, it is reasonable to assume that this is not the case since the intensity combined with frequency used is below the cavitation threshold (the higher the frequency is the higher the threshold is), not to mention the higher viscosity of Phastgels and membranes compared to aqueous solutions [the higher the viscosity is the higher the threshold is [40]]. Thus, the increased transfer rate can probably be attributed to the enhanced convection resulting from ultrasound effects such as acoustic streaming.

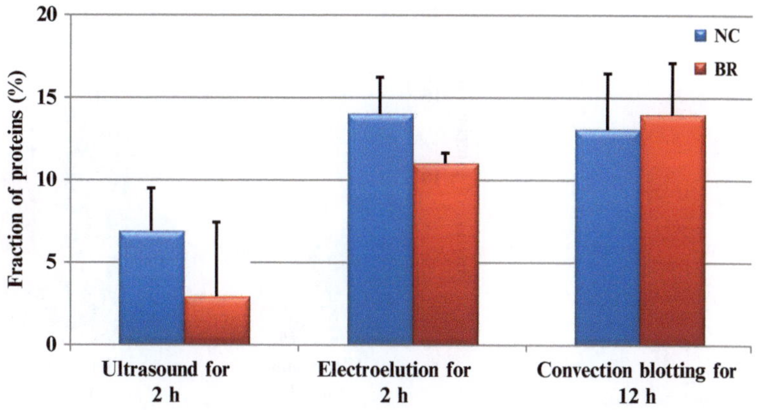

Fig. 3 Fraction of proteins, out of the initial amount loaded to the PhastGel, transferred and retained in BR and NC membranes, following blotting in transfer solution TB-1 by ultrasound for 2 h (*vertical lines*), electroelution (*open bars*) and 12 h of convection blotting (*tilted lines*) [19]

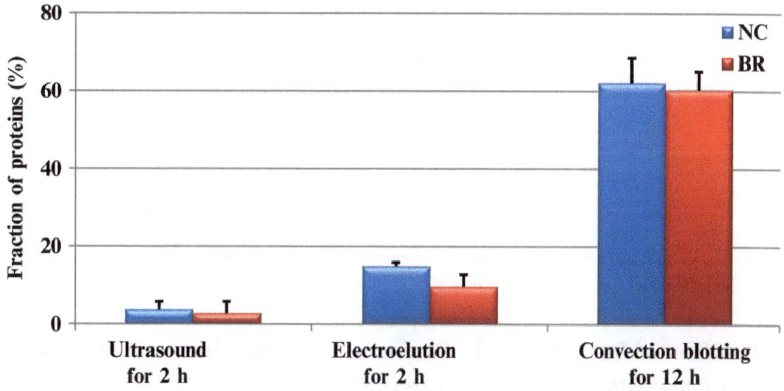

Fig. 4 Fraction of proteins, out of the initial amounts loaded to the PhastGel, retained in PhastGel following blotting in transfer solution TB-1 by ultrasound for 2 h (*vertical lines*), electroelution for 2 h (*open bars*), and 12 h of convection blotting (*tilted lines*) [19]

2 Materials

1. Transfer buffer –1 (TB-1): 3.05 g/L of Tris(hydroxymethyl) aminomethane (Tris), 19.2 g/L of glycine, and 200 mL/L of methanol.
2. Transfer buffer –2 (TB-2): 3.05 g/L of Tris and 19.2 g/L of glycine.
3. Transfer buffer-3 (TB-3): 3 M sodium chloride and 0.3 M sodium citrate.
4. Waterproof tape (Scotch Super 33+, 3 M, Hutchinson, MN, USA).

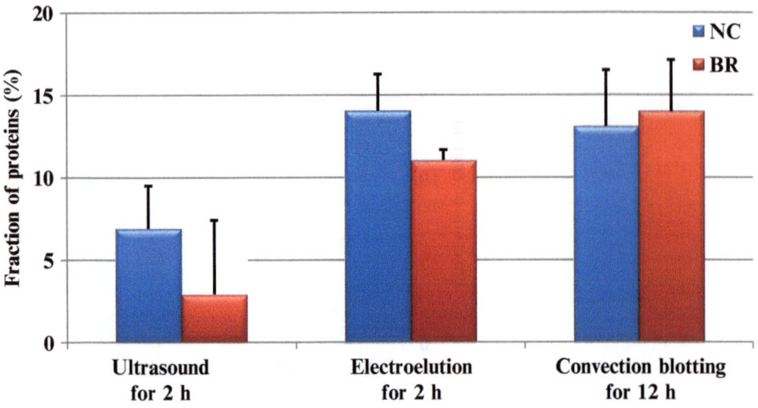

Fig. 5 Fraction of proteins, out of the initial amount loaded to the PhastGel, transferred from the gel and lost following blotting in transfer solution TB-1 by ultrasound for 2 h (*vertical lines*), electroelution for 2 h (*open bars*), and 12 h of convection blotting (*tilted lines*) [19]

5. Filter paper.
6. Aquasonic gel 100 (Parker Laboratories, Fairfield, NJ, USA).
7. Ultrasound (Sonopuls 434, Enraf Nonius, The Netherlands).
8. Nitrocellulose membranes (0.45 μm).
9. Nylon 66 membranes (0.45 μm).

3 Methods

3.1 Effect of Transfer Buffer Composition on BSA Blotting by Ultrasound Irradiation

1. Perform BSA blotting with ultrasound and electroelution in different transfer buffer solutions (Fig. 6) in order to evaluate the effect of transfer buffer solution composition on blotting.
2. No effect of transfer buffer composition on BSA blotting by ultrasound irradiation could be detected, while for BSA electroelution, TB-1 was more efficient than TB-2 (*see* **Note 1**).
3. Although the therapeutic ultrasound applied in these experiments is routinely used on humans in physiotherapy, its possible effects on the protein biological activity were examined. Figure 7 shows that 10 min exposure of phospholipase A2 (PLA2), from porcine pancreas, to ultrasound at intensities comparable to those used for ultrasonic blotting, did not affect its biological activity (*see* **Note 2**).

3.2 Protocol for Protein Transfer Using Ultrasonic Irradiation

Protein transfer from PhastGels to membranes by ultrasonic irradiation was performed as follows:

1. Place a sheet of dry filter paper (equal in size to the PhastGel used in electrophoresis) on a glass plate.
2. Place protein-loaded PhastGel on top of the filter paper.

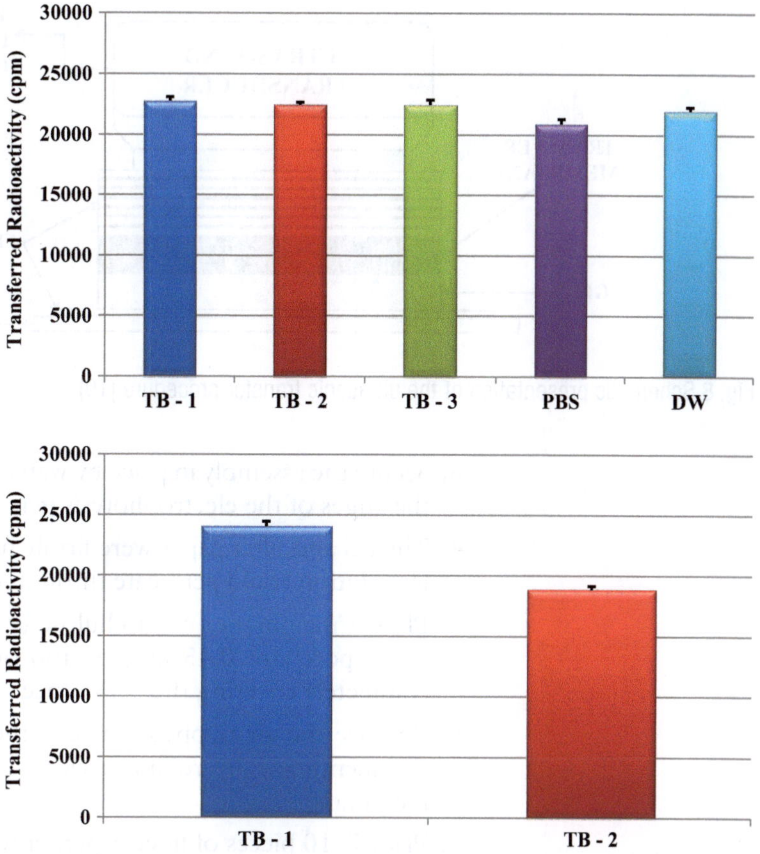

Fig. 6 Amount of ^{125}I-BSA transferred to NC membranes (**a**) ultrasonic blotting for 2 h in TB-1, TB-2, TB-3, PBS, and distilled water (**b**) Electroelution for 2 h in TB-1 and TB-2 [19]

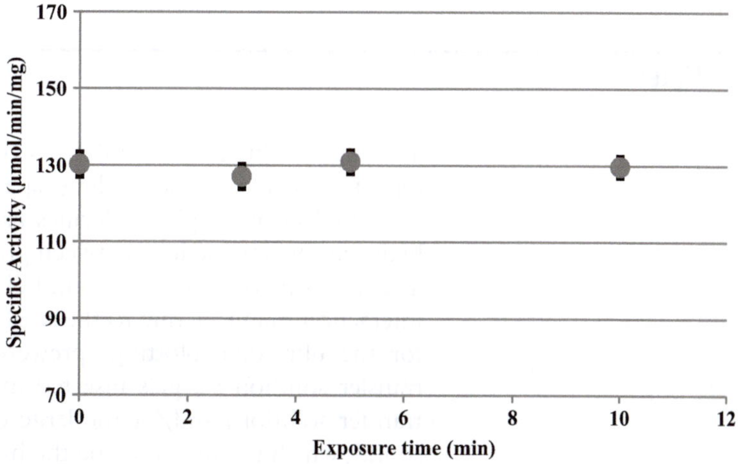

Fig. 7 Specific activity of phospholipase A2 solution subjected to ultrasound at 1 MHz, 2.5 W/cm^2 [19]

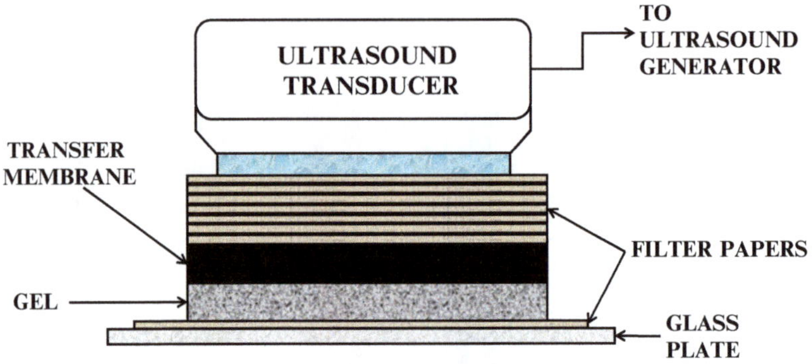

Fig. 8 Schematic presentation of the ultrasonic transfer procedure [19]

3. Secure the assembly in place by waterproof tape, placed around the edges of the electrophoresis gel, taped to the glass plate.
4. The gel and filter paper were firmly held together so the transfer solution could permeate through the gel and not around it.
5. Place a membrane (nitrocellulose (NC) or nylon 66 Biotrans (BR), pore size 0.45 μm) on top of the electrophoresis gel, completely covering the gel surface.
6. Remove the air trapped between the electrophoresis gel and the membrane by rolling a clean metallic cylinder over the membrane.
7. Place 7–10 pieces of filter paper, prewetted in the transfer buffer, sheet by sheet on the membrane.
8. Place aquasonic gel 100 (about 5 mL) on top of the filter papers as a coupling medium for ultrasound transmission.
9. Apply ultrasound at 1 MHz, 2.5 W/cm^2, placing the ultrasonic probe on top of the Aquasonic gel (Fig. 8).

4 Notes

1. The higher efficiency of TB-1 was expected as methanol was reported to increase the binding capacity of proteins, with moderate molecular weight molecules like BSA, strengthening the hydrophobic interactions between protein and membrane [41]. The insensitivity of the ultrasound blotting to this hydrophobic interaction might be due to the smaller amounts of buffer used for the ultrasonic blotting (prewetted filter papers with the transfer solution vs. gels inserted into buffer tank containing transfer solution) and/or the large enhancement of the ultrasound, which might overcome the hydrophobic interaction.
2. The procedure is fast and not as charge specific as electroblotting. We believe that after optimization, the required times could even be shorter and potentially applied to all proteins.

References

1. Fenster A et al (2013) Three-dimensional ultrasound imaging. In: Zhang S (ed), Handbook of 3D machine vision: optical metrology and imaging, vol 16, CRC Press, p 285
2. Polat BE et al (2011) Ultrasound-mediated transdermal drug delivery: mechanisms, scope, and emerging trends. J Control Release 152(3): 330–348
3. Suzuki R et al (2011) Progress in the development of ultrasound-mediated gene delivery systems utilizing nano- and microbubbles. J Control Release 149(1):36–41
4. Jia X et al (2010) Using sonochemistry for the fabrication of hollow ZnO microspheres. Ultrason Sonochem 17(2):284–287
5. Lu Y et al (2012) Research on ultrasonic online cleaning system for MBR. New Technol New Proc 1:013
6. Hodgson WJ, DelGuercio LR (1984) Preliminary experience in liver surgery using the ultrasonic scalpel. Surgery 95(2):230–234
7. Sanborn MR et al (2011) Safety and efficacy of a novel ultrasonic osteotome device in an ovine model. J Clin Neurosci 18(11):1528–1533
8. Chemat F, Khan MK (2011) Applications of ultrasound in food technology: processing, preservation and extraction. Ultrason Sonochem 18(4):813–835
9. Wan CPL et al (2012) Increased accumulation and retention of micellar paclitaxel in drug-sensitive and p-glycoprotein-expressing cell lines following ultrasound exposure. Ultrasound Med Biol 38(5):736–744
10. Suzuki R, Maruyama K (2010) Effective in vitro and in vivo gene delivery by the combination of liposomal bubbles (bubble liposomes) and ultrasound exposure. Methods Mol Biol 605:473–486
11. Anderson CD et al (2013) Ultrasound directs a transposase system for durable hepatic gene delivery in mice. Ultrasound Med Biol 39(12): 2351–2361
12. Song S et al (2011) Explorations of high-intensity therapeutic ultrasound and microbubble-mediated gene delivery in mouse liver. Gene Ther 18(10):1006–1014
13. Qin P et al (2012) Ultrasound-microbubble mediated cavitation of plant cells: effects on morphology and viability. Ultrasound Med Biol 38(6):1085–1096
14. Schroeder A et al (2009) Ultrasound, liposomes, and drug delivery: principles for using ultrasound to control the release of drugs from liposomes. Chem Phys Lipids 162(1):1–16
15. Giustetto P et al (2013) Release of a paramagnetic magnetic resonance imaging agent from liposomes triggered by low intensity non-focused ultrasound. J Med Imag Health Inform 3(3):356–366
16. Kost J, Langer R (2012) Responsive polymeric delivery systems. Adv Drug Deliv Rev 64: 327–341
17. Souza J et al (2013) Effect of phonophoresis on skin permeation of commercial anti-inflammatory gels: sodium Diclofenac and Ketoprofen. Ultrasound Med Biol 39(9):1623–1630
18. Kost J et al (2000) Non-invasive blood glucose measurement using ultrasound. Nat Med 6:347–350
19. Khoo X et al (2012) Formulations for trans-tympanic antibiotic delivery. Biomaterials 34(4):1281–1288
20. Maryam A, Mark EB (2008) Review: challenges and solutions in topical ocular drug-delivery systems. Exp Rev Clin Pharm 1(1):145–161
21. Sakuma S et al (2012) Performance of cell-penetrating peptide-linked polymers physically mixed with poorly membrane-permeable molecules on cell membranes. Eur J Pharm Biopharm 81:64–73
22. Meher JG et al (2012) Buccal drug delivery system and penetration enhancer: a review. Pharm Rev 10(55):113–117
23. Nabili M et al (2013) Ultrasound-enhanced delivery of antibiotics and anti-inflammatory drugs into the eye. Ultrasound Med Biol 39(4): 638–646
24. Chisti Y, Moo-Young M (1998) Disruption of microbial cells for intracellular products. Enzyme Microb Technol 8:194–204
25. Bar R (1998) Ultrasound-enhanced bioprocesses: cholesterol oxidation by Rhodococcus erythropolis. Biotechnol Bioeng 32:655–663
26. Nyborg WL (1998) Ultrasonic microstreaming and related phenomena. Br J Cancer 45: 156–160
27. Zabaneh M, Bar R (1991) Ultrasound-enhanced bioprocess. II: dehydrogenation of hydrocortisone by Arthrobacter simplex. Biotechnol Bioeng 37:998–1003
28. Matysik F-M et al (1997) Ultrasound-enhanced anodic stripping voltammetry using perfluorosulfonated ionomer-coated mercury thin-film electrodes. Anal Chem 69(8):1651–1656
29. Krasovitski B et al (2011) Intramembrane cavitation as a unifying mechanism for ultrasound-induced bioeffects. Proc Natl Acad Sci 108(8):3258–3263

30. Ogura M et al (2008) Low-frequency sonophoresis: current status and future prospects. Adv Drug Deliv Rev 60(10):1218–1223
31. Mitragotri S, Kost J (2004) Low-frequency sonophoresis: a review. Adv Drug Deliv Rev 56(5):589–601
32. Park D et al (2014) Sonophoresis in transdermal drug deliveries. Ultrasonics 54(1):56–65
33. Polat BE et al (2010) Low-frequency sonophoresis: application to the transdermal delivery of macromolecules and hydrophilic drugs. Expert Opin Drug Deliv 7(12):1415–1432
34. Lavon I, Kost J (1998) Mass transport enhancement by ultrasound in non-degradable polymeric controlled release systems. J Control Release 54(1):1–7
35. Aschkenasy C, Kost J (2005) On-demand release by ultrasound from osmotically swollen hydrophobic matrices. J Control Release 110(1):58–66
36. Chisti Y (2003) Sonobioreactors: using ultrasound for enhanced microbial productivity. Trends Biotechnol 21:89–93
37. Kost J et al (1994) Enhanced protein blotting from PhastGel media to membranes by irradiation of low-intensity. Anal Biochem 216:27–32
38. Kost J (2009) Blotting from PhastGel to membranes by ultrasound. In: Kurien B, Scofield H (eds) Protein blotting and detection, Humana Press. Methods Mol Biol 536:173–179
39. Aoki K et al (1969) Heat denaturation of bovine serum albumin. I: Analysis by acrylamide-gel electrophoresis (commemoration issue dedicated to professor Rempei Gotoh on the occasion of his retirement). Bull Inst Chem Res Kyoto Univ 47(4):274–282
40. Suslick KS (1988) Ultrasound: its chemical, physical, and biological effects. VCH Publishers, New York, NY
41. Issaq HJ et al (1991) The effect of electric field strength, buffer type and concentration on separation parameters in capillary zone electrophoresis. Chromatographia 32(3–4): 155–161

Chapter 26

Western Blotting of High and Low Molecular Weight Proteins Using Heat

Biji T. Kurien and R. Hal Scofield

Abstract

A method for the electrophoretic transfer of high and low molecular weight proteins to nitrocellulose membranes following sodium dodecyl sulfate (SDS) polyacrylamide gel is described here. The transfer was performed with heated (70–75 °C) normal transfer buffer from which methanol had been omitted. Complete transfer of high and low molecular weight antigens (molecular weight protein standards, a purified protein, and proteins from a human tissue extract) could be carried out in 10 min for a 7 % (0.75 mm) SDS polyacrylamide gel. For 10 and 12.5 % gels (0.75 mm) the corresponding time was 15 min. A complete transfer could be carried out in 20 min for 7, 10, and 12.5 % gels (1.5 mm gels). The permeability of the gel is increased by heat, such that the proteins trapped in the polyacrylamide gel matrix can be easily transferred to the membrane. The heat mediated transfer method was compared with a conventional transfer protocol, under similar conditions. The conventional method transferred minimal low molecular weight proteins while retaining most of the high molecular weight proteins in the gel. In summary, this procedure is particularly useful for the transfer of high molecular weight proteins, very rapid, and avoids the use of methanol.

Key words SDS-PAGE, Western blotting, Nitrocellulose, High molecular weight protein

1 Introduction

Protein transfer from sodium dodecyl sulfate polyacrylamide gel electrophoresis (SDS-PAGE) gels to nitrocellulose or polyvinylidene difluoride membranes have been achieved by (a) simple diffusion [1], (b) vacuum-assisted solvent flow [2], and (c) electrophoretic elution [3]. There is considerable interest in diffusion mediated transfer of proteins, since it was showed that [1] multiple immunoblots can be generated following non-electrophoretic bidirectional transfer of a single SDS-PAGE gel with multiple antigens. The lifts from SDS-PAGE gels, for immunoblotting, using this method are particularly useful in identification of proteins by mass spectrometry [4, 5]. However, electrophoretic elution is

Biji T. Kurien and R. Hal Scofield (eds.), *Western Blotting: Methods and Protocols*, Methods in Molecular Biology, vol. 1312, DOI 10.1007/978-1-4939-2694-7_26, © Springer Science+Business Media New York 2015

still the method of choice for protein transfer to membranes in most laboratories.

Electrophoretic transfer of proteins, resolved by SDS-PAGE, to nitrocellulose is a fundamental step prior to detection of specific proteins with specific antibodies [6–8]. The protein transfer procedure normally takes about 2–4 h at about 70 V or an overnight transfer at 30 V. High molecular weight proteins are often stubbornly resistant to transfer [9] in spite of these prolonged runs and this problem is accentuated when higher percentage gels are used. Prolonged electrotransfer (16–20 h) at high current density coupled with inclusion of 0.01 % sodium dodecyl sulfate, to enhance protein elution, has been used to efficiently transfer high-molecular weight proteins [9].

In order to determine the efficiency of heat-mediated electroblotting, we have used purified Ro 60 autoantigen, prestained molecular weight standards and a human cell extract using gels of two different thicknesses (0.75 and 1.5 mm) and gels with three different amounts of acrylamide (7, 10, and 12.5 %). In addition, transfer of proteins from a 4–20 % gradient SDS-PAGE gel and a 12.5 % (1.5 mm) regular gel was also investigated and compared to that obtained using a conventional transfer method under similar conditions.

Prestained high and low molecular weight protein standards (Figs. 1 and 2; Lane 4) and bovine Ro 60 (Figs. 1 and 2; Lane 2)

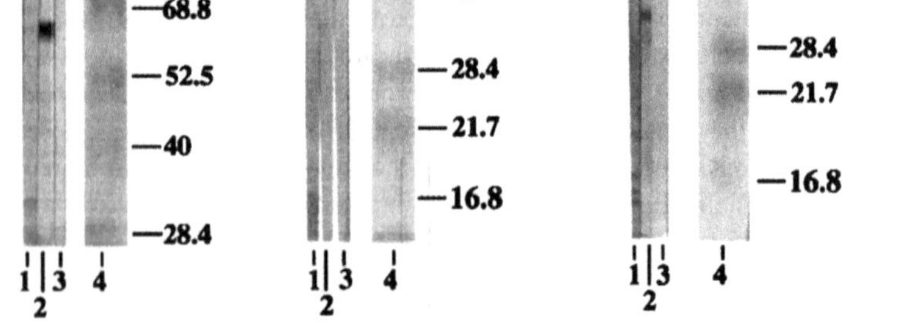

Fig. 1 Western blot transfer and immunoblotting of bovine Ro 60 and prestained molecular weight standards using 7, 10, and 12.5 % gels (*0.75 mm gels*). (**a–c**) Proteins on a 7, 10, and 12.5 % SDS-PAGE gels respectively transferred using heat. *Lane 1*—Anti-Ro 60 negative control; *Lane 2*—Anti-Ro 60 positive control; *Lane 3*—Conjugate control; *Lane 4*—Prestained molecular weight standards. (Reproduced from ref. 8 with permission from Elsevier)

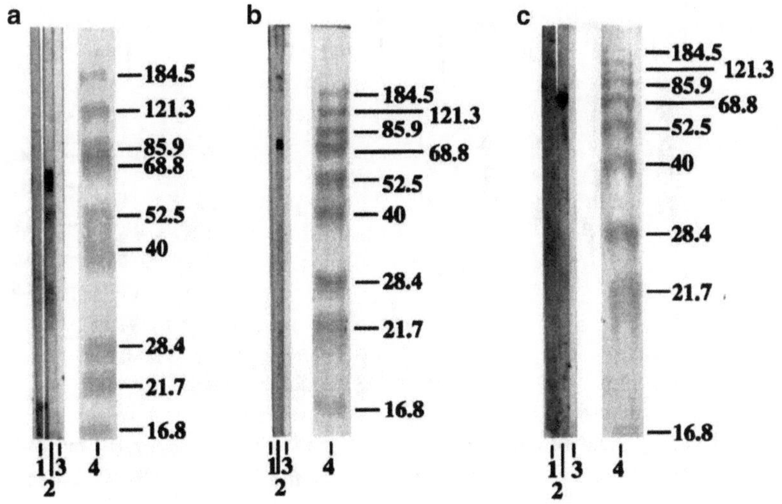

Fig. 2 Western blot transfer and immunoblotting of bovine Ro 60 and prestained molecular weight standards using 7, 10, and 12.5 % gels (*1.5 mm gels*). (**a**–**c**) Proteins on a 7, 10, and 12.5 % SDS-PAGE gels transferred using heat. *Lane 1*—Anti-Ro 60 negative control; *Lane 2*—Anti-Ro 60 positive control; *Lane 3*—Conjugate control; *Lane 4*—Prestained molecular weight standards. (Reproduced from ref. 8 with permission from Elsevier)

could be transferred completely to nitrocellulose. All the protein markers (5 μL of the marker) could be transferred in about 10 min from a 0.75 mm, 7 % gel (Fig. 1a, Lane 4). All the protein markers, ranging from 184 to 9 kDa, could be transferred to membranes from 7, 10, and 12.5 % gels (Fig. 1b, c—Lane 4) in 15 min. The post-transfer polyacrylamide gels were clean, without any sign of residual non-transferred protein markers. It took 20 min to transfer all the protein markers in the case of the 1.5 mm gels (7, 10, and 12.5 % gels) (Fig. 2, Lane 4).

Figure 3 shows the fast green stained nitrocellulose membranes following transfer of protein using our method (Fig. 3, right panel) and a conventional method (Fig. 3, left panel). The protein transfer was found to be efficient with heat mediated transfer, while even the low molecular weight proteins can be barely seen following 20 min of transfer using a conventional method. Similarly only low levels of proteins could be detected immunologically following a conventional electrotransfer compared to that obtained following heat mediated transfer (Fig. 4).

This method clearly demonstrates that both low and high molecular weight proteins can be transferred very efficiently to nitrocellulose membranes in a very short time using heated transfer buffer without methanol [10].

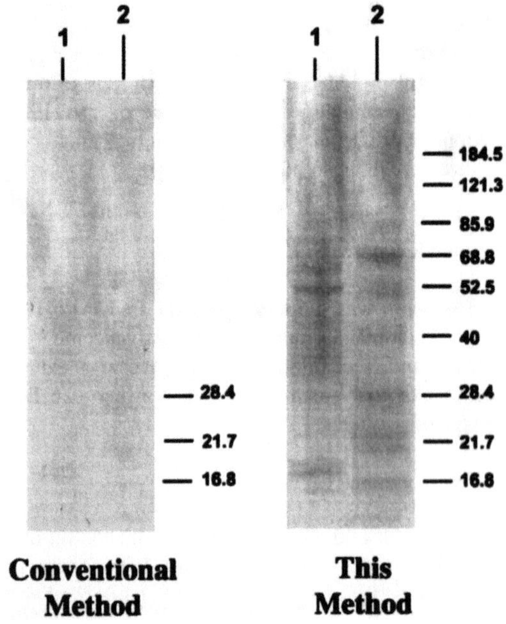

Fig. 3 Fast Green staining of proteins from a HeLa cell extract transferred to nitrocellulose membrane by a conventional method (*left*) and using the heat transfer method (*right*) from a 4–20 % SDS-PAGE gradient gel. *Lane 1*—HeLa cell extract; *Lane 2*—10 μL of prestained molecular weight marker. (Reproduced from ref. 8 with permission from Elsevier)

2 Materials

Prepare all solutions using ultrapure water (prepared by purifying deionized water, to attain a sensitivity of 18 MΩ-cm at 25 °C) and analytical grade reagents. Prepare and store all reagents at room temperature (unless indicated otherwise). Diligently follow all waste disposal regulations when disposing waste materials. We do not add sodium azide to our reagents.

1. 10 % SDS-PAGE precast gels (10-well).
2. SDS lysis buffer (5×): 0.3 M Tris–HCl (pH 6.8), 10 % SDS, 25 % β-mercaptoethanol, 0.1 % bromophenol blue, 45 % glycerol. Leave one aliquot at 4 °C for current use and store remaining aliquots at −20 °C (*see* **Note 1**).
3. SDS-PAGE running buffer: 0.025 M Tris–HCl, pH 8.3, 0.192 M glycine, 0.1 % SDS.
4. Phosphate buffered saline (PBS), pH 7.4.
5. Purified bovine Ro 60 autoantigen [11, 12] was a gift from Immunovision, Springdale, AK, USA.
6. BenchMark pre-stained molecular weight standards.

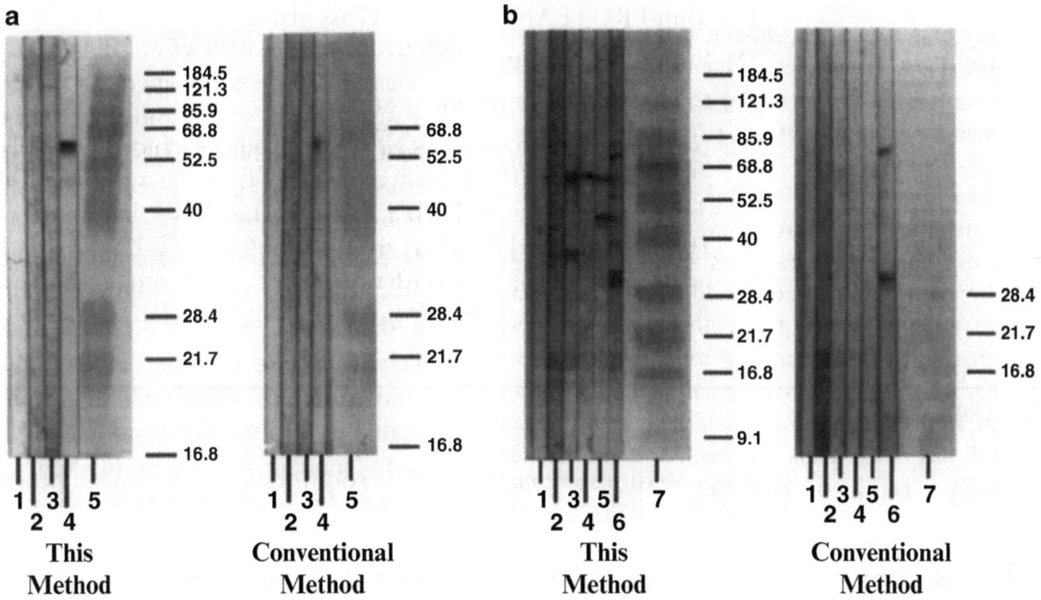

Fig. 4 Immunoblots of purified Ro 60 and proteins derived from a HeLa cell extract transferred to nitrocellulose membrane using the heat transfer method (*left*) and by a conventional method (*right*). (**a**) Purified Ro 60 autoantigen immunoblot obtained from a 12.5 % SDS-PAGE gel using the heat transfer method (*left*) and by a conventional method (*right*). *Lane 1*—conjugate control; *lanes 2* and *3*—normal controls; *lane 4*—anti-Ro 60 SLE sera; *lane 5*—prestained protein molecular weight standards. (**b**) HeLa cell extract immunoblot obtained from a 4–20 % gradient SDS-PAGE gel using the heat transfer method (*left*) and a conventional transfer method (*right*). *Lane 1*—conjugate control; *lane 2*—normal control; *lane 3*—anti-Ro 60 SLE sera; *lane 4*—anti-La sera; *lane 5*—anti-Ro 52 sera; *lane 6*—anti-Sm/nRNP sera; *lane 7*—prestained protein molecular weight markers (10 μL). (Reproduced from ref. 8 with permission from Elsevier)

7. Nitrocellulose membranes.
8. Transfer buffer (with methanol): 0.025 mM Tris, 192 mM glycine, 20 % methanol.
9. Transfer buffer (without methanol): 0.025 mM Tris, 192 mM glycine.
10. Western blot transfer apparatus with capability of circulating hot or cold water at the base.
11. Refrigerator bath/circulating hot water bath.
12. Branson Sonifier Cell Disruptor 185.
13. Tris buffered saline (TBS; 10×): 1.5 M NaCl, 0.1 M Tris–HCl, pH 7.4.
14. TBS containing 0.05 % Tween 20 (TBST).
15. Blocking solution: 5 % milk in TBS, pH 7.4. Store at 4 °C (*see* **Note 2**).
16. Diluent solution: 5 % milk in TBS, pH 7.4 containing 0.05 % Tween. Store at 4 °C (*see* **Note 2**).

17. Mini PROTEAN® 3 System Glass plates.
18. FB300 power supply.
19. Corning PC-351 magnetic stirrer.
20. Nitro blue tetrazolium (NBT)/5-bromo-4-chloro-3-indolyl phosphate (BCIP): Dissolve 1 g NBT in 20 mL of 70 % dimethylformamide (DMF). Dissolve 1 g BCIP in 20 mL of 100 % DMF. Add 33 μL of BCIP and 66 μL of NBT to 10 mL of alkaline phosphatase buffer just before adding to membrane. Alternatively, use 1-Step™ premixed NBT/BCIP.

3 Methods

All procedures are carried out at room temperature unless otherwise specified.

3.1 Preparation of HeLa Cell Lysate

1. Harvest freshly cultured HeLa cells by centrifuging at 800 × *g* and wash twice with PBS.
2. Lyse cells by sonication in SDS-PAGE lysis buffer using a Branson sonicator (setting 4) and spin at 10,000 × *g* for 10 min.
3. Use an aliquot of the supernatant for SDS-PAGE.

3.2 SDS-PAGE

1. Carry out SDS-PAGE [13] (*see* Chapters 11 and 34) on 7, 10 or 12.5 % gels (0.75 or 1.5 mm thickness).

3.3 Conventional Electrophoretic Transfer

1. Carry out conventional electrotransfer of proteins [3] separated on gradient and regular gels (*see* Chapters 22 and 34) at 4 °C using standard transfer buffer (with methanol) for 20 min.
2. Save the nitrocellulose membrane for use with the membrane to be obtained following heat mediated electrophoretic transfer for 20 min also. Stain the polyacrylamide gel with Coomassie brilliant blue stain to ensure the efficiency of transfer.

3.4 Heat Mediated Electrophoretic Transfer

1. Turn on the circulating hot water bath and set temperature at 70 °C.
2. Following SDS PAGE, pry the gel plates open with the use of a spatula. The gel remains on one of the glass plates. Rinse the gel gently with deionized water and transfer carefully to a container with transfer buffer (without methanol).
3. Heat the transfer buffer (without methanol) to about 70–75 °C in a 1-L glass beaker. Cover the beaker with clear plastic wrap.
4. Cut a nitrocellulose membrane to the size of the gel and immerse in transfer buffer (*see* **Note 3**).

5. Cut four sheets of Whatman 3MM filter paper to the size of the gel and transfer to the transfer buffer (*see* **Note 4**). Place two adsorbent pads also in the buffer and remove air bubbles by pressing down on it with help of fingers.
6. Place clear plastic wrap (12 in. in length) on the work-bench. Place two filter papers on top of the plastic wrap. Position the membrane on top of the filter papers. Transfer the gel to the top of the membrane in such a way that there are no air bubbles between the gel and the membrane (*see* **Note 5**). Place the remaining two filter papers on top of the gel. Place in transfer cassette. Ensure that the nitrocellulose membrane is between the gel and the anode (*see* **Note 6**).
7. Place the transfer apparatus on a magnetic stirrer and connect it to the hot water bath maintained at 70 °C. Transfer the hot transfer buffer into the apparatus (*see* **Note 7**). Place a magnetic stir bar inside the transfer apparatus to circulate the buffer.
8. Carry out the transfer at 40 V (*see* **Note 8**) for periods ranging from 10 to 20 min, depending upon the type of SDS polyacrylamide gel used (*see* **Note 9**).
9. Disconnect power supply and disassemble the sandwich. Save the membrane along with one obtained using conventional electrotransfer.
10. Stain the polyacrylamide gel with Coomassie brilliant blue stain to ensure the efficiency of transfer.

3.5 Immunoblotting

1. Carry out standard immunoblotting [3] (*see* Chapters 16 and 25) for both sets of membranes.
2. Use human systemic lupus erythematosus (SLE) sera with autoantibodies against 60 kDa Ro, 48 kDa La, 52 kDa Ro, Sm or nuclear ribonucleoprotein autoantigens [11, 12] to identify the respective antigens transferred to nitrocellulose from various gels (*see* **Notes 9** and **10**).

4 Notes

1. SDS precipitates at 4 °C. Therefore, the lysis buffer needs to be warmed prior to use.
2. Add 100 mL of 10× TBS to a 1 L graduated cylinder and make it to about 800 mL with water. Transfer 50 g skim milk powder into the cylinder and mix stir until dissolved. Make to 1 L with water. Separate 500 mL as the blocking solution. To the remaining 500 mL add 250 μL of Tween 20 (cut end of blue tip to aspirate Tween 20 easily), dissolve and use it as the diluent.

3. Moistening the nitrocellulose completely with the transfer buffer (without methanol) takes about 10 min. Alternately, the nitrocellulose membrane can be moistened almost instantaneously with regular transfer buffer that contains methanol. The membrane can then be rinsed with water and the transfer buffer (without methanol).
4. Preferably, the buffer used to assemble the sandwich should be at room temperature or just slightly warm. Using the hot buffer to assemble the sandwich causes the 3MM filter paper, used as a part of the gel sandwich, to disintegrate partially.
5. Hold the two top corners of the gel with each hand. Lower the bottom part of the gel first of the membrane and gently release the gel little by little to lay the complete gel on the membrane. This will prevent trapping of bubbles in between the gel and the membrane. We use clear plastic wrap. Part of the wrap was folded over the sandwich. A 10 mL pipette was used to roll out the air bubbles from the gel membrane sandwich prior to placing in transfer cassette.
6. In the Bio-Rad transfer apparatus, we place the gel side of the transfer cassette facing the black side of the transfer cassette holder and the membrane side facing the red side.
7. Exercise caution in handling the hot contents. Use thermo gloves to hold the hot beaker and pour carefully.
8. Forty volts was close to the upper limit possible using the FB300 power supply. The current was about 500 mA at this point.
9. Transfer from a 7.5 % SDS polyacrylamide gel (0.75 mm thick) can be carried out in 10 min. However, it takes longer (20 min) for a 12.5 % gel (1.5 mm thick). If a circulating water bath is not available, it is still possible to transfer using just the heated buffer, by increasing the time of transfer. The main advantage of this procedure is that it is possible to transfer high molecular weight protein, in addition to saving time.
10. Exercise universal precaution when handling human sera. Treat each serum sample as potentially dangerous.

Acknowledgement

This work was supported by NIH grant ARO1844 and Oklahoma Center for the Advancement of Science and Technology to RHS. We also express our thanks to Samantha Ganick for her excellent technical assistance.

References

1. Kurien BT, Scofield RH (1997) Multiple immunoblots after non-electrophoretic bidirectional transfer of a single SDS-PAGE gel with multiple antigens. J Immunol Methods 205:91–94
2. Peferoen M, Huybrechts R, De Loof A (1982) Vacuum-blotting: a new simple and efficient transfer of proteins from sodium dodecyl sulfate-polyacrylamide gels to NC. FEBS Lett 145:369–372
3. Towbin H, Staehelin T, Gordon J (1979) Electrophoretic transfer of proteins from polyacrylamide gels to NC sheets: procedure and applications. Proc Natl Acad Sci U S A 76:4350–4354
4. Kurien BT, Scofield RH (2000) Association of neutropenia in systemic lupus erythematosus with anti-Ro and binding of an immunologically cross-reactive neutrophil membrane antigen. Clin Exp Immunol 120:209–217
5. Kurien BT, Matsumoto H, Scofield RH (2001) Purification of tryptic peptides for mass spectrometry using polyvinylidene fluoride membrane. Indian J Biochem Biophys 38:274–276
6. Bolt MW, Mahoney PA (1997) High efficiency blotting of proteins of diverse sizes following sodium dodecyl sulfate-polyacrylamide gel electrophoresis. Anal Biochem 247:185–192
7. MacPhee DJ (2010) Methodological considerations for improving Western blot analysis. J Pharmacol Toxicol Methods 61:171–177, Review. Methods Mol Biol 536:181–190
8. Kurien BT, Scofield RH (2000) Ultrarapid electrophoretic transfer of high and low molecular weight proteins using heat. Methods Mol Biol 536:181–190
9. Otter T, King SM, Witman GB (1987) A two-step procedure for efficient electrotransfer of both high-molecular weight (greater than 400,000) and low-molecular weight (less than 20,000) proteins. Anal Biochem 162:370–377
10. Kurien BT, Scofield RH (2002) Heat mediated, ultra-rapid electrophoretic transfer of high and low molecular weight proteins to NC membranes. J Immunol Methods 266: 127–133
11. Deutscher SL, Harley JB, Keene JD (1988) Molecular analysis of the 60 kD human Ro ribonucleoprotein. Proc Natl Acad Sci U S A 85:9479–9483
12. Ben-Chetrit E, Gandy BJ, Tan EM, Sullivan KF (1989) Isolation and characterization of a cDNA clone encoding the 60 kDa component of the human SS-A/Ro ribonucleoprotein autoantigen. J Clin Invest 83:1284–1292
13. Laemmli UK (1970) Cleavage of structural proteins during assembly of the head of bacteriophage T4. Nature 227:680–685

Chapter 27

Membrane Strip Affinity Purification of Autoantibodies

Biji T. Kurien

Abstract

A method for affinity purification of autoantibodies from a membrane strip using small volumes of human sera is described. The membrane strip is excised from a western blot containing a target antigen electrophoretically transferred from a sodium dodecyl sulfate (SDS) polyacrylamide gel. This method is a very useful alternative for affinity column chromatography, particularly when the antigen of interest is of low abundance in a HeLa cell extract. The protein mixture is resolved on a preparative SDS polyacrylamide gel and transferred to nitrocellulose membrane. A couple of strips are excised from either side of the blotted membrane and immunoblotted with specific antisera to identify the target band. Then the target band is excised horizontally and used for affinity purification. We have used this procedure to affinity purify antibodies to a 70,000 molecular weight protein derived from HeLa cell extract. A sham band, excised away from the target antigen, was used as a control for sham purification of autoantibodies. The autoantibodies purified in this manner reproduced the multiple nuclear dot anti-nuclear antibody pattern obtained using crude sera from 21 patients without primary biliary cirrhosis or anti-mitochondrial antibody.

Key words Anti-nuclear antibody (ANA), Multiple nuclear dot ANA, Autoimmunity, Affinity purification, Nitrocellulose, Autoantibodies, Autoantigens

1 Introduction

Anti-nuclear antibodies (ANA), directed against a variety of nuclear antigens, are a common characteristic of autoimmune diseases [1]. Its detection depends on the evaluation of immunofluorescence of tissue culture cells. ANA positive sera, typically, exhibit nuclear homogeneous, nuclear speckled or nucleolar immunofluorescence patterns on Hep-2 cells. However, in certain instances a distinct set of fluorescent spots called multiple nuclear dot (MND) ANA occur all over the nucleus except in the nucleoli [2, 3]. The MND ANA is similar to the anti-centromere pattern but differs in that the discrete fluorescent dots are larger, fewer in number and seen only in the interphase cell, whereas the anti-centromere antibodies (ACA) give a speckled or punctate staining of the chromosomes in metaphase.

We undertook a study to determine the antigen bound by sera from a group of 21 patients with MND ANA but no ACA.

Biji T. Kurien and R. Hal Scofield (eds.), *Western Blotting: Methods and Protocols*, Methods in Molecular Biology, vol. 1312, DOI 10.1007/978-1-4939-2694-7_27, © Springer Science+Business Media New York 2015

We found that the crude sera from these patients showed a MND ANA pattern on Hep-2 cells. These sera were also found to bind to a 70,000 molecular weight antigen on a HeLa cell extract immunoblot. This study was undertaken to see whether autoantibodies affinity purified off this 70,000 molecular weight protein could reproduce the pattern that we found using the crude sera.

Purified antibodies are essential for a number of techniques, such as immunoblots, immunoassays, or cell staining (e.g., Hep-2 cells). There are several methods for purifying antibodies, e.g., precipitation with ammonium sulfate, use of hydroxyapatite column, gel filtration, protein A/B beads and column affinity chromatography [4]. Ammonium sulfate precipitation is advantageous in that it is cheap and convenient to work with large volumes. However, the antibody yields are impure and therefore this procedure has to be coupled with other methods to obtain pure antibodies. Gel filtration is appropriate for IgM derived from all sources, since it can efficiently separate IgM from other antibody in polyclonal sera [4]. Sample dilution, impure antibody yields, low sample capacity are some of the disadvantages of this method. Protein A beads are useful for IgG that bind to it from various sources and results in pure antibody providing high yield in a single step. It is disadvantageous in that it is expensive and also because it is not suitable for all species and classes [4]. Affinity column chromatography has been commonly used for purifying antibodies from polyclonal sera. This procedure requires pure antigen for coupling to the column, is expensive and involves multiple steps. However, it is possible to obtain pure and specific antibody. The main disadvantage of this procedure is that it requires large amounts of pure antigen.

However, when the antigen of interest is of low abundance in a mixture of proteins (like in HeLa cell extract) and antibodies to that antigen is to be characterized from limited amounts of patient sera, affinity purification using a western blot strip containing that antigen becomes very useful. The protein mixture is electrophoresed on a SDS polyacrylamide gel, transferred to nitrocellulose membrane and stained with fast green. Two thin strips, one from each side of the blotted membrane, are excised and immunoblotted with specific antisera to identify the target band. Then the target band is excised horizontally and used for affinity purification of autoantibodies from patient sera. A sham band of the nitrocellulose, approximately similar in size to the excised target band, was also cut from an area away from the 70 kDa band. We have used affinity purification using nitrocellulose membrane strips to purify antibodies to the 70,000 molecular weight protein derived from HeLa cell extract to reproduce binding to the 70 kDa protein on immunoblot (Fig. 1) as well as the multiple nuclear dot antinuclear antibody pattern (Fig. 2) detected using crude sera from 21 patients without primary biliary cirrhosis or anti-mitochondrial antibody [5].

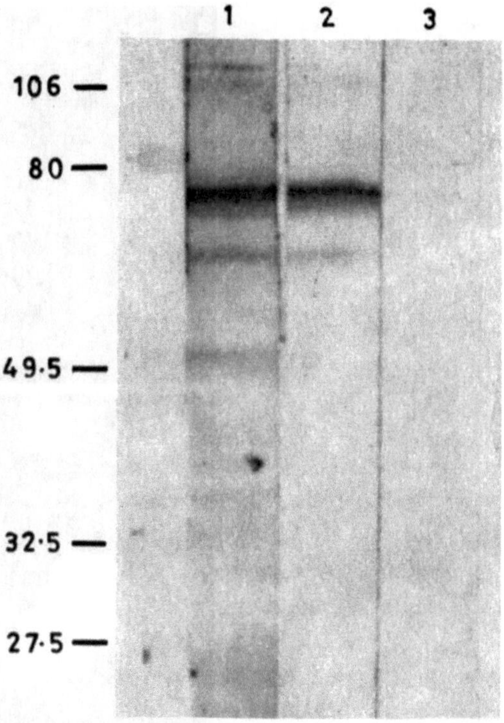

Fig. 1 Demonstration of immunoglobulin purified from the 70 kDa band binding back to the 70 kDa band on immunoblot. *Lane 1*—Crude MND ANA prototype human sera prior to affinity purification of anti-70 kDa antibody; *lane 2*—prototype human sera partially depleted of the anti-70 kDa antibody following affinity purification (crude sera minus affinity purified antibodies); *lane 3*—affinity purified anti-70 kDa immunoglobulin. (Reproduced from ref. 5 with permission from publisher)

2 Materials

Prepare all solutions using ultrapure water (prepared by purifying deionized water, to attain a sensitivity of 18 MΩ-cm at 25 °C) and analytical grade reagents. Prepare and store all reagents at room temperature (unless indicated otherwise). Diligently follow all waste disposal regulations when disposing waste materials. We do not add sodium azide to reagents.

1. Resolving gel buffer: 1.5 M Tris–HCl, pH 8.8. Add about 100 mL water to a 1 L graduated cylinder or a glass beaker (*see* **Note 1**). Weigh 181.7 g Tris and transfer to the cylinder. Add water to a volume of 900 mL. Mix and adjust pH with HCl (*see* **Note 2**). Make up to 1 L with water.
2. Stacking gel buffer: 0.5 M Tris–HCl, pH 6.8. Weigh 60.6 g Tris and prepare a 1 L solution as in previous step.

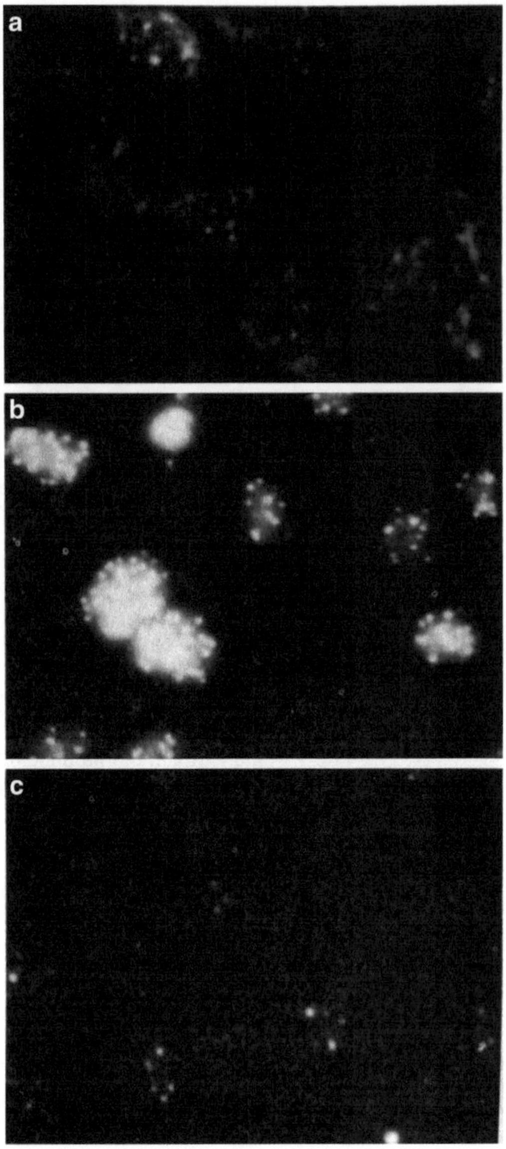

Fig. 2 Indirect immunofluourescence showing that affinity anti-70 kDa antibody reproduces the pattern obtained with crude sera. *Top panel*—MND sera that bound to a 70 kDa antigen in immunoblot; *bottom panel*—antibody affinity purified off the 70 kDa band from the MND sera. HEp-2 slides are incubated with crude or purified anti-70 kDa IgG, washed with PBS and further incubated with FITC labeled anti-human IgG secondary antibody. The slides are washed with PBS and photographed. (Reproduced from ref. 5 with permission from publisher)

3. Thirty percent acrylamide/bis solution (29.2:0.8) acrylamide:bis: Weigh 29.2 g of acrylamide monomer and 0.8 g bis (cross-linker) and transfer to a 100 mL graduated cylinder containing about 40 mL of water. Add a spatula of mixed-resin beads and mix for about 30 min. Make up to

100 mL with water and filter through a 0.45 μm Corning filter (*see* **Note 3**). Store at 4 °C, with bottle wrapped with aluminum foil (*see* **Note 4**).

4. Ammonium persulfate: 10 % solution in water (*see* **Note 5**).
5. *N*,*N*,*N*′,*N*′-Tetramethyl-ethylenediamine. Store at 4 °C (*see* **Note 6**).
6. SDS-PAGE running buffer: 0.025 M Tris–HCl, pH 8.3, 0.192 M glycine, 0.1 % SDS (*see* **Note 7**).
7. SDS lysis buffer (5×): 0.3 M Tris–HCl (pH 6.8), 10 % SDS, 25 % β-mercaptoethanol, 0.1 % bromophenol blue, 45 % glycerol. Leave one aliquot at 4 °C for current use and store remaining aliquots at −20 °C (*see* **Note 8**).
8. Bromophenol blue (BPB) solution: Dissolve 0.1 g BPB in 100 mL water.
9. Phosphate buffered saline (PBS), pH 7.4. Dissolve five PBS tablets in 1 L water to obtain PBS solution, pH 7.4 at 25 °C (0.01 M phosphate, 0.00027 potassium chloride, 0.137 M sodium chloride).
10. Human lymphocytes: Purify lymphocytes from normal human peripheral blood using Lymphoprep according to manufacturer's instruction.
11. BenchMark pre-stained molecular weight standards.
12. Nitrocellulose membranes.
13. Western blot transfer buffer: 0.025 M Tris, 0.192 M glycine, 20 % methanol (*see* **Note 9**).
14. Western blot transfer apparatus.
15. Tris buffered saline (TBS; 10×): 1.5 M NaCl, 0.1 M Tris–HCl, pH 7.4.
16. Blocking solution: 5 % milk in TBS (*see* **Note 10**). Store at 4 °C.
17. Diluent solution: 5 % milk in TBST (*see* **Note 10**). Store at 4 °C.
18. Alkaline phosphatase (AP) buffer: Weigh 6.1 g of Tris, 2.9 g sodium chloride, and 0.51 g magnesium chloride-6H_2O and make it to 500 mL with water after adjusting pH to 9.3 with HCl. Store at 4 °C (*see* **Note 1**).
19. Nitro blue tetrazolium (NBT)/5-bromo-4-chloro-3-indolyl phosphate (BCIP): Add 33 μL of BCIP and 66 μL of NBT to 10 mL of AP buffer at the time of assay. Alternatively, use 1-Step™ premixed NBT/BCIP solution.
20. Glycine elution buffer (Glycine Buffered Saline-Tween 20; GBST): 0.2 M glycine, 0.15 M NaCl, 0.05 % Tween 20, pH 2.7 (GBST). Store at 4 °C.

21. Electrophoresis system: Mighty small vertical slab gel unit (Hoeffer Scientific) (*see* **Note 11**).
22. HEp-2 slides.
23. Centricon 30 micro-concentrators.
24. Alkaline phosphatase conjugated affinity purified goat anti-human IgG.
25. Nitro blue tetrazolium (NBT)/5-bromo-4-chloro-3-indolyl phosphate (BCIP): Dissolve 1 g NBT in 20 mL of 70 % dimethylformamide (DMF). Dissolve 1 g BCIP in 20 mL of 100 % DMF. Add 33 μL of BCIP and 66 μL of NBT to 10 mL of alkaline phosphatase buffer just before adding to membrane. Alternatively, use 1-Step™ premixed NBT/BCIP solution.
26. Fluorescent isothiocyanate conjugated affinity purified goat anti-human IgG.
27. 3.5 mL tubes (Sarstedt).
28. 0.009″ single edge razor blades–Smith Brand.
29. Membrane strip mini-incubation container (Bio-Rad).
30. Helium.

3 Methods

All procedures are carried out at room temperature unless otherwise specified.

3.1 Preparation of Human Lymphocyte Extract

1. Lyse the lymphocytes by sonication in PBS using a Branson sonicator (setting # 4) and spin at 10,000 × *g* for 10 min (*see* **Note 12**). Analyze an aliquot of the supernatant by SDS-PAGE.

3.2 10 % Preparative SDS Polyacrylamide Gel

1. Mix 2.5 mL of resolving buffer, 3.33 mL of acrylamide mixture, and 4 mL water in a 50 mL conical flask. Degas with helium for 10 min. Add 100 μL of SDS, 80 μL of ammonium persulfate, and 10 μL of TEMED and cast gel within a 7.25 cm × 10 cm × 1.5 mm gel cassette. Allow space for stacking gel and gently overlay with isobutanol or water (*see* **Note 13**).
2. Prepare the stacking gel by mixing 1.25 mL of resolving buffer, 0.67 mL of acrylamide mixture, and 3 mL water in a 50 mL conical flask. Degas with helium for 10 min. Add 100 μL of SDS, 40 μL of ammonium persulfate, and 5 μL of TEMED. Insert a preparative gel comb immediately without introducing air bubbles.
3. Heat an aliquot of the lymphocyte extract at 95 °C for 5 min. Do not add lysis buffer to the prestained protein standard or subject it to heat (manufacturer instructions). The heated

lymphocyte extract was centrifuged at 3,000×g for 30 s to bring down the condensate. Load the sample and the protein standard (10 μL/well) on the gel. Electrophorese at 15 mA until the sample enters the gel and then continue at 25 mA till the dye front (from the BPB dye in the samples) reached the bottom of the gel (*see* **Note 14**).

3.3 Immunoblotting

1. Transfer proteins resolved on SDS PAGE to nitrocellulose membrane by Towbin's electrophoretic transfer method [5] (*see* Chapters 22 and 34).
2. Stain with fast green to visualize proteins transferred to the membrane (*see* **Note 15**).
3. Trim the edges of the membrane. Wrap the fast green stained membrane in cling wrap and make a photocopy for record keeping purposes.
4. Excise the lane with molecular weight standards from the main nitrocellulose sheet containing the transferred protein. Excise a tiny wedge from the bottom left side of the marker lane and the main membrane sheet for orientation purposes (*see* **Note 16**).
5. Excise two strips, each about 2 mm in width, from either ends of the main nitrocellulose membrane sheet. Place them in a 12-lane mini-incubation container.
6. Destain strips with TBST.
7. Block for 1 h with 1 mL each of blocking solution.
8. Add 1 mL primary sera at 1:100 dilution and incubate for 2 h. Wash 5× with TBST, 5 min each time. Dilute anti-human IgG conjugated to alkaline phosphatase 1:5,000 with diluent and add 1 mL to each strip. Incubate for 1 h. Wash 5× with TBST, 5 min each time. Add 0.5 mL substrate and develop bands.
9. Align the two strips (with developed bands) on either side of the main nitrocellulose membrane and excise the desired antigen strip (horizontally this time). Use this strip, containing the antigen of interest, for the affinity purification of autoantibodies (*see* **Note 17**).
10. Cut a sham strip away from the target antigen strip and also similar to the size of the target antigen strip. This is used as a sham purification control.

3.4 Purification of Autoantibodies [6, 7]

1. Block the tubes for collecting purified antibodies, the tubes for incubating the membrane strips with sera and the Centriprep 30 concentrator with 1 % milk/TBST solution for 30 min at 4 °C (*see* **Note 18**).
2. Wash the tubes adequately with TBST to get rid of unbound milk protein.

3. Cut the nitrocellulose membrane strip with the desired antigen into smaller pieces and transfer to Sarstedt 3.5 mL 'B' tube (for blocking). Incubate these pieces with 2.5 mL 5 % milk/TBST for 30 min on an orbital shaker.
4. Wash well with several changes of TBST.
5. Transfer pieces from the 'Blocking tube' ('B') to a serum Incubation ('I') previously blocked with 1 % milk/TBST.
6. Incubate nitrocellulose pieces with 2.5 mL of a 1:50 sera dilution in 5 % milk/TBST for 1 h on an orbital shaker (*see* **Note 19**).
7. At the end of the incubation, pipet off and save the diluted sera (sera diluted 1:50 in 5 % milk/TBST) into another tube for use in later repeat sera incubations with the membrane pieces.
8. Transfer the membrane pieces to a TBST 'Wash' tube ('W') and wash five times (10 min each time) on the orbital shaker.
9. After the wash, transfer the membrane pieces to a 'Glycine Elution tube' ('E').
10. Add 2.5 mL glycine elution buffer (GBST) and shake vigorously on an orbital shaker for 2 min. Then pipette the solution to a previously blocked collection tube.
11. Repeat with 2.5 mL of GBST and pipette into collection tube.
12. Add 2.5 mL of TBST and shake for approximately 1 min and pipette this solution into the collection tube.
13. Neutralize GBST with 500 μL of 1 M Tris.
14. Transfer pieces back to Blocking tube ('B'). Block for 30 min in 5 % milk/TBST. Wash with TBST.
15. Repeat **steps 4–10** for two more elution cycles. However, care need to be taken to use a 1:100 dilution of sera in 5 % milk/TBST saved in **step 7**.
16. Upon completion of three elution cycles, transfer nitrocellulose pieces back to Blocking tube ('B'). Block in 5 % milk/TBST for 30 min. The nitrocellulose can be stored at 4 °C for 1–2 weeks if further elutions are needed.
17. Concentrate the eluted antibody solution to required volume (*see* **Note 20**). Antibody is now ready for western blot assay (*see* **Note 21**) carried out as in Subheading 3.3 or ANA testing.
18. Arrange the immunoblotted strips on card board inserts (*see* **Note 22**).

4 Notes

1. Having water at the bottom of the cylinder helps to dissolve the Tris much faster, allowing the magnetic stirrer to go to work immediately. If using a glass beaker, the Tris can also be

dissolved faster if the water is warmed to about 37 °C. However, the downside is that care should be taken to bring the solution to room temperature before adjusting pH.

2. Concentrated HCl (12N) can be used at first to narrow the gap from the starting pH to the required pH. From then on it would be better to use a series of HCl (e.g., 6N and 1N) with lower ionic strengths to avoid a sudden drop in pH below the required pH.
3. Wear a mask when weighing acrylamide. To avoid exposing acrylamide to co-workers, cover the weigh boat containing the weighed acrylamide with another weigh boat (similar size to the original weigh boat containing the weighed acrylamide) when transporting it to the fume hood. Transfer the weighed acrylamide to the cylinder inside the fume hood and mix on a mixer inside the hood. Unpolymerized acrylamide is a neurotoxin and care should be exercised to avoid skin contact.
4. The acrylamide solution can be stored at 4 °C for 1 month. The acrylamide mixture, buffer and water can be prepared in large batches, frozen in aliquots (for greater day-to-day reproducibility) and used indefinitely (*see* ref. 4). Remove the required amount, bring to room temperature and add the other ingredients for polymerization.
5. We find it is best to prepare this fresh each time.
6. We find that storing at 4 °C reduces its pungent smell.
7. Simple method of preparing running buffer: Prepare 10× native buffer (0.25 M Tris, 1.92 M glycine). Weigh 30.3 g Tris and 144 g glycine, mix and make it to 1 L with water. Dilute 100 mL of 10× native buffer to 990 mL with water and add 10 mL of 10 % SDS. Care should be taken to add SDS solution last, since it makes bubbles.
8. SDS precipitates at 4 °C. Therefore, the lysis buffer needs to be warmed prior to use.
9. Dilute 100 mL of 10× native buffer to 800 mL with water and add 200 mL of methanol. Avoid adding methanol directly to the 10× buffer, since it precipitates its ingredients. Even in such a scenario the precipitate can be redissolved by the addition of 800 mL water.
10. Add 100 mL of 10× TBS to a 1 L graduated cylinder and make it to about 800 mL with water. Transfer 50 g skim milk powder into the cylinder and mix stir until dissolved. Make to 1 L with water. Separate 500 mL as the blocking solution. To the remaining 500 mL add 250 μL of Tween 20 (cut end of blue tip to aspirate Tween easily), dissolve and use it as diluent.
11. Hoeffer Scientific was taken over by Pharmacia LB and now by GE Healthcare Piscataway, NJ, USA.

12. Cells were lysed four times, 15 s each time. The tube with the lysate is chilled on ice in between sonication steps to prevent the contents from heating up. Do this in a fume hood with an ear protector.
13. The gel cassette was sealed at the base using 1 % agarose. Overlay the resolving gel with water for gels having acrylamide concentration lower than 8 % and use isobutanol (or isobutanol saturated with water) for gels of 10 % or greater (*see* ref. 4). This overlay prevents contact with atmospheric oxygen, (which inhibits acrylamide polymerization) in addition to helping to level the resolving gel solution.
14. Centrifuging the samples prior to the run helps remove insoluble debris, which could produce streaks in the protein lanes (revealed when stained with Coomassie blue). Add a drop of 0.1 % BPB to the upper chamber buffer. This helps to form a much stronger dye front during the electrophoretic run.
15. Using a dilute fast green solution help to prevent overstaining of the membrane. Dilute stock fast green solution in gel destaining solution (25 % methanol, 10 % acetic acid). Use TBST to destain the membrane if it is overstained with fast green.
16. Strips can be excised very nicely using a razor blade. A Mini PROTEAN® 3 System glass plate is placed at an angle on the nitrocellulose at a distance of 2 mm from the edge and the razor is used to cut the strip (pull the razor blade along the sides of the glass plate to cut).
17. The fast green stained proteins bands in the main nitrocellulose membrane should be still visible. After aligning the two strips on either side of the membrane the developed band on the strips will line up with one of the bands on the membrane. This will permit the easy excision of the desired band from the main membrane.
18. This will prevent antibody loss as a consequence of adherence to the tubes.
19. Exercise universal serum handling precaution when handling human sera. Treat each serum sample as potentially hazardous.
20. It is possible to concentrate up to 500 μL with Centri-Prep 30 concentrator. If there is a need to reduce the volume further, use a 10,000 or 30,000 molecular weight cut off microcentrifuge tube.
21. If assaying for antibody purity (of the purified antibody) on western Blot, best results are obtained by incubating on nitrocellulose strips overnight at room temperature on the orbital shaker. This allows better binding of low titer antibody recovery and also washes should be longer (10 min instead of 5 min) and more frequent (five versus normally four washes).

22. We use paper boards placed in between stacks of ELISA plates in packages of ELISA plates (Costar, Cambridge, MA, USA) for this purpose.

Acknowledgement

This work was supported by NIH grant ARO1844 and Oklahoma Center for the Advancement of Science and Technology to RHS.

References

1. Kurien BT, Scofield RH (2006) Autoantibody determination in the diagnosis of systemic lupus erythematosus. Scand J Immunol 64:227–235, Review
2. Hansen BU, Eriksson S, Lindgren S (1991) High prevalence of autoimmune liver disease in patients with multiple nuclear dot, anticentromere, and mitotic spindle antibodies. Scand J Gastroenterol 26:707–713
3. Bernstein RM, Neuberger JM, Bunn CC, Callender ME, Hughes GR, Williams R (1984) Diversity of autoantibodies in primary biliary cirrhosis and chronic active hepatitis. Clin Exp Immunol 55:553–560
4. Harlow E, Lane D (1988) Electrophoresis. Appendix I. In: Harlow E, Lane D (eds) Antibodies. A laboratory manual. Cold Spring Harbor Laboratory Press, Cold Spring Harbor, NY, pp 283–318
5. Kurien BT, Scofield RH (1998) Multiple nuclear dot antinuclear antibody in patients without primary biliary cirrhosis. Indian J Biochem Biophys 35:41–47
6. Towbin H, Staehelin T, Gordon J (1979) Electrophoretic transfer of proteins from polyacrylamide gels to NC sheets: procedure and applications. Proc Natl Acad Sci U S A 76: 4350–4354
7. Olmsted JB (1981) Affinity purification of antibodies from diazotized paper blots of heterogeneous protein samples. J Biol Chem 256: 11955–11957

Chapter 28

Strip Immunoblotting of Multiple Antigenic Peptides

Biji T. Kurien

Abstract

Sodium dodecyl sulfate (SDS) polyacrylamide gel electrophoresis can be employed to efficiently separate multiple antigenic peptides (MAPs). Moreover, the electrophoresed MAPs are amenable for transfer to nitrocellulose membrane for immunoblotting. MAPs involve a hepta lysine core with end groups for anchoring multiple copies of the same synthetic peptide. MAPs are amenable to staining with Coomassie and silver on SDS polyacrylamide gels as well as by Fast Green on a blotted nitrocellulose membrane. They lend themselves to analysis on an immunoblot as they behave like low molecular weight proteins. Affinity immunoblotting for analysis of antibody clonotype distribution has also been carried out using these peptides.

Key words Multiple antigenic peptides, SDS-PAGE, Immunoblotting, Nitrocellulose

1 Introduction

MAPs (multiple antigenic peptides) consist of a hepta lysine backbone with end groups for anchoring multiple copies of the same synthetic peptide. Ever since MAPs were first introduced in 1988 [1] they have been used by several investigators for eliciting anti-peptide antibodies [2–7], antibodies to ribonucleoprotein complexes [8] and in enzyme linked immunosorbent assay [9, 10]. The potential of MAPs as vaccines [1, 11] has also been investigated. In addition to these studies, MAPs have also been used in surface plasmon resonance studies, double immunodiffusion, affinity column purification [12], and calcium binding assays [13].

We demonstrate in this report that multiple antigenic peptides (MAPs) can be efficiently separated on sodium dodecyl sulfate (SDS) polyacrylamide gels and transferred to a nitrocellulose membrane for subsequent use in western blot. MAPs lend themselves to Coomassie and silver on SDS-PAGE as well as by Fast Green on an immunoblot. Affinity immunoblotting for analysis of antibody clonotype distribution has been carried out using MAPs [14].

Biji T. Kurien and R. Hal Scofield (eds.), *Western Blotting: Methods and Protocols*, Methods in Molecular Biology, vol. 1312, DOI 10.1007/978-1-4939-2694-7_28, © Springer Science+Business Media New York 2015

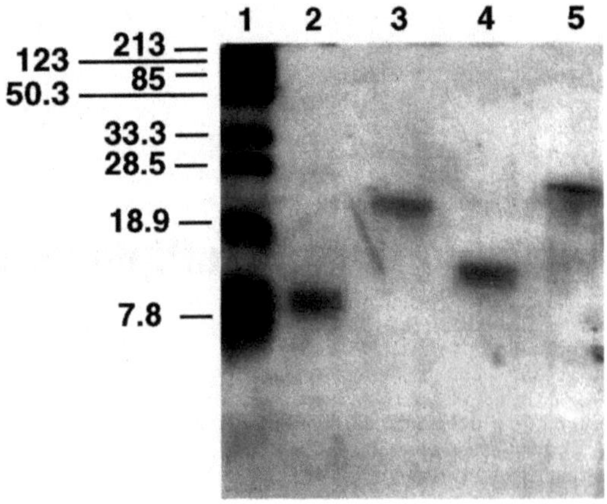

Fig. 1 MAPs from the human Ro 60 sequence analyzed on a 15 % SDS-PAGE and stained with Coomassie blue. *Lane 2, 3, 4* and *5* show the MAP's 1, 2, 3 and 4 respectively. *Lane 1* corresponds to prestained high range molecular weight standards. MAPs at a concentration of 5 μg/well (for Coomassie stain) and about a third of this for silver staining were used in SDS-PAGE. Immediately following electrophoresis the gel was stained with Coomassie brilliant blue or silver according to standard procedures. (Reproduced from ref. 21 with permission from Wiley-VCH)

Other investigators have used MAPs previously in SDS-PAGE [15, 16]. MAPs appear to migrate as smears in these studies. However, the MAPs used in this study migrated like low molecular weight proteins, as distinct bands. The difference most likely is on account of the fact that the MAPs used in our studies were purified by HPLC and shown to be homogeneous by mass spectroscopy. The data presented herein indicate that MAPs can be used in SDS-PAGE (Fig. 1) and immunoblot (Fig. 2) without any further modification.

2 Materials

Prepare all solutions using ultrapure water (prepared by purifying deionized water, to attain a sensitivity of 18 MΩ-cm at 25 °C) and analytical grade reagents. Prepare and store all reagents at room temperature (unless indicated otherwise). Diligently follow all waste disposal regulations when disposing waste materials. We do not use sodium azide in our reagents.

1. Ro 60 multiple antigenic peptides (MAPs): Twenty one Ro 60 MAPs were synthesized (*see* **Note 1**) from the sequence of the Ro 60 autoantigen [17, 18] (Molecular Biology Resource Facility, University of Oklahoma Health Sciences Center, Oklahoma City, OK, USA) by a manual stepwise solid phase procedure. An unrelated MAP with the sequence PPPGRRPP from the Sm autoantigen [8] was also synthesized (*see* Table 1).

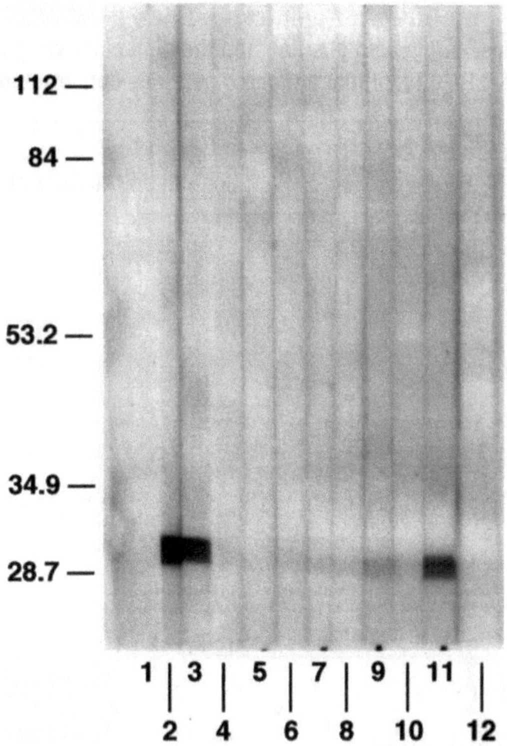

Fig. 2 Ro MAP 482–495 from the sequence of the human Ro autoantigen analyzed on an immunoblot after transfer from a 15 % SDS-PAGE. *Lane 1* corresponds to low range molecular weight standards. *Lanes 2* and *3*—probed with sera from mice immunized with 479–494 60 kDa Ro linear, peptide mouse sera binding to 482–495 Ro MAP (Table 1, MAP #4); *lanes 4* and *5*—negative peptide mouse sera; *lanes 6* and *7*—Freund's adjuvant only immunized mouse sera; *lane 8*—saline only injected mouse sera; *lane 9*—anti 52 kDa Ro human sera; *lane 10*—normal human sera; *lane 11*—anti 60 kDa Ro human sera and *lane 12*—conjugate control. Sera: DBA mice were immunized with a linear peptide synthesized from the 60 kDa human Ro sequence (Ro 479–494). Groups of mice were immunized with an unrelated peptide (a peptide from the sequence of the Human Lymphocyte Antigen-A2), with Freund's Aadjuvant only and with saline only. The peptides (100 μg) made up in 0.89 % saline was mixed well with Freund's complete adjuvant and injected subcutaneously and intraperitoneally. Subsequent boosts were in Incomplete Freunds adjuvant. The mice were bled from their tail veins. Blood was centrifuged and the antisera used in the immunoblots. Anti-Ro sera from systemic lupus erythematosus patients and normal sera were also used in immunoblots. (Reproduced from ref. 21 with permission from Wiley-VCH)

2. 15 % SDS polyacrylamide gels (*see* **Note 2**).
3. SDS PAGE running buffer: 0.025 M Tris, pH 8.3, 0.192 M glycine, 0.1 % SDS.
4. SDS PAGE lysis buffer: 0.3 M Tris–HCl (pH 6.8), 10 % SDS, 25 % β-mercaptoethanol, 0.1 % bromophenol blue, 45 % glycerol.

Table 1
The sequences, amino acid position and the molecular weight (as measured by mass spectrometry) of the different Ro 60 multiple antigenic peptides constructed from the sequence of the Ro 60 protein

Amino acid sequence of 60 kDa Ro-MAPs used	Location on the Ro protein	M. Wt. by mass spec. (in kDa)
1 TYYIKEQKLGL	45–55	11.69
2 SQEGRTTKQ	81–89	9.12
3 STKQAAFKAV	106–115	9.25
4 TFIQFKKDLKES	126–137	12.7
5 MKCGMWGRA	139–147	9.2
6 MWGRALRKAIA	143–153	11.02
7 LAVTKYKQRNGWSHK	166–180	15.37
8 LRLSHLKPS	183–191	9.25
9 VTKYITKGWKEVH	198–210	13.55
10 LYKEKALS	212–219	8.45
11 TEKLLKYL	222–229	8.9
12 EAVEKVKRTKDELE	230–243	14.23
13 HLLTNHLKSKEVWKAL	257–272	16.18
14 ALLRNLGKMTA	280–290	10.34
15 NEKLLKKARIHPFH	310–323	14.69
16 YKTGHGLRGKLKWRP	331–345	15.22
17 AAFYKTFKTV	355–364	10.25
18 VEPTGKRFL	364–372	9.21
19 MVVTRTEKDSY	401–411	11.4
20 LPMIWAQKTNTP	449–460	12.04
21 ALREYRKKMDIPAK	482–495	14.59
22 PPPGRRPP	Sm	7.63

HPLC purification and characterization by mass spectrophotometry confirmed the homogeneity and purity of these MAPs

Leave one aliquot at 4 °C for current use and store remaining aliquots at −20 °C (*see* **Note 3**).

5. BenchMark pre-stained high and low molecular weight standards.
6. 0.1 % bromophenol blue.
7. Nitrocellulose membranes.
8. Western blot transfer buffer: 0.025 M Tris, 0.192 M glycine, 20 % methanol.

9. Western blot transfer apparatus: Mini Trans-Blot apparatus.
10. Tris buffered saline (TBS)—10× solution: 1.5 M NaCl, 0.1 M Tris–HCl, pH 7.4.
11. Blocking solution: 5 % milk in TBS (*see* **Note 4**). Store at 4 °C.
12. Diluent solution: 5 % milk in TBST (*see* **Note 4**). Store at 4 °C.
13. Membrane strip mini-incubation containers.
14. Alkaline phosphatase (AP) buffer: 100 mM Tris–HCl, pH 9.5, 100 mM NaCl, 5 mM $MgCl_2$.
15. Nitro blue tetrazolium/5-bromo-4-chloro-3-indolyl phosphate, *p*-toluidine salt (NBT/BCIP) substrate stock: Dissolve 1 g NBT in 20 mL of 70 % dimethylformamide (DMF). Dissolve 1 g BCIP in 20 mL of 100 % DMF (*see* **Note 5**). Store at 4 °C.
16. Working substrate solution: Add 66 μL of NBT to 5 mL of AP buffer and then add 33 μL BCIP. Make up to 10 mL with AP buffer. Prepare this just prior to adding to membrane (*see* **Note 5**).
17. 0.1 % Fast green in 25 % methanol and 10 % acetic acid.

3 Methods

3.1 Immunoblotting

1. Perform regular 15 % SDS PAGE (preparative gel or 10-well gels) (*see* **Note 6**) [19] (*see* Chapters 11 and 34) and transfer to nitrocellulose membrane by Towbin's electrophoretic transfer method [20] (*see* Chapters 22 and 34; *see* **Note 7**).
2. Stain with fast green to visualize the MAPs transferred to the membrane (*see* **Note 8**).
3. Trim the edges of the membrane. Wrap the fast green stained membrane in cling wrap and make a photocopy for record keeping purposes.
4. Excise the lane (for strip immunoblotting) with molecular weight standards from the main nitrocellulose sheet containing the transferred protein. Excise a tiny wedge from the bottom left side of the marker lane and the main membrane sheet for orientation purposes (*see* **Note 9**).
5. Excise thin strips, about 2 mm in width, from the main nitrocellulose membrane sheet. Cut a tiny wedge from the bottom left side of the strips for orientation purposes. Place them in numbered 12-lane plastic trays with the antigen side up.
6. Destain strips with TBST. Decant TBST (*see* **Note 10**) once it becomes colored and replace with fresh TBST if necessary.
7. Block for 1 h with 1 mL each of blocking solution. Decant blocking.

8. Add 1 mL primary human or mouse sera at 1:100 dilution and incubate for 2 h. Decant (*see* **Note 11**).
9. Rinse strips with deionized water two to three times (*see* **Note 12**).
10. Wash 5× with TBST, 5 min each time. Dilute anti-human or anti-mouse IgG conjugated to alkaline phosphatase 1:5,000 with diluent and add 1 mL to each strip. Incubate for 1 h. Wash 5× with TBST, 5 min each time. Add 0.5 mL substrate and develop bands.
11. Align the strips (with developed bands) on immunoblotted strips on card board insert (*see* **Note 13**).

4 Notes

1. Twenty one MAPs, from the Ro 60 sequence (*see* Table 1) were synthesized at the Molecular Biology Resource Facility, University of Oklahoma Health Sciences Center, Oklahoma City. OK, USA. The peptide sequences were designed based on the regions on Ro 60 bound by anti-Ro 60 human sera, obtained from systemic lupus erythematosus patients using, in a PIN ELISA. The MAPs were quick HPLC purified and showed a single peak on mass spectrometric analysis [21]. In experiments varying the amount of MAP, as little as 250 ng of MAP could be detected on an immunoblot.
2. Four to 20 % gradient gels can also be used.
3. SDS precipitates at 4 °C. Therefore, the lysis buffer needs to be warmed prior to use.
4. Add 100 mL of 10× TBS to a 1 L graduated cylinder and make it to about 800 mL with water. Transfer 50 g skim milk powder into the cylinder and mix stir until dissolved. Make to 1 L with water. Separate 500 mL as the blocking solution. To the remaining 500 mL add 250 μL of Tween 20 (cut end of blue tip to aspirate Tween 20 easily), dissolve and use it as diluent.
5. Dissolve NBT in the bottle it came in. Likewise for BCIP. We started using stabilized ready-to-use BCIP/NBT substrate (Kirkegaard and Perry Laboratories, Gaithersburg, Maryland, USA) solution prior to switching to ECL plus chemiluminescence (Amersham/GE Healthcare, Piscataway, NJ, USA).
6. Do not heat MAPs with the lysis buffer, since it makes the MAPs to migrate as doublets at about 100,000 molecular weight on SDS-PAGE. Since they contain short sequences, there is no need to denature them with heat.
7. Good contact between membrane and gel is essential for good transfer. Add an extra foam if contact is insufficient.

8. Using a dilute fast green solution help to prevent overstaining of the membrane. Dilute stock fast green solution in gel destaining solution (25 % methanol, 10 % acetic acid). Use TBST to destain the membrane if it is overstained with fast green.
9. Strips can be excised very nicely using a razor blade. A Mini PROTEAN® 3 System glass plate is placed at an angle on the nitrocellulose at a distance of 2 mm from the edge and the razor is used to cut the strip (pull the razor blade along the sides of the glass plate to cut).
10. The tray can be gently slanted and the entire buffer can be discarded into the sink. The membrane strips remain stuck to the plates.
11. Decant diluted human sera into container having the diluted germicidal detergent Vesphene (Steris Corporation, St. Louis, MO, USA). Let sit in this for a couple of hours and then discard into sink.
12. Rinsing the membrane strips with deionized water two to three times will help remove a bulk of the nonspecific antibodies and other contaminants. This will help reduce the amount of TBST used subsequently and also reduce the number of washes. This wash helps to reduce nonspecific binding of NBT/BCIP to the strip. The water, owing to its low ionic strength compared to TBST, will be able to remove contaminants much better than TBST. Water is much cheaper compared to TBST, in terms of money and labor. Other investigators have found no reduction in detection of specific signals due to washing with water [22].
13. We use paper boards placed in between stacks of ELISA plates in packages of ELISA plates (Costar, Cambridge, MA, USA) for this purpose.

Acknowledgement

This work was supported by NIH grant ARO1844 and Oklahoma Center for the Advancement of Science and Technology to RHS.

References

1. Tam JP (1988) Synthetic peptide vaccine design: synthesis and properties of high-density multiple antigenic peptide system. Proc Natl Acad Sci U S A 85:5409–5413
2. Posnett DN, McGrath H, Tam JP (1988) A novel method for producing anti-peptide antibodies. Production of site specific antibodies to the cell antigen receptor beta-chain. J Biol Chem 263:1719–1725
3. Troalen F, Razafindratsita A, Puisieux A, Voeltzel T, Bohuon C, Bellet D et al (1990) Structural probing of human lutropin using antibodies raised against synthetic peptides constructed by classical and multiple antigen peptide system approaches. Mol Immunol 27: 363–368
4. Munesinghe DY, Clavijo P, Calle MC, Nuzzenzweig RS, Nardin E (1991)

Immunogenicity of multiple antigen peptides (MAP) containing T and B cell epitopes of the repeat region of the *P. falciparum* circumsporazoite protein. Eur J Immunol 21:3015–3020

5. McLean GW, Owsianka AM, Subak-Sharpe JH, Marsden HS (1991) Generation of anti-peptide and anti-protein sera. Effect of peptide presentation on immunogenicity. J Immunol Methods 137:149–157
6. Habluetzel A, Pessi A, Bianchi E, Rotigliano G, Esposito F (1991) Multiple antigen peptides for specific detection of antibodies to a malaria antigen in human sera. Immunol Lett 30:75–80
7. Francis MJ, Hastings GZ, Brown F, McDermed J, Lu Y-A, Tam JP (1991) Immunological evaluation of the multiple antigen peptide (MAP) system using the major immunogenic site of foot-and-mouth disease virus. Immunology 73: 249–254
8. James JA, Gross T, Scofield RH, Harley JB (1995) Immunoglobulin epitope spreading and autoimmune disease after peptide immunization: Sm B/B′-derived PPPGMRPP and PPGIRPP induce spliceosome autoimmunity. J Exp Med 181:453–461
9. Tam JP, Zavala F (1989) A novel approach to increase detection sensitivity of synthetic peptides in solid-phase immunoassays. J Immunol Methods 124:53–61
10. Briand J-P, Barin C, Regenmortel MHVV, Muller S (1992) Application and limitations of the multiple antigenic peptide (MAP) system in the production and evaluation of anti-peptide and anti-protein antibodies. J Immunol Methods 156:255–265
11. Tam JP, Lu Y-A (1989) Vaccine engineering: enhancement of immunogenicity of synthetic peptide vaccines related to hepatitis in chemically defined models consisting of T- and B-cell epitopes. Proc Natl Acad Sci U S A 86: 9084–9088
12. Scofield RH, Kurien BT, Zhang F, Mehta P, Kaufman K, Gross T et al (1999) Protein-protein interaction of the Ro-ribonucleoprotein particle using multiple antigenic peptides. Mol Immunol 36:1093–1106
13. Scofield RH, Kurien BT, Gross JK, Reed NB, Taylor AK, Dominguez N et al (2004) Interaction of calcium and Ro60: increase of antigenicity. Mol Immunol 41:809–816
14. Scofield RH, Kurien BT, Reichlin M (1997) Immunologically restricted and inhibitory anti-Ro/SSA in monozygotic twins. Lupus 6: 395–398
15. McLean GW, Cross AM, Munns MS, Marsden HS (1992) Rapid attachment of a helper T cell epitope to branched peptides by fragment condensation to give enhanced immunogenicity. J Immunol Methods 155:113–120
16. Yao ZJ, Kao MCC, Loh KC, Chung MCM (1994) Antibody against branched epitope as an affinity ligand to separate the parent protein. J Chromatogr A 679:190–194
17. Deutscher SL, Harley JB, Keene JD (1988) Molecular analysis of the 60 kD human Ro ribonucleoprotein. Proc Natl Acad Sci U S A 85:9479–9483
18. Ben-Chetrit E, Gandy BJ, Tan EM, Sullivan KF (1989) Isolation and characterization of a cDNA clone encoding the 60 kDa component of the human SS-A/Ro ribonucleoprotein autoantigen. J Clin Invest 83:1284–1292
19. Laemmli UK (1970) Cleavage of structural proteins during assembly of the head of bacteriophage T4. Nature 227:680–685
20. Towbin H, Staehelin T, Gordon J (1979) Electrophoretic transfer of proteins from polyacrylamide gels to nitrocellulose sheets: procedure and applications. Proc Natl Acad Sci U S A 76:4350–4354
21. Kurien BT, Jackson K, Scofield RH (1998) Immunoblotting of multiple antigenic peptides. Electrophoresis 19:1659–1661
22. Wu M, Stockley PG, Martin WJ II (2002) An improved Western blotting technique effectively reduces background. Electrophoresis 23: 2373–2376

Chapter 29

Double-Blotting: A Solution to the Problem of Nonspecific Binding of Secondary Antibodies in Immunoblotting Procedures

Françoise Lasne

Abstract

Nonspecific interactions between blotted proteins and unrelated secondary antibodies generate false positives in immunoblotting techniques. Some procedures have been developed to reduce this adsorption, but they may work in specific applications and be ineffective in others. "Double-blotting" has been developed to overcome this problem. It consists of interpolating a second blotting step between the usual probings of the blot membrane with the primary antibody and the secondary antibodies. This step, by isolating the primary antibody from the interfering proteins, guarantees the specificity of the probing with the secondary antibody. This method has been developed for the study of erythropoietin in concentrated urine since a strong nonspecific binding of biotinylated secondary antibodies to some urinary proteins is observed using classical immunoblotting protocols. However, its concept makes it usable in other applications that come up against this kind of problem. This method is expected to be especially useful for investigating proteins that are present in minute amounts in complex biological media.

Key words Double-blotting, Immunoblotting false positive secondary antibodies, Erythropoietin

1 Introduction

The development of a test for anti-doping control of erythropoietin (EPO), a hormone used in endurance sport to stimulate red blood cell production, has been a long and exacting task. The method is based on differentiation of natural and recombinant (used in case of doping) hormones in urine by their isoelectric profiles [1]. For this, urine is first submitted to ultra-filtration to concentrate EPO in retentates that are then subjected to isoelectric focusing. Immunoblotting of EPO is then performed using primary monoclonal anti-human EPO antibodies and secondary biotinylated goat anti-mouse IgG antibodies [2]. The major drawback of the ultrafiltration step is the resulting very high protein content of the retentates that are then subjected to the next step of isoelectric focusing. This is particularly true for urine samples taken

Biji T. Kurien and R. Hal Scofield (eds.), *Western Blotting: Methods and Protocols*, Methods in Molecular Biology, vol. 1312, DOI 10.1007/978-1-4939-2694-7_29, © Springer Science+Business Media New York 2015

at the end of a competition, due to proteinuria induced by physical exercise. This results in retentates with huge total protein contents (about 50 g/L for samples taken at rest and up to 200 g/L for samples taken after a physical exercise) for an EPO concentration generally no more than 4 μg/L. Such a situation is a real challenge for the classical immunoblotting procedures. Indeed, a strong nonspecific binding of secondary antibodies to some proteins present in the retentates was observed, completely masking the detection of EPO. All attempts to prevent or reduce this nonspecific binding were ineffective when working directly on the blotting membrane. The problem was solved by isolating the primary antibody from the interfering proteins on a second membrane that was then probed by the secondary antibody without any risk of nonspecific binding [3–6]. Gershoni has emphasized the difficulty of resolving such a problem [7]. For this, after it has been probed by the primary antibody, the membrane with the blotted proteins is assembled with a second blank membrane and submitted to a second blotting under acidic conditions. The primary antibody molecules are thus desorbed from their corresponding antigen and transferred onto the second membrane, whereas the antigen and the interfering proteins remain bound to the first one. The second membrane can then be probed by the secondary antibodies without the risk of nonspecific binding.

2 Materials

Since double blotting (DB) takes place after probing of the blotting membrane (B membrane) with a primary antibody and before probing with the secondary one (Fig. 1), the reagents and materials for these steps (blotting membrane, blocking and washing buffers, primary and secondary antibody solutions, possibly amplification system, development reagents) will not be indicated here, being specific for the application in which DB is introduced (*see* **Note 1**).

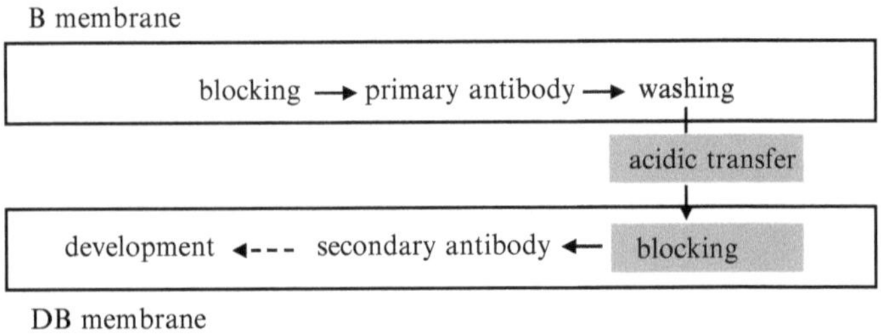

Fig. 1 Positioning of the double-blotting process into an immunoblotting procedure. The additional steps specifically related to DB are grey colored

Only the materials used for the DB step itself will be detailed.

1. Semidry transfer unit.
2. Roller.
3. Immobilon P polyvinylidene fluoride membranes (PVDF), 0.45 μm pore size.
4. Durapore hydrophilic PVDF membranes, 0.65 μm pore size
5. Filter paper sheets: Electrode paper Novablot.
6. 0.7 % (v/v) acetic acid solution.
7. Phosphate buffered saline (PBS) pH 7.4.
8. Sealing film, Parafilm.

3 Methods

1. Proceed to the usual blotting, blocking, primary antibody probing and washing steps of your B membrane according to your application. DB is performed after the last wash of your B membrane (*see* Fig. 1 and **Note 2**).
2. Cut two stacks of nine filter paper sheets, a Durapore (intermediate membrane) and an Immobilon P (DB membrane) membrane to the dimensions of the blotting membrane.
3. Condition them in 0.7 % acetic acid: just immerse the Durapore in the acidic solution for at least 10 min, pre-wet the Immobilon P membrane in methanol for 3 s, rinse in water for 2 min before equilibration in acidic solution for 10 min. The stacks of filter paper are moistened in acetic acid solution by capillary action.
4. During the same time, perform a rinsing of the B membrane in two changes of PBS.
5. Layer the B membrane of your application onto a first stack of filter paper with the blotted proteins facing up and cover it with the intermediate and DB membranes successively. Quickly put the second stack of filter paper onto the DB membrane to prevent the membranes from drying (*see* **Note 3**).
6. Position this sandwich on the anode plate of the semidry electrophoretic blotting instrument so that the B and the DB membranes will face the anode and the cathode, respectively (Fig. 2).
7. Place a sealing film onto the sandwich and carefully press out the air bubbles with the roller. Remove the film.
8. Place the cathode plate on the sandwich and connect the blotting instrument to the power supply.
9. Apply a constant intensity of 0.8 mA per cm^2 for 10 min (*see* **Notes 4–6**).

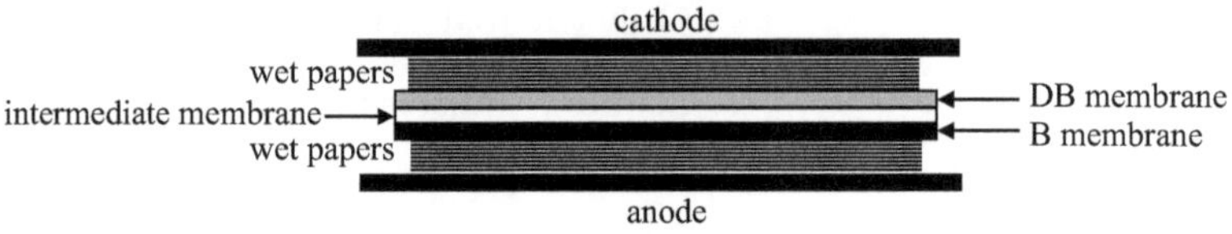

Fig. 2 Experimental set-up for DB

10. Disconnect the blotting instrument and disassemble the membranes (*see* **Note 3**).
11. Keep the B membrane in PBS at 4 °C (*see* **Note 7**).
12. Rinse the DB membrane quickly in two changes of PBS (*see* **Note 3**).
13. Proceed to the blocking of the DB membrane according to your usual procedure.
14. Proceed to the usual steps of your application from the probing with the secondary antibody to the final development, on the DB membrane.

4 Notes

1. Reagents for detection of erythropoietin. The method is illustrated by its application to immunodetection of erythropoietin. The specific reagents used in the case of EPO are:
   (a) Primary antibodies: monoclonal mouse anti-human EPO (clone AE7A5 from R&D, Abingdon, England).
   (b) Secondary antibodies: biotinylated goat anti-mouse IgG (H + L).
   (c) Amplifying systems: Streptavidin:biotinylated peroxidase complexes.
   (d) Development system: Chemiluminescence.
   (e) Blocking buffer: 5 % (w/v) nonfat milk in PBS buffer.
2. DB has been developed using PVDF as blotting membrane and has not been tested with other types of membranes.
3. Be aware that the DB membrane is very sensitive to drying when not saturated. This, while being not visually perceptible during the handling of the membrane, will produce a high background in the final image. It is thus extremely important to quickly perform the steps in which the membrane is handled without any liquid contact.
4. The acidic pH of this step induces dissociation of the primary antibody molecules from their corresponding antigen. The released antibodies, being positively charged due to the acidic pH, migrate towards the cathode, passing through

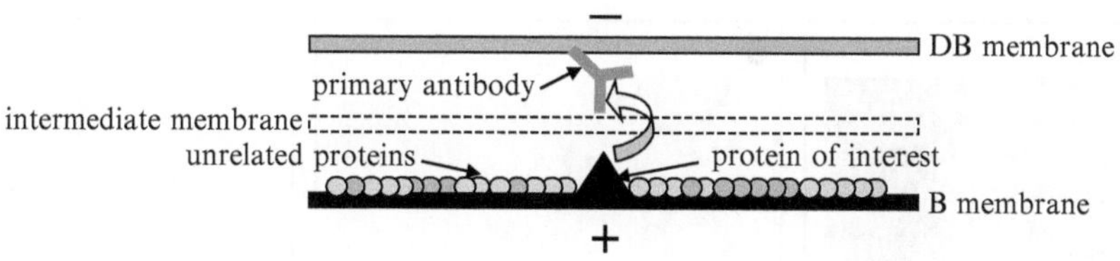

Fig. 3 Principle of double-blotting

the intermediate membrane and thus are transferred onto the DB membrane. Since the acidity does not affect hydrophobic interactions with PVDF, the antigen and the unrelated proteins (interfering proteins and proteins used for blocking the B membrane) are retained on the B membrane (Fig. 3). It should be noted that the result is actually an "image" of the probed antigen since it is only the antibody on the second membrane and not the antigen that gives rise to the final signal. However, this image is quite representative of the probed antigen (Fig. 4).

5. The interposition of the intermediate membrane between the B and the DB was very useful in our application (EPO analysis) which uses nonfat milk as a blocking agent, some "holes" in the final image obtained with the DB membrane being sometimes observed when this intermediate membrane was omitted (Fig. 3). Though the explanation for this is purely hypothetical (local releases of clumps of casein precipitated by the acidic pH), the interpolated microporous Durapore membrane worked as a barrier definitely remedying this problem. It is possible that this membrane is not necessary in other applications but its use is strongly advised in case of "holes" in the final image.
6. The use of an electric field speeds up the transfer of the primary antibody from the blotting to the DB membrane. However, a simple contact between the membranes (passive transfer) without applying an electric field for a prolonged time (30 min) has been tested in dot blot experiments and proved to be usable too (data not shown).
7. Storage of the B membrane (on which the blotted proteins are retained) in PBS may be useful. If some problem is observed on the final image, e.g., high background (*see* **Note 3**), it is possible to reuse the stored B membrane to perform a second DB. In this case, the B membrane is re-incubated in the primary antibody of the application and DB is then performed as described above. If desired, it is possible to probe the stored B membrane with a different primary antibody prior to performing a second DB. This enables obtaining an image of another antigen from the same sample.

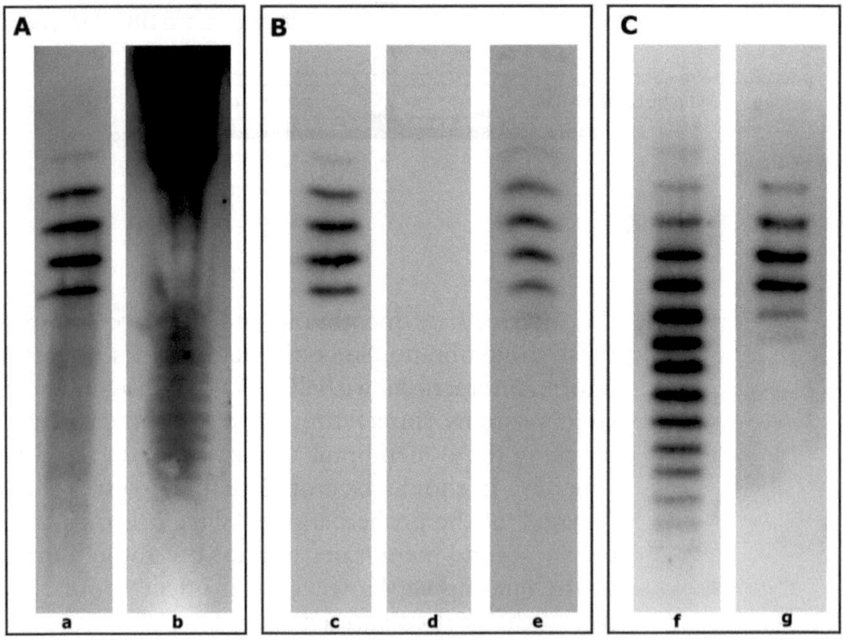

Fig. 4 Isoelectric patterns of EPO (**A**) images obtained without DB of (**a**) pure recombinant human EPO (rHuEPO) (Epoetin α), (**b**) a retentate obtained from ultrafiltration of a urine sample (chosen because it is devoid of endogenous EPO); (**B**) images obtained with DB of (**c**) pure rHuEPO (Epoetin α), (**d**) the same retentate as in (**b**), (**e**) rHuEPO introduced into this same retentate; (**C**) images obtained with DB of (**f**) a retentate from urine containing natural endogenous EPO, (**g**) a retentate from urine containing excreted rHuEPO. In the case of pure rHuEPO, comparison of the images obtained without (**a**) and with (**c**) DB shows that no significant change in the isoelectric pattern is induced by the DB process. In the case of a retentate, due to the binding of the secondary antibody on the urinary proteins, a very strong EPO nonspecific signal of is observed with the classical immunoblotting process (**b**). DB totally eliminates this nonspecific signal (**d**) and enables to specifically detect the introduced rHuEPO (**e**). An illustration of the use of DB in anti-doping control analysis is given in (**C**), natural endogenous urinary EPO (**f**) and administered rHuEPO excreted in urine (**g**) are differentiated by their isoelectric patterns

Acknowledgement

The DB process has been patented (2 786 273) by "Hospices Civils de Lyon", and by "Laboratoire National de Dépistage du Dopage" with "Hospices Civils de Lyon" as PCT/FR01/01331.

References

1. Lasne F, de Ceaurriz J (2000) Recombinant erythropoietin in urine. Nature 405:635
2. Lasne F, Martin L, Crepin N, de Ceaurriz J (2002) Detection of isoelectric profiles of erythropoietin in urine: differentiation of natural and administered recombinant hormones. Anal Biochem 311:119–126
3. Lasne F (2001) Double-blotting: a solution to the problem of non-specific binding of secondary antibodies in immunoblotting procedures. J Immunol Methods 253:125–131
4. Lasne F (2003) Double-blotting: a solution to the problem of non-specific binding of secondary antibodies in immunoblotting

procedures (protocol). J Immunol Methods 276:223–226

5. Kurien BT, Scofield RH (eds) (2009) Protein blotting and detection, vol 536, Methods in molecular biology series. Humana, New York, NY
6. Lasne F (2009) Double-blotting: a solution to the problem of nonspecific binding of secondary antibodies in immunoblotting procedures. Methods Mol Biol 536:213–219
7. Gershoni JM (1988) Protein blotting: a manual. In: Glick D (ed) Methods of biochemical analysis, vol 33. Wiley, New York, pp 1–58

Chapter 30

Method for Resolution and Western Blotting of Very Large Proteins Using Agarose Electrophoresis

Marion L. Greaser and Chad M. Warren

Abstract

Proteins larger than 200 kDa are difficult to separate electrophoretically using polyacrylamide gels, and their transfer during western blotting is typically incomplete. A vertical SDS agarose gel system was developed that has vastly improved resolving power for very large proteins. Complete transfer of proteins as large as titin (Mr 3,000–3,700 kDa) onto blots can be achieved. The addition of a sulfhydryl reducing agent in the upper reservoir buffer and transfer buffer markedly improves the blotting of large proteins.

Key words SeaKem agarose, Titin, DATD, Large protein blotting

1 Introduction

Proteins with large subunit size (~>200 kDa) are difficult to separate by electrophoresis because of their poor penetration into gels with the widely used Laemmli SDS (sodium dodecyl sulfate) polyacrylamide system [1]. Protein migration in SDS gels has been found to be linear with the log of the molecular weight [2], so the larger the protein, the more poorly it is resolved from other big proteins. Others have attempted to solve this problem by using very low concentration acrylamide gels [3], acrylamide mixed with agarose [4], or acrylamide gradients [5] to better separate large proteins. Low concentration acrylamide gels are mechanically fragile and distort easily during handling; these problems become magnified when blotting is attempted. An additional difficulty in blotting very large proteins is their poor transfer to the membrane. Inclusion of 2-mercaptoethanol in the transfer buffer improves transfer efficiency, but acrylamide gels stained after transfer typically still contain most of the giant muscle protein titin [6].

An electrophoresis system using SDS and agarose for protein electrophoresis and blotting has been described [7]. An example showing the resolution for several muscle samples containing large

Biji T. Kurien and R. Hal Scofield (eds.), *Western Blotting: Methods and Protocols*, Methods in Molecular Biology, vol. 1312, DOI 10.1007/978-1-4939-2694-7_30, © Springer Science+Business Media New York 2015

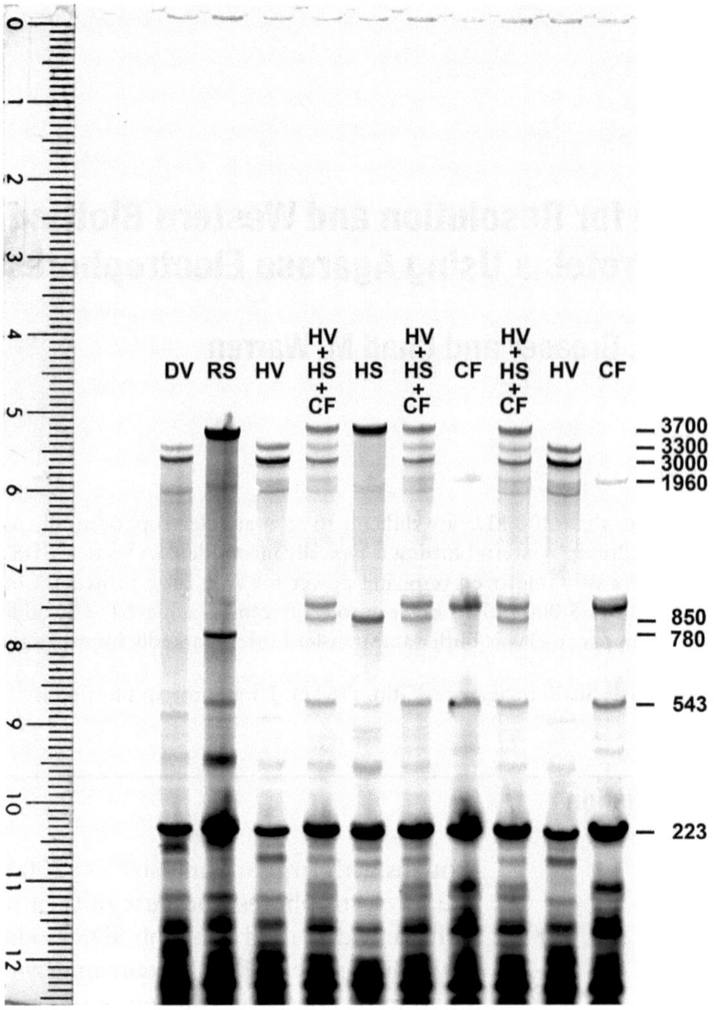

Fig. 1 SDS 1 % agarose gel stained with silver. A centimeter ruler is shown on the *left* and the sizes of the various protein bands in kDa are listed on the *right*. *DV* dog ventricle, *RS* rat soleus, *HV* human ventricle, *HS* human soleus, *CF* crayfish claw muscle. Human soleus titin is 3,700 kDa and human ventricle has two titin bands of 3,300 and 3,000 kDa. The bands at 780 and 850 kDa are rat and human nebulin, respectively. The myosin heavy chain is 223 kDa. Blotting proteins this size from acrylamide gels usually results in incomplete transfer, but full transfer can be achieved with agarose [7]

proteins is shown in Fig. 1. Migration distance shows a linear relationship with the log of the molecular weight [7]. This system allows reproducible and quantitative transfer of proteins from the gel in contrast to methods using low percentage acrylamide. The method has been widely adapted for studying muscle proteins, but it also has been used to study protein aggregates of the von Willenbrand Factor [8, 9] and Huntingtin [10].

2 Materials

2.1 Apparatus

1. SE 600 Slab Gel Unit with 16 × 18 cm glass plates (Hoefer) or a similar commercial gel unit (*see* **Note 1**).
2. 65 °C Oven.
3. A constant current power supply.
4. Circulating cooler.
5. TE62 Tank Blotting Unit, Hoefer.

2.2 Stock Solutions

1. Acrylamide gel for plug: 38.5 % acrylamide. Weigh 37.5 g of acrylamide and 1 g DATD (*N,N′*-diallyl-tartardiamide) into a beaker, add about 50 mL of water, stir till dissolved, dilute to 100 mL. Filter through a 0.45 μm filter. The solution should be stored in a brown bottle in the cold room (4 °C). Danger! Avoid skin contact.
2. Reservoir and agarose gel buffer concentrate (5×): 0.25 M Trizma base—1.92 M glycine—0.5 % SDS. Buffer concentrate can be stored at room temperature.
3. Ammonium persulfate; Prepare a 100 mg/mL solution in water; store frozen in 0.5 mL aliquots (stable indefinitely at -20 °C).
4. Sample buffer: 8 M urea, 2 M thiourea, 0.05 M Tris–HCl (pH 6.8), 75 mM DTT, 3 % SDS, 0.05 % bromophenol blue (adapted from ref. 11). (Dissolve urea and thiourea and treat with mixed bed resin to remove ionic constituents; then add remaining ingredients. Store at −20 °C).
5. 50 % v/v glycerol.
6. Transfer buffer: 20 mM Trizma base, 150 mM glycine, 20 % v/v methanol [7, 11] or 10 mM CAPS (*N*-cyclohexyl-3-aminopropanesulfonic acid), pH 11 [12]. For high molecular weight proteins, add SDS and 2-mercaptoethanol to 0.1 % and 10 mM, respectively to the transfer buffer.

3 Methods

3.1 Gel Preparation

1. Volumes listed will provide enough solution for two 16 × 18 cm gels with 1.5 mm spacers. One is used for staining either with Coomassie blue or with silver with a special procedure for agarose [7], the other for blotting.
2. Clean plates and spacers with soap, rinse with distilled water and finally with ethanol.
3. Assemble gel plates. Place plate on clean bench top. Place spacers hanging half the way off each side of plate. Place second

plate on top. Stand up plates and place one side into the clamp. Align spacer with side of plates and clamp and push spacer down so that bottom is flush with the glass plates (top buffer will leak if spacers are not flush with plates).

4. Pour acrylamide plugs in bottom of gel plate assembly (*see* **Note 2**): In a 15 mL plastic beaker add: 1.924 mL deionized water, 1.7 mL 50 % glycerol, 2.12 mL 3 M Tris (pH 9.3), 2.72 mL acrylamide (40 %), 24 μL 10 % ammonium persulfate, and 13 μL TEMED (tetramethylethylenediamine) (*see* **Note 2**). Mix by pipeting a few times. Immediately add 2.5 mL to each gel assembly. Add a small amount of water on top of each plug to level the upper surface and provide an oxygen barrier. Allow gel to polymerize for 20–30 min. Drain off water layer by inverting gel plate assembly on a paper towel.
5. Place assembly, 20 lane sample combs, and 60 mL plastic syringe in a 65 °C oven for 10 min (*see* **Note 3**).
6. Weigh 0.8 g of SeaKem Gold agarose SeaKem Gold Agarose (Lonza Group Ltd) (*see* **Note 4**) into a 600 mL beaker (*see* **Note 5**). To a 100 mL graduated cylinder add 48 mL of 50 % v/v glycerol (*see* **Note 6**), 16 mL 5× electrophoresis buffer, and bring volume up to 80 mL with deionized water. Place Parafilm over top of the graduated cylinder, mix by inverting a few times, and pour solution into the 600 mL beaker containing the agarose. Place Saran wrap over top of beaker and poke a few holes in the Saran wrap. Weigh beaker with contents. Place beaker in a microwave oven along with a separate beaker of deionized water. Heat for a total of 2 min (stop every 30 s to swirl—protect hand with an insulated glove) (*see* **Note** 7).
7. Allow agarose to cool for a few minutes at room temperature. Re-weigh, and add sufficient heated deionized water to replace that lost by evaporation.
8. Draw up about 40 mL of agarose in the pre-warmed 60 mL Luer-Lock syringe and pour each gel slowly until it just overflows the top of the plates. Try to avoid formation of bubbles (if bubbles present, bring them to the top of the gel and pinch them with the sample comb). Insert sample combs and allow unit to cool at room temperature for about 45 min (*see* **Note 8**).

3.2 Electrophoresis Setup and Sample Loading

1. Add 4 L of buffer to lower chamber (3,200 mL deionized water plus 800 mL 5× electrophoresis buffer). Start cooling unit and stir bar (gels run at 6 °C).
2. Prepare 600 mL upper chamber electrophoresis buffer (same concentration as lower chamber buffer). Add 2-mercaptoethanol (final concentration of 10 mM). Buffer will be poured into top chamber after samples are loaded and assembly placed in unit.

3. Take combs out of gels by bending them back and forth to detach from gel and slowly pull them up. Pour a small amount of upper chamber buffer into a 15 mL beaker and pipette buffer into first and last wells (the rest will fill over). Add buffer to remove any trapped bubbles. Insert pipette tip to deposit sample in bottom of the sample well. Skip the first and last lanes (*see* **Note 9**).
4. Running gels. Once samples are loaded, put upper chamber on the assembly. Pour upper chamber buffer into upper chamber from corners (don't pour buffer directly over wells). Place lid on unit, and connect to power supply. Turn electrophoresis unit on and run at 30 mA (2 gels) for 3 h.

3.3 Staining and Western Blotting

1. After tracking dye reaches the bottom of the acrylamide plug, turn off the power and disassemble the plates. Cut off sample wells and acrylamide plug and discard. Soak the remaining agarose gel in 10 mM CAPS (pH 11.0), 0.1 % SDS, and 10 mM 2-mercaptoethanol for 30 min with gentle shaking.
2. The gel is then placed on top of either a sheet of PVDF (polyvinylidene difluoride) or nitrocellulose, assembled into the transfer unit, and the protein electrophoretically transferred using 40 V constant voltage for 2–3 h (*see* **Note 10**).
3. Blotted proteins can then be treated using conventional procedures with either colorimetric (horseradish peroxidase or alkaline phosphatase substrates) or ECL (enhanced chemiluminescence) methods.

4 Notes

1. The agarose gel procedure works equally well with small format gels (i.e., 8 × 10 cm).
2. The acrylamide plug is used to prevent the agarose from slipping out of the vertical gel plate assembly. Use of DATD as the cross-linker provides a stickier bond of the acrylamide to the glass plates than if a conventional bisacrylamide cross-linker is used. Plugs can be poured a day before making the gel (place tape or Parafilm over the top of the plates to prevent drying and store in cold room).
3. Preheating the glass plate assembly, well comb, and syringe prevents premature agarose gelling when the solution touches the colder surfaces. In addition the plates are less likely to crack during pouring if they are closer to the temperature of the hot agarose.
4. The supplier for SeaKem Gold agarose has changed twice since 2003. Biowhittaker was succeeded by Cambrex who was

followed by Lonza Group Ltd, Muenchensteinerstrasse 38, CH-4002 Basel, Switzerland.

5. It is essential to use SeaKem Gold agarose for optimal migration of high molecular weight proteins. This type has large pore size and excellent mechanical stability. Other types of agarose may be used, but the protein mobility will be significantly reduced.
6. Glycerol is included in the mixture to increase the solution viscosity inside the gel and thus sharpen the protein bands.
7. Periodic swirling during the heating step eliminates non-hydrated agarose granules in the final gel.
8. Sample combs should extend no longer than 1 cm into agarose; otherwise they may be difficult to remove. Gels can be used right away or stored overnight in a cold room.
9. Conventional sample buffers may not be dense enough for the sample to stay at the bottom of the well. If necessary add additional glycerol (up to 30 % v/v final concentration) to increase sample density.
10. The disulfide bond formation of large proteins during electrophoresis also retards their migration out of the gel onto blots during transfer. Thus inclusion of 2-mercaptoethanol in the transfer buffer improves efficiency of transfer of high molecular weight proteins. The use of the agarose electrophoresis system with inclusion of 2-mercaptoethanol in the transfer buffer results in complete transfer of all high molecular weight proteins out of the gel, including titin (Mr 3,000–3,700 kDa subunit size) [7]. Alternatively, protein can be alkylated to prevent disulfide bond formation during the transfer process [13].

Acknowledgements

This work was supported by the College of Agricultural and Life Sciences, University of Wisconsin-Madison and from grants (MLG-NIH HL77196 and Hatch NC1184).

References

1. Laemmli UK (1970) Cleavage of structural proteins during the assembly of the head of bacteriophage T4. Nature 227:680–685
2. Weber K, Osborn M (1969) The reliability of molecular weight determinations by dodecyl sulfate-polyacrylamide gel electrophoresis. J Biol Chem 244:4406–4412
3. Granzier H, Wang K (1993) Gel electrophoresis of giant proteins: solubilization and silver-staining of titin and nebulin from single muscle fiber segments. Electrophoresis 14:56–64
4. Tatsumi R, Hattori A (1995) Detection of giant myofibrillar proteins connectin and nebulin by electrophoresis in 2 % polyacrylamide slab gels strengthened with agarose. Anal Biochem 224:28–31
5. Cazorla O, Freiburg A, Helmes M, Centner T, McNabb M, Wu Y, Trombitas K, Labeit S,

Granzier H (2000) Differential expression of cardiac titin isoforms and modulation of cellular stiffness. Circ Res 86:59–67

6. Fritz JD, Swartz DR, Greaser ML (1989) Factors affecting polyacrylamide gel electrophoresis and electroblotting of high-molecular-weight myofibrillar proteins. Anal Biochem 180:205–210
7. Warren CM, Krzesinski PR, Greaser ML (2003) Vertical agarose gel electrophoresis and electroblotting of high-molecular-weight proteins. Electrophoresis 24:1695–1702
8. Ott HW, Griesmache A, Schnapka-Koepf M, Golderer G, Sieberer A, Spannagl M, Scheibe B, Perkhofer S, Will K, Budde U (2010) Analysis of von Willebrand Factor multimers by simultaneous high- and low-resolution vertical SDS-agarose gel electrophoresis and Cy5-labeled antibody high-sensitivity fluorescence detection. Am J Clin Pathol 133:322–330
9. Akiyama R, Komori I, Hiramoto R, Isonishi A, Matsumoto M, Fujimura Y (2011) H1N1 influenza (swine flu)-associated thrombotic microangiopathy with a markedly high plasma Ratio of von Willebrand Factor to ADAMTS13. Intern Med 50:643–647
10. Hoffner G, Island M-L, Djian P (2005) Purification of neuronal inclusions of patients with Huntington's disease reveals a broad range of N-terminal fragments of expanded huntingtin and insoluble polymers. J Neurochem 95:125–136
11. Yates LD, Greaser ML (1983) Quantitative determination of myosin and actin in rabbit skeletal muscle. J Mol Biol 168:123–141
12. Matsudaira P (1987) Sequence from picomole quantities of proteins electroblotted onto polyvinylidene difluoride membranes. J Biol Chem 262:10035–10038
13. Sechi S, Chait BT (1998) Modification of cysteine residues by alkylation. A tool in peptide mapping and protein identification. Anal Chem 70:5150–5158

Chapter 31

Immunodetection of P-Selectin Using an Antibody to Its C-Terminal Tag

Padmaja Mehta-D'souza

Abstract

P-selectin is a multi-domain glycoprotein expressed on activated endothelial cells and activated platelets. We previously expressed a recombinant form of P-selectin containing only its N-terminal lectin and EGF domains in CHO-Kl cells and showed that these two domains are sufficient to mediate ligand binding. We have now expressed the same construct in CHO-Lecl cells that make truncated glycans. The uniform glycosylation in these cells should make it easier to crystallize this protein.

Key words SDS-PAGE, Western blotting, Chromogenic detection, Immobilon-P, P-selectin, P-selectin glycoprotein ligand-1 (PSGL-1)

1 Introduction

Selectins are multi-domain glycoproteins that play an important role both in inflammation and tissue injury [1]. P-selectin is expressed on activated endothelial cells and platelets, E-selectin is expressed on activated endothelial cells and L-selectin is expressed constitutively on leukocytes. The tetrasaccharide sialyl Lewis x forms the core structure recognized by all selectins [2]. In addition, P-selectin and L-selectin require sulfation of one or more N-terminal tyrosine residues in their cognate ligand P-selectin glycoprotein ligand-1 (PSGL-1). L-selectin also requires sulfation of one or more GlcNAc residues in its ligands on HEV mucins on lymph nodes [3].

All selectins contain an N-terminal lectin domain, an EGF domain, a variable number of consensus repeats, a transmembrane domain and a cytoplasmic tail. Using a recombinant truncated form of soluble P-selectin expressed in CHO-Kl cells, we have shown that the lectin and EGF domains of P-selectin are sufficient to bind its ligand, PSGL-1 [4].

Biji T. Kurien and R. Hal Scofield (eds.), *Western Blotting: Methods and Protocols*, Methods in Molecular Biology, vol. 1312, DOI 10.1007/978-1-4939-2694-7_31, © Springer Science+Business Media New York 2015

We also generated a recombinant, soluble, monomeric form of P-selectin in the CHO-Lec1 cell line for further structure–function studies. The recombinant protein was purified by affinity chromatography over an HPC4 antibody column [5] and its purity checked by SDS-PAGE followed by silver-staining and western blotting.

2 Materials

2.1 P-selectin lec-EGF

1. P-selectin lec-EGF was purified by HPC4 affinity chromatography from the conditioned medium of transfected CHO-Lec1 cells.

2.2 SDS-Polyacrylamide Gel Electrophoresis (SDS-PAGE)

1. 30 % Acrylamide/bis solution (37.5:1 with 2.6 % C) (*see* **Note 1**).
2. Separating gel buffer (3×); 1 M Tris–HCl, pH 8.8. Store at 4 °C.
3. Stacking gel buffer (5×); 1 M Tris–HCl, pH 6.8. Store at 4 °C.
4. Gel loading buffer: 62.5 mM Tris–HCl, pH 6.8, 5 % β-mercaptoethanol, 10 % glycerol, 2 % SDS. A 3× version of the gel-loading buffer was also made by reducing the amount of water used to make up the final volume (*see* **Note 6**).
5. 10 % SDS solution. (Exercise care when using this chemical in the dry form.) Store at room temperature.
6. *N,N,N,N′* Tetramethyl-ethylenediamine (TEMED): Make a 10 % solution in water. This should be made fresh every time to ensure good polymerization of gels.
7. Ammonium persulfate (APS): Make a 10 % solution in water. This should also be made fresh every time to ensure good polymerization (*see* **Note 2**).
8. Running Buffer for SDS-PAGE: 25 mM Tris, 190 mM glycine, 0.2 % (w/v) SDS. This buffer was made as a 10× solution and diluted with water before use. Store at room temperature.
9. β-Mercaptoethanol.
10. Rainbow Molecular Weight Markers.

2.3 Western Blotting

1. Setup Buffer: 25 mM Tris, 190 mM Glycine, 20 % (v/v) Methanol. Do not adjust the pH of the Tris solution.
2. Immobilon-P membrane.
3. Methanol.
4. Thick filter paper.

2.4 Immunostaining

1. Equilibration buffer: 25 mM Tris, 190 mM Glycine, 20 % Methanol. Do not adjust the pH of the Tris solution.
2. 20 mM Tris–HCl pH 7.5, 100 mM NaCl, 1 mM $CaCl_2$ (TBS) (*see* **Note 3**).

3. Wash buffer: TBS with 1 % Tween 20 (*see* **Note 4**).
4. Blocking Solution: Nonfat dry milk, 5 % in TBS (*see* **Note 5**).
5. Primary antibody: Diluted to 5 μg/mL in blocking solution.
6. Secondary antibody (biotinylated): 5 μg/mL in TBS containing 0.1 % BSA.
7. Vectastain kit (ABC) (peroxidase conjugate).
8. Chromogenic substrate: 4-chloro-1-naphthol.

3 Methods

Over the years, several antibodies have been raised against human P-selectin [6, 7]. Many of these antibodies have been shown to have epitopes that lie mainly within the lectin domain. Most of these antibodies recognize three-dimensional (conformational) epitopes and hence will stain only gels that have been run without treating the samples with β-mercaptoethanol (i.e., under non-reducing conditions). This general principle is applicable to other proteins as well, hence when staining any protein, the antibody used for the western blot should be selected carefully. The antibody used for staining here is an antibody to its C-terminal epitope tag, and will stain under reducing conditions as well (*see* **Note 6**).

3.1 P-selectin lec-EGF (Sample for Western Blotting)

P-selectin lec-EGF was purified by HPC4 affinity chromatography from conditioned medium of transfected CHO-Lec1 cells as previously described [4].

3.2 SDS-Polyacrylamide Gel Electrophoresis (SDS-PAGE)

A minigel apparatus, made by Atto (Tokyo, Japan) was used. The proteins were electrophoresed in buffer containing 2-mercaptoethanol essentially as described by LaemmLi [8], using a discontinuous SDS-polyacrylamide gel system with a 12.5 % separating gel.

3.3 Western Blotting

This technique is used to electrophoretically transfer the samples run on SDS-PAGE, to a membrane support, like Immobilon-P, which is a PVDF membrane [9]. The directions below are specifically applicable to the Bio-Rad Trans-Blot SD Semi-dry Transfer cell. The gel equilibration steps and the assembly of the transfer cassette will be similar even when using other types of transfer apparatus, including a transfer tank system. However, the volume of buffer required and the length of run will differ.

1. After the dye front in the separating gel reaches the base of the gel, turn the power off and disconnect the gel unit. Pour out the running buffer, remove the gel cassette from the unit and disassemble. Very carefully, pry the two plates open, allowing the gel to remain on the straight plate. The straight plate is laid

down on the bench with the gel face up. Cut out the stacking gel and make a small mark in the left corner at the base of the separating gel near the dye front.

2. Place the separating gel in a small flat trough (the clean cover of a tip box is a good option) containing western blotting setup buffer. Allow the gel to equilibrate in this buffer on a rocker for 5–10 min.
3. Fill two more flat troughs with the setup buffer, and one trough each with water and methanol.
4. Cut a piece of Immobilon-P the same size as the separating gel (*see* **Note** 7). Place it in the methanol container for 1 min, drip dry, and place it in the trough of water for 2 min. Place this container on the rocker and ensure that the membrane is completely submerged in the water. After 2 min, transfer the membrane to a fresh trough containing setup buffer.
5. Take two pieces of extra thick filter paper and wet them in a fresh trough containing setup buffer. Place one wet filter paper on the transfer apparatus surface. Layer this with the piece of Immobilon that is soaked in buffer. Place the equilibrated gel over the Immobilon and complete the assembly by placing the second price of moist filter paper over the sandwich.
6. Place the lid of the transfer apparatus on this sandwich and connect the electrodes to the power supply. Run the transfer at 15 V for 20 min with the current limit at 350 mA per gel (increase the time if transferring two gels, but do not increase the voltage beyond 25 V).

3.4 Immunostaining

1. After the transfer is complete, remove the lid of the transfer apparatus carefully. Separate the gel sandwich and remove the Immobilon membrane (*see* **Note 8**).
2. Place the Immobilon membrane in the blocking solution in a flat trough and keep this on a rocker at room temperature for 1 h (*see* **Note 9**).
3. Dilute the primary antibody (HPC4, an anti Protein C antibody which reacts with its epitope fused to the C-terminus of P-selectin) to a concentration of 5 μg/mL into blocking solution (add one drop of normal horse serum) and incubate the proteins on the membrane with the primary antibody for one hour at room temperature. This incubation can also be performed at 4 °C overnight).
4. Wash the membrane with four changes of transfer wash buffer over 20 min, on a rocker.
5. Transfer the membrane to a 5 μg/mL solution of the biotinylated secondary antibody, made in dilution buffer (add 2 drops antibody for every 20 mL buffer). Incubate the membrane

with the secondary antibody at room temperature for 45 min, on a rocker.

6. Wash the membrane with four changes of transfer wash buffer over 20 min, on a rocker.
7. In the meantime, prepare the Vectastain ABC reagent by adding four drops of reagent A and four drops of reagent B to 20 mL of the dilution buffer. Mix well and let it stand at room temperature for 30 min.
8. Transfer the membrane to the Vectastain ABC reagent prepared above and incubate on a rocker for 30 min at room temperature.
9. Wash the membrane with three changes of transfer wash buffer over 15 min, on a rocker. Wash once with TBS.
10. In the meantime prepare the peroxidase substrate solution containing 0.5 mg/mL of 4-chloro-1-naphthol as follows: dissolve 30 mg of 4-chloro-1-naphthol in 10 mL of chilled methanol. To this, add 50 mL TBS and 30 μL of 30 % hydrogen peroxide (*see* **Note 10**).
11. Incubate the membrane with the substrate solution at room temperature. Color development should occur within 5–10 min (*see* **Note 11**).
12. After sufficient color has developed and the bands are clearly visible as shown in Fig. 1, wash the membrane twice in deionized water and air dry for storage. After the membrane is dry, store it wrapped in Saran wrap.

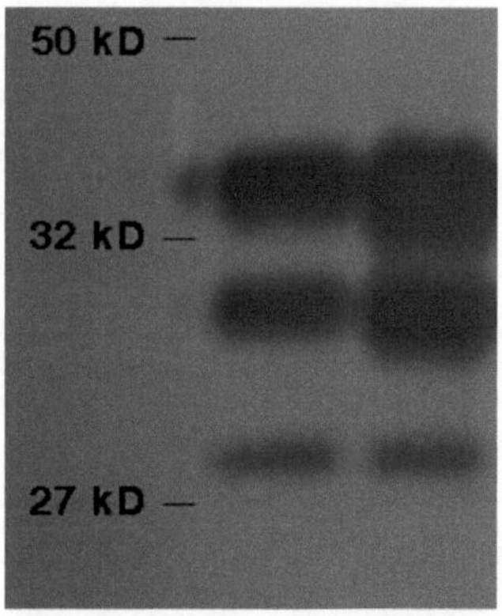

Fig. 1 Electrophoresis of P-selectin lec-EGF expressed in CHO-Lec1-cells. After SDS-PAGE, the proteins were transferred to Immobilon-P and stained with the HPC4 antibody. P-lec-EGF has three potential *N*-glycosylation sites and the three bands seen on the Western blot represent three different glycosylated forms of the protein

4 Notes

1. This is a neurotoxin, so great care should be taken while handling this chemical. Use gloves and do not breathe fumes.
2. For the sake of convenience, this can be made in a larger volume and then stored in single use (200 μL) aliquots at −20 °C.
3. The calcium chloride is not required for the technique itself and its addition is optional. We add it for the stability of our protein.
4. We have observed that a wash buffer containing Triton-X-100 instead of Tween 20 is not as effective.
5. Nonfat dry milk contains carbohydrates. When using detection reagents that stain carbohydrates (such as lectins and certain antibodies), a different blocking solution (such as 3 % BSA) should be used, or else you will get a high background.
6. When running samples under non-reducing conditions, the β-mercaptoethanol should be omitted from the buffer.
7. Always use gloves when handling the gel and the Immobilon membrane.
8. After the transfer, the gel can be stained with a protein stain to assess the quality of the transfer so that it can be optimized for subsequent runs. If a large amount of higher molecular weight bands are still visible on the gel, this indicates that the proteins should be transferred for a longer time.
9. The amount of blocking or antibody solution used should be sufficient to keep the membrane completely submerged even when the container is on the rocker.
10. The conjugated enzyme will vary depending on the ABC kit used. For peroxidase conjugates, other substrates that may be used include TMB and DAB-Ni. The entire immunostaining protocol will remain the same, only the substrate preparation and incubation times will vary.
11. It is recommended that for color development, a fresh container be used for incubation with substrate.

Acknowledgement

The author thanks Dr. Rodger P. McEver for the soluble, recombinant P-selectin and Dr. Charles Esmon for the HPC4 antibody.

References

1. McEver RP, Zhu C (2010) Rolling cell adhesion. Thromb Hemost Annu Rev Cell Dev Biol 26: 363–396
2. McEver RP (2002) Selectins: lectins that initiate cell adhesion under flow. Curr Opin Cell Biol 14:581–586
3. McEver RP (2005) A sulfated address for lymphocyte homing. Nat Immunol 6:1067–1069
4. Mehta P, Patel KD, Laue T, Erickson HP, McEver RP (1997) Soluble monomeric P-selectin containing only the lectin and epidermal growth factor domains binds to P-selectin Glycoprotein Ligand-1 on leukocytes. Blood 90:2381–2389
5. Stearns DJ, Kurosawa S, Sims PJ, Esmon NL, Esmon CT (1988) The interaction of a Ca^{2+}-dependent monoclonal antibody with the protein C activation peptide region. Evidence for obligatory Ca^{2+} binding to both antigen and antibody. J Biol Chem 263:826–832
6. Geng JG, Bevilacqua MP, Moore KL, McIntyre TM, Prescott SM, Kim JM, Bliss GA, Zimmerman GA, McEver RP (1990) Rapid neutrophil adhesion to activated endothelium mediated by GMP-140. Nature 343:757–760
7. Boukerche H, Ruchaud-Sparagano MH, Rouen C, Brochler J, Kaplan C, McGregor JL (1996) A monoclonal antibody directed against a granule membrane protein (GMP-140/PADGEM, P-selectin, CD62P) inhibits ristocetin-induced platelet aggregation. Br J Haematol 92: 442–451
8. LaemmLi UK (1970) Cleavage of structural proteins during the assembly of the head of bacteriophage T4. Nature 227:680–685
9. Towbin H, Staehelin T, Gordon J (1979) Electrophoretic transfer of proteins from polyacrylamide gels to nitrocellulose sheets: procedure and some applications. Proc Natl Acad Sci U S A 76:4350–4354

Chapter 32

Improvements and Variants of the Multiple Antigen Blot Assay-MABA: An Immunoenzymatic Technique for Simultaneous Antigen and Antibody Screening

Oscar Noya, Sandra Losada, Marilyan Toledo, Adriana Gauna, María Angelita Lorenzo, Henry Bermúdez, and Belkisyolé Alarcón de Noya

Abstract

This simple, versatile, reliable, reproducible, multipurpose, and inexpensive technique is based on the adhesion of different antigens to a single nitrocellulose strip using, as template, an acrylic device containing 28 parallel channels. The inclusion of channels containing normal human serum improves the quality control of this assay. Antigen-sensitized nitrocellulose strips are cut perpendicularly to the antigen-rows, exposed to immune sera followed by the appropriate conjugate. Positive signals are recorded using chemiluminescent or precipitable colorimetric substrates. This assay allows the simultaneous qualitative demonstration of antigenicity and immunogenicity of antigens obtained as synthetic peptides, recombinant molecules, or crude preparations, with high sensitivity and specificity. Its major value is based on the rapid and simultaneous comparative evaluation of various antigenic preparations allowing the diagnosis of a variety of infectious, allergic, and autoimmune diseases. It can in general be used to detect any type of antibody or circulating antigen. Some improvements and variants of the original technique are included.

Key words Multiple, Blot, Antigen, MABA, Immunochromatographic, Multiscreen enzyme-immunoassay, Immunodiagnosis, Luminescent, Colorimetric

1 Introduction

Three diagnostic techniques comply with the concept of simultaneous multidiagnosis: the antigen/antibody microarrays [1], the multiple laser detection beads (Multiplex®) [2] and the multiple antigen blot assay (MABA) [3, 4]. The first two have been used for the detection of antigens, antibodies or nucleic acids, while the third one only for antibodies, so far.

The occurrence of simultaneous infections is not an infrequent finding worldwide; therefore it would be ideal for health personnel to be aware of a patient's current or past history of contact with the most

Biji T. Kurien and R. Hal Scofield (eds.), *Western Blotting: Methods and Protocols*, Methods in Molecular Biology, vol. 1312, DOI 10.1007/978-1-4939-2694-7_32, © Springer Science+Business Media New York 2015

important pathogens, as well as his/her immune status, in order to have a more integral knowledge of each patient for a more rationale medical practice. To this end, MABA can be implemented and adapted for particular populations and institutions, such as blood banks, pediatric diseases, immunocompromised patients, vaccine evaluation, this being a major line in research of our laboratory. It is also a practical, rapid, reproducible, sensitive, and cheap technique for the identification and evaluation of antigens (antigenicity and immunogenicity) and sera. This can be done with crude, chemically synthesized, or recombinant antigens [5–9]. Allergic conditions detecting specific IgE have been successfully evaluated with this technique [10]. Another advantage is that a single MABA assay allows the evaluation of at least 26 different sera against 26 different antigens this being equivalent to 13–14 conventional ELISA plates. The cost of evaluation of each antigen per patient is \$0.006 by MABA versus \$0.129 by conventional ELISA. For this technique, an acrylic device with grooves is needed, so nitrocellulose paper can be sensitized with different antigens. Moreover, this device can be made with the appropriate acrylic plastic by a skilled artisan in various formats depending on the requirements (number and width of channels, etc.).

This technique has been used for other purposes such as the identification and selection of the best candidate antigens [11] as well as the identification of a specific molecule from eluent fractions during a chromatographic separation [12]. Schematic diagram of the whole protocol of MABA is presented in Fig. 28.

2 Materials

The quality of the water is critical. We recommend the use of Type 1 ultrapure water (18.2 MΩ).

1. Miniblotter® Immunetics Inc, Cambridge, MA, USA or Mini-PROTEAN® Multiscreen Apparatus Bio-Rad.
2. Nitrocellulose Membrane Trans-Blot.
3. Carbonate–bicarbonate buffer, 0.05 M, pH 9.6: 1.59 g sodium carbonate (Na_2CO_3); 2.52 g sodium bicarbonate ($NaHCO_3$). Adjust pH to 9.6 and complete to a final volume of 1,000 mL with distilled water. Store at 2–8 °C.
4. Phosphate Buffered Saline (PBS)-0.05 % Tween 20 (PBST), pH 7.5; Solution A (0.2 M NaH_2PO_4): 24 g NaH_2PO_4 in 1,000 mL distilled water; solution B (0.2 M Na_2HPO_4): 5.68 g Na_2HPO_4 in 200 mL distilled water; 13 mL solution A + 87 mL solution B + 8.76 g NaCl + up to 800 mL distilled water. Adjust pH to 7.5 with solutions A or B. Then, complete to 1,000 mL with distilled water and add 0.5 mL of Tween 20. Store at 4 °C.
5. Tris Buffered Saline Buffer (TBS)-0.05 % Tween20; 6.05 g Tris (50 mM); 8.76 g NaCl (150 mM), pH 7.5. Dissolve in 800 mL

of H_2O. Adjust pH to 7.5 with 1 M HCl and make volume up to 1 L with H_2O. Add 0.5 mL Tween20. Store at 4 °C.

6. Blocking solution (5 % nonfat milk in PBST): 5 g nonfat milk in 100 mL PBST. Prepare fresh.
7. Anti-human IgG (Fc specific) horseradish peroxidase antibody produced in goat. For continuous use, store at 2–8 °C up to 1 month. For extended storage, the solution may be frozen in working aliquots at −20 °C, since repeated freezing and thawing is not recommended. Dilute to 1:30,000 in blocking solution. Standardization is necessary for each conjugate and each brand. Prepare fresh. Anti-human IgG alkaline phosphatase (AKP) conjugated antibody should be used at approximately 1:25,000. If MABA is performed with AKP conjugate, TBST rather than PBST should be used. Other immunoglobulins (IgM, IgE, IgA, IgG1, etc.) may also be evaluated.
8. SuperSignal® (West Pico Chemiluminescent Peroxidase Substrate, Pierce, USA,). The two solutions should be mixed 1:1. Prepare fresh.
9. TMB membrane peroxidase substrate. This substrate contains 3,3′, 5,5′- tetramethylbenzidine in acidic buffer and is a very sensitive chromogenic substrate for peroxidase detection, developing a deep blue color in the presence of the horseradish peroxidase conjugates. The substrate is ready to use. Store at 2–8 °C.
10. BCIP/NBT substrate for alkaline phosphatase, ready to use.
11. Acetate buffer 0.05 M, pH 4.8: 110 mL Solution A + 15 mL Solution B + distilled water to a final volume of 500 mL. Solution A: Sodium acetate 0.2 M: 5.44 g sodium acetate·$3H_2O$ in 200 mL distilled water; solution B: Acetic acid 0.2 M: 2 mL acetic acid + 172 mL distilled water. The buffer may be stored at 2–8 °C for up to 1 month.
12. Sodium metaperiodate (SMP), stock solution 0.02 M in acetate buffer: 85.6 mg SMP in 20 mL acetate buffer. Prepare fresh.
13. 0.05 M sodium borohydride ($NaBH_4$) in PBS: 0.189 g $NaBH_4$ in 100 mL PBS. Prepare fresh.
14. Hyperfilm®.

3 Methods

3.1 Luminescent MABA [3, 4]

1. A rectangle of nitrocellulose (NC) paper (7.5 × 10 cm), is cut (Fig. 1) and labeled with an indicator line along one of the borders with a fine ball point-water resistant pen (Fig. 2).
2. The paper is soaked in distilled water for 5 min (Fig. 3).
3. The wet paper is aligned in the upper part of the Miniblotter®, with the reference line parallel to channel 1 (Fig. 4). Then, a plastic cushion is placed over the nitrocellulose sheet (Fig. 5).

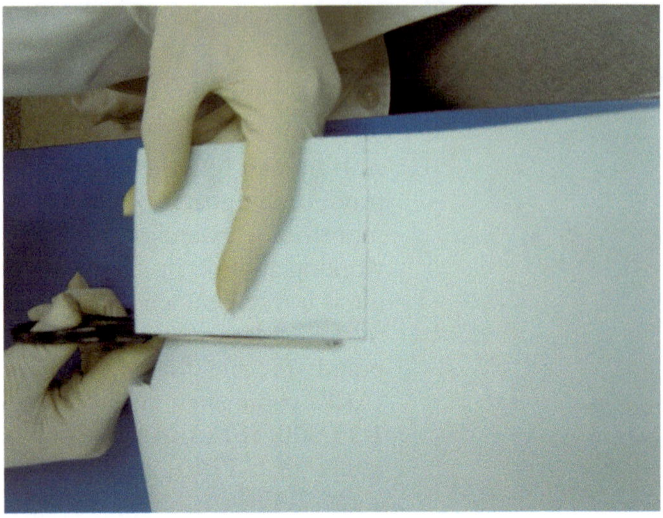

Fig. 1 Cut a rectangle of NC sheet 7.5 × 10 cm

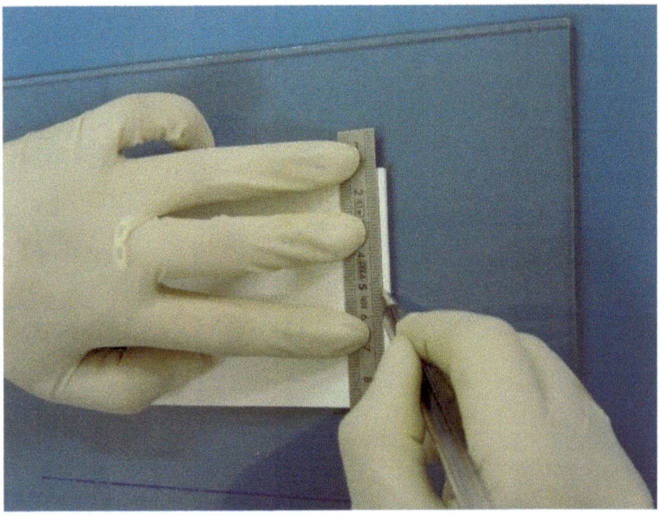

Fig. 2 Reference line in one of the borders

4. The Miniblotter® is closed with the lower piece and hand tightened with the two provided screws (Fig. 6). Avoid excessive tightening. The remaining water in the parallel channels is removed with pipette (Fig. 7). The next steps should be carried out in less than 30 min, to prevent drying and possible cracking of the paper.
5. The different antigen preparations (crude, synthetic peptides or recombinant molecules) up to a maximum of 26 can be analyzed on each NC sheet, at a concentration between 10 and 50 μg/mL in carbonate–bicarbonate buffer, pH 9.6. Inclusion

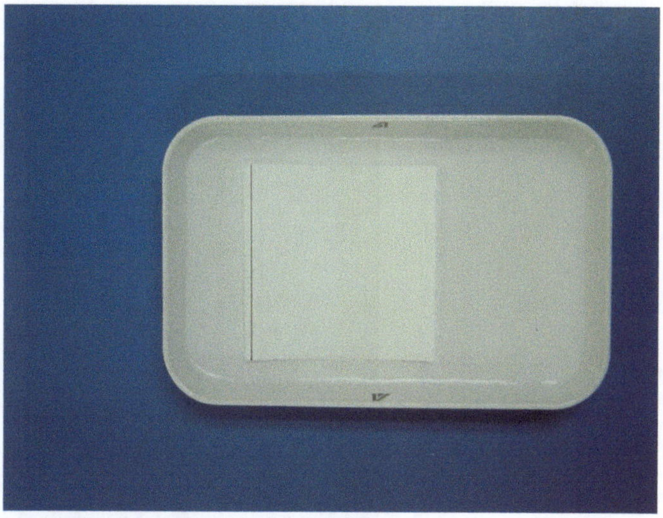

Fig. 3 NC paper is soaked in distilled water for 5 min

Fig. 4 NC is aligned to the upper part of Miniblotter® with the reference line parallel to channel 1

of a normal human serum on the membrane of one or two of the parallel lanes is strongly recommended as an internal control, as human serum should always give a positive reaction in the presence of the secondary anti-human antibody. If nonhuman samples are to be assayed (mice, rabbit, etc.), the control row should be sensitized with the corresponding normal serum.

6. To sensitize the NC sheet, 60 μL of each antigen are introduced in each groove of the Miniblotter® (Fig. 8), avoiding the formation of bubbles, and incubated for 60 min on an orbital

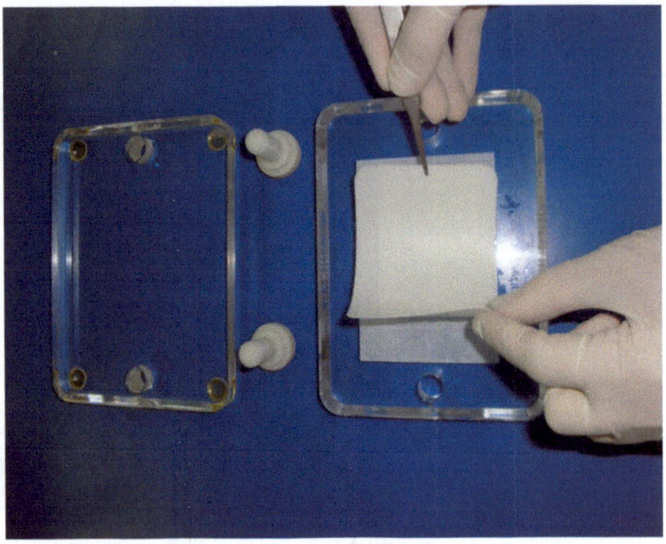

Fig. 5 A plastic cushion is placed over the NC

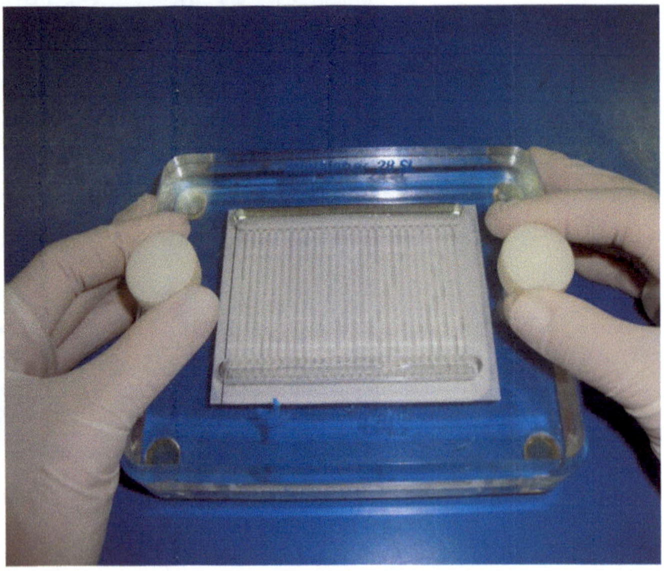

Fig. 6 The Miniblotter® is closed with the lower piece and hand tightened

or horizontal shaker, at room temperature (RT) (Fig. 9). The volume may need to be adjusted depending on the apparatus/ size of groove used).

7. The antigen solutions are removed by washing with 40 mL PBST, using a washing bottle (Fig. 10) or the manifold provided with the equipment (Fig. 11).

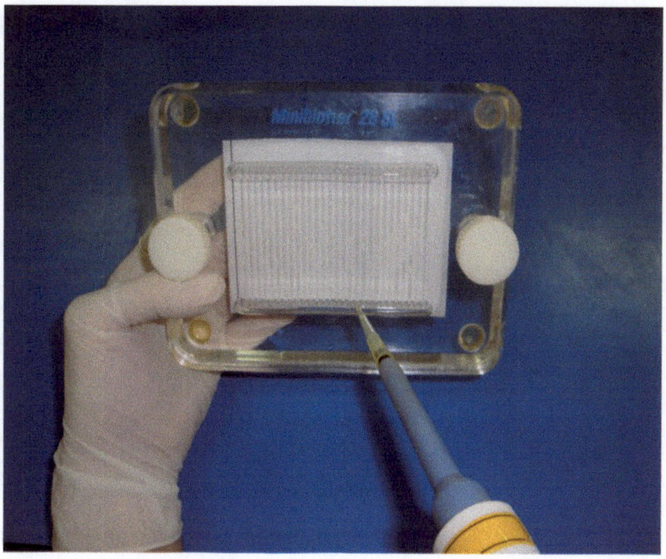

Fig. 7 Remaining water in the channels is removed by aspiration

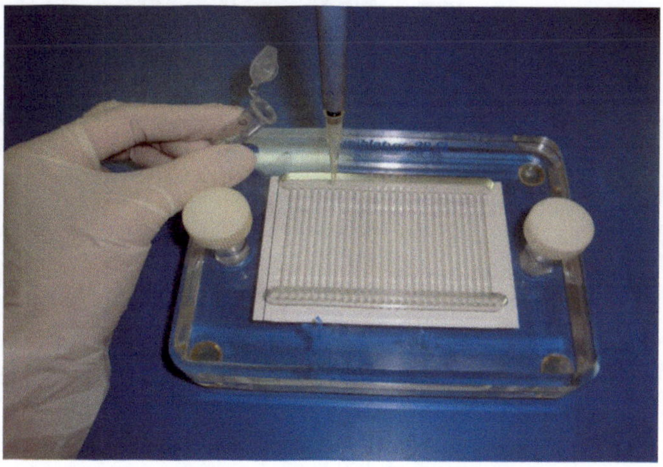

Fig. 8 Sixty microliter of each antigen are introduced into each groove

8. Then, the Miniblotter® is opened and the NC sheet is washed three times in a tray with 10 mL PBST on a shaker for 10 min each.
9. Blocking is achieved by immersing the NC in 5 % nonfat milk in PBST for 2 h on a shaker, at RT (Fig. 12).
10. Once blocked, the NC sheet may be processed immediately or stored wrapped in filter paper inside a plastic bag at –20 °C.

Fig. 9 Incubation for 60 min on an orbital or horizontal shaker

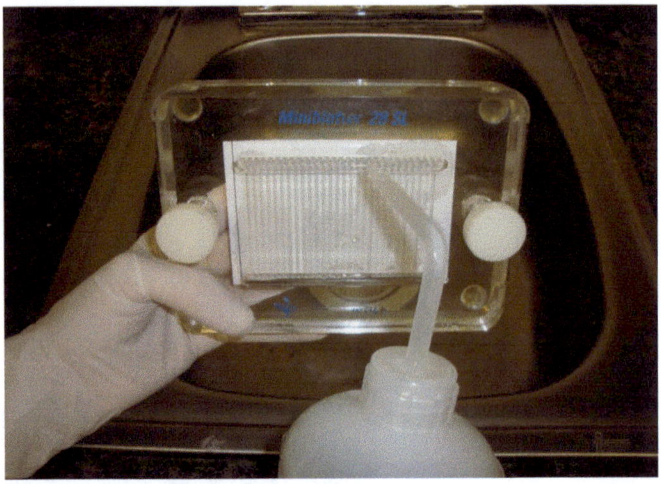

Fig. 10 Antigen excess is removed by washing with 40 mL PBST using a wash bottle

11. Numbered 2 mm-width-strips are cut perpendicular to the channels and the reference line, with a scalpel (Fig. 13). Each strip will thus contain a row of square spots corresponding to a maximum of 26 different antigens. Shorter strips with less spot rows may be used.
12. Strips are immersed individually in the troughs of an incubation tray, in serum diluted—1:100 or 1:200—in blocking solution (Figs. 14, 15, and 16) and incubated for 90 min at RT on a shaker (Fig. 17). Positive and negative serum controls and a

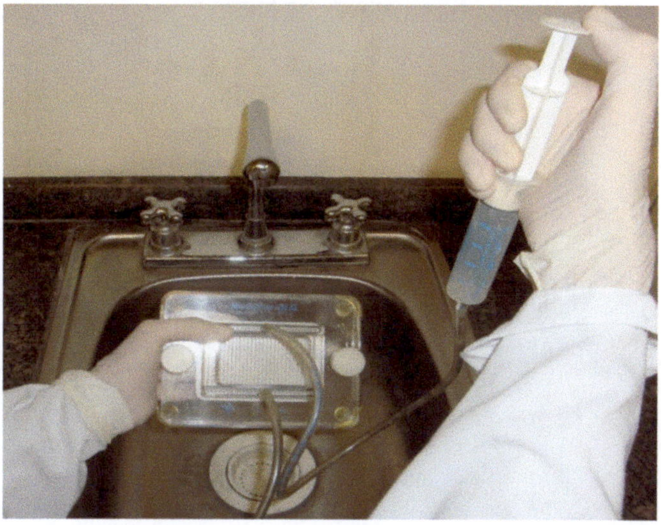

Fig. 11 Antigen excess can also be removed by washing with 40 mL PBST using the manifold provided with the equipment

Fig. 12 Blocking is achieved by immersing the NC sheet in 5 % nonfat milk in PBST for 2 h on a shaker

conjugate control (which consists in a strip without serum) should be included.

13. Strips are washed three to five times depending of the background of each system, for 10 min each, with PBST (Fig. 18), and then incubated for 90 min with the corresponding anti-Ig horseradish peroxidase labeled conjugate (Fig. 19), diluted in blocking solution at RT, on a shaker.
14. The strips are washed three times for 10 min each with PBST, as in Fig. 18.

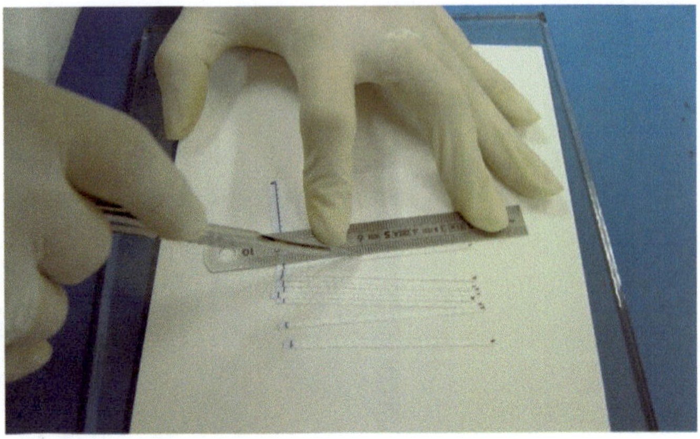

Fig. 13 Numbered 2 mm wide strips are cut perpendicular to the channels and the reference line, with a scalpel

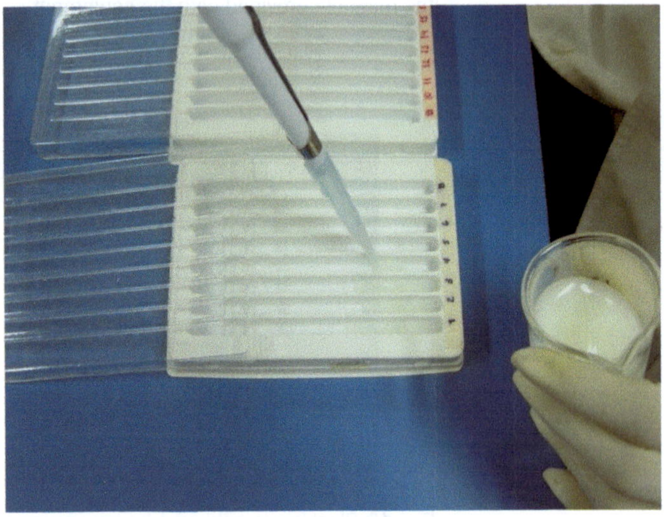

Fig. 14 Add 600 μL of blocking solution to troughs of an incubation tray

15. All strips are put together in a single tray. When using a chemiluminescent substrate, strips are immersed in approximately 1 mL of SuperSignal® solution for at least 1 min (Fig. 20), arranged in parallel onto a glass sheet (Fig. 21) and covered with a plastic wrap (Figs. 22 and 23).
16. Finally, the strips are exposed to Hyperfilm®, in darkness and developed photographically (Fig. 24). Routinely, two different exposure times are carried out: 5 and 20 s, depending on the intensity of the signal for each condition. If a luminescent

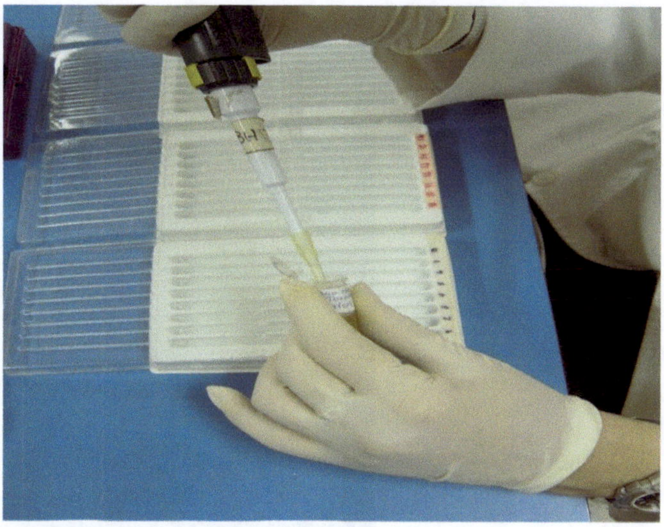

Fig. 15 Add 3–6 μL of sera (dilution 1:100 or 1:200, respectively) in blocking solution

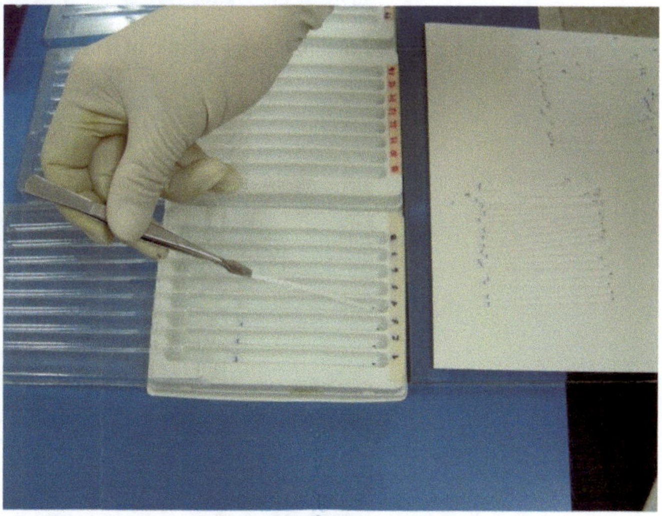

Fig. 16 Strips are immersed individually in the troughs

detection system (ChemiDoc™, BIO-RAD Laboratories, Inc. CA, USA) is available, the equipment registers electronically the luminescent signal. The signal density can also be quantified.

3.2 Colorimetric MABA [4]

The use of colorimetric substrates avoids the need of a dark room or expensive detection equipment (*see* Subheading 3.1, **steps 1–12** are the same for luminescent and colorimetric MABA).

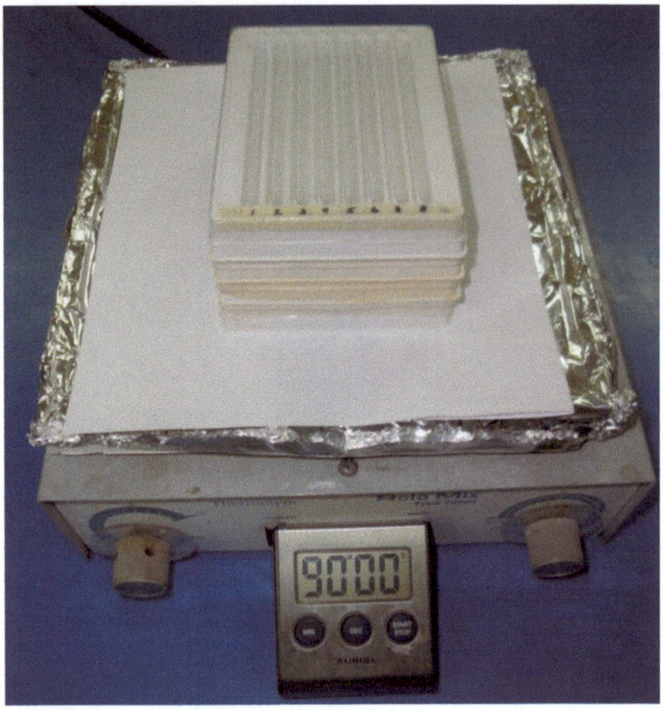

Fig. 17 Incubation for 90 min at RT on a shaker

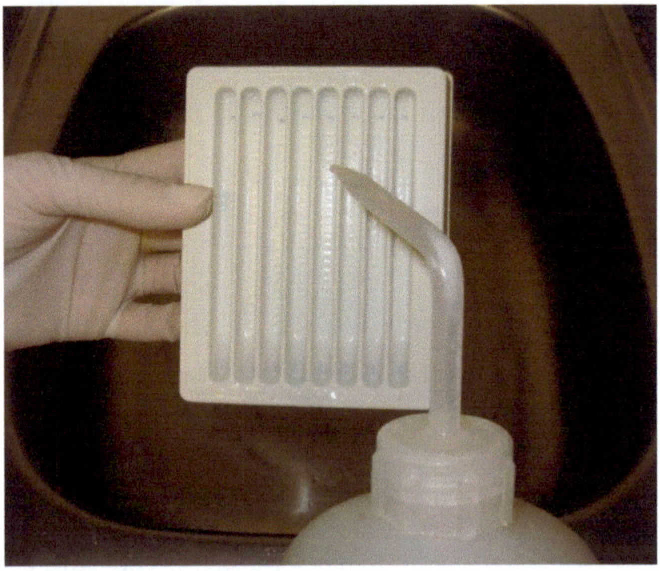

Fig. 18 Strips are washed three to five times for 10 min each, with PBST on a shaker

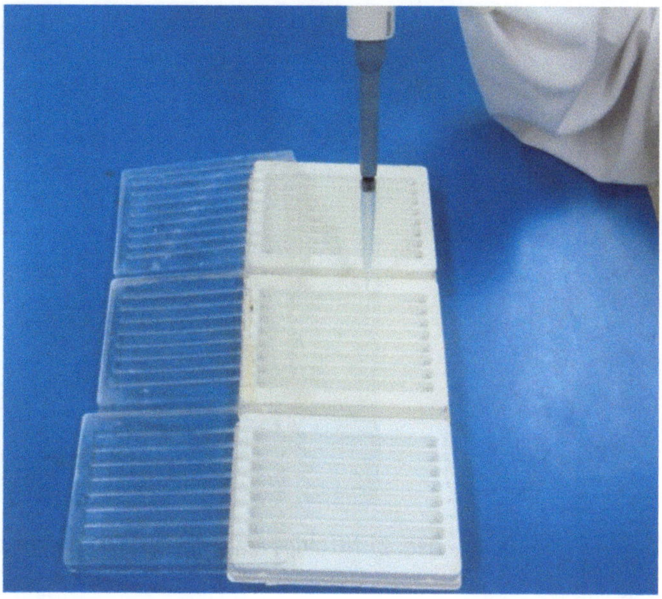

Fig. 19 Incubation for 90 min with the corresponding anti-Ig enzyme-labeled conjugate

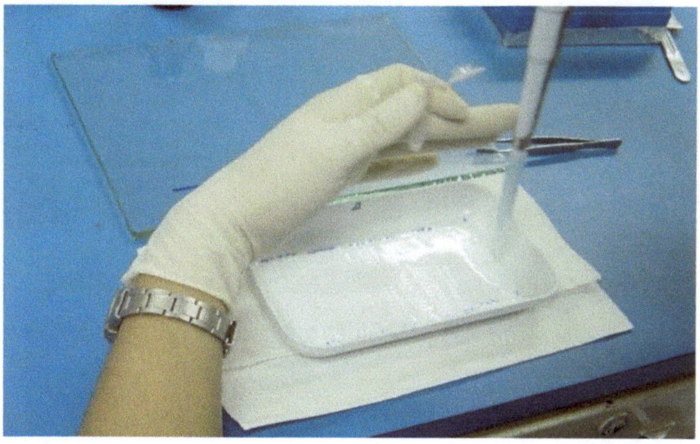

Fig. 20 Strips are immersed in approximately in 1 mL Super Signal® for at least 1 min

1. To develop with a precipitable colorimetric substrate, the peroxidase-labeled conjugate should be diluted in blocking solution and incubated for 90 min at room temperature with continuous gentle agitation on a shaker.
2. Wash the strips three times for 10 min with PBST.

Fig. 21 NC strips are arranged in parallel on a glass sheet

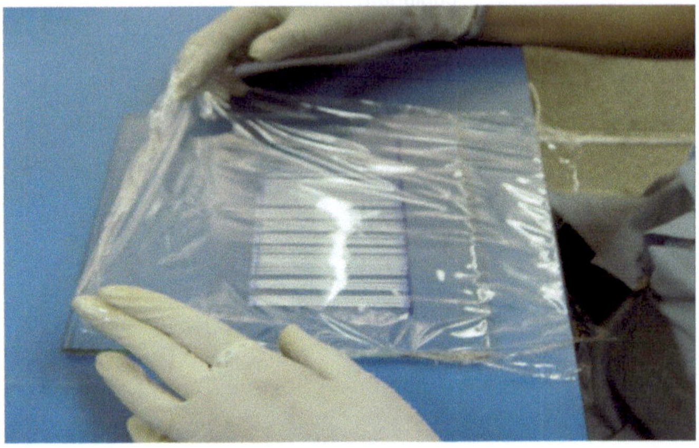

Fig. 22 Strips are covered with a plastic wrap

3. Remove the PBST and immerse strips in TMB® peroxidase substrate solution substrate for 5–15 min, or until the desired color is achieved.
4. Stop the reaction by immersing the membrane in distilled water for 20–30 s. The reaction should be stopped before background color becomes too intense (Fig. 25). For storage, dry the membrane, seal with plastic wrap and store in the dark to minimize fading.

Fig. 23 Strips are exposed to Hyperfilm® in darkness

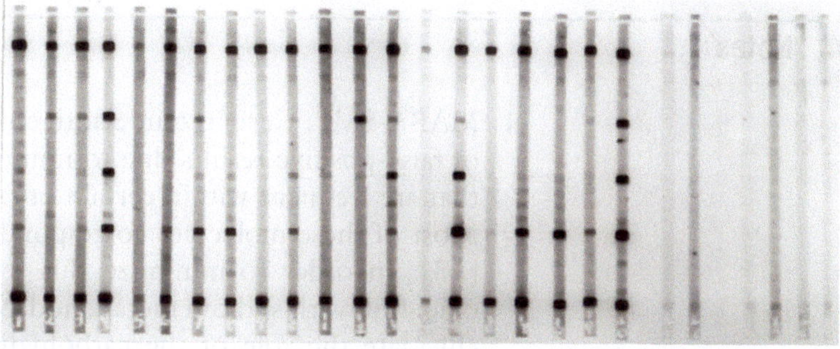

Fig. 24 Representative MABA of antigenicity of crude extracts and synthetic peptides from different infectious agents recognized by patient's sera developed with chemiluminescent substrate

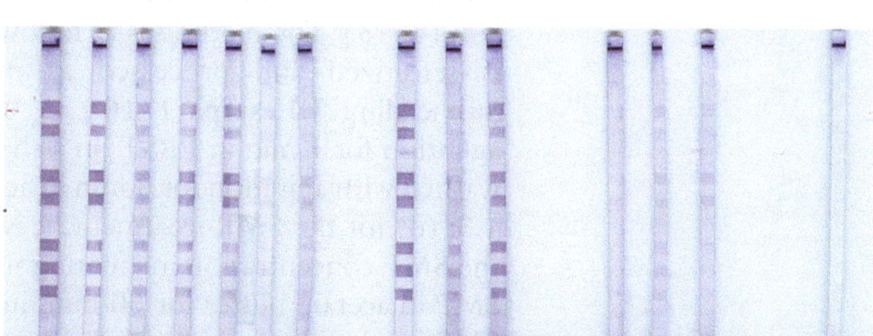

Fig. 25 Representative MABA of antigenicity of crude extracts and synthetic peptides from different infectious agents recognized by patient's sera developed with colorimetric substrate

5. If an AKP conjugate is used, strips should be immersed in enough volume of ready-to-use BCIP/NBT substrate at RT. Wait 5–15 min, or until the desired purple color is achieved and stop the reaction by immersing the membrane in distilled water. For storage, dry the membrane, seal with plastic wrap and store in the dark to minimize fading.

For each antigen and antibody system, all steps must be standardized.

3.3 SMP Treatment in MABA

In case of nonspecific reactivity due to cross-reacting crude antigens, especially those containing carbohydrates, treatment of the antigen preparation with sodium metaperiodate is recommended (*see* **Note 1**).

3.4 Cross-MABA

A variation in the technique, the Cross-MABA, avoids cutting, numbering or manipulation of strips (*see* **Note 2**).

4 Notes

1. MABA-SMP: Certain crude antigens have limitations because of false-positive results due to immunological cross-reactions that are frequent within certain organisms such as helminths. Most of these molecules correspond to glycosylated epitopes [13]. In order to minimize this serological cross-reactivity, oxidation with sodium metaperiodate (SMP) has been used to elucidate the role of the carbohydrate portion of antigenic glycoproteins in the reactivity by serology [14, 15]. After treatment of crude antigens with SMP, the hydroxyl groups of sugars are oxidized to aldehydes under acidic pH. In the presence of a reducing agent such as sodium borohydride ($NaBH_4$), aldehydes become alcohols, which are not recognized by the cross-reactive anti-carbohydrate epitope antibodies [14, 15]. The method is as follows: (a) the NC membrane is sensitized and processed as previously described (*see* Subheading 3.1, **steps 1–10**); (b) the sheet is washed with agitation for 1 min in PBST on a shaker; (c) the membrane is washed with agitation for 3 min in acetate buffer (50 mM) pH 4.8; (d) for the SMP treatment, it is necessary to standardize the SMP concentration incubating with between 1 and 3 mM SMP in acetate buffer for 60 min, in the dark, at RT. Prepare SMP just before use; (e) the sheet is washed with 0.05 M acetate buffer for 3 min; (f) add 0.05 M sodium borohydride ($NaBH_4$) in PBS and incubate for 30 min at RT, with continuous gentle agitation on a shaker. The reagent should be prepared just before use. Bubbles should be apparent. If the reagent is excessively hydrated and does not bubble, the treatment will not work properly; (g) wash for 3 min with PBS

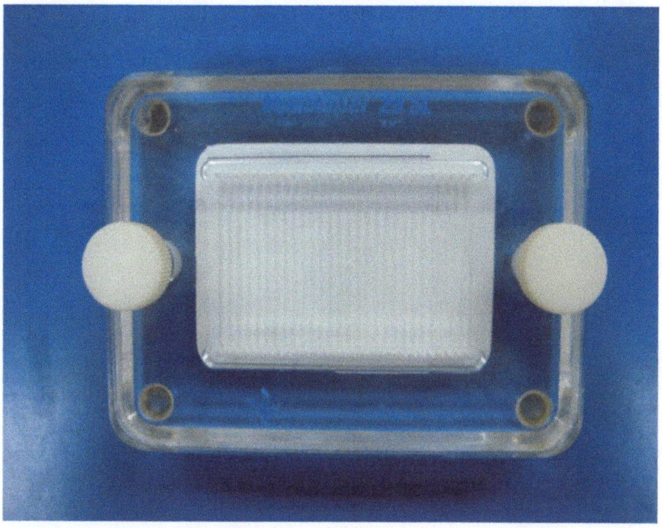

Fig. 26 A square 7.5 × 7.5 NC membrane is sensitized with antigens. After blocking, NC is placed again inside the Miniblotter® device positioned 90° respect to the original position. The reference line should be perpendicular to the channels, aligning perfectly the previous tracks in the paper with the grooves in the apparatus

with agitation; (h) continue the procedure as described previously (*see* Subheading 3.1, **steps 11–16**).

2. Cross-MABA: This is a simplification of the original technique, since no cutting, numbering or manipulation of strips is necessary. This variant of the technique is as follows: (a) a square 7.5 × 7.5 NC membrane is sensitized with antigens as described in Subheading 3.1, **steps 1–10**; (b) After blocking, the wet NC is placed again inside the Miniblotter® device positioned at an angle of 90° with respect to the original position, i.e., with the reference line perpendicular to the channels (Fig. 26), aligning perfectly the previous tracks in the paper with the grooves in the apparatus; (c) The Miniblotter® is closed tightly with the plastic cushion and water is removed; (d) patient sera are poured into the grooves, avoiding bubbles (1:100 or 1:200 in blocking solution) and incubation proceeds for 90 min; (e) Sera are removed with PBST, the membrane is placed in a tray and three washings are performed continuing the procedure as described previously in Subheading 3.1, **steps 13–16**. The whole NC paper is incubated with a secondary antibody and with the respective substrate. The resulting image can be analyzed with a grid that helps in the interpretation (Fig. 27). There are two limiting factors with this technique: only 17 antigens can be evaluated and only 16 sera.
3. Schematic diagram of the whole protocol of MABA is represented in Fig. 28.

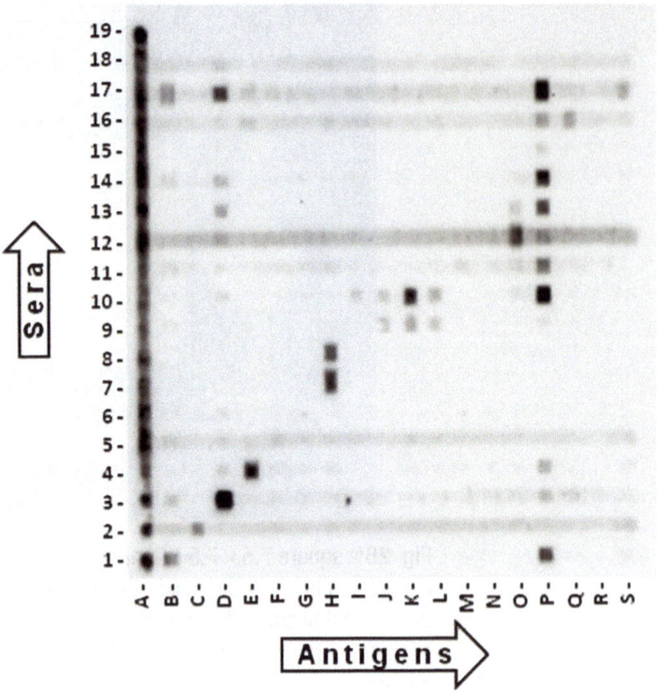

Fig. 27 Representative Cross-MABA

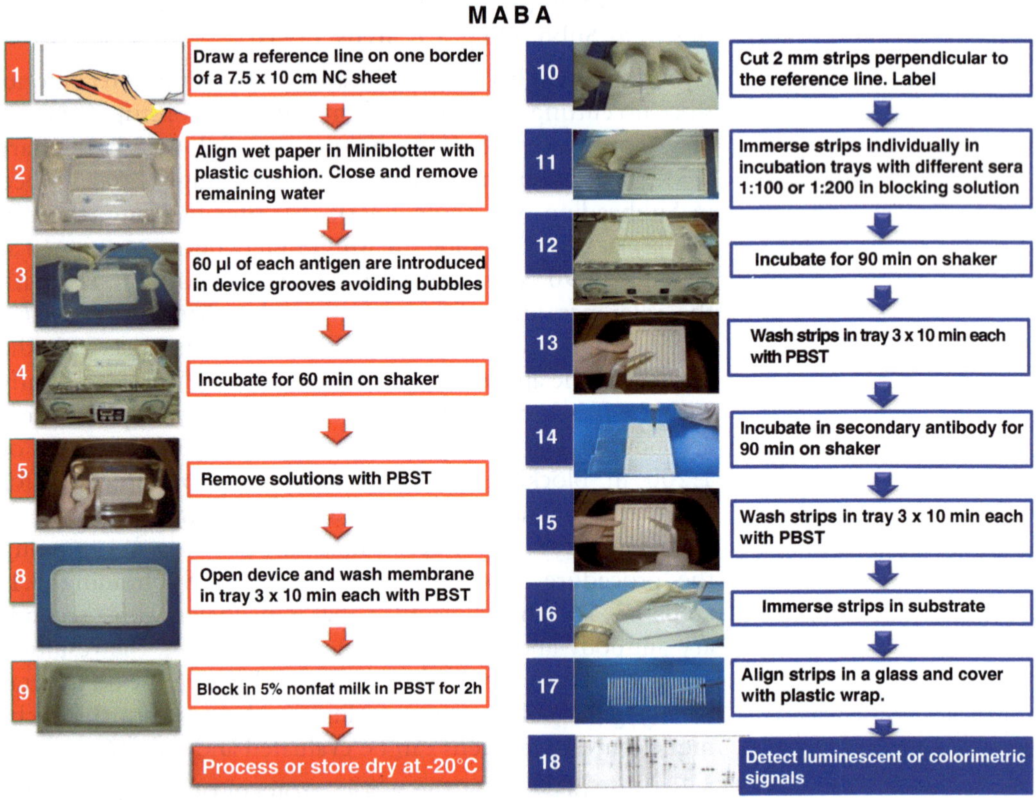

Fig. 28 Schematic diagram of the whole protocol of MABA is represented

Acknowledgment

This work has been partially supported by grants N° G-2005000387, 201100413, 2007001425, F-2005000122 and 20122001311 from FONACIT-Venezuela. We also thank the Consejo de Desarrollo Científico y Humanístico de la UCV for financing our participation in the "XVIII International Congress of Tropical Medicine and Malaria" in Brazil, with the conference "The simultaneous multidiagnosis of infectious agents as a routine laboratory screening test for the near future" (ICC09-0142-2012). We thank Dr. Peter Taylor, for help with the manuscript.

References

1. Schena M, Heller RA, Theriault TP, Konrad K, Lachenmeier E, Davis RW (1998) Microarrays: biotechnology's discovery platform for functional genomics. Trends Biotechnol 16: 301–306
2. Dunbar SA, Vander Zee C, Oliver K, Karem K, Jacobson J (2003) Quantitative, multiplexed detections of bacterial pathogens: DNA and protein applications of the Luminex LabMAP™ system. J Microbiol Methods 53:245–252
3. Noya O, Alarcón de Noya B (1998) The multiple antigen blot assay (MABA): a simple immunoenzymatic technique for simultaneous screening of multiple antigens. Immunol Lett 63:53–56
4. Noya O, Losada S, Toledo M, Alarcón de Noya B (2009) The multiple antigen blot assay-MABA, a simple, versatile and multi-purpose immunoenzymatic technique. Chapter 26. In: Protein blotting and detection, The Humana Press INC, USA. Methods Mol Biol 536:237–251
5. Noya O, Patarroyo ME, Guzmán F, Alarcón de Noya B (2003) Immunodiagnosis of parasitic diseases with synthetic peptides. Curr Protein Pept Sci 4:299–308
6. Noya O, Alarcón de Noya B, Guzmán F, Bermúdez H (2003) Immunogenicity of Sm32 synthetic peptides derived from the *Schistosoma mansoni* adult worm. Immunol Lett 88:211–219
7. Chacón N, Losada S, Bermúdez H, Cesari IM, Hoebeke J, Noya O (2003) Immunogenicity of polymerizable synthetic peptides derived from a vaccine candidate against schistosomiasis: the asparaginyl endopeptidase (Sm32). Immunol Lett 88:199–210
8. Wide A, Capaldo J, Zerpa N, Pabón R, Noda A, Noya B, González J, Noya O (2006) Sharing of antigens between *Plasmodium falciparum* and *Anopheles albimanus*. Rev Inst Med Trop Sao Paulo 48:327–332
9. Losada S, Chacón N, Colmenares C, Bermúdez H, Lorenzo A, Pointier JP, Theron A, Alarcón de Noya B, Noya O (2005) *Schistosoma*: cross-reactivity and antigenic community among different species. Exp Parasitol 111:182–190
10. Puccio FC, Lynch NR, Noya O, Noda A, Hagel I, López E, López R, Caraballo L, Mercado D, DiPrisco MC (2004) Importance of including *Blomia tropicalis* in the routine diagnosis of Venezuelan patients with persistent allergic symptoms. Allergy 59:753–757
11. Zerpa N, Wide A, Noda JA, Bermúdez H, Pabón R, Noya O (2006) Immunogenicity of synthetic peptides derived from *Plasmodium falciparum* proteins. Exp Parasitol 113: 227–234
12. Sulbarán G, Noya O, Brito B, Ballén DE, Cesari IM (2013) Immunoprotection of mice against schistosomiasis mansoni using solubilized membrane antigens. PLoS Negl Trop Dis 7:1–13
13. Alarcón de Noya B, Colmenares C, Lanz H, Caracciolo MA, Losada S, Noya O (2000) *Schistosoma mansoni*: immunodiagnosis is improved by sodium metaperiodate which reduces cross-reactivity due to glycosylated epitopes of soluble egg antigen. Exp Parasitol 95:106–112
14. van Lenten L, Ashwell G (1971) Studies on chemical and enzymatic modification of glycoproteins. J Biol Chem 246:1889–1894
15. Woodward M, Young W, Bloodgood RA (1985) Detection of monoclonal antibodies specific carbohydrate epitopes using periodate oxidation. J Immunol Methods 78:143–153

Chapter 33

Blotting from Immobilized pH Gradient Gels: Application to Total Cell Lysates

Harry Towbin

Abstract

Isoelectric focusing as used in the first dimension of two-dimensional gel electrophoresis separates protein isoforms such as those due to phosphorylation and acetylation. The immunoblotting method described here reveals this diversity by a one-dimensional separation. Using commercially available immobilized pH gradient plates or strips, the resolved proteins are transferred to PVDF membranes by diffusion and are probed with protein-specific antibody. The system is useful for monitoring changes of banding patterns and permits parallel processing of samples. Since the effect of posttranslational modifications on the isoelectric point can be predicted, inferring the number and extent of modifications is possible.

Key words Isoelectric focusing, Posttranslational modifications, Phosphorylation, Immunoblotting

1 Introduction

When looked at closely, proteins, both intracellular and secreted, display a bewildering number of modifications. Because of the importance of posttranslational modifications for regulating biological activities of proteins there is a corresponding effort aiming at detecting and monitoring their occurrence. An increasing number of antibodies are becoming available that allow specific detection of modifications on selected proteins, mostly well-characterized components of signaling cascades. Such antibodies are, however, often not available for other proteins. The method described in this chapter can be used to detect and track protein modifications that lead to alterations of their charge and consequently of a change of the isoelectric point. Such charge isoforms can be separated by isoelectric focusing and are evident from two-dimensional separations where trains of spots are seen that are frequently due to different phosphorylation states. Especially high resolution is achieved by IEF on immobilized pH gradient gels. Although SDS-PAGE is sometimes capable of resolving modified proteins, the extent or even direction of the changed migration is difficult to predict.

Biji T. Kurien and R. Hal Scofield (eds.), *Western Blotting: Methods and Protocols*, Methods in Molecular Biology, vol. 1312, DOI 10.1007/978-1-4939-2694-7_33, © Springer Science+Business Media New York 2015

The IEF blotting method described here was developed for monitoring alterations occurring on intracellular proteins that require denaturing conditions for extraction and separation [1]. The method allows sensitive detection and estimation of the relative abundance of isoforms with antibodies. The latter is a quantity that is not easily measured by modification-specific reagents. Quantification is facilitated by running samples side by side under the same conditions. In principle, similar information can be obtained from two-dimensional separations by running a series of westerns. Results from such experiments are, however, difficult to quantify because of unavoidable technical variability. The separation method used here is very similar to the first dimension of two-dimensional separation systems [2, 3]. It uses commercially available precast gels and allows direct comparison with two-dimensional separations. Transfer from the plastic-supported gels is achieved by diffusion blotting.

2 Materials

1. IPG plates (immobilized pH gradient gels), Immobiline DryPlates (GE Healthcare or other manufacturers), or immobilized pH gradient strips (*see* **Note 1**). Various pH ranges are available.
2. R-Buffer [4]: 4 % CHAPS, 7 M urea, 2 M thiourea, 10 mg/mL dithiothreitol, and 1 % carrier ampholytes (pH 3–10) (Pharmalytes; Amersham Pharmacia Biotech, Uppsala, Sweden), 0.02 mg/mL bromophenol blue. Store at −80 °C.
3. Reswelling buffer: R-buffer without carrier ampholytes and without bromophenol blue. Store at −80 °C.
4. TCA fixative: 12 % trichloroacetic acid, 3.5 % sulfosalicylic acid.
5. Multiphor II Electrophoresis System.
6. Sample applicator for IEF (*see* **Note 2**).
7. Transfer buffer. 4 M guanidinium chloride, 0.05 M Tris–HCl pH 8, 1 mg/mL DTT (added shortly before use).
8. Filter paper, e.g., Whatman 1Chr.
9. Polyvinylidene difluoride (PVDF) membrane.

3 Methods

3.1 Preparation of Extracts

1. Collect cells by centrifugation, wash once with PBS, suspend the pellet, transfer into a tared microcentrifuge tube, centrifuge once more, and remove as much of the supernate as possible (*see* **Note 3**). Estimate the volume by weighing, add R-buffer corresponding to five volumes of the cell pellet, suspend,

and leave on ice for about 1 h. Centrifuge at 1,000×*g* for 15 min in the cold. Store the sample at −80 °C. Protein concentration may be determined by a Coomassie binding assay [5] (*see* **Note 4**). Protein concentrations are in the range of 3–10 mg/mL.

3.2 Isoelectric Focusing on IEF Immobiline Plates

1. Reswell polyacrylamide sheets in R-buffer devoid of ampholines (*see* **Note 5**). The procedure is otherwise according to instructions of the manufacturer. Cut required size if only part of a sheet is used, and mark the polarity of the pH gradient by cutting off a corner from the sheet. Place the volume of buffer required for the sheet in the reswelling tray and place the sheet face down on the liquid. Avoid trapping air bubbles and spilling of buffer over the top side of the sheet. Leave overnight at room temperature in a box containing wet paper towels for achieving a water-saturated atmosphere. Instead of Immobiline plates, it is also possible to use individual IEF strips (*see* **Note 1**).
2. Place about 2 mL of kerosene on the cooling stage of the Multiphor apparatus and position the IPG plate paying attention to match the IPG plate to the required polarity of the electrodes. Remove excess buffer with a filter paper (provided for that purpose by the manufacturer). Position electrode filter paper strips and sample applicator on the gel surface (about 2 cm from the anode). The position may vary according to the proteins of interest.
3. Using a micropipette apply the samples (1 μL or up to 4 μL for the larger sample comb), which will be held by capillary force underneath the indentations (*see* **Note 6**).
4. Running conditions: 15 °C, from 0 to 300 V in 30 min, then to 900 V in 9 h, keep at 900 V for 6 h, raise to 3,500 V within 1 h, and keep at 3,500 V for a final 4 h. The entire run takes 20 h with an accumulated Volt-hour product of approx. 27,000 Vh. We avoided focusing at the highest voltages without attention. Adhere to the safety instructions of the IEF manufacturer. Depending on the pH range and type of protein, different focusing conditions may be needed.

3.3 Transfer to PVDF Membrane

1. Incubate the gel with TCA fixative for 1 h. Wash three times with water, 10 min each (*see* **Note 7**).
2. Cut a piece of PVDF to the size of the gel, and moisten it by soaking in ethanol followed by washing in water.
3. Prepare ten pieces of filter paper (e.g., Whatman 1Chr), 1 cm larger than the membrane, and soak in transfer buffer (about 35 mL is required for half a gel, approx. 13 cm×13 cm).
4. Assemble the sandwich by placing the gel on a glass plate, followed by the PVDF membrane, and a stack of filter papers

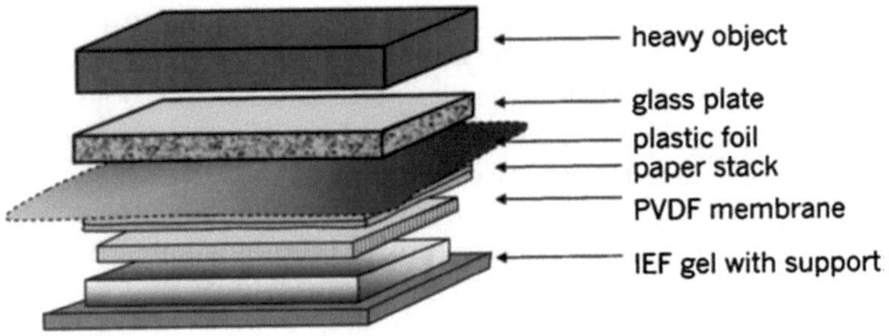

Fig. 1 Setup used for diffusion transfer from plastic-backed Immobiline DryPlates

soaked in the transfer buffer. Overall thickness of the paper stack will be about 2 mm. Wrap the assembly with a plastic foil to prevent drying out and cover with glass plate (Fig. 1). Ensure good contact by compressing with a heavy object and leave overnight at room temperature.

5. Wash the membrane with water to remove guanidinium chloride.
6. At this stage, the membrane is processed like any western blot.

3.4 Visualization of Total Protein Pattern (Optional)

In order to assess any disturbances or differences between samples we find it useful to visualize and photograph the total protein pattern on the PVDF membrane using stains that are compatible with subsequent immunostaining (e.g., SyproRuby). Also, if needed, IEF standards may be added in adjacent lanes and thus be visualized on the membrane.

3.5 Interpretation

It is useful to judge whether an assumed modification can account for the observed shift of the isoelectric point of the protein under investigation. Thus, for proteins with known amino acid composition, the isoelectric point can be predicted [6]. The effect of phosphorylation can also be calculated [7]. For acetylation on lysine groups, it is possible to simply omit one lysine from the sequence before computation of the pI. Similarly, other modifications may be incorporated by considering their effect on the charge of the protein. Figure 2 shows an example of an application of the IEF blotting method.

4 Notes

1. Instead of Immobiline DryPlates, individual strips may also be used. Here, the procedures for the first dimension of 2D electrophoresis are closely adhered to. Thus, reswelling of the

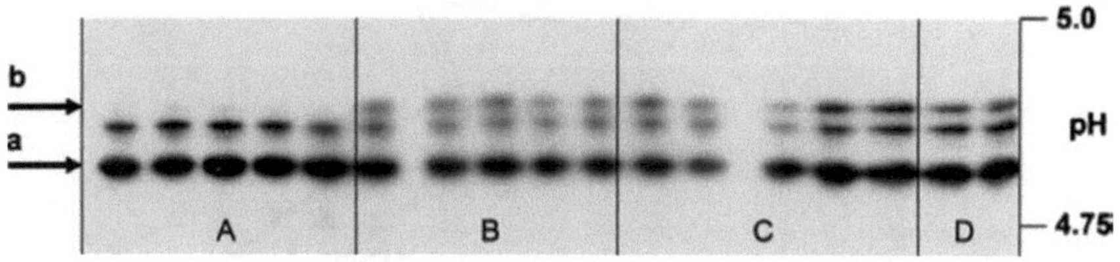

Fig. 2 Example of an IEF blotting experiment with extracts from rat peripheral blood white blood cells. The protein extract loaded (4 μL) corresponds to cells obtained from 80 μL of blood. The separation was carried out on Immobiline plates, pH range 4.5–5.4. Transfer to PVDF membranes was achieved by diffusion blotting. An antibody against 14-3-3 gamma protein detects two bands in untreated animals and three bands in animals that had been treated with bengamide. This compound inhibits methionine aminopeptidase [9]. As a consequence, acetylation of the penultimate N-terminal amino acid is also inhibited and a slightly more basic protein accumulates. The position of normal acetylated protein is marked as "a," and the unprocessed form as "b." The nature of the intermediate form is not known. *A*: untreated; *B*, *C*, and *D*: animals treated with increasing doses of bengamide

gel strips proceeds in the presence of the sample, thereby allowing application of higher sample volumes. After electrophoresis, the strips are processed exactly as described for the plates. For the blotting step, the strips can be arranged side by side. A large choice of strips covering wide and narrow pH ranges is commercially available. For strips, an electrophoretic transfer method, which requires removal of plastic backings, has been described [8].

2. The applicator comb may be built by cutting a disposable applicator intended for the PhastGel system (PhastGel™ Sample Applicators, 8 × 1 μL or 6 × 4 μL). The comb needs to be cut horizontally to fit to the limited height available on the Multiphor II. A rectangular piece of the plastic material is glued to one end of the comb with a drop of dichloromethane (this solvent is toxic and irritant, work under a fume hood) (*see* Fig. 3).
3. It is important to keep the salt concentration as low as possible as high conductivity leads to disturbances during running.
4. We find it convenient to carry out the assay out in 96-well microtiter plate by adding 0.5 and 1 μL of the sample to 10 μL water and then adding 150 μL prediluted Bradford reagent to the wells. This amount of R-buffer is tolerated in this assay.
5. We found that ampholines at the relatively high concentrations used for focusing as well as CHAPS have a tendency to bind to the blotting membrane. Hydrophilic proteins may poorly bind in the presence of these components.
6. Other methods of sample application may also be useful, such as cup loading or paper pieces.

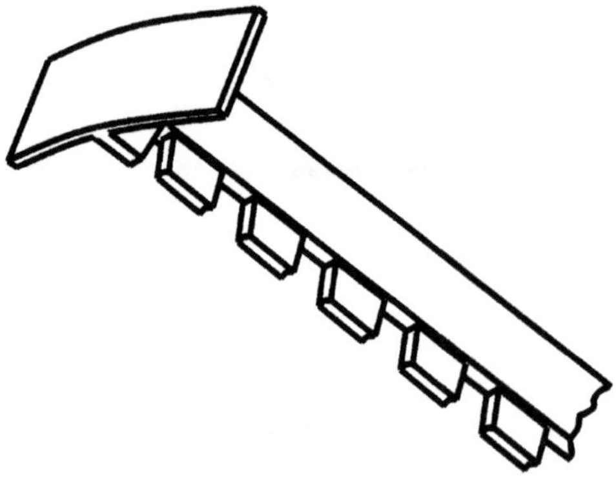

Fig. 3 T-shaped sample applicator

7. This step serves to wash out CHAPS, which interferes with adsorption of many proteins. Large or hydrophobic proteins may not require the fixation and CHAPS washout step, one example being tubulins [1]. If IEF blotting is to be carried out repeatedly, it is worthwhile to test simplified transfer procedures.

References

1. Towbin H, Ozbey O, Zingel O (2001) An immunoblotting method for high-resolution isoelectric focusing of protein isoforms on immobilized *p*H gradients. Electrophoresis 22:1887–1893
2. Görg A, Weiss W, Dunn MJ (2004) Current two-dimensional electrophoresis technology for proteomics. Proteomics 4:3665–3685
3. Hoving S, Voshol H, van Oostrum J (2005) Using ultra-zoom gels for high-resolution two-dimensional polyacrylamide gel electrophoresis. In: Walker JM (ed) The proteomics protocols handbook. Humana Press Inc., Totowa, NJ, pp 151–166
4. Rabilloud T (1998) Use of thiourea to increase the solubility of membrane proteins in two-dimensional electrophoresis. Electrophoresis 19:758–760
5. Bradford M (1976) A rapid and sensitive method for the quantitation of microgram quantities of protein utilizing the principle of protein-dye binding. Anal Biochem 72:248–254
6. Bjellqvist B, Hughes GJ, Pasquali C, Paquet N, Ravier F, Sanchez J-C et al (1993) The focusing positions of polypeptides in immobilized *p*H gradients can be predicted from their amino acid sequences. Electrophoresis 14: 1023–1031
7. Obenauer JC, Cantley LC, Yaffe MB (2003) Scansite 2.0: Proteome-wide prediction of cell signaling interactions using short sequence motifs. Nucleic Acids Res 31: 3635–3641
8. Poland J, Bohme A, Schubert K, Sinha P (2002) Revisiting electroblotting of immobilized pH gradient gels: a new protocol for studying post-translational modification of proteins. Electrophoresis 23:4067–4071
9. Towbin H, Bair KW, DeCaprio JA, Eck MJ, Kim S, Kinder FR et al (2003) Proteomics-based target identification: bengamides as a new class of methionine aminopeptidase inhibitors. J Biol Chem 278:52964–52971

Chapter 34

Immunoprecipitation: Western Blot for Proteins of Low Abundance

Edward P. Trieu and Ira N. Targoff

Abstract

Combining the procedures of immunoprecipitation and immunoblotting can help overcome some of the limitations of each separate procedure. Immunoblotting can identify immunoprecipitated proteins more specifically and with higher sensitivity than nonspecific protein stains or autoradiography. Immunoprecipitation can enrich proteins of interest to improve sensitivity for detection when compared with immunoblotting of whole cell extracts. Recently, immunoprecipitation-blotting helped us characterize a new autoantibody, anti-p155, and to test for the presence of the autoantibody in patient sera to study its clinical associations. The procedure for immunoprecipitation-blotting, with specific reference to this autoantibody test ("reverse" immunoprecipitation-blotting), is reported here in detail.

Key words Immunoprecipitation, Immunoblot (western blot), Autoantibody, Autoantigen, Dimethyl pimelimidate

1 Introduction

Immunoprecipitation has been a useful method for purification of specific proteins and protein complexes from cell extracts or other mixtures, and has been used extensively to characterize autoantigens and autoantibodies in autoimmune diseases. Usually the analysis of the immunoprecipitated material is by gel electrophoresis, but sometimes, more specificity is required. For autoantibodies, although the band pattern by electrophoresis is often enough to determine identity between the autoantibodies in different sera, this band pattern may not be specific, particularly when only a single protein band is immunoprecipitated.

Combining immunoprecipitation with immunoblotting can often solve these problems. It can provide greater specificity for protein identification, and, by using different sera for immunoprecipitation and blotting, can help determine immunologic identity. Immunoblotting can also help identify the antigenic protein within

Biji T. Kurien and R. Hal Scofield (eds.), *Western Blotting: Methods and Protocols*, Methods in Molecular Biology, vol. 1312, DOI 10.1007/978-1-4939-2694-7_34, © Springer Science+Business Media New York 2015

an immunoprecipitated multi-protein complex. The high sensitivity of immunoblots can also enhance the sensitivity of detection of immunoprecipitated proteins, particularly when present in complex mixtures or crude extracts. In turn, immunoblotting of immunoprecipitates can solve some of the problems that arise with immunoblotting of crude extracts. It can often enhance detection and recognition of low-abundance protein. For autoantibodies, it can improve the sensitivity and specificity of autoantibody recognition when compared with immunoblotting of whole extract.

The method we describe here facilitates the detection of low-abundance proteins, and increases the ability to detect antibodies to these proteins in a specific manner. For autoantibody testing, immunoprecipitation is used to enrich the low-abundance proteins, which can then be electrophoresed, transferred, and used to test other sera for autoantibodies to these proteins. Some autoantibodies do not recognize denatured antigens, a limitation that can be overcome by reversing this approach. For reverse immunoprecipitation-blotting, immunoprecipitates are prepared using the test sera, and blotted with known positive reference sera. The use of this method for detection of a recently described autoantibody found in patients with dermatomyositis, anti-p155, is used as an example. Immunoprecipitation-western blots could be used to detect p155 autoantibody even in sera that could not recognize the denatured form of p155 antigen after electrophoresis and western blotting (*see* Figs. 1 and 2) [1]. Thus, immunoprecipitation-western blot is a basic tool to provide useful data about a new antibody that other methods could not.

This procedure consisted of five different stages:

1. *Preparation of antibody-protein A gel*: The antibody is allowed to bind to protein A by incubating the antibody with protein A gel. After incubation, the column is washed with buffer to wash away the unbound proteins. The antibody is cross-linked to the protein A gel with a cross-linking solution to avoid IgG contaminating the eluate [2]. Following cross-linking, block the remaining unbound sites with an amine group from a chemical such as Tris or glycine. Next, a mild eluting agent removes the unbound antibodies that remain. Then, equilibrate the column with antigen-binding buffer.
2. *Immunoprecipitation*: Incubate the cell extract with the antibody-protein A gel. The antibody immunoprecipitates the antigen. Rinsing with buffer removes the unbound proteins. After washing, eluted protein from the beads with SDS sample buffer is the purified antigen.
3. *Gel electrophoresis of immunoprecipitate*: Run the purified antigen on SDS PAGE to resolve the proteins according to size by the Laemmli method.

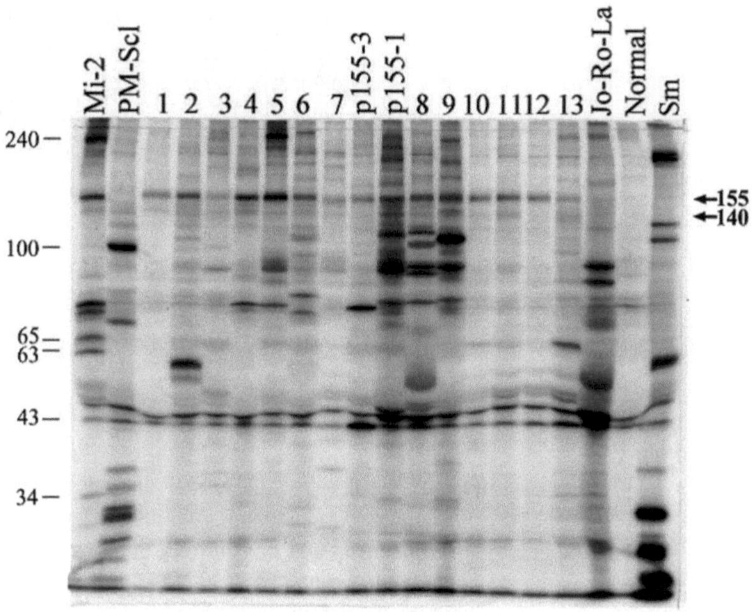

Fig. 1 Before proceeding to the immunoprecipitation-western blot, 13 anti-p155 autoantibody-positive sera (*lanes 1–13*) were screened by an immunoprecipitation of S35-labeled Hela cell extract. This is an autoradiograph of the polyacrylamide gel electrophoresis of the immunoprecipitates. *Lanes 1–13* were sera that precipitated a similar appearing protein to that of the prototype sera, p155-1 and p155-3; however, several sera show additional coexistent autoantibodies in the same region. To identify positively the p155 band, the sera are next screened with the immunoprecipitation western blot. Control autoantibodies used in the other *lanes* were Mi-2, PM-Scl, Jo-Ro-La, and Sm. Normal is serum from a normal subject. The *numbers* on the *left side* of the gel are molecular weight markers in kDa (reproduced from [1] with permission from John Wiley & Sons)

4. *Western blot*: The antigen separated via discontinuous SDS PAGE is electrophoretically transferred onto a nitrocellulose membrane [3].
5. *Antibody detection*: Cut the nitrocellulose paper with purified antigen into strips to develop with different sera for the detection of antibody.

2 Materials

2.1 Preparation of Antibody-Protein A Gel

1. Protein A sepharose.
2. Micro Bio-Spin Chromatography Columns, caps, and tip enclosures (Bio-Rad, Hercules, CA, USA).
3. Sera or monoclonal antibodies: Mouse IgG subclass 2a, 2b, and 3 can bind to protein A but they are weak. Protein G can bind to all mouse IgG sub class 1–3.

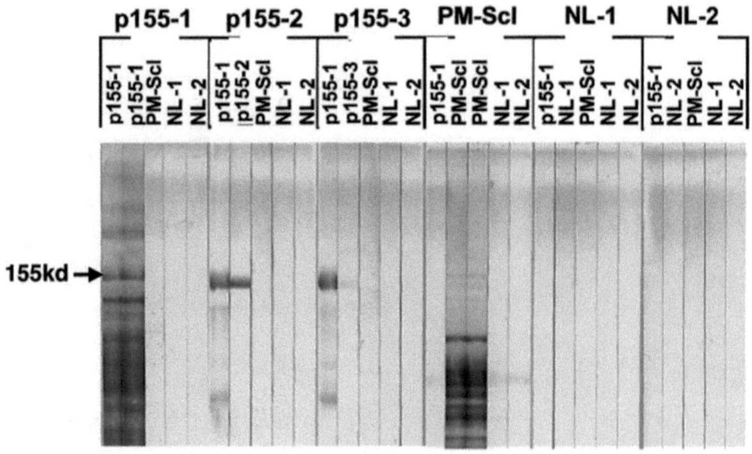

Fig. 2 Results from a typical immunoprecipitation-western blot. Immunoprecipitates were prepared from the six sera listed across the *top*, separated via electrophoresis and transferred to nitrocellulose. The sera as indicated on the figure blotted each of the *30 lanes*. In this experiment, the immunoprecipitates were the immunoprecipitation of test serums and the sera in the blotting phase were the reference sera. The result shows that the three anti-p155-positive sera, p155-1, p155-2, and p155-3, recognized the native form of p155 antigen, precipitated it and p155-1 reference serum detected it on the blot. In the antibody development panel, p155-1 blots the first strip in each set and it stained the p155 band with all three p155-positive sera. The *second strip* in each set, stained with the serum used for immunoprecipitation. P155-1 and p155-2 recognized the denatured form of the antigen but p155-3 did not [1]. Thus, this "reverse" immunoprecipitation-blotting procedure allowed the detection of the anti-p155 antibody in serum p155-3, whereas direct immunoprecipitation-western blotting (immunoprecipitation using reference serum and blotting using test serum) would not have detected this antibody. Thus, the serum used in the blotting phase must be able to recognize the denatured form of the antigen. The control sera PM-Scl, NL-1, and NL-2 were all negative for p155 antigen (reproduced from [1] with permission from John Wiley & Sons)

4. Antibody Binding Buffer (ABB): 0.2 M triethanolamine, pH 8.9, 0.5 M NaCl. Store at 4 °C.
5. Cross-linking solution: 30 mg/mL of dimethyl pimelimidate (DMP) in antibody binding buffer. Store DMP at 4 °C or lower since it is unstable and hygroscopic. Prepare the solution immediately before use.
6. Quenching buffer: 0.1 M Tris–HCl, pH 8.5, 0.5 M NaCl. Store at 4 °C.
7. Unbound antibody elution solution: 1.0 M acetic acid, 0.15 M NaCl. Store at RT.
8. Immunoprecipitation buffer (IPP): 10 mM Tris–HCl, pH 8.0, 0.5 M NaCl, 0.1 % NP-40 or Igepal CA-630. Store at 4 °C.
9. 8-Channel Nunc-Immuno* washer (*see* **Note 1**).

10. Sonicator (Ultra Sonic Processor).
11. Protein-free T20 (PBS) blocking buffer: Used for blocking of protein A sepharose beads or for the dilution of the serum for the first antibody development.

2.2 Immunoprecipitation

1. K562 cell culture pellets in log phase stored frozen at −80 °C in 15 mL conical tubes.
2. SDS sample buffer, 4×: 40 mM Tris–HCl, pH 6.8, 40 % glycerol, 8 % SDS, 0.04 bromophenol blue, 4 % mercaptoethanol. Store frozen at −20 °C (*see* **Note 2**).
3. 4× stacking buffer: 0.5 M Tris–HCl, pH 6.8. Stored frozen at −20 °C in 50 mL aliquots.

2.3 Gel Electrophoresis of Immunoprecipitate

1. Protean II xi gel apparatus from Bio-Rad, 16 cm glass plates, 25-well combs, and 1.5 mm spacers.
2. EZ Run Rec Protein Ladder: Dilute 1–10 with SDS sample buffer and make 50 μL aliquots and store at −80 °C. Use 25–50 μL per well for molecular weight markers.

2.4 Western Blot

1. Trans-Blot cell.
2. A small power supply with a current up to 450 mA.
3. Refrigerated circulator, or alternatively a refrigerator with an electrical outlet inside.
4. Filter paper or Pellon sheet from fabric, reusable. Pellon sheets come with many different thicknesses. Pellon # 806 Stitch N-Tear—White has the same thickness as filter paper. Pellon sheet is made with plastic fiber; it will not absorb water readily but will absorb alcohol such as ethanol. Before using, wet the Pellon sheet with 70 % ethanol and rinse with distilled water.
5. Nitrocellulose sheets with 0.2 μm pore size.
6. Pre-cut cellophane sheet.
7. Protein transfer buffer: 25 mM Tris–HCl, 192 mM glycine, 0.0005 % SDS, 20 mM NaOH, 20 % methanol (pH 8.3). Methanol is very dangerous. Read MSDS before using it.

2.5 Antibody Development

1. Mini incubation trays.
2. Antibody development panel:
   (a) Sera to be used for immunoprecipitation, including "test" or positive sera and "control" sera.
   (b) Sera to be used for blotting, including "test" or "positive" sera and "control" sera.
   (c) Control sera can include those with antibodies to other antigens and normal sera.

3. Nitro Blue Tetrazolium (NBT): 50 mg/mL in 70 % dimethylformamide (DMF). Light sensitive. To avoid loss of activity prepare in a small amount. Store at 4 to −20 °C in a dark bottle.
4. 5-5-Bromo-4-chloro-3-indolyl phosphate (BCIP): 25 mg/mL in DMF. Light and temperature sensitive. Store at −20 °C in a dark bottle.
5. Alkaline phosphatase (AP) buffer: 100 mM Tris–HCl, pH 9.5, 100 mM NaCl, and 5 mM $MgCl_2$ ($MgCl_2$ is optional; tends to precipitate).
6. AP substrate solution: Add 15 μL each of NBT and BCIP for every 5 mL AP buffer. Use within 1 h and discard any unused solution.
7. High-precision stainless steel forceps.
8. Scalpel handles No. 3 and Scalpel blade No. 10.
9. Nalgene reusable plastic utility box, dimension: 71/2×6×1 1/2 in.
10. Colloidal gold staining solution (optional).
11. Rolling ruler, 12 om., part No. 961-812 (Staedtler).
12. 10× TBST: 0.1 M Tris–HCl, pH 7.4, 1.5 M NaCl, 0.5 % Tween 20.
13. Instant nonfat dry milk: Dissolve and use immediately with TBS-T at RT.
14. Fast Green FCF (Fisher Scientific): 0.2 % in 5 % acetic acid.
15. White tulle from fabric or craft stores or Spectra/Mesh woven nylon filters (Spectrum Laboratories, Inc, Rancho Dominguez, CA, USA).

3 Methods

3.1 Preparation of Antibody-Protein A Gel

3.1.1 Monoclonal Antibody or Sera for a Gel Protocol

1. For column with monoclonal antibody: Measure the amount of antibody available to add to protein A gel. To get an optimal amount of monoclonal antibody, first use 100–200 μg of mouse monoclonal IgG of subclass 2a, 2b or 3 per column. Usually 1 mm of protein A gel can bind 20–40 mg of IgG.
2. For column with human serum: The amount of each serum used for preparation of column is 20–40 μL based on estimated 15–20 mg of IgG per milliliter of human serum of which specific antibody of interest is very little. For one gel, we recommend preparing 12 protein A columns simultaneously. The control sera usually include at least one that is positive for the antibody of interest and one that is normal or negative, and often sera with other antibodies that are positive.

3.1.2 Equilibration of Protein A-Sepharose

1. Reconstitution of protein A gel may be necessary. The gel can be reconstituted with ABB buffer before use. For 12 columns add 0.3 g protein A gel to 10 mL of ABB. Usually 1 mL of pre-swelled gel consists of 0.2 g dry beads.
2. Cap, vortex and mix on a rotator for 5 min.
3. Centrifuge at low speed about 500 × *g* without brake.
4. Wash beads one time with ABB to remove preservatives that prevent antibody binding.
5. Aspirate the buffer to gel level. Add exactly the amount of ABB to equal the amount of gel to make a 1:1 ratio.

3.1.3 Preparation of Antibody-Protein A-Sepharose

1. To each Micro Bio-Spin Column add 50 or 100 μL of protein A gel from **step 5** above (*see* **Note 3**).
2. Add 25 or 50 μL of each serum into the columns (*see* **Note 4**).
3. Add 25 or 50 μL of ABB to the columns for a better suspension.
4. Cap and place on a rotator for 20 min at RT or overnight at 4 °C.
5. After 20 min, the binding to protein A gel is complete. To bring all the beads to the bottom of the column, add 0.9 mL of ABB to each of the column, cap, invert to mix and centrifuge. After centrifuging, remove the lids and snap off the column tips. If no manifold is available, put the columns into a 2 mL 96 deep well plate or put the columns into tubes (13 × 100 mm) in a rack to keep the columns upright for washing. Wash the beads thrice with ABB (*see* **Note 5**). Remove and put back the tip enclosures on the columns after the washing.
6. Add 100 μL of fresh cross-linking solution.
7. Shake the columns horizontally to mix. Let them sit for 15 min with momentary shaking.
8. Add 200 μL of blocking solution.
9. Cap and place on a rotator for 10 min at RT.
10. Wash the columns with ABB buffer once and add 300–400 μL of unbound antibody elution buffer.
11. Wash columns with PBS thrice or more to remove any trace of acetic acid.
12. Place the tip enclosures back on the columns after the washing and add 200 μL of protein-free blocking solution and leave it overnight at 4 °C for blocking.

3.2 Immunoprecipitation

1. Estimate the volume of K562 frozen cell pellet to make the cell extract. The volume of dry cell pellet necessary per column is roughly equal from one to one and a half volume of serum loaded on the protein A column (e.g., for 12 columns which

were loaded with 50 μL sera one would need 0.6–0.9 mL of pelleted cells (*see* **Note 6**).

2. Take the cell pellet out from the freezer and add PMSF for a final concentration of 1 mM. Allow it to thaw in a beaker of water at RT. Vortex to loosen the pellet/s and to speed thawing.
3. Add six volumes of thawed cell pellet with lysis buffer (0.3 % NP-40 and 1 mM PMSF in IPP buffer) and vortex to mix.
4. Sonicate the cells in a beaker with ice and water at power of 30 W with 20 s. Repeat this six times with a 10 s rest between each sonication.
5. Transfer the cell extract into an ultracentrifuge tube or microcentrifuge tubes. Centrifuge at 10,000 ×*g* for 40 min. Remove cell extract into a new tube with a pipet (avoid aspirating any material from the cell debris) and discard the cell debris pellet.
6. Remove tip enclosures from columns to let the protein free blocking solution come down to gel level.
7. Put the tip enclosures back on the columns, and then cell extract. If 20 μL of serum is used then add 100 μL of cell extract; if 40 μL of serum is used then add 200 μL of cell extract to each column.
8. Place columns on a rotator and incubate them for 3 h at 4 °C.
9. When incubation is complete, wash the beads 10× with ice-cold IPP buffer.
10. Wash each column twice with 1 mL of ice cold 10 mM Tris, pH 6.8 (from 1/50 dilution of stacking buffer).
11. To remove the remaining water that is still inside the columns, shake the water off into a sink.
12. Place columns into new, labeled 1.5 mL tubes.
13. To each column, add 4× SDS sample buffer: If 20 μL of serum is used, add 10 μL of sample buffer. If 40 μL of serum is used, then add 20 μL of sample buffer.
14. Centrifuge briefly, about 20 s.
15. Take the same sample from each 1.5 mL microcentrifuge tube and put it back into the same column. Centrifuge again briefly about 20 s. Remove column from the tube and cap the tube.
16. Store the samples at −80 °C until gel electrophoresis.

3.3 Gel Electrophoresis of Immunoprecipitate

1. Prepare a discontinuous 7 % polyacrylamide SDS gel, 3.3 % C with 25-well comb, 1.5 mm thick spacer. The procedure is based on the Laemmli method [4] but with some modifications: The pH of running gel buffer was 8.7 for better separation of proteins because the gel runs a little bit slower. The stacking

gel was 3.9 % with pH 6.8. Cover the top with Parafilm and leave it overnight at RT.

2. Remove the comb the next day. Combine two wells by removing one gel finger (well wall) in between every two wells to make the lanes wider (*see* **Note 8**). To remove gel finger, insert two strips of plastic (*see* **Note 9**) on either side of a gel finger, close the end to cut and pull it up. Leave the last well for the molecular weight marker.
3. Heat-denature the samples at 95 °C for 5 min to reduce disulfide bonds (*see* **Note 7**).
4. Load the samples and molecular weight markers into the gel (*see* **Note 10**).

Check manufacturer's instruction for the best current setting for the gel apparatus. In general, with Protean II xi, start the current at 26 mA per gel and adjust it up to 32 mA per gel when the blue dye enters the running gel. Turn off current when the blue dye come down to 2–3 cm from the bottom of gel.

3.4 Western Blot

1. Before gel electrophoresis is complete, make up the transfer buffer. Degas to remove bubbles that may have formed from reaction of methanol with water.
2. Remove clamps from gel assembly. Separate the two glass plates. Leave gel on one glass plate. Cut off the stacking gel. Mark the gel for orientation, for example, cut off a small piece of corner of the molecular weight marker lane without cutting into the sample lanes.
3. Cut a piece of nitrocellulose paper and two pieces of 3MM filter paper. The nitrocellulose paper should be 5–10 mm bigger than the size of gel.
4. Pour some transfer buffer into another container and put one of the 3MM filter paper or Pellon sheet into the tray. Push it down to the bottom of the container. Put nitrocellulose paper into the transfer buffer. Start from one corner of membrane onto the surface of the solution. Let the nitrocellulose membrane absorb buffer and then gradually lay the whole piece down. Orient nitrocellulose paper onto the center of the first filter paper (*see* **Note 11**).
5. Invert the glass plate and drop the gel into tray. Place the gel with molecular weight markers side (a corner marked with a cut). Now the gel is on the middle of nitrocellulose paper with 3MM filter paper or Pellon sheet at the bottom. Put a second 3MM filter paper or Pellon sheet on top of gel with a cut one corner to recognize the gel side. Remove any bubbles below the second filter paper. Align the filter papers on top of each other.

6. Place the whole gel sandwich onto a glass plate with the nitrocellulose paper side up.
7. Using a roller or a long tube, lightly roll out any excess water or bubbles between the gel and nitrocellulose paper. Repeat a few times (*see* **Note 12**).
8. Transfer gel sandwich with membrane side toward the anode (clear or red). Assemble with Scotch Brite pads supports on both sides of the Trans-Blot cassette.
9. Close the cassette and put it into a slot of Trans-Blot cell that holds the cassette (*see* **Note 13**).
10. Pour transfer buffer into Trans-Blot cell to cover the gel inside the cassette. Put the cover with the leads on and connect them to power supply.
11. Set the current at 200 mA.
12. Transfer the protein overnight at 4 °C about 16–22 h in a cold room or using refrigerated circulator (*see* **Note 14**). Transfer at 4 °C in a refrigerator is much more heat transfer effective than using a refrigerated circulator.

3.5 Antibody Development

1. Turn off the power when the protein transfer is complete. Remove the leads that connect to power supply. Take out protein transfer cassette and put it onto a tray.
2. Remove membrane from gel and filter papers with fine point forceps.
3. Stain with Fast Green solution.
4. De-stain with ddH_2O.
5. When de-staining is complete, air-dry the membrane on a 3MM filter paper or Pellon sheet until completely dry. The drying of the membrane could reduce the background [4].
6. Place the membrane on a clean glass plate, use a sharp scalpel on a handle and a ruler to trim off excess nitrocellulose paper.
7. Cut the lane with molecular weight and keep it separately on a 3MM filter paper. Do not wet it with water; keep it as it is for later use when putting the strips together.
8. Wet the remaining membrane with ddH_2O.
9. It may be necessary to use background rather than the protein of interest itself to identify the lanes, since the latter may not be visible.
10. Set-up to cut membrane into strips: Lay the membrane on a clean glass plate. Make sure the membrane is always wet. Put a rolling ruler on the membrane with a cutting strip template underneath the glass plate. The line guide to cut the strips should be visible through both top and bottom of membrane. The strip template should be wide enough within or inside the protein lane.

11. After cutting the nitrocellulose membrane into strips and placing into the wells of the mini incubation trays, add ddH_2O into the wells.

12. Set up washing device: Prepare 1× TBST and pour it into a bottle. Use one of the 8-channel multi-well plate washer/dispenser manifold to dispense the TBST buffer and a second port to aspirate the buffer with vacuum. Adjust the amount of washing buffer from a syringe that connects to a port with a three-way Luer stop cock to switch from dispensing to loading or vice versa. Manifold tips will contact the membrane strips and leave marks on them. To avoid the marks of needles on the strips, cover the tips of manifold with some silicone tubing size 18 GA extended out from the needle about 1–2 mm to cover it.

13. Destain Fast green on membrane strips with 1 mL of 2 mM NaOH. Destain the molecular weight marker lane with 10–20 mL. Put them on the shaker for a few min.

14. Block the molecular weight marker strip with 0.3 % Tween 20 for 1 h at RT. Wash with ddH_2O and add enough colloidal gold staining solution to cover the strip. Put on shaker for a few hours or overnight at 4 °C. When staining is complete keep it in ddH_2O until antibody development is finished. More often, the molecular weight markers are visible well enough with Fast Green staining and can use as it is. Sometimes, the other staining such as colloidal gold staining, provides a better visualization of the markers, Colloidal gold stain is as sensitive as silver stain and it does not shrink the membrane [5].

15. When most of Fast Green comes off from membrane strips, aspirate 2 mM NaOH. Remaining small amount of Fast Green will come off at the end of development. Add into each well 1 mL of blot blocking solution at 37 °C.

16. Block the membranes with 5 % instant nonfat dry milk in TBST on a shaker for 1–3 h at 22–37 °C or at 4 °C overnight [6].

17. Add primary antibodies for blotting (*see* Subheading 2.5, **item 2**). Dilute the serum 1:100 in blocking solution by adding 5 μL of each antibody panel serum into each well of mini incubation tray that contain 0.5 mL of 1 % instant nonfat dry milk in TBST with membrane strip [7]. An eight multichannel pipettor will shorten the time of adding antibody panel sera except the serum used to prepare column.

18. Incubate first antibody on shaker at 4 °C overnight or at least 8 h.

19. When first antibody incubation time is over, put mini incubation trays on a tray large enough for one set of 12 mini incubation trays or less. To aspirate the buffer, raise one side of tray with an object.

20. Wash the membrane strips with TBST thrice (5–10 min each on shaker at RT to allow antibodies inside of membrane to equilibrate with the buffer outside).
21. Add second antibody in 5 % milk blocking solution.
22. Incubate second antibody on shaker at 4 °C overnight or at least 8 h.
23. Wash membrane strips with TBST thrice (5–10 min for each wash on a shaker at RT).
24. Set-up to develop the strips with substrate solution (*see* **Note 15**). Place on bench top in the following order:
    (a) A piece of steel or stainless steel sheet larger than the membrane size.
    (b) A piece of saran wraps large enough to cover the inside of Nalgene Reusable Plastic Utility Box.
    (c) A piece of nylon filter or white tulle about the same size with a piece of steel or stainless steel (*see* Fig. 3). Wet the nylon filter or white tulle with TBST.
25. Aspirate the wash buffer inside the incubation trays.
26. Transfer the strips on the incubation trays with fine tip forceps onto the wetted nylon filter or white tulle according to their original order of lane number. Use fingers to move the strips next to each other to make room for the next strips.

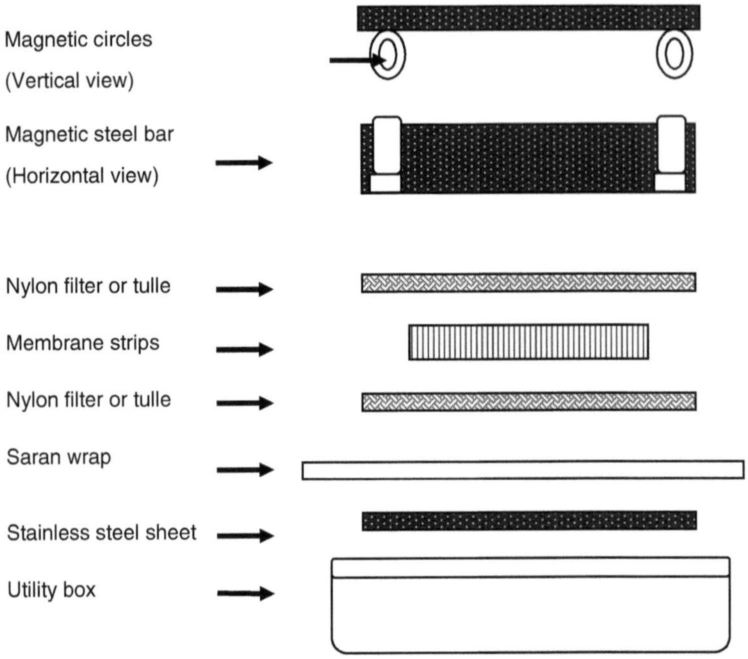

Fig. 3 The membrane strips sandwich used in substrate development

Fig. 4 A homemade device secures the membrane strips together and facilitates the development of all strips simultaneously in one tray [1]. Make two strong magnetic circles (The Magnetic Source) and permanently attach by glue or screw on either side of a piece of plastic or metal bar [2]. Cut a piece of stainless steel sheet 1-mm thick about the size of the inside development tray (5¼ × 6 in.). Round off the corners and smooth out the rough edges to avoid cuts [3]. Two pieces of nylon filter or white tulle (5 × 6 in.) aid in holding the membrane strips together and in order, while letting the substrate solution go through during development

27. Use two rulers; align the top and bottom of strips to line them straight.
28. Cover the membranes with a second nylon filter or white tulle.
29. Put on each top and bottom edge of membrane a stainless steel bar or plastic bar with two pieces of super magnet circles attached on either side (*see* Fig. 4).
30. Put the whole set up of membrane sandwich between two pieces of nylon filter or white tulles on stainless steel sheet into a Nalgene Reusable Plastic Utility Box.
31. Prepare 50 mL of the alkaline phosphatase substrate solution and pour it into the utility box that contains the membrane strips.
32. Color changes rapidly. Stop the reaction with dH_2O before it become too dark. After development is complete, scan the membrane strips for analysis.
33. Often things can turn out wrong unexpectedly (*see* **Note 16** for trouble shooting western blot results).

4 Notes

1. Lock the tips of an 8-channel Nunc-immuno* Washer to fit the pattern of columns on a vacuum manifold. The ports of dispenser manifolds come in different gauges. Sometimes it will be possible to use the rubber cover of a Vacutainer needle to plug some of the ports. Other times it may be necessary to insert thin walled plastic tubing (18 GA) into the rubber covers for a tight fit.
2. SDS sample buffer usually contains bromophenol blue (BPB) to aid sample visualization when loading and running the gel. A small amount of this dye in the sample buffer will not affect protein migration during gel electrophoresis, but it may affect it if there is too much. The amount of BPB can be reduced from 0.05 to 0.01 % for better gel electrophoresis separation.
3. Protein A-sepharose beads equilibrated 1:1 with another buffer, such as immunoprecipitation buffer, is an option since there are three washes of antibody-binding buffer that would remove any buffers that might interfere with the cross-linking reaction. Allow the wash buffer to come down to the gel level before adding new buffer.
4. If there is precipitate present in the serum, centrifuge it at 10,000 × g for 5 min. Avoid pipetting the pellet or lipid.
5. If there are many columns to wash, use a vacuum manifold in conjunction with a dispenser manifold. When a column is attached to a manifold, the water comes down into the manifold tubing by gravity. The weight of water inside the tubing creates a vacuum; therefore, if the tubing is longer, the vacuum is greater. Conveniently, capillary action inside the tubing will prevent the column from running dry. A homemade manifold can be made from any source, such as any polyethylene micro test tube or 4 mm syringe filter that can attach to a column.
6. Our lab keeps a stock of frozen cell pellets at −70 °C and routinely uses pliers to pool frozen cell pellets by squeezing from one end of a 15 mL tube.
7. Microwaving the samples for 90 s is a quick way of heat denaturing the samples [8]. Sample volumes less than 100 μL will not boil or explode while microwaving. Sample volumes greater than 150 μL or more require a test run in the microwave first with sample buffer to avoid sample loss. This technique is not recommended for samples having high salt content, concentrated samples from a large sample volume or samples containing a lot of glycerol because when microwaved they can be unstable. Tubes should be uncapped.
8. Cutting off the gel fingers allows for more versatility by creating a wider lane to accommodate a larger sample.

9. Tools to make wider wells: Cut out two thin strips of polyethylene plastic, the same thickness as the spacer. Make one strip with a narrow sharp hook to cut the finger and the other strip with no hook for support. The tools should be thin, strong and long enough to hold the gel fingers while inserting them into the upper gel chamber. To make the new well, cut the gel finger at the bottom of the well and remove it.
10. Prepare one lane of unstained molecular weight marker: One lane of a 25-well comb requires about 2–2.5 μg of unstained molecular weight markers. Reconstitute the markers at 0.1 g/μL and add 25–50 μL of this to 20–25 μL of sample buffer to bring the volume of markers equal to the volume of the sample wells.
11. The nitrocellulose paper wets by capillary action inside the nitrocellulose membrane. After long-term storage, nitrocellulose paper will become brittle, will no longer be wettable, and should be replaced with a new one.
12. If there are two or more gels, separate them with cellophane membrane. The cellophane membrane prevents proteins from depositing on the membrane belonging to the other gel.
13. Orient the membrane toward the positive electrode (red color) and the gel toward the negative electrode (black color). If the membrane is not in the path of protein migration, the protein will not deposit onto the membrane.
14. Protein transfer buffer will be partially exhausted after 16–22 h. Most power supplies are not equipped to detect this change, but some for semidry transfer systems can detect the conductivity of the protein transfer buffer during electrophoresis. To regain conductivity, add 10–20 mM NaOH to the buffer after the first use; it will last several times until discarded [9]. To avoid unwanted artifacts depositing onto the membrane from a small amount of protein leaking through the membrane in the old buffer, add cellophane membrane in front of the gel.
15. Development of membrane strips with substrate solution on mini incubation trays has a disadvantage in that if there are many incubation trays to develop at the same time, the reaction of substrate solution may happen too rapidly. This may lead to undesirable high background. A new method to develop all of the membrane strips simultaneously is to put all of the membrane strips for one gel side by side on one tray (*see* Fig. 4).
16. Troubleshooting western blot results: (a) High background, no signal: Bad cell pellet. Use a new cell pellet at the log phase; (b) High background, with signals: the concentration of secondary antibody is too high, causing high background even if extensively washed. Dilute the conjugate further from 1:10,000 to 1:30,000. (c) Net like appearance after substrate development: Tulle or nylon filter may cause this appearance. When

developing strips with substrate, place tray on a shaker at slow speed or lightly agitate by hand. The substrate solution makes the top tulle or nylon filter float a little above the membrane strips, but the strips will still be secure. After stopping development with ddH_2O keep the strips in ddH_2O in the tray for at least 10 min to prevent further development; (d) A strong shadow of IgG and artifacts: Insufficient blocking of membrane causes these artifacts. Block the membrane strips longer and at higher temperature with 5 % instant nonfat dry milk in TBST at 42 °C for 3–4 h. Instant nonfat dry milk may not dissolve completely if it has been stored dry for too long; (e) The image of the protein band turns light when stained by strong antibodies but dark when stained by normal serum: The pH of the blocking solution may be lower because the milk in solution has soured; (f) High background within the lane: Cell extract or the antigen source, such as recombinant protein, is too concentrated or needs further centrifugation; (g) Bubbles appear on the membrane: Gas in the transfer buffer released during the protein transfer or failure to carefully roll out the water between the membrane and the gel. Degas the transfer buffer before use or prepare it a few hours ahead; (h) High molecular weight proteins are still in the gel after transfer: The higher molecular weight proteins are harder to transfer and require a longer electrophoresis time. Add 0.01 % SDS to the transfer buffer to facilitate the transfer of the largest proteins by solubilizing them [10]. Use no more than 0.01 % of SDS because it has the reverse effect.

References

1. Targoff IN, Mamyrova G, Trieu EP, Perurena O, Koneru B, O'Hanlon TP et al (2006) A novel autoantibody to a 155-kDa protein is associated with dermatomyositis. Arthritis Rheum 54:3682–3689
2. Schrieider C, Newman RA, Sutherland DR, Asser U, Greaves MF (1982) A one-step purification of membrane proteins using a high efficiency immunomatrix. J Biol Chem 257:10766–10769
3. Towbin H, Staehelin T, Gordon J (1979) Electrophoretic transfer of protein from polyacrylamide gels to nitrocellulose sheets: procedure and some applications. Proc Natl Acad Sci U S A 76:4350–4354
4. Laemmli UK (1970) Cleavage of structural proteins during the assembly of the head of bacteriophage T4. Nature 227:680–685
5. Moeremans M, Daneels G, De Mey J (1985) Sensitive colloidal metal (gold or silver) staining of protein blots on nitrocellulose membranes. Anal Biochem 145:315–321
6. Johnson DA, Gautsch JW, Sportsman JR, Elder JH (1984) Improved technique utilizing nonfat dry milk for analysis of proteins and nucleic acids transferred to nitrocellulose. Gene Anal Technol 1:3–8
7. Sheng S, Schuster SM (1992) Simple modifications of a protein immunoblotting protocol reduce nonspecific background. Biotechniques 13:704–708
8. Horscroft NJ, Roy P (1997) Thermal denaturation of proteins for SDS-PAGE analysis by microwave irradiation. Biotechniques 22: 224–226
9. Brody JR, Kern SE (2004) Sodium boric acid: a Tris-free, cooler conductive medium for DNA electrophoresis. Biotechniques 36:214–216
10. Wang K, Fanger BO, Guyer CA, Staros JV (1989) Electrophoretic transfer of high-molecular-weight proteins for immunostaining. Meth Enzymol 172:687–696

Chapter 35

Native Electrophoresis and Western Blot Analysis (NEWeB): Methods and Applications

Ioannis N. Manoussopoulos and Mina Tsagris

Abstract

Native Electrophoresis and Western Blot Analysis (NEWeB) has been developed for the study of plant virus characteristics, among others, virus particle–protein interactions, electrophorotype formation, and strain separation. The method is based on the property of electrophoretic mobility of virus particles (VP) and proteins and combines the analytical capacity of electrophoresis with the specificity of western blot. One of its advantages is that it deals with entire VP that can be studied in cause and effect or in time-interval experiments. Some of the most interesting approaches include VP structural studies, VP interaction with host or viral proteins, and also the characterization of VP-protein complexes. In this protocol, NEWeB is used to demonstrate the interaction of *Plum pox virus* particles with the helper component, a virus encoded protein. It is expected that the method could be used in analogous studies of other viruses or large protein complexes, where similar principles apply.

Key words Native electrophoresis, NEWeB, Virus particle–protein interaction, Helper component, HCPro, Coat protein, *Plum pox virus*, *Potyvirus*

1 Introduction

Most viruses consist of particles with helical (rod or filamentous) or icosahedral (roughly spherical) architecture. Virus particles in their simplest form, are made up of an infectious nucleic acid (the genetic material) encapsidated in a protective coat (the capsid). The capsid is constructed by the spatial arrangement of multiple copies of one or more types (depending on the virus) of virus encoded protein subunits. VPs are thus large entities, having masses in the MD range compared to single proteins that have molecular masses usually in the kilodalton range.

One important physicochemical property of VPs is their net charge, resulting from the ionization of the side groups of the basic and acidic amino acids in the surface of their capsid proteins [1]. Thus, at low pH the amino groups of lysine and arginine and the nitrogen atoms in the imidazole ring of histidine are charged,

Biji T. Kurien and R. Hal Scofield (eds.), *Western Blotting: Methods and Protocols*, Methods in Molecular Biology, vol. 1312, DOI 10.1007/978-1-4939-2694-7_35, © Springer Science+Business Media New York 2015

whereas at high pH the carboxyl groups of aspartic and glutamic acids are charged [2]. At a specific pH, known as the isoelectric point (p*I*), characteristic for each protein, negative and positive charges are balanced, and net charge equals zero. Virus particles or proteins will carry a negative net charge when the pH of the solution is above their p*I*, while they will be positively charged when the pH is below it. When in an electric field such negatively or positively charged VPs or proteins will move towards the anode or the cathode, respectively, whereas uncharged particles will be immobile. This property is known as the "electrophoretic mobility" (EMb) and the rate of migration of the virus particles or proteins is mainly depended on both their size and net charge [2]. As a result, the EMb of free capsid or other proteins is different from that of large VPs. Typically, such proteins will move much faster than VPs, covering longer distances in the gel during electrophoresis.

NEWeB exploits the EMb of VPs and proteins to allow the study of several properties of viruses. Its main advantage is that it deals with whole VPs that can be treated and detected as such, enabling a number of approaches and studies that cannot be designed otherwise.

Some of the most interesting NEWeB approaches are outlined below:

1. Study of electrophorotype formation without the need of virus purification (electrophorotypes are virus particle populations of the same virus strain with different EMb)
2. Characterization of protein–VP interaction and stability. Proteins could be either of viral (structural or other proteins depending on the virus genus) or host origin. Unknown host proteins interacting with VPs could be further explored by proteomic approaches.
3. Time interval or "cause and effect" experiments, as for example a follow-up of the infection kinetics and the formation of viral genome-protein complexes or the effect of different factors on VP, respectively.

In this protocol some of the above approaches are demonstrated using a Greek stain of *Plum pox virus* (PPV-Lar) [3].

Plum pox virus belongs to genus *Potyvirus* having members with flexuous filamentous particles of about 680–900 nm length and 12 nm width. The genetic material is an ssRNA of positive polarity, coding for at least eight mature proteins, most having multifunctional properties. The capsid consists of about 2,000 protein subunits of the virus encoded coat protein (CP) molecule. Members of the *Potyvirus* genus are transmitted by aphids in a mechanism involving the interaction of VPs with another virus encoded protein, the helper component (HC) [4–9].

Although much progress has been made in the subject, the exact mechanism of transmission has not yet been elucidated. Using NEWeB it was shown that particles of PPV-Lar were in association with HC in extracts of diseased plants [6, 8], implying a VP–HC interaction already in the plant cell (Fig. 1). Using the same method, it was found that a small population of VPs associated with HC had survived ultracentrifugation at high salt concentrations [8, 9] suggesting a strong interaction between VP

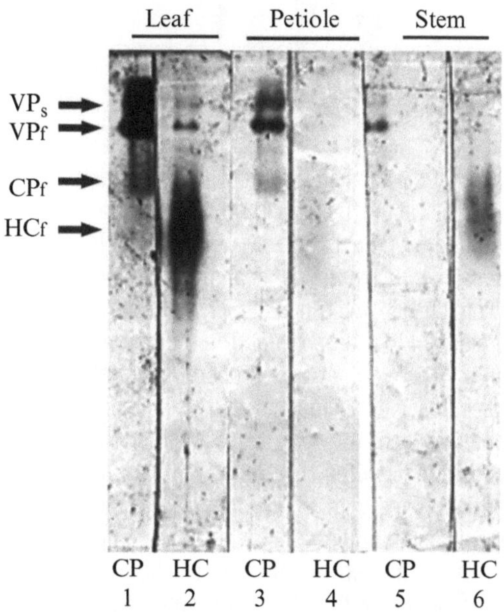

Fig. 1 Electrophorotype formation and VP-HC complex distribution in different tissues of a PPV-Lar infected *Nicotiana benthamiana* plant. Sap from a leaf, its petiole and a piece of the stem was extracted separately in TBE (pH 8.3) about 15 days post infection. After centrifugation, 100 μL of the supernatant was mixed in equal volumes with loading buffer. Two groups of three wells each in the same gel were loaded with 25 μl of each sample. Samples were loaded and electrophoresed in a 0.8 % agarose gel which was blotted on a nitrocellulose membrane. After transferring, the membrane was cut into two identical sheets (bearing the same biological material). The one sheet was probed with coat protein specific antibodies (CP) and the other with antibodies specific for PPV helper component (HC), as described in the protocol. Both membranes were incubated with secondary antibodies conjugated to alkaline phosphatase. After NBT/BCIP incubation and color development, the membranes were cut into strips corresponding to the above lanes for comparison. Two electrophorotypes (VP_s and VP_f) were detected in the leaf the petiole and the stem (traces) (*lanes 1*, *3*, and *5*). Particles of both electrophorotypes were in association with HC in the leaf (compare *lanes 1* and *2*), but not in the other tissues (compare *lanes 3* with *4* and *5* with *6*). Free HC aggregates (HC_f) were detected in the leaf and the stem but not in the petiole (*lanes 2*, *4*, and *6*), whereas free coat protein aggregates (CP_f) were detected in the leaf and the petiole but not in the stem (*lanes 1*, *3*, and *5*)

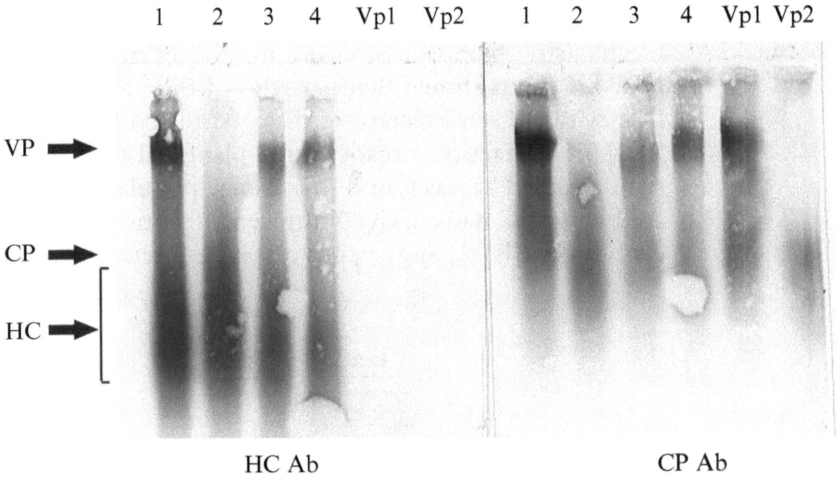

Fig. 2 Incubation of PPV-Lar with different concentrations of urea. Leaves from a PPV-Lar *N. benthamiana* infected plant were extracted in TBE (pH 8.3) and sap was incubated, after low speed centrifugation, in aliquots with 0, 1.25, 2.5, and 5 M (final concentration) of urea for 15 min at room temperature. Samples were mixed with 10 % sucrose, loaded in two identical groups of wells (four wells per group) in the same gel (0.6 % agarose–20 % acrylamide-bis) and run in TBE pH 8.3. After electrophoresis, the gel was blotted on a nitrocellulose membrane overnight, in the same buffer. The membranes were cut into two sheets (containing the same biological material) and each sheet was probed separately with virus (CP Ab) and helper component (HC Ab) specific antibodies, followed by incubation with secondary antibodies conjugated with alkaline phosphatase, as described in the protocol. After substrate addition (NBT/BCIP) and color development the membranes were washed in tap water. At low urea concentrations (*lane 4*) all complexes were stable. VPs detected with the CP Abs (CP Ab) and incubated with 2.5 M urea (*lane 3*) appeared more unstable than those detected with HC (compare *lane 3* in each membrane). However, all complexes and VPs incubated with 5 M urea were dissociated (compare *lane 2* in each membrane). HC Ab = membrane incubated with HC primary antibodies; CP Ab = membrane incubated with CP primary antibodies; 1, 2, 3, 4 = samples incubated with 0, 5, 2.5, and 1.25 M urea, respectively; VP = level of virus particles; VP1, VP2 = purified PPV-Lar virus particles incubated with 0 and 5 M urea, respectively; CP_f = level of free coat protein aggregates; HC_f = level of free helper component aggregates

and HC in these populations. In support to these observations, "subpopulations" of VPs associated with HC have been recently discovered for two potyviruses, namely *Potato virus Y* (PVY) and *Potato virus A* [10]. Furthermore, it was shown by NEWeB that VP-HC complexes from plant extracts were stable under low urea concentrations (Fig. 2).

As concluded from the above synopsis, NEWeB could be a simple and useful tool for further characterization of *potyvirus* VP-protein associations and could also be helpful in the study of similar phenomena of other viruses, where similar principles apply.

An outline of the procedure is given below followed by the detailed protocol:

VPs are extracted from infected tissue in a buffer adjusted to the appropriate pH, loaded in agarose or mixed agarose-acrylamide gels, electrophoresed under native conditions and blotted onto nitrocellulose or other suitable membranes.

When VP-protein complexes are studied, two or more identical gels are prepared, each blotted to a different membrane. However, two or more replicas (groups) of a few wells in the same gel, loaded with the same samples, will be adequate for most studies.

After blotting, each membrane is treated with the antibodies (Ab) of interest, and following probing with secondary Abs, substrate incubation and color development, membranes are compared against each other for signal positioning. If a VP-protein complex exists, a signal coming from the (host or viral) protein under study will be spotted at the same level with that of VPs detected in the second membrane (protein shift) by the coat protein specific antibody (detecting VPs), suggesting a VP–protein interaction (Figs. 1 and 2).

2 Materials

2.1 Virus Isolates

Virus particles can be used either from purified preparations diluted to the nanogram or picogram range in extraction buffer, or detected directly in extracts of diseased plants.

2.2 Solutions

2.2.1 Electrophoresis and Blotting

1. Extraction and electrophoresis buffer: 0.2 M sodium phosphate pH 7.5, or TBE pH 8.3 (89 mM Tris base, 89 mM boric acid, 2.5 mM EDTA).
2. Loading buffer: 10 % sucrose in extraction buffer. Sucrose is added to increase the density of the sample thus enabling smooth precipitation into the well during loading. It is inert and will not influence the integrity of the complexes.
3. Tracing solution (1×): 0.25 % bromophenol blue, 0.25 % xylene cyanol FF, 40 % (w/v) sucrose in water. Weigh 0.25 g bromophenol blue, 0.25 g xylene cyanol FF and 40 g sucrose and dissolve thoroughly in 100 mL dH_2O by continuous steering. Store at 4 °C.
4. Acrylamide solution: 19 % acrylamide 1 % *N*, *N′*-methylene-bis-acrylamide (bis). Weigh 19 g acrylamide and 1 g bis and make up to 100 mL with distilled water (dH_2O) (*see* **Note 1**). Adjust a heated magnetic stirrer to about 37 °C and allow the solution to mix by continuous stirring in a fume cupboard. Store at 4 °C in dark bottles.
5. Ammonium persulfate (10 %): Weigh 1 g ammonium persulfate and make up to 10 mL with dH_2O. Mix in a magnetic stirrer. Make this solution always fresh.
6. Transfer buffer: Same as extraction buffer.

2.2.2 Immunodetection

1. Tris buffered saline (TBS)–Tween (TBST): 10 mM Tris, 0.9 % NaCl, 0.05 % Tween 20, pH 7.4. Add 1.21 g Tris–HCl in 800 mL dH_2O and adjust the pH to 7.4 with HCl. Add 9 g NaCl and 0.5 mL Tween 20 and make up to 1 L with dH_2O.
2. TBS/T/Milk (TBS/T/M): Dissolve 10 g skim milk powder in 150 mL TBS/T. Stir in a magnetic stirrer for 10 min. After homogenization make up to 200 mL with TBS/T.
3. TBST/M/Sap (TBS/T/M/S): Extract 5 g of leaves from a healthy plant, if possible the same species as that of the diseased one, in 50 mL TBST/M. Centrifuge for 10 min at 10,000 × *g* in a bench centrifuge to get rid of debris and collect the supernatant.
4. Alkaline phosphatase buffer (100 mM Tris–HCl, 100 mM NaCl, 5 mM $MgCl_2$ pH 9.5): Weigh 2.4 g Tris, 1.16 g NaCl, and 0.1 g $MgCl_2$. Dissolve in 100 mL dH_2O and adjust the pH to 9.5 with HCl. Make up to 200 mL with dH_2O.
5. Substrate solution:
   (a) NBT (nitro blue tetrazolium). Dissolve 1 g NBT in 20 mL of 70 % dimethylformamide.
   (b) BCIP (5-bromo-4-chloro-3-indolyl phosphate): Dissolve 1 g BCIP in 20 mL of 100 % dimethylformamide. Store at 4 °C in dark vials.

3 Methods

3.1 Extraction

The pH of the extraction and electrophoresis buffer should be selected so that the VPs and the proteins of interest have the desired charge. In the case of PPV both VP and HC were negatively charged at pH above 7.3. However, when both, positively and negatively charged proteins and VPs are examined in the same gel, a modified version called two-directional NEWeB (td-NEWeB) should be applied. In td-NEWeB, gels are prepared as described in the protocol but the comb is placed in the middle of the tray. After electrophoresis, each part of the gel upwards and downwards the loading points (wells) is blotted according to the polarity of the particles or molecules it contains (Fig. 3). Thus, the blotted half part of the gel containing the negative particles will be facing the anode during transfer, while the half blotted part containing the positive particles will be facing the cathode (Fig. 3). The blotting membranes should be cut to the exact size of each gel part to be blotted. The pieces of the Whatman paper should be cut to the exact size of the whole gel side, and applied accordingly along with the pads as described in the protocol.

1. Extract tissue with a pestle and mortar in extraction buffer (1 g in 5 mL).

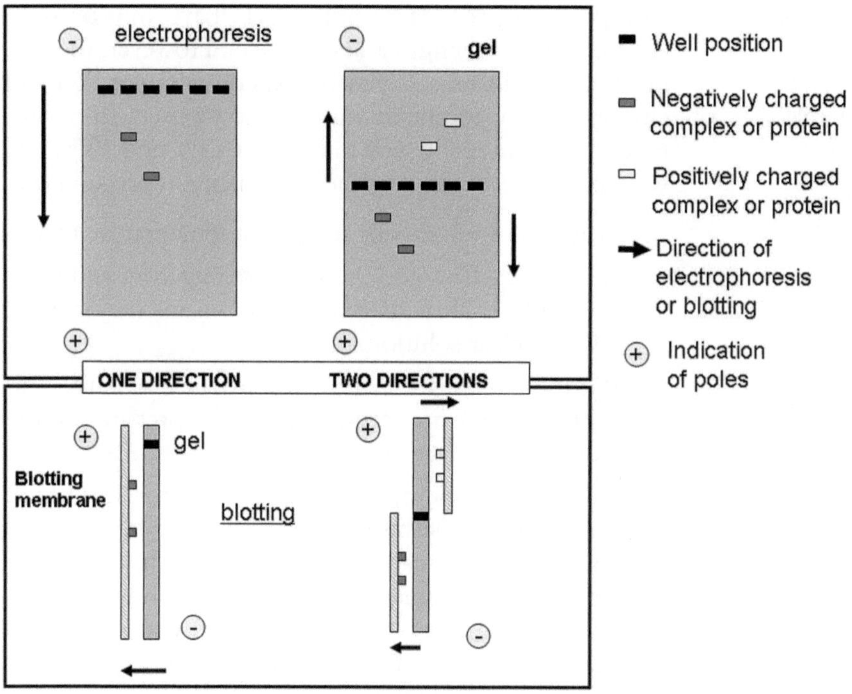

Fig. 3 One and two directional NEWEB: In one directional NEWeB (*left*) samples are loaded near the top of the gel and complexes and proteins with the same polarity are separated moving according to their net charge towards the opposite pole. After electrophoresis, the gel is blotted and treated as described in the protocol. In two directional (td) NEWeB (*right*) samples are loaded in the middle of the gel and VPs and proteins of different polarities are separated according to their net charge in the two gel directions moving to the opposite poles. After electrophoresis, the parts of the gel containing molecules and particles of opposite polarity are blotted accordingly on nitrocellulose membranes cut to the exact size of each gel part to be blotted. Gels are placed in the sandwich so that the blotted part containing positive molecules faces the cathode and the blotted part containing negative molecules faces the anode. Following transfer, the membranes are treated as described in the protocol for signal development and analysis. The gel is shown in gray and the membranes in downward diagonal line pattern. Black arrows indicate the direction of particle migration during electrophoresis or blotting

2. Centrifuge for 10 min at 10,000 × *g* in a bench centrifuge to separate cell debris and use the supernatant.

3.2 Gel Preparation

3.2.1 Agarose (0.8 %)

1. Weigh 0.8 g agarose and dissolve in 100 mL extraction buffer (*see* **Note 2**) in a conical flask.
2. Heat in a microwave oven until the agarose dissolves giving a clear solution. Cool the solution to about 55 °C.
3. Pour the solution in an appropriately assembled gel tray.
4. Fix the comb at the appropriate position, either on top or in the middle of the tray according to the expected particle charge and the desired migration direction.
5. After setting, place the gel in the electrophoresis unit and cover with electrophoresis buffer.

3.2.2 Mixed Agarose-Acrylamide (0.6 % in Agarose, 0.76 % in Acrylamide, 100 mL Total Volume)

Composite agarose/acrylamide gels have improved mechanical characteristics (strength and elasticity). However, the type of gel to be used should be selected by experimentation. Yet, other acrylamide/agarose combinations can also be used. In our hands, both types worked equally well for at least PPV and PVY. Agarose gels were preferred because of their simplicity in preparation.

1. Switch on a water bath and adjust temperature to about 60 °C.
2. In a conical flask add 0.6 g powder agarose and 60 mL extraction buffer. Heat in a microwave until the agarose dissolves giving a clear solution.
3. Put the conical flask in the already heated bath. This will prevent agarose from setting during the preparation of the acrylamide solution.
4. In another flask mix 24 mL dH_2O, 6 mL TBE (10×), and 4 mL acrylamide-bis solution (19 % acrylamide, 1 % bis), and place it in the already warmed water bath.
   When the acrylamide solution is warm enough, mix it with the agarose solution by gentle shaking. The acrylamide solution should be warm enough for avoiding premature solidification of agarose during mixing.
5. Add 6 μL ammonium persulfate (10 %), and 50 μL TEMED. Mix by gentle shaking.
6. Pour the solution in an appropriately assembled gel tray.
7. Adjust the combs as needed.
8. Allow about half an hour for the gel to set.
9. Place the gel in the electrophoresis unit and cover it with electrophoresis buffer.

3.3 Sample Loading

1. Prepare the samples by mixing an equal amount of the extracted preparations (e.g., 100 μL) with an equal volume of loading buffer. Mix thoroughly and give a quick centrifugation in a bench centrifuge.
2. Load in sequence 20–25 μL of each sample in each well. Samples are usually colorless or slightly greenish. Fixing a red tape below the electrophoresis tray at the level of the wells will increase contrast, making them visible and facilitate loading. When a VP-complex is examined load one or more identical groups of wells, in the same gel. Let one or two wells empty between the groups.
3. Add 20 μl tracker solution in the outside wells. Dyes of the tracker solution may interfere with electrophoretic behavior, therefore it is better to load them separately.
4. Run at 35–45 V until bromophenol blue reaches the end of the gel (about 5 h).

3.4 Blotting

1. Mark the gel by cutting the corner indicating the starting point of loading.
2. Wearing gloves cut four pieces of 3MM Whatman paper, in the exact dimensions of the gel and soak them in transfer buffer. Cut a piece of nitrocellulose membrane at the exact size of the gel. Allow the piece of nitrocellulose membrane to wet by capillary action in transfer buffer.
3. The following steps are described for one-directional NeWEB. For the two-directional NeWEB, another set up of the blotting array is needed (Fig. 3). Slide the gel from the electrophoresis tray onto the supporting plate of the blotting sandwich, the one that will be facing the anode during transferring. Lay the supporting plate with the gel on an open tray containing some amount of transfer buffer. Put the two pieces of the wetted paper on the gel so that they will fit exactly. Roll out any air bubbles with a glass pipette and add a porous pad over the papers. Cover with the second supporting plate.
4. Holding the two supporting plates firmly, turn the sandwich up-side-down. Remove the upper supporting plate to expose the side of the gel to be blotted.
5. Align the moistened nitrocellulose membrane on the gel and mark, with a pen, the corner corresponding to that of the marked gel. Mark gently, with a suitable pen, the positions of the wells in which the samples have been loaded. Squeeze out air bubbles by gently rolling a glass pipette on the membrane. Cover with two pieces of wet paper and add a porous pad.
6. Fix the sandwich with rubber bands, and place it in the tank with the blotted part facing the anode. Fill the tank with transfer buffer.
7. Run at 15–40 V 120 mA overnight keeping the buffer cooled (*see* **Note 3**).

3.5 Immunodetection

1. Switch off the blotting apparatus.
2. Disassemble the sandwich and wearing gloves carefully remove the membrane.
3. Incubate the membrane in 15 mL TBS/T/M by gentle shaking for 1 h.
4. Wash the membranes by shaking in TBS/T three times for 5 min each.
5. If complexes are studied, cut the membrane appropriately so that two (or more) identical sheets are obtained (each will be treated with different primary Abs). Mark each membrane according to the Ab it will be treated.
6. Incubate the membranes separately with Abs specific for each of the antigens examined, diluted in TBS/T/M/S at predetermined

ratios defined by experimentation. This is a necessary step for blocking (inactivating) nonspecific Abs that may have been raised against host proteins co-purified with the virus particles used for Ab production. However, when monoclonal or Ab raised against recombinant proteins are used, incubation could be done in TBS/T/M. In the above examples shown in Figs. 1 and 2 one membrane was incubated in CP particle specific Abs and the other in HC specific ones.

7. Repeat washing **step 4**.
8. Incubate each membrane separately for 1 h in appropriately conjugated secondary Abs suspended in TBS/T/M at a predetermined titration dilution depending on manufacturer.
9. Repeat washing **step 4**.
10. For each membrane mix 66 μL NBT in 10 mL alkaline phosphatase buffer and then add 33 μL BCIP. Incubate each membrane for appropriate time (usually 5–60 min) by gentle shaking (*see* **Note 4**).
11. After satisfactory signal development, immerse the membranes in tap water for 20 min and let them dry between two sheets of dry Whatman paper. Take pictures of the membranes and compare the signals, examining VP and protein positions as revealed in each membrane.

4 Notes

1. Acrylamide is a potent neurotoxin. Weighing and mixing should be done in a fume cupboard, whereas a lab coat, gloves and a mask must be worn during handling. Even after setting, gels should be handled with care because of the possibility of containing small amounts of unpolymerized acrylamide.
2. This amount has been calculated for gels with approximate dimensions of 16 cm in length, 12 cm in width and 0.5 cm in height. Smaller gels can be prepared by reducing the ingredients proportionally.
3. Some blotting apparatuses have incorporated buffer cooling systems. Blotting can also be performed in a cold room, but precautions should be taken that the power supply apparatus be compatible for operation in such environments.
4. NBT/BCIP and alkaline phosphatase conjugated antibody give good sensitivity. However, secondary Abs conjugated to fluorophores (e.g., Qdots) could also be tested for further increasing sensitivity. In that case, membranes appropriate for fluorescence detection should be used (e.g., Immobilon FL (Millipore)), whereas a documentation system will also be needed for best signal detection.

Acknowledgment

We thank Prof. E. Maiss for providing PPV CP and HC specific antibodies and M. Bastaki for help during her rotation. This work was partly supported by the Greek Ministry of Development (Project EPET II, 453) and by the INTAS 989 program of the EC.

References

1. Levy AJ, Fraenkel-Conrat H, Owens AR (1994) Virology. Prentice Hall International Limited, London
2. Gibbs A, Harrison BD (1976) Plant virology: the Principles. Edward Arnold, London
3. Tsagris M, Liu B, Tzortzakaki S, Tabler M (1993) Construction of a full length cDNA clone of an aphid transmissible PPV strain from Greece. IX International Congress of Virology
4. Govier DA, Kassanis B (1974) Evidence that a component other than the virus particle is needed for aphid transmission of potato virus Y. Virology 57:285–286
5. Govier DA, Kassanis BA (1974) Virus-induced component of plant sap needed when aphids acquire potato virus Y from purified preparations. Virology 61:420–426
6. Manoussopoulos IN, Maiss E, Tsagris E (1996) Detection of virus particle-helper component complexes in *Nicotiana benthamiana* plants infected with plum pox virus. PW37-8. Jerusalem, Israel, Xth International Congress of Virology
7. Blanc S, Lopez-Moya JJ, Wang RY, Garcia-Lampasona S, Thornbury DW, Pirone TP (1997) A specific interaction between coat protein and helper component correlates with aphid transmission of a potyvirus. Virology 231:141–147
8. Manoussopoulos IN, Maiss E, Tsagris M (2000) Native electrophoresis and Western blot analysis (NEWeB): a method for characterization of different forms of potyvirus particles and similar nucleoprotein complexes in extracts of infected plant tissues. J Gen Virol 81:2295–2298
9. Manoussopoulos IN, Tsagris M (2009) Native electrophoresis and western blot analysis: method and applications. Methods Mol Biol 536:277–287
10. Torrance L, Andreev IA, Gabrenaite-Verhovskaya R, Cowan G, Makinen K, Taliansky ME (2006) An unusual structure at one end of potato potyvirus particles. J Mol Biol 357:1–8

Chapter 36

Shift-Western Blotting: Separate Analysis of Protein and DNA from Protein–DNA Complexes

Matthias Harbers

Abstract

The electrophoretic mobility shift assay (EMSA) is the most frequently used experiment for studying protein–DNA interactions and to identify DNA-binding proteins. Protein–DNA complexes formed during EMSA experiments can be further analyzed by shift-western blotting, where the protein and DNA components contained in a polyacrylamide gel are transferred to stacked membranes: First a nitrocellulose membrane retains the proteins while double-stranded DNA passes through the nitrocellulose membrane and binds only to a charged membrane placed below. Immobilized proteins can then be stained with specific antibodies while the DNA can be detected by a radioactive label or a nonradioactive detection system. Shift-western blotting can overcome many limitations of supershift experiments and allows for the analysis of complex protein–DNA complexes containing multiple protein factors. Moreover, proteins and/or DNA may be recovered from membranes after the blotting step for further analysis by other means.

Key words Electrophoretic mobility shift assay, EMSA, Band shift assay, Gel shift assay, Shift-western blot, Transcription factor, DNA-binding protein, Protein–DNA interaction, Immunostaining, Response element

1 Introduction

The sequence-specific binding of transcription factors to their response elements is of fundamental importance for the regulation of gene expression and the utilization of genetic information [1, 2]. While DNA-binding proteins and their response elements have been studied individually for decades, only the advent of new high-speed sequencing methods has enabled genome-wide views on regulatory regions in the genome. For example, the cap analysis gene expression (CAGE) method [3, 4] obtains short sequencing reads from the 5′-end of mRNAs at a very high throughput. Those short sequences, or "CAGE tags," can be mapped to a reference genome identifying transcription start sites (TSSs) and their utilization. The results from the FANTOM 5 project [5] show that the activity of human TSSs is accurately regulated depending on the biological context. Moreover, information on gene-specific TSSs

Biji T. Kurien and R. Hal Scofield (eds.), *Western Blotting: Methods and Protocols*, Methods in Molecular Biology, vol. 1312, DOI 10.1007/978-1-4939-2694-7_36, © Springer Science+Business Media New York 2015

allowed identification of promoter [5, 6] and enhancer regions [7] in the human genome, which are largely regulated by interaction with protein factors. The occupation of regulatory regions by such protein factors can be further analyzed by chromatin immunoprecipitation followed by sequencing (ChIP-seq) experiments [8, 9], where protein–DNA complexes are isolated from cells with specific antibodies against selected proteins. From the antibody-precipitated protein–DNA complexes the DNA component is isolated and then sequenced on a high-speed sequencer. Mapping back the sequencing reads from ChIP-seq experiments to a reference genome provides genome-wide profiles on where the selected protein factor occupied its binding sites in the genome. While approaches like CAGE and ChIP-seq identify regulatory regions in the genome, these methods are limited in their ability to further analyze protein–DNA complexes and specific response elements. CAGE cannot directly tell which factors regulate an individual TSS activity, and features of TSSs and surrounding promoter regions are only inferred by computational sequence analysis. Similarly, while ChIP-seq can be very specific in its focus on the specified transcription factor, the DNA fragments sequenced in the experiments are commonly too long to identify the actual binding site occupied by the protein. In addition, ChIP-seq is mostly limited to the analysis of known factors for which good antibodies are available. Therefore other methods are needed to study protein–DNA interactions in more detail. Different strategies for the identification of DNA-binding proteins have been described in the literature [10], which can be very useful to follow up on individual observations made during genome-wide sequencing studies by CAGE or ChIP-seq.

The electrophoretic mobility shift assay (EMSA), also known as "band shift assay" or "gel shift assay," is the most frequently used method for characterizing protein–DNA interactions [10–12]. After incubation of a DNA fragment with proteins, the reaction mixture is subjected to polyacrylamide gel electrophoresis under native conditions. Under these conditions the free DNA migrates fast at the front of the gel, while the larger protein–DNA complexes are retained and migrate slower into the gel. Commonly EMSA experiments are analyzed by detection of the DNA, which can be easily labeled using radioactivity. The free DNA is found at the bottom of the gel, whereas in the protein–DNA complexes the DNA is "shifted" up to give rise to new signals higher up in the gel. Because of the high sequence specificity of many DNA-binding proteins, this rather simple method can accurately detect individual transcription factors even starting from crude nuclear extracts. The identity of the transcription factor can be confirmed by adding an antibody to the binding reaction that is directed against the protein in question. Such an antibody can confirm the specificity of an EMSA assay by either disrupting the protein–DNA complex (the shifted band will be lost or reduced) or shifting up the position of

Table 1
Large-scale cloning projects and collections

Project	Resources	URL
Broad Institute	Human ORF clones	http://www.broadinstitute.org/rnai/orfs
Drosophila Gene Collection	Drosophila	http://www.fruitfly.org:9005/DGC/index.htmL
Human Gene and Protein Database	Human	http://hgpd.lifesciencedb.jp/cgi/
I.M.A.G.E Consortium	Cow, zebrafish, pufferfish, human, mouse, rat, primate, frog	http://www.imageconsortium.org/
Mammalian Gene Collection	Human, mouse, rat, bovine	http://mgc.nci.nih.gov/
National Bioresource Project	Different plants	http://www.nbrp.jp/report/reportProject.jsp?project=arabidopsis#availableResources
ORFeome Collaboration	Human ORF clones	http://www.orfeomecollaboration.org/
RIKEN FANTOM	Mouse	http://fantom.gsc.riken.jp/

the protein–DNA complex in the gel (the so-called supershift experiment). Various versions of the basic EMSA format have been developed over the years that allow for qualitative and quantitative analysis of protein–DNA complexes including determination of binding stoichiometry, affinity constants, and binding kinetics. EMSA also has been combined with other methods such as western blotting and two-dimensional gel electrophoresis, and was used to enrich proteins from extracts. All these different applications show the great potential of the EMSA method for many studies on DNA-binding proteins and regulatory sequences in the genome (*see* Table 1 in [12] for more details).

While working on the thyroid hormone receptor (TR) [13], a nuclear hormone receptor [14–17], and the related oncogene v-erbA, we used EMSA to study various response elements and protein complexes. In those experiments, TR could bind as a monomer, a homodimer, or as a heterodimer together with the retinoic acid receptor (RAR) or retinoid X receptor (RXR) to DNA oligonucleotides comprising two binding sites arranged in different orientations and distances. To better analyze the complex patterns in those EMSA experiments we were seeking for alternatives to the commonly used supershift method. Moreover, we wanted not only to better characterize the nature of the different complexes separated in the gels, but also to have the further possibility to isolate separately the DNA and proteins from those protein–DNA complexes. In those days, we had focused on identifying optimal binding sites selected from pools of random sequences followed by DNA sequencing [18, 19], but the identification of cofactors binding to TR and other nuclear receptors was considered a very important aspect for future directions as well. These considerations let to the development of the "shift-western" blot method [20]. During a shift-western blot experiment the free DNA, free protein, and protein–DNA complexes are transferred to stacked membranes using an electric field: The first membrane is commonly a nitrocellulose membrane that can bind proteins but not double-stranded DNA (single-stranded DNA is not used in regular EMSA experiments). Therefore the free DNA and DNA from protein–DNA complexes migrate further and are then retained on a second charged filter, e.g., a DE 81 paper or a polyvinylidene difluoride (PVDF) membrane. After the transfer the proteins and the DNA can be separately detected. Also, free proteins that did not bind to DNA can be detected in the assays. Those commonly stayed at the top of the gels during gel electrophoresis and could be easily separated from the protein–DNA complexes. In our experiments, we detected Jun, TR, gag-v-erbA, and RAR on the nitrocellulose filters by immunostaining with specific antibodies. However, there are many more examples for the use of shift-western blot experiments in the literature from other laboratories working on different protein factors. To name only a few, Sp1 and NF-κB [21], p50, Rel, p65, RelB and p52 [22], or acetylated histone H4 [23] have been analyzed by shift-western blotting. We worked with radioactively labeled oligonucleotides that could easily be detected on the DE 81 paper by autoradiography or by using a phosphorimager. However, since the oligonucleotides were transferred onto a membrane, nonradioactive detection methods also have been used for DNA detection (e.g., digoxigenin-end labeling) [20]. Because both membranes relate to the same gel, protein and DNA are located at the same positions, and the signals can be overlaid to identify the location of specific protein factors within different protein–DNA complexes. The shift-western blot method may likewise be used to study the effect of cofactors on the DNA binding of certain proteins. In our original publication for example, we showed that 3,5,3′-L-triiodothyronine (T3) binds mainly to monomeric TR but not to TR homodimers [20].

While we normally used overexpressed proteins in our studies, one of the key advantages of EMSA is that proteins in crude nuclear extracts can be identified only by their binding to specific DNA sequences. This can be used to identify unknown transcription factors that bind to a regulatory region characterized in other experiments [11] using CAGE or ChIP-seq data for instance. Great progress has been made for the use of mass spectrometry to identify and characterize proteins as was recently shown by the publication of a mass-spectrometry-based draft of the human proteome [24]. Such methods have also been used to study active transcription factors [25, 26], suggesting that the combination of EMSA experiments with protein mass spectrometry is a successful way to study proteins binding to genomic regions. Here, blotting methods like the shift-western blotting are very useful to transfer proteins from protein–DNA complexes formed during EMSA experiments directly onto nitrocellulose. Proteins bound to nitrocellulose membranes can be characterized for example by blotting and removal of nitrocellulose (BARN) followed by mass spectrometry [27, 28]. We showed in our original publication that depending on membranes used in shift-western blot experiments, the DNA and proteins can be recovered for further analysis [20] enabling the use of other protocols for protein and DNA identification. Therefore, it is hoped that the shift-western blot method will be used more in the future not only to directly analyze protein–DNA complexes, but also to enrich those proteins and/or DNA for identification by mass spectrometry or DNA sequencing. Combining those methods could be a good way to better follow up on the results from genome-wide sequencing experiments and to characterize in more detail regulatory important protein–DNA interactions.

2 Materials

Optimal conditions for EMSA experiments must be individually determined for each study. Therefore reagents and instruments listed here are focusing on those that are required for electroblotting of the proteins and DNA onto staged membranes. For the other steps general directions are given.

Use deionized water to prepare buffers, wear gloves, and use forceps when working with membranes or filter papers. Do not touch membranes or filters with your hands. Use a clean glass plate and scalpel to cut membranes and filters into the desired size. Avoid damage of the membranes; some of the materials are fragile (e.g., nitrocellulose filter breaks easily).

Take special precautions when working with radioactive materials as for example when using labeled oligonucleotides or DNA fragments. Follow regular laboratory procedures and safety measures while doing the experiment. Note that methanol, bisacrylamide, and acrylamide are toxic and have to be handled with special care and safety precautions.

2.1 Antibody

Select an optimal antibody against your target protein(s) (*see* **Note 1**).

2.2 Buffers

1. Reaction buffer: 4 % w/v Ficoll 400, 80 mM KCl, 5 mM $MgCl_2$, and 10 mM Hepes pH 7.9 (*see* **Note 2**).
2. Competing DNA: Poly d(I-C) 1 mg/mL stock solution in water (*see* **Note 3**).
3. Loading buffer (10×): 50 % v/v glycerol, 0.001 % w/v bromophenol blue, 0.001 % w/v xylene cyanol (*see* **Note 4**).
4. PAGE: Prepare a 6 % polyacrylamide gel (*see* **Note 5**).
5. Transfer buffer: 48 mM Tris, 39 mM glycine, 20 % methanol, pH 8.5.

2.3 Filters

Different filters and membranes can be used in shift-western blot experiments (*see* **Note 6**).

1. Nitrocellulose BA85.
2. DE 81 paper.
3. Nylon membrane.
4. PVDF membrane.
5. Filter paper (Whatman paper).
6. Paper towels.
7. Plastic film: Saran wrap or equivalent.

2.4 Instruments

1. Electrophoresis chambers: Use a vertical electrophoresis apparatus with matching glass plates, spacers, and comb.
2. Power supply: Up to 250 V and 200 mA capacity.
3. Glass plate.
4. Forceps: Select flat forceps with a smooth surface so as not to damage membranes.
5. Scalpel.
6. Transfer unit: Multiphor II NovaBlot.
7. Gel dryer.
8. Autoradiography cassette.

3 Methods

Protein–DNA interactions can be studied with high specificity in EMSA experiments. However, this specificity is entirely dependent on accurately defining the optimal experimental conditions and control experiments; refer to Table 2 in [12] for more details. It is beyond the scope of this protocol to list all options, and the conditions given here relate to our own work on nuclear receptors.

Figure 1 summarizes the different considerations necessary to plan a shift-western blot experiment. All those parameters should be taken seriously to make sure that meaningful results can be obtained.

Table 2
Public clone depositories and clone providers

Clone provider	Resources	URL
Addgene	Nonprofit plasmid repository	https://www.addgene.org/
DGRC	Drosophila clones	https://dgrc.cgb.indiana.edu/Home
DNAFORM	Clone distributor	http://www.cloneresources.com/cgi-bin/DNAFORM/welcome_en
DNASU	Central repository	https://dnasu.org/DNASU/
GE Dharmacon	Clone distributor	http://dharmacon.gelifesciences.com/geneius-product-search/
Genecopoeia	Clone provider	http://www.genecopoeia.com/
Invitrogen	Gateway entry clone provider	http://www.lifetechnologies.com/us/en/home/life-science/cloning/gateway-cloning/entry-clones.htmL
Kazusa DNA Research Institute	Human, mouse, plant	http://www.biosupport.kazusa.or.jp/english/sub_center1/index.htmL#human_clone
NBRC	Nonprofit repository for human clones	http://www.nbrc.nite.go.jp/e/hgentry-e.htmL
NIAS DNA Bank	Nonprofit repository for rice and animal resources	http://www.dna.affrc.go.jp/
Origene	Human clone provider	http://www.origene.com/
RIKEN BRC	Nonprofit plasmid repository	http://en.brc.riken.jp/
Source Biosciences	Clone distributor	http://www.lifesciences.sourcebioscience.com/genomecube/

3.1 Considerations Before Starting Shift-Western Blot Experiments

3.1.1 Protein Source

Studying protein–DNA complexes in a biological context, nuclear extracts are the first choice to see whether a certain DNA-binding protein is present. Such experiments can be extended to identify protein factors that bind to individual promoter or enhancer regions. Nuclear extracts are best prepared by the classical protocol from Digman [29, 30], although other methods have been described in the literature (*see* ref. 10 for more details). Alternatively individual transcription factors can be used in the binding reactions. Because transcription factors are often difficult to express in bacterial systems, EMSA experiments often use proteins prepared by in vitro translation or cell-free protein expression systems. Such systems allow for rapid preparation of proteins directly from DNA templates within a few hours. The cell-free expression systems most commonly used are based on extracts from *E. coli* [31], wheat germ [32], rabbit reticulocyte [33], and insect [34] or human cell lines [35]. All those expression systems can be purchased from different

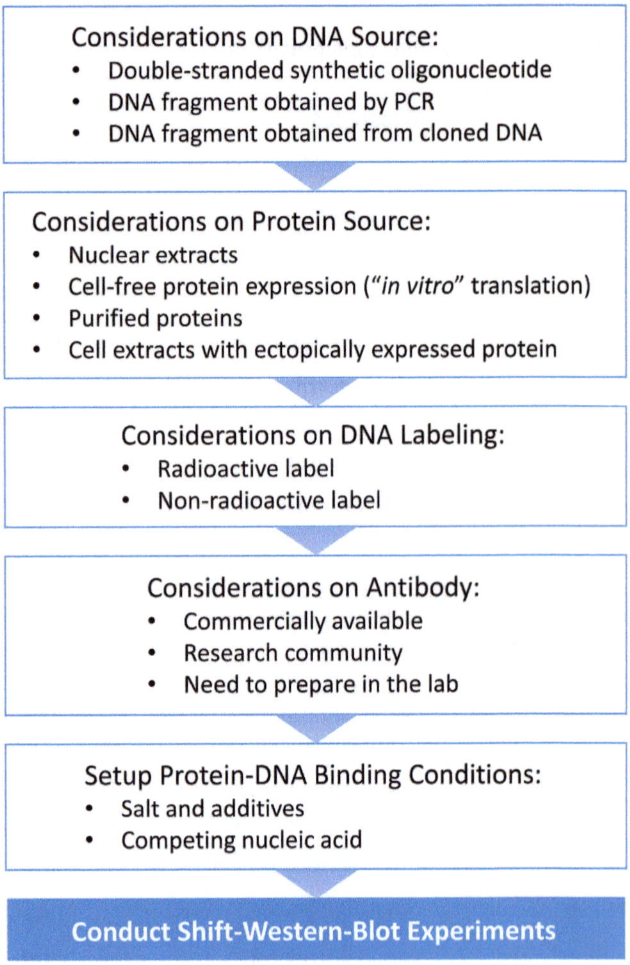

Fig. 1 Considerations for planning a shift-western blot experiment

providers. Some providers also sell purified transcription factors that may be suitable for EMSA experiments. Transcription factors may also be ectopically expressed in cell lines before preparing cell extracts enriched for the protein or proteins of interest.

It has been estimated that there could be in the range of 2,600 DNA-binding proteins encoded by the human genome, most of which are likely to be transcription factors. Therefore, it can be meaningful to search also public cDNA collections from genome-wide cloning projects [36] for clones encoding proteins of interest. Refer to Table 1 for a list important clone collections and projects; refer to Table 2 for a list of public clone depositories and clone providers.

3.1.2 DNA Source

Double-stranded DNA fragments of some 20–300 bp length can be used in EMSA experiments. The lower limit is set by the length of the binding site and some additional sequences that should be present at both ends to stabilize protein binding (5–10 bp at each end). The upper length is rather set by the limitation of polyacrylamide gels to separate free DNA from protein–DNA complexes. Because long DNA fragments migrate slower in the gel, binding of additional protein factors may lead only to a small shift in the DNA band. Moreover, working with long DNA fragments and crude proteins from nuclear extracts can lead to the formation of multiple complexes further complicating the analysis. In general, short DNA oligonucleotide probes are preferable (20–25 bp), which have to be double-stranded DNA for studying factors binding to genomic DNA. Synthetic oligonucleotides can be annealed to form the double-stranded DNA needed for the binding assays. Single-stranded DNA regions should be excluded to avoid nonspecific binding by unrelated proteins.

When studying known transcription factors the oligonucleotide sequences may be selected from known binding sites published in the literature or taken from online databases for DNA-binding sites like TRANSFAC (http://www.biobase-international.com/product/transcription-factor-binding-sites) or JASPER (http://jaspar.genereg.net/). There is further the "DBD: Transcription factor prediction database" (www.transcriptionfactor.org) that holds information on predicted transcription factors found in completely sequenced genomes.

However, individual binding sites in the genome do not have always the same sequence as the consensus sequence given in a reference database. Therefore, one has to consider whether the presence of a transcription factor in a biological sample should be detected using a consensus sequence or whether a specific binding site in the genome should be confirmed for binding to a certain transcription factor. Biologically active response elements do not necessarily match consensus sequences assembled from many naturally occurring motifs or obtained in a selection experiment [37].

When longer DNA regions are studied for potential binding proteins, the DNA fragment is preferably prepared by PCR or from cloned genomic regions. PCR also provides means to label DNA fragments during the amplification reaction.

3.1.3 DNA Labeling

We had normally used radioactive labeling for our EMSA and shift-western blot experiments. Those labels can be easily introduced by standard methods. Radioactive labels are highly sensitive and very easy to detect after the transfer from the polyacrylamide gel to a filter. We found DE 81 paper the most suitable and cost-effective solution for working with radioactively labeled oligonucleotides in shift-western blot experiments.

Alternatively, nonradioactive labeling methods can be easily used in shift-western blot experiments, because the DNA is already transferred onto a membrane that can be used in the staining experiments. There are different commercial kits for working with nonradioactive labels, for example the DIG-labeling kit from Roche we have been using in our studies. Alternative nonradioactive labeling and detection kits for EMSA experiments are offered by Active Motif (Gelshift™ Chemiluminescent EMSA), or Thermo Scientific (LightShift Chemiluminescent EMSA Kit).

For safety reasons, nonradioactive methods are preferable. Such experiments can be conducted without need for an isotope laboratory and should also be suitable for teaching classes.

3.1.4 Test Protein–DNA Binding Conditions

The most critical part of setting up EMSA and shift-western blot experiments is finding appropriate binding conditions for studying the protein–DNA interactions of interest. Therefore for every protein factor and DNA probe parameters like mono- and divalent salt concentrations, pH, competitor DNA, and buffering system have to be tested. Weak protein–DNA interactions can be further stabilized by additives, increasing the density of the buffer. Similarly, adding carrier proteins (e.g., BSA) is recommended when working with low concentrations of purified transcription factors. When working with cell extracts, protease, nuclease, and phosphatase inhibitors are useful additives. Some transcription factors could require specific cofactors that may have to be considered.

In particular, when working with cell extracts containing many different DNA-binding proteins, an excess of unlabeled competing nucleic acids must be added to the reaction mixture. The most commonly used nonspecific competitor is poly d(I-C), but poly d(A-T) and salmon-sperm genomic DNA also have been used. However, genomic DNA can cause nonspecific signals when working with cell extracts.

The specificity of a protein–DNA interaction can be further confirmed by competition experiments using cold oligonucleotides having the same binding site or mutated versions of the binding site. Such experiments not only confirm the specificity of the binding reaction, but could also be further used for Scatchard analysis to obtain dissociation constants and so on.

3.2 Shift-Western Blot Experiment

The principal workflow of a shift-western blot experiment is outlined in Fig. 2. For conducting successful shift-western blot experiments, the foregoing setup experiments should be made to assure reliable conditions for the EMSA experiments; refer to Hellman and Fried [12] on detailed information on EMSA reaction conditions.

3.2.1 Preparation of Labeled DNA Probe

We commonly used ~0.3 ng of ^{32}P-labeled double-stranded DNA (*see* **Note 7**), where synthetic oligonucleotides were labeled using T4 polynucleotide kinase and [γ-^{32}P] ATP (*see* **Note 8**). PCR products were labeled by using a ^{32}P-labeled primer in the amplification reaction. After titration of an optimal DNA probe concentration, we kept the DNA amount per assay usually constant in our experiments while changing the protein concentration.

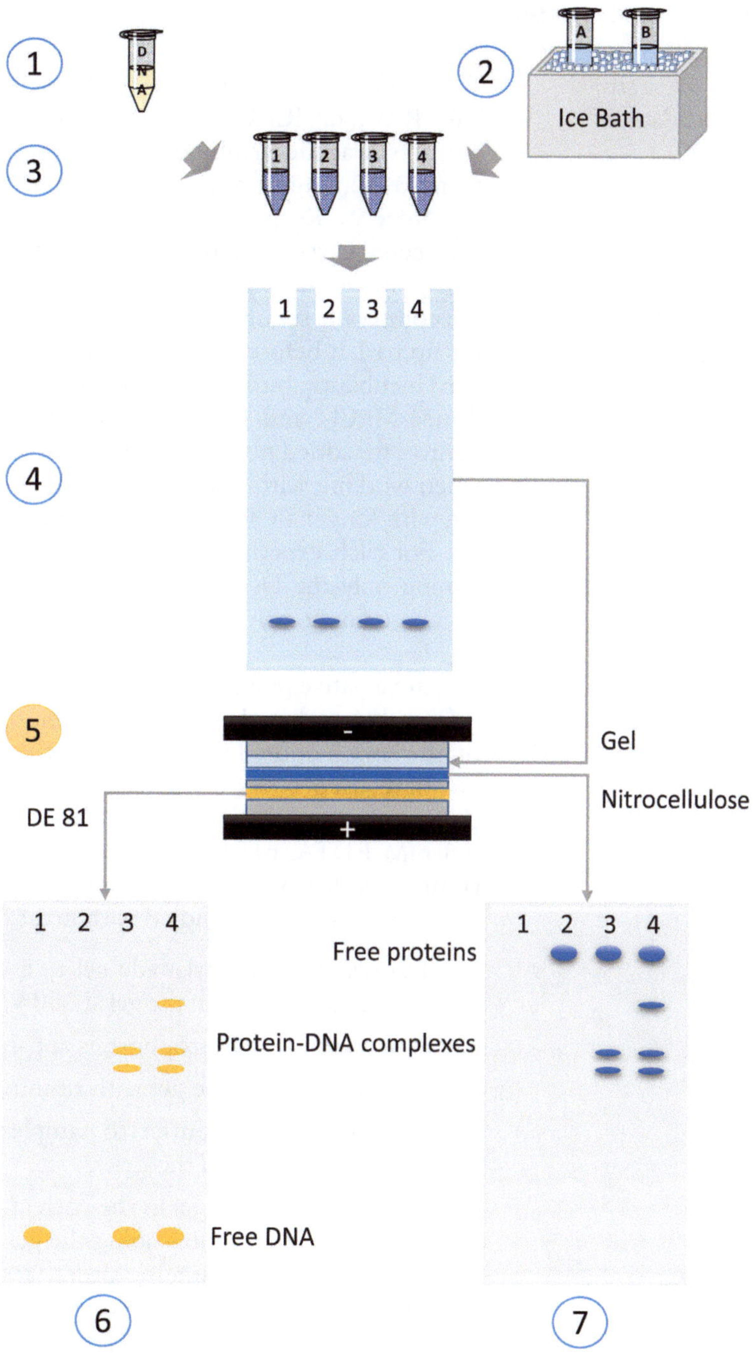

Fig. 2 Principal workflow of a shift-western blot experiment. Shift-western blot experiments comprise the following steps: *1*. Preparation of labeled DNA probe. *2*. Preparation of protein sample. *3*. Protein–DNA binding reactions. *4*. Polyacrylamide gel electrophoresis to separate free DNA and proteins from protein–DNA complexes. *5*. Blotting of proteins and DNA onto stacked membranes (polyacrylamide gel = *light blue*; nitrocellulose = *dark blue*; DE81 paper = *orange*). *6*. DNA detection on DE 81 paper. *7*. Protein detection on nitrocellulose membrane. The schematic representation of the results from a shift-western blot experiment shows in *lane 1*: negative control with a DNA oligonucleotide only, *lane 2*: negative control with a protein sample only, and *lanes 3* and *4*: samples containing free DNA oligonucleotides, free proteins, and protein–DNA complexes

3.2.2 Preparation of Protein Sample

We commonly used nuclear extracts from HeLa cells overexpressing TR, RAR, or RXR. The protein concentration of the nuclear extracts was determined before use by Bradford assay, and some 3–12.5 μg of nuclear protein was used per binding reaction (*see* **Note 9**). Keep all protein samples on ice at all times and avoid unnecessary freeze/thawing of proteins.

3.2.3 Protein–DNA Binding Reaction

We commonly set up 20 μL binding reactions and kept them on ice for up to 1 h before loading onto a polyacrylamide gel. The standard incubation buffer contained 4 % w/v Ficoll 400, 80 mM KCl, 5 mM $MgCl_2$, and 10 mM Hepes pH 7.9. As an unspecific competitor, we added poly d(I-C) at a final concentration of 0.1 μg/μL when working with oligonucleotides, and 0.2 μg/μL when working with longer PCR products (*see* **Note 3**).

For each experiment, set up a negative control reaction containing only the DNA probe, and another negative control reaction containing only the protein sample.

3.2.4 Gel Electrophoresis to Separate Free DNA, Free Proteins, and Protein–DNA Complexes

Prepare a native polyacrylamide gel following standard procedures. Adjust the acrylamide and bisacrylamide concentration depending on the length of the DNA probe and the size of the protein factors. We have used 6 % polyacrylamide gels in our experiments. We commonly used 0.25× TBE (22.5 mM Tris, 22.5 mM borate, 0.5 mM EDTA, pH 8.3) buffer for our polyacrylamide gels and running buffer. Note that acrylamide and bisacrylamide are neurotoxic and have to be handled with great care!

1. Transfer the polyacrylamide gel to a vertical gel electrophoresis apparatus and pre-run the gel at 250 V for at least 30 min at 4 °C.
2. Disconnect gel electrophoresis apparatus from power supply.
3. Wash the slots in the gel with running buffer.
4. Add 10× loading buffer to samples to increase the density before loading onto the gel.
5. Load samples onto gel in the correct order. Avoid any air bubbles in the slots while loading the gel. Use negative controls to load gel in an asymmetric order (*see* **Note 10**) to avoid mistakes by flipping over the gel or membranes.
6. Run the gel at 250 V at 4 °C. Reduce voltage if the gel is getting warm during electrophoresis. Monitor the migration of the dyes in the gel to determine when the electrophoresis is completed. Bromophenol blue will run at the front, and electrophoresis should be stopped before bromophenol blue reaches the end of the gel.

3.2.5 Transferring Proteins and DNA onto Stacked Membranes

Prepare membranes of the size of the polyacrylamide gel before starting the electroblotting step. During the shift-western blot, two stacked membranes are used to separately retrieve the protein and DNA from the polyacrylamide gel. We commonly used a nitrocellulose membrane to capture the proteins and a DE 81 paper to capture the DNA (*see* **Note 6**). Keep the order of the membranes; otherwise proteins and DNA will not be separated during the transfer reaction.

We used a semidry blotting method to transfer the proteins and DNA. This is commonly faster and more efficient than other electroblotting methods. In our experiments we observed a nearly complete separation of protein–DNA complexes during the transfer reaction (measured by monitoring the presence of radioactive DNA on either membrane).

Prepare a chamber with transfer buffer to soak papers and membranes in transfer buffer before assembling the transfer reaction. The transfer buffer in the papers and membranes is the only buffer reservoir to maintain the currency flow. Wear gloves while working with the gel, membranes, and transfer buffer. Note: The transfer buffer contains methanol to avoid shrinking of the polyacrylamide gel. Methanol is toxic and safety precautions must be taken.

Keep strictly to the order of the membranes while preparing the transfer reaction (compare Fig. 2, Step 5). Avoid air bubbles while assembling the transfer reaction; there will be no protein and DNA transfer where air bubbles block the electricity flow. Mark the orientation of the gel and membranes, e.g., by cutting of one corner. The order given in this protocol assumes that the cathode is at the bottom of the transfer unit. Assure that the same applies to the unit used in your experiments.

Assemble the transfer reaction in the following order:

1. Prepare transfer unit.
2. The cathode is at the bottom of the transfer unit.
3. Place some six layers of paper towels soaked with transfer buffer onto cathode.
4. Place one layer of Whatman paper soaked in transfer buffer on the paper towels.
5. Place one layer of DE81 paper soaked in transfer buffer on the Whatman paper.
6. Place one layer of Whatman paper soaked in transfer buffer on the DE81 paper.
7. Place one layer of nitrocellulose soaked in transfer buffer on the Whatman paper.
8. Rinse polyacrylamide gel in transfer buffer.
9. Place the polyacrylamide gel onto the nitrocellulose.
10. Place one layer of Whatman paper soaked in transfer buffer on the polyacrylamide gel.

11. Place some six layers of paper towels soaked with transfer buffer on top of the Whatman paper.
12. Place anode on top of the sandwich to begin electroblotting.
13. Proteins and DNA are blotted at room temperature at a fixed current of 0.8 mA/cm^2.
14. The transfer is usually completed after 90 min.

3.2.6 Detection of DNA Fragments

Radioactively labeled DNA is detected by autoradiography after drying the DE 81 paper on a gel dryer.

1. Remove the DE 81 paper together with the Whatman paper below from the transfer reaction.
2. Place the Whatman paper with the DE 81 paper on top onto the gel dryer.
3. Cover the DE 81 paper with a plastic film before closing the gel dryer.
4. DE 81 paper is commonly dry after some 5–10 min of drying on the gel dryer.
5. Wrap the DE 81 paper and the Whatman paper into a plastic film.
6. Forward the DE 81 paper on the Whatman paper and the plastic cover to an autoradiography cassette or use a phosphorimager to detect the signals. The exposure time will depend on the signal strength. It can be useful to make more than one exposure if the signal for the free DNA is very strong as compared to the shifted bands from the protein–DNA complexes.

When working with a nonradioactive label, do not use DE 81 paper (it is not stable in aqueous solutions) but a suitable membrane for DNA binding like DEAE, PVDF, or nylon. Note that some membranes need special treatments for binding DNA (*see* Table 3). After the transfer reaction, follow the steps indicated in the kit manual to detect the signals. Those steps may vary depending on which labeling methods have been used, and some kits may advise on the use of a specific membrane.

For recovery of DNA from membrane *see* **Note 11**.

3.2.7 Detection of Proteins

Proteins bound to the nitrocellulose membrane are detected by immunostaining. We used a standard protocol with phosphate-buffered saline (PBS) supplemented with 0.35 M NaCl, 1 % BSA, and 0.2 % Tween 20. In brief, nonspecific binding to the nitrocellulose membrane was blocked by incubation with 3 % skim milk powder, followed by incubation with the first antibody (directed against the protein target) for 1 h. After washing the membrane three times, the membrane was incubated with the second antibody (directed against the first antibody) for 1 h. After washing the membrane three times, protein signals were detected by a peroxidase reaction leading to enhanced chemiluminescence.

For recovery of proteins from membrane *see* **Note 12**.

Table 3
Membranes and filters tested for use in shift-western blot experiments

Membrane	Protein binding	DNA binding	Comment
Nitrocellulose	++	–	Used for protein detection, does not bind double-stranded DNA
PVDF (Immobilon P)	++	++	Must be activated with methanol before use, used for detection of nonradioactively labeled DNA
Nylon (Hybond)	++	–/+	May require cross-linking to bind DNA, used for detection of nonradioactively labeled DNA
DEAE membrane	++	++	High background for protein detection, used to recover DNA
DE 81 paper	++	++	Used to detect radioactive DNA, not suitable for detection of proteins and nonradioactively labeled DNA

++ = good binding, –/+ = poor binding, – = no binding

4 Notes

1. A specific antibody is required to detect proteins after transfer onto a nitrocellulose membrane. The proteins detected in shift-western blot experiments are not of the same quality as proteins studied by western blotting: While EMSA experiments use native proteins and gel conditions, proteins are denatured during SDS-PAGE experiments. Still it is useful to make sure whether the selected antibody is suitable for western blot experiments. For commercial antibodies, this is often indicated in the product description, but we recommend testing new antibodies before use in shift-western blot experiments by doing a western blot experiment on the protein source used in the shift-western blot experiment. These western blot experiments can be further used to optimize immunostaining conditions, where optimal antibody concentrations may be tested and where the specificity of the antibody is confirmed, e.g., by testing a nuclear extract. If an antibody works in western blotting but not in the shift-western blot experiment, look for a different antibody that can recognize the native protein. These are commonly antibodies that also work in immunoprecipitation experiments. Large antibody collections are available from different providers, which also cover many transcription factors (e.g., Abcam, Abnova, Active Motif, or Sigma-Aldrich). Moreover, there are large antibody collections for antibody proteomic projects [38]. Use the Antibodypedia (http://www.antibodypedia.com/), an open-access database of publicly available antibodies against human protein targets, to find

suitable antibodies for working on human transcription factors. Search the literature for antibody resources in the research community. Antibodies may also be prepared in the lab using standard procedures. In-house-made antibodies should be thoroughly characterized before use in shift-western blot experiments to ensure their specificity. Note that antibodies working in western blotting experiments do not necessarily work also in other detection assays. In principle, proteins blotted onto nitrocellulose membranes may also be detected by other methods, such as southwestern blotting [39] or different staining methods [40].

2. Reaction buffer conditions have to be tested for each protein–DNA binding reaction. Make sure that the reaction buffer is suitable for your studies.
3. Different competitor nucleic acids may be used in shift-western blotting experiments, but poly d(I-C) is the most commonly used competitor. Confirm the length of poly d(I-C) before use. It may be preferable to reduce the length of poly d(I-C) to some 200–600 bp by sonication before use in EMSA experiments.
4. The 50 % v/v glycerol is used to increase the density of the buffer for loading onto the gel; 0.001 % w/v bromophenol blue and 0.001 % w/v xylene cyanol are migration markers for gel electrophoresis. Other buffer components may be added in line with the reaction conditions to stabilize the protein–DNA complex.
5. Commonly 4–6 % polyacrylamide gels are used in EMSA or shift-western blotting experiments, and it may be necessary to test optimal gel electrophoresis conditions for each protein–DNA complex.
6. Table 3 provides more information on the different membranes that had been tested for use in shift-western blot experiments. The choice of membranes depends on the experimental needs, but we commonly worked with nitrocellulose membranes to capture the proteins, and DE 81 paper to capture radioactively labeled oligonucleotides. For use of nonradioactively labeled oligonucleotides, we preferred a nylon or PVDF membrane, because DE 81 paper is not stable during the staining experiments. Proteins and DNA can be recovered when bound to charged filters like DE 81 paper or a DEAE membrane.
7. Oligonucleotides can be annealed by mixing equal ratios of the upper and lower strand in 180 mM NaCl, denaturing at 95 °C in a water bath, followed by slowly cooling down the reaction mixture to room temperature overnight. This can be achieved by just turning off the water bath after the denaturation step.

8. Oligonucleotides may be designed to have single-stranded overhangs after the annealing step. Those overhangs can be used to label the oligonucleotides in a fill-in reaction with the Klenow fragment of DNA polymerase I. Working with oligonucleotides having a single-stranded overhang can be very useful if the same sequences are to be cloned into a reporter gene vector to test the activity of the binding site in subsequent transfection experiments.
9. We used recombinant *Vaccinia* viruses to overexpress nuclear receptors in HeLa cells. Nuclear extracts from infected cells can be directly used in shift-western blot experiments, where approximately about 60 nM of the receptors were used per binding assay.
10. Use the negative control reactions to asymmetrically load the gel. The negative control reaction containing only the DNA probe should not show any shifted bands and should give no signal during protein staining. The negative control reaction containing only the protein sample should give no signal for the DNA probe. Free proteins are commonly found at the top of the gel, since they do not migrate much into the gel without the additional charges from the DNA probe.
11. DNA can be extracted from charged membranes by the following steps:
    (a) Excise region with the DNA of interest.
    (b) Wash membrane piece 3× in low-salt buffer (150 mM NaCl, 20 mM Tris–HCl, pH 8.0, 0.05 mM EDTA).
    (c) Elute DNA by incubation at 65 °C in high-salt buffer (1 M NaCl, 20 mM Tris–HCl, pH 8.0, 0.05 mM EDTA) for 45 min.
    (d) DNA can be recovered from the high-salt buffer by ethanol precipitation.
12. Proteins can be recovered from shift-western blot experiments, when the nitrocellulose membrane is replaced by a charged membrane that can bind protein–DNA complexes. Proteins are extracted by the following steps:
    (a) Excise region with the protein of interest.
    (b) Elute protein by incubation at 65 °C in 0.4 M acetic acid and 1 M NaCl for 45 min.
    (c) Proteins are precipitated from the high-salt buffer by adding 10 % trichloroacetic acid (TCA). Protein pellets can be washed with cold acetone followed by cold ethanol.
    (d) Proteins can also be recovered from nitrocellulose membranes by dissolving the membrane in acetone [41].

Acknowledgement

I want to express my special thanks to Stephen Demczuk and Björn Vennström, with whom I had originally developed the shift-western blotting method, and Piero Carninci for his help to make this protocol possible. I further want to thank Mitch Dushay for critically reading the manuscript and his suggestions to improve the text. This work was supported by a research grant from the Japanese Ministry of Education, Culture, Sports, Science and Technology (MEXT) to the RIKEN Center for Life Science Technologies (No. 95).

References

1. Spitz F, Furlong EE (2012) Transcription factors: from enhancer binding to developmental control. Nat Rev Genet 13:613–626
2. Woringer M, Darzacq X, Izeddin I (2014) Geometry of the nucleus: a perspective on gene expression regulation. Curr Opin Chem Biol 20C:112–119
3. Kodzius R, Kojima M, Nishiyori H, Nakamura M, Fukuda S, Tagami M, Sasaki D, Imamura K, Kai C, Harbers M, Hayashizaki Y, Carninci P (2006) CAGE: cap analysis of gene expression. Nat Methods 3:211–222
4. Shiraki T, Kondo S, Katayama S, Waki K, Kasukawa T, Kawaji H, Kodzius R, Watahiki A, Nakamura M, Arakawa T, Fukuda S, Sasaki D, Podhajska A, Harbers M, Kawai J, Carninci P, Hayashizaki Y (2003) Cap analysis gene expression for high-throughput analysis of transcriptional starting point and identification of promoter usage. Proc Natl Acad Sci U S A 100:15776–15781
5. Forrest AR, Kawaji H, Rehli M, Baillie JK, de Hoon MJ, Lassmann T, Itoh M, Summers KM, Suzuki H, Daub CO, Kawai J, Heutink P, Hide W, Freeman TC, Lenhard B, Bajic VB, Taylor MS, Makeev VJ, Sandelin A, Hume DA, Carninci P, Hayashizaki Y (2014) A promoter-level mammalian expression atlas. Nature 507: 462–470
6. (2012) An integrated encyclopedia of DNA elements in the human genome. Nature. 489: 57–74
7. Andersson R, Gebhard C, Miguel-Escalada I, Hoof I, Bornholdt J, Boyd M, Chen Y, Zhao X, Schmidl C, Suzuki T, Ntini E, Arner E, Valen E, Li K, Schwarzfischer L, Glatz D, Raithel J, Lilje B, Rapin N, Bagger FO, Jorgensen M, Andersen PR, Bertin N, Rackham O, Burroughs AM, Baillie JK, Ishizu Y, Shimizu Y, Furuhata E, Maeda S, Negishi Y, Mungall CJ, Meehan TF, Lassmann T, Itoh M, Kawaji H, Kondo N, Kawai J, Lennartsson A, Daub CO, Heutink P, Hume DA, Jensen TH, Suzuki H, Hayashizaki Y, Muller F, Forrest AR, Carninci P, Rehli M, Sandelin A (2014) An atlas of active enhancers across human cell types and tissues. Nature 507:455–461
8. Furey TS (2012) ChIP-seq and beyond: new and improved methodologies to detect and characterize protein-DNA interactions. Nat Rev Genet 13:840–852
9. Sims D, Sudbery I, Ilott NE, Heger A, Ponting CP (2014) Sequencing depth and coverage: key considerations in genomic analyses. Nat Rev Genet 15:121–132
10. Carey MF, Peterson CL, Smale ST (2012) Experimental strategies for the identification of DNA-binding proteins. Cold Spring Harb Protoc 2012:18–33
11. Jiang D, Jarrett HW, Haskins WE (2009) Methods for proteomic analysis of transcription factors. J Chromatogr A 1216:6881–6889
12. Hellman LM, Fried MG (2007) Electrophoretic mobility shift assay (EMSA) for detecting protein-nucleic acid interactions. Nat Protoc 2:1849–1861
13. Aranda A, Alonso-Merino E, Zambrano A (2013) Receptors of thyroid hormones. Pediatr Endocrinol Rev 11:2–13
14. Evans RM, Mangelsdorf DJ (2014) Nuclear receptors, RXR, and the big bang. Cell 157:255–266
15. Mendoza-Parra MA, Gronemeyer H (2013) Genome-wide studies of nuclear receptors in cell fate decisions. Semin Cell Dev Biol 24: 706–715
16. Rastinejad F, Huang P, Chandra V, Khorasanizadeh S (2013) Understanding nuclear receptor form and function using structural biology. J Mol Endocrinol 51:T1–T21

17. Xiao X, Wang P, Chou KC (2013) Recent progresses in identifying nuclear receptors and their families. Curr Top Med Chem 13: 1192–1200
18. Harbers M, Wahlstrom GM, Vennstrom B (1996) Transactivation by the thyroid hormone receptor is dependent on the spacer sequence in hormone response elements containing directly repeated half-sites. Nucleic Acids Res 24:2252–2259
19. Harbers M, Wahlstrom GM, Vennstrom B (1998) Identification of DNA binding sites for the V-erbA oncoprotein, the viral homolog to thyroid hormone receptor alpha. J Steroid Biochem Mol Biol 67:181–191
20. Demczuk S, Harbers M, Vennstrom B (1993) Identification and analysis of all components of a gel retardation assay by combination with immunoblotting. Proc Natl Acad Sci U S A 90:2574–2578
21. Hirano F, Tanaka H, Hirano Y, Hiramoto M, Handa H, Makino I, Scheidereit C (1998) Functional interference of Sp1 and NF-kappaB through the same DNA binding site. Mol Cell Biol 18:1266–1274
22. Kontgen F, Grumont RJ, Strasser A, Metcalf D, Li R, Tarlinton D, Gerondakis S (1995) Mice lacking the c-rel proto-oncogene exhibit defects in lymphocyte proliferation, humoral immunity, and interleukin-2 expression. Genes Dev 9:1965–1977
23. Vettese-Dadey M, Grant PA, Hebbes TR, Crane-Robinson C, Allis CD, Workman JL (1996) Acetylation of histone H4 plays a primary role in enhancing transcription factor binding to nucleosomal DNA in vitro. Embo J 15:2508–2518
24. Wilhelm M, Schlegl J, Hahne H, Moghaddas GA, Lieberenz M, Savitski MM, Ziegler E, Butzmann L, Gessulat S, Marx H, Mathieson T, Lemeer S, Schnatbaum K, Reimer U, Wenschuh H, Mollenhauer M, Slotta-Huspenina J, Boese JH, Bantscheff M, Gerstmair A, Faerber F, Kuster B (2014) Mass-spectrometry-based draft of the human proteome. Nature 509: 582–587
25. Ding C, Chan DW, Liu W, Liu M, Li D, Song L, Li C, Jin J, Malovannaya A, Jung SY, Zhen B, Wang Y, Qin J (2013) Proteome-wide profiling of activated transcription factors with a concatenated tandem array of transcription factor response elements. Proc Natl Acad Sci U S A 110:6771–6776
26. Zhu G, Cai G, Liu Y, Tan H, Yu C, Huang M, Wei M, Li S, Cui X, Huang D, Tian Y, Zhang X (2014) Quantitative iTRAQ LC-MS/MS proteomics reveals transcription factor crosstalk and regulatory networks in hypopharyngeal squamous cell carcinoma. J Cancer 5:525–536
27. Luque-Garcia JL, Zhou G, Spellman DS, Sun TT, Neubert TA (2008) Analysis of electroblotted proteins by mass spectrometry: protein identification after Western blotting. Mol Cell Proteomics 7:308–314
28. Luque-Garcia JL, Zhou G, Sun TT, Neubert TA (2006) Use of nitrocellulose membranes for protein characterization by matrix-assisted laser desorption/ionization mass spectrometry. Anal Chem 78:5102–5108
29. Dignam JD, Lebovitz RM, Roeder RG (1983) Accurate transcription initiation by RNA polymerase II in a soluble extract from isolated mammalian nuclei. Nucleic Acids Res 11:1475–1489
30. Carey MF, Peterson CL, Smale ST (2009) Dignam and Roeder nuclear extract preparation. Cold Spring Harb Protoc. 2009: pdb prot5330
31. Kigawa T, Yabuki T, Matsuda N, Matsuda T, Nakajima R, Tanaka A, Yokoyama S (2004) Preparation of Escherichia coli cell extract for highly productive cell-free protein expression. J Struct Funct Genomics 5:63–68
32. Harbers M (2014) Wheat germ systems for cell-free protein expression. FEBS Lett 588(17):2762–2773
33. Jackson RJ, Hunt T (1983) Preparation and use of nuclease-treated rabbit reticulocyte lysates for the translation of eukaryotic messenger RNA. Methods Enzymol 96:50–74
34. Ezure T, Suzuki T, Higashide S, Shintani E, Endo K, Kobayashi S, Shikata M, Ito M, Tanimizu K, Nishimura O (2006) Cell-free protein synthesis system prepared from insect cells by freeze-thawing. Biotechnol Prog 22:1570–1577
35. Mikami S, Masutani M, Sonenberg N, Yokoyama S, Imataka H (2006) An efficient mammalian cell-free translation system supplemented with translation factors. Protein Expr Purif 46:348–357
36. Harbers M (2008) The current status of cDNA cloning. Genomics 91:232–242
37. Liu J, Stormo GD (2005) Combining SELEX with quantitative assays to rapidly obtain accurate models of protein-DNA interactions. Nucleic Acids Res 33:e141
38. Taussig MJ, Schmidt R, Cook EA, Stoevesandt O (2013) Development of proteome-wide binding reagents for research and diagnostics. Proteomics Clin Appl 7:756–766
39. Siu FK, Lee LT, Chow BK (2008) Southwestern blotting in investigating transcriptional regulation. Nat Protoc 3:51–58
40. Harper S, Speicher DW (2001) Detection of proteins on blot membranes. Curr Protoc Protein Sci. Chapter 10, Unit 10 18
41. Anderson PJ (1985) The recovery of nitrocellulose-bound protein. Anal Biochem 148:105–110

Chapter 37

Grid-Immunoblotting

Olga Yeliosof and R. Hal Scofield

Abstract

Grid-immunoblotting is a fast, simple, and efficient method for simultaneously testing multiple allergens utilizing small amount of antibody.

Key words Allergens, Antibodies, Nitrocellulose

1 Introduction

Grid-immunoblotting is a technique developed by Reese et al. [1, 2] that could alleviate many problems often encountered in allergy diagnosis and research. This method is particularly valuable for it requires only about 150–200 μL of the serum sample compared to ten times more serum required for enzyme-linked immunosorbant assay (ELISA). Furthermore, this technique permits simultaneous testing of up to 20 different antibodies against 20 different allergens. In clinical or laboratory settings, the amount of sera drawn from allergic subjects is often limited. This limitation really becomes a barrier when blood is drawn from allergic children or patients with complicated preexisting conditions who are unable to give sufficient amounts of blood.

The grid-immunoblotting procedure consists of three basic steps. First, the protein is immobilized on a carrier nitrocellulose membrane. This membrane is placed onto a multichannel manifold (Surf-blot apparatus) and various proteins are applied to the channels. After a 1-h incubation period, the membrane is washed and blocked. Second, the blot is incubated with the primary antibody. Finally, specific binding is detected using a detection system [1–3].

Biji T. Kurien and R. Hal Scofield (eds.), *Western Blotting: Methods and Protocols*, Methods in Molecular Biology, vol. 1312, DOI 10.1007/978-1-4939-2694-7_37, © Springer Science+Business Media New York 2015

2 Materials

2.1 Protein Immobilization and Blocking

1. 12.5 × 12.5 cm reinforced nitrocellulose membrane (pore size 0.45 μm).
2. Cyanogen bromide (CNBr) used to activate the nitrocellulose membrane and improve protein binding.
3. Tris-buffered saline (TBS): 100 mM Tris–HCl, 100 mM NaCl, 2.5 mM $MgCl_2$, pH 7.4.
4. Surf-blot apparatus (Idea Scientific, Minneapolis, MN, USA).
5. ELISA carbonate/bicarbonate coating buffer: 60 mM Na_2CO_3, 140 mM $NaHCO_3$, pH 9.6.
6. Allergen solutions (200 μL/channel).
7. 0.01 % Pyronin Y solution (200 μL).
8. Washing buffer (TBST): TBS supplemented with 0.05 % Tween-20.
9. Blocking solution: 200 mL of TBST containing 1 % (w/v) nonfat dry milk powder.

2.2 Blot Incubation with Primary Antibody

1. Antibodies specific to the allergens.
2. Detection antibody (*see* **Note 1**).

2.3 Detection of Specific Binding with a Detection System (See Note 4)

1. Alkaline phosphatase (AP) buffer: 100 mM Tris–HCl, 100 mM NaCl, 5 mM $MgCl_2$, pH 9.5.
2. Substrate/chromogen mixture for alkaline phosphatase at 37 °C (450 μM 5-bromo-4-chloro-indonyl-phosphate disodium salt (BCIP) and 400 μM nitroblue tetrazolium chloride (NBT) solubilized in AP buffer).

3 Methods

3.1 Protein Immobilization and Blocking

1. Activate reinforced nitrocellulose membranes with CNBr according to the technique used by Demeulemester and colleagues [4] (can store this membrane at 4 °C for extended periods of time).
2. Soak the membranes for 1 min in Tris-buffered saline.
3. Place the soaked membranes onto the Surf-blot apparatus and assemble the apparatus according to the manufacturer's instructions (*see* **Note 2**).
4. Dilute the allergens in standard ELISA carbonate/bicarbonate coating buffer with a concentration ranging from 100 μg to 1,000 μg/mL.
5. Pipette 200 μL of allergen solution into each channel of the assembled Surf-blot apparatus except for the last channel.

Coat the last channel with 200 μL of a 0.01 % Pyronin Y solution instead of the allergen in order to avoid confusion about the orientation of the membrane. This will result in a permanently stained pink lane on the blot.

6. Incubate the Surf-blot apparatus for 1 h at room temperature (RT) with end-over-end rotation (*see* **Note 3**).
7. Perform three washing steps (5 min each) with 2 mL per channel of washing TBST.
8. Disassemble the apparatus using a regular 1 mL pipette and block the unoccupied binding sites on the membrane with 200 mL of TBST containing 1 % (w/v) nonfat dry milk for 1 h at RT.
9. Rinse the membranes twice for 10 min in 200 mL of the washing buffer TBST and air-dry until usage.

3.2 Blot Incubation with Primary Antibody

1. For detection of specific antibody reactivity, soak the membrane in TBST for 1 min.
2. Arrange the membranes in such a way that the antigen-coated lanes on the membrane are perpendicular to the incubation channels of the manifold, thus permitting each antibody to interact with each of the 20 antigens.
3. Remove the excess buffer and pipette 150–200 μL of diluted (1:100) antibody-containing solution into the lanes.
4. Incubate the blot for 1 h with end-over-end rotation.
5. Perform another washing procedure (3×) (*see* Subheading 3.1, **step 2**) with TBST and incubate the membrane in 100 mL of TBST containing 1 % nonfat dry milk powder and detection antibody for 1 h (*see* **Note 3**).

3.3 Detection of Specific Binding with a Detection System

1. Use calorimetric detection method to make the antibody binding visible (*see* **Note 4**).

3.3.1 Calorimetric Detection

1. Wash the membranes with 100 mL of TBST and 50 mL TBS-AP buffer for 5 min each.
2. Incubate the blots in substrate/chromogen mixture for alkaline phosphatase at 37 °C till spots appear.
3. Stop color development by washing with TBST.

4 Notes

1. For the detection of mouse antibodies, use 1:20,000 diluted alkaline phosphatase-conjugated goat anti-mouse IgG + IgM antibody (Jackson ImmunoResearch, West Groove, PA).

For the detection of IgE antibodies, incubate the blot with 1:1,000 diluted alkaline phosphatase-conjugated monoclonal anti-human-IgE.

2. Aspirate any TBS remaining in the channels before the membrane is coated with allergen.
3. Make sure that the antigen evenly coats the membrane and that no bubbles are trapped in the channels.
4. As an alternative to colorimetric detection, one can use a chemiluminescence substrate for alkaline phosphatase. The following materials (*see* (a)–(c)) and methods (*see* (d)–(g)) need to be used for this procedure.
   (a) Washing assay buffer (100 mM diethanolamine/HCl, 1.0 mM $MgCl_2$, pH 10).
   (b) 1:20 diluted nitroblock chemiluminescence enhancer (Tropix, Bedford, MA, USA).
   (c) 250 μM CSPD (disodium 3-(4-methoxy-spiro{dioxetane-3, 2′-(5′ chloro) tricyclo [3.3.1.1.^{3,7}] decan}-4-yl)phenyl phosphate; Tropix).
   (d) Wash the blots with freshly prepared assay buffer (100 mM diethanolamine/HCl, 1 mM $MgCl_2$, pH 10).
   (e) Incubate the blots in the 1:20 diluted Nitroblock chemiluminescense enhancer for 5 min.
   (f) Incubate the blots in 250 μM CSPD (disodium 3-(4-methoxy-spiro{dioxetane-3,2′(5′ chloro) tricyclo [3.3.1.1.^{3,7}] decan}-4-yl)phenyl phosphate; Tropix) for 5 min and drain any excessive liquid.
   (g) Expose the blots to autoradiography film for 15, 30, 60, and 120 s sealing the blots between transparencies.

References

1. Reese G, Schmechel D, Ayuso R, Lehrer SB (2001) Grid-immunoblotting: a fast and simple technique to test multiple allergens with small amounts of antibody for cross-reactivity. J Chromatogr B 756:151–156
2. Yeliosof O, Scofield RH (2009) Grid-immunoblotting. Methods Mol Biol 536:289–292
3. Kurien BT, Scofield RH (2003) Protein blotting: a review. J Immunol Methods 274:1–15
4. Demeulemester C, Peltre G, Laurente M, Panheleux D, David B (1987) Cyanogen bromide-activated nitrocellulose membranes: a new tool for immunoprint techniques. Electrophoresis 8:71

Chapter 38

Detection and Quantification of Protein–Protein Interactions by Far-Western Blotting

Joshua A. Jadwin, Bruce J. Mayer, and Kazuya Machida

Abstract

Far-western blotting is a convenient method to characterize protein–protein interactions, in which protein samples of interest are immobilized on a membrane and then probed with a non-antibody protein. In contrast to western blotting, which uses specific antibodies to detect target proteins, far-western blotting detects proteins on the basis of the presence or absence of binding sites for the protein probe. When specific modular protein binding domains are used as probes, this approach allows characterization of protein–protein interactions involved in biological processes such as signal transduction, including interactions regulated by posttranslational modification. We here describe a rapid and simple protocol for far-western blotting, in which GST-tagged Src homology 2 (SH2) domains are used to probe cellular proteins in a phosphorylation-dependent manner. We also present a batch quantification method that allows for the direct comparison of probe binding patterns.

Key words Protein–protein interaction, Far-western blotting, GST fusion protein, Affinity purification, SH2 domain, Tyrosine phosphorylation, Reverse-phase assay

1 Introduction

Far-western blotting is a method of characterizing protein–protein interactions, in which protein samples of interest are separated by gel electrophoresis, immobilized on a membrane, and then probed with a non-antibody protein [1]. The term "far-western" was derived from western blotting, a similar method in which membranes are probed directly with specific antibodies, and is also referred to as a west-western or blot overlay assay [2–4]. Non-antibody proteins have been also used as a means to screen phage-based expression libraries [2, 5–7].

Far-western blotting is very different from other commonly used methods to detect and characterize protein–protein interactions, and therefore complements these other approaches. Because the probe protein directly binds to denatured/separated proteins immobilized on a membrane, far-western blotting detects only

Biji T. Kurien and R. Hal Scofield (eds.), *Western Blotting: Methods and Protocols*, Methods in Molecular Biology, vol. 1312, DOI 10.1007/978-1-4939-2694-7_38, © Springer Science+Business Media New York 2015

direct interactions; by contrast, most non-far-western protein binding assays, such as immunoprecipitation and pull-down assays, may detect both direct association (two proteins make contact directly) and indirect association (two proteins do not make contact, and another molecule in the ternary complex mediates the association). Thus, the far-western assay is often used to confirm direct interaction following immunoprecipitation or pull-down assays.

The ability of far-western blotting to detect direct interactions is offset by limitations in the types of protein–protein interactions that can be detected. Because target proteins in a cell lysate are usually denatured in the process of gel electrophoresis, it may be difficult or impossible to detect interactions that require the native, folded conformation of the target protein. For this reason, far-western blotting has been particularly useful in characterizing the binding partners of modular protein binding domains that bind to short, linear peptide motifs. It is now apparent that many signaling proteins interact with their partners via such modular binding domain–peptide interactions, and thus, the far-western approach is quite useful for analysis of signaling networks. However, these differences highlight the importance of using multiple approaches to assess specific protein–protein interactions.

In far-western blotting, either a whole protein or fragment of a protein containing a suspected binding interface is used to probe interaction partners immobilized on a membrane. The interaction is visualized by direct labeling of the probe or by its subsequent detection with antibodies (Fig. 1a). There are a number of considerations in selecting the specific probe. First, ease of growth and purification of the probe must be considered. For the sake of cost and convenience, expressing probe proteins in bacteria is advantageous. However, only relatively small proteins (less than ~100 kDa) tend to remain soluble when grown in bacteria, so in general a fragment of a protein containing only the known or suspected binding domain will be easier to work with than the full-length protein. Second, it is useful to fuse the probe protein or domain to a tag sequence for ease of purification and detection. We routinely make probe proteins as glutathione S-transferase (GST) fusions, which has the dual advantage of allowing purification of proteins on glutathione columns, and allowing detection of bound probe with glutathione conjugates or with anti-GST antibodies. A further advantage of GST fusions is that GST exists as a stable dimer in solution. As in the case of antibodies, a dimeric probe binds with much greater avidity compared to a monomer to targets containing multiple binding sites, such as a membrane surface bearing many molecules of a target protein.

In contrast to western blotting where a target protein is usually known in advance, far-western blotting can detect proteins on the basis of presence or absence of binding sites without any previous knowledge about their identities (Fig. 1a, b). From the intensity of

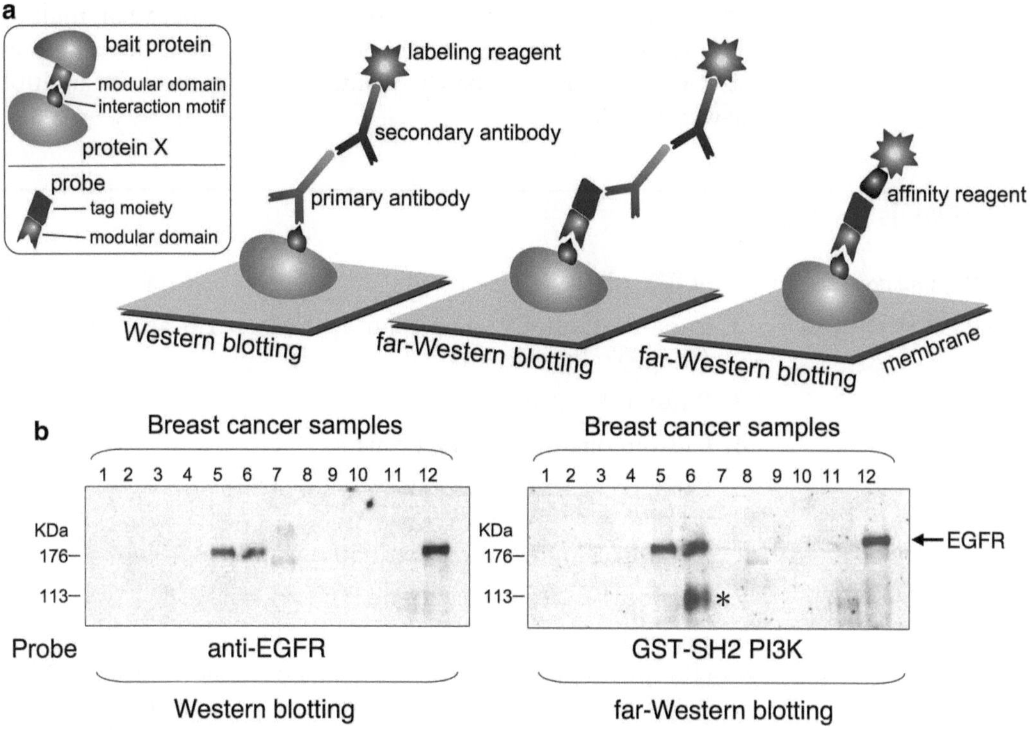

Fig. 1 Comparison of western and far-western blotting. (**a**) Schematic representation of both methods. Box, bait protein and protein X interact via modular interaction domain and interaction motif. *Left*, western blotting uses antibodies raised against a target protein X to detect its presence. *Middle* and *right*, far-western blotting uses a protein probe containing a modular interaction domain or a short interaction motif, fused to a tag moiety (e.g., GST), to detect protein X based on presence of its binding sites. Specific interaction of far-western blotting is visualized by antibody-based detection (*middle*) or by direct labeling using isotope or an enzyme-conjugated affinity reagent, e.g., glutathione-HRP (*right*). (**b**) Application of both methods. Breast cancer patient-derived samples (tumor 1–12) are immobilized on a membrane, and then identical membranes were probed with anti-EGFR antibody (*left*, western blotting) or GST-PI3K SH2 domain fusion (*right*, far-western blotting). Whereas both results look similar, interpretation of the results is somewhat different; the western result indicates a presence of EGFR in sample *lane 5*, *6*, and *12*, while the far-western result indicates a presence of PI3K SH2 domain recognition sites on a protein which is likely to be EGFR (*arrow*). In addition, the far-western probe also detected an uncharacterized PI3K SH2-binding protein (*asterisk*) that can potentially distinguish tumor 6 from 5 and 12, not otherwise detected by western blotting

bands observed on a far-western blot of a complex mixture of proteins, one can gain insight into both the number and relative affinity of binding partners for the probe in that sample. Furthermore, since some protein binding domains recognize their targets only after specific post-transcriptional modifications, far-western blotting can be used to assess the modification status of multiple proteins in a sample [8–10]. In this chapter, we will present a specific example of the utility of far-western blotting methods, in which GST-tagged Src homology 2 (SH2) domains, which bind specifically to tyrosine-phosphorylated target proteins, are used to probe the state of tyrosine phosphorylation of cellular proteins.

Since quantitative comparison of interactions between multiple SH2s or other signaling proteins and their ligands is often of interest, we also provide a batch quantification method for multiple far-western blots.

2 Materials

2.1 Subcloning of GST-SH2 Construct

1. pGEX-6P1.
2. Luria-Bertani (LB)-ampicillin agar plate: LB agar plate with 100 μg/ml ampicillin.
3. Phusion DNA polymerase.
4. Custom oligonucleotide primers.
5. Competent bacteria (strain NB42 or DH5α).

2.2 Evaluation of GST-SH2 Clones

1. LB-ampicillin: LB broth with 50 μg/mL ampicillin.
2. Isopropyl-β-D-thiogalactoside (IPTG).
3. Bacteria Triton X-lysis buffer (BXB): Phosphate buffered saline (PBS) with 100 mM ethylene diamine tetraacetic acid (EDTA), 1 % Triton X-100; add phenylmethyl sulphonyl fluoride (PMSF) to 1 mM, aprotinin to 1 % v/v (3 trypsin international units (TIU)/mL), dithiothreitol (DTT) to 1 mM just before use.
4. Sonicator with microtip probe (e.g., Branson Sonifier 450 or equivalent).
5. 5× sample buffer: 0.3 M Tris–HCl pH 6.8, 10 % sodium dodecyl sulfate (SDS), 25 % β-mercaptoethanol, 0.1 mg/mL bromophenol blue, 45 % glycerol.
6. Glutathione Sepharose 4B.
7. 12 % SDS-polyacrylamide gel electrophoresis (PAGE) mini gel (*see* Subheading 2, **item 4**).
8. Control lysates (*see* Subheading 2, **item 4**).
9. Anti-phosphotyrosine antibody.
10. Coomassie blue solution: 40 % methanol, 10 % acetic acid, 0.25 % Coomassie blue R-250.
11. Fixing solution: 20 % methanol, 10 % acetic acid.
12. Bacteria stock solution: 50 % glycerol (autoclaved).
13. Cryogenic tubes.

2.3 Large-scale Preparation of GST-SH2 Probe

1. Tris-NaCl-EDTA (TNE) buffer: 50 mM Tris–HCl 7.4, 150 mM NaCl, 10 mM EDTA; add aprotinin to 1 % (3 TIU/mL), PMSF to 1 mM just before use.
2. Chromatography column (poly-prep, 0.8×4 cm).
3. Elution buffer: 20 mM reduced glutathione, 100 mM Tris–HCl, pH 8.0; add aprotinin to 1 % (3 TIU/mL), PMSF to 1 mM just before use.

4. Sephadex G-25 PD-10 column.
5. Dialysis membrane tubing (molecular weight cut-off 3,500).
6. PBS with 10 % glycerol.
7. Bio-Rad Bradford dye reagent.
8. Ultrafiltration membrane.

2.4 Far-Western Blotting

1. Kinase lysis buffer (KLB) : 150 mM NaCl, 25 mM Tris–HCl pH 7.4, 5 mM EDTA, 1 % Triton X-100, 10 % glycerol, 0.1 % sodium pyrophosphate, 10 mM β-glycerophosphate, 10 mM sodium fluoride (NaF); add aprotinin to 1 % (3 TIU/mL), PMSF to 1 mM, pervanadate (50 mM orthovanadate, 4 % hydrogen peroxide) to 50 μM just before use.
2. Sodium orthovanadate: Dissolve powdered sodium orthovanadate (final concentration will be 50 mM, but leave some extra volume for multiple rounds of pH adjustment); adjust with NaOH to pH of 10 (solution will turn bright yellow); boil in microwave until colorless, then stir until cooled to room temperature; adjust pH once again to 10.0, and repeat boiling; continue boiling and adjusting pH as above until pH stays at 10.0 after boiling (usually three rounds total); adjust volume for 50 mM, filter, and store at room temperature (RT).
3. Pervanadate solution: Mix 16 μL 30 % (w/w) hydrogen peroxide and 100 μL 50 mM sodium orthovanadate; incubate at room temperature for 30 min (not stable, needs to be freshly prepared before use).
4. Positive control lysates: Combine equal amounts of KLB lysate from pervanadate-treated NIH 3T3, HepG2, A431, and MR20 cells [9].
5. Negative control lysate: Prepare lysates of each cell line in the absence of vanadate, combined, and treated with tyrosine phosphatase PTP-1B for 1 h at RT.
6. 12 % SDS-PAGE gel (1 mm thickness, for three mini gels).

Stock solutions	Resolving gel	Stacking gel
acrylamide/bis-acrylamide	20 mL	2.2 mL
(30 %/0.8 %, w/v)		
1 M Tris–HCl (pH 8.8)	18.6 mL	—
1 M Tris–HCl (pH 6.8)	—	2.1 mL
Distilled water	10.5 mL	12.2 mL
10 % SDS	500 μL	167 μL
10 % ammonium persulfate	500 μL	125 μL
TEMED	16.7 μL	16.7 μL

7. Electrophoresis buffer: 25 mM Tris, 192 mM glycine, 0.1 % SDS.
8. Nitrocellulose membrane (0.2 μm pore size).
9. Transfer buffer: 10 mM 3-(cyclohexylamino)-1-propanesulfonic acid (CAPS), pH 11.0, 20 % methanol, kept at 4 °C.
10. Tris buffered saline-Tween 20 (TBST): 25 mM Tris–HCl, pH 8.0, 150 mM NaCl, and 0.05 % (v/v) Tween-20.
11. Blocking solution: 10 % fat-free milk in TBST, 1 mM EDTA, 1 mM sodium orthovanadate.
12. Labeling reagent: Glutathione-HRP conjugate or anti-GST-HRP conjugate.
13. Chemiluminescence kit:

Product	Maker	Sensitivity	Background
RPN2106	GE Healthcare	Medium	Low
NEL103	PerkinElmer	High	Medium
#34079	Pierce	Very high	High
RPN2132	GE Healthcare	Very high	High
We routinely use NEL103			

14. Imaging: X-ray film; CCD detection system (Kodak Image Station 4000 MM PRO).

2.5 Stripping and Reprobing

1. Acidic stripping buffer: 100 mM Glycine-HCl pH 2.0.
2. SDS-ME stripping buffer: 2 % SDS, 100 mM β-Mercaptoethanol, 62.5 mM Tris–HCl pH 6.8.

2.6 Probing of Replica Membranes

Same as Subheadings 2.4 and 2.5.

2.7 Image Adjustment

1. Photoshop CS.

2.8 Batch Quantification

1. ImageJ (National Institutes of Health, Bethesda, MD).
2. Probe Reader plug-in (http://sites.imagej.net/Kazy/plugins/).

3 Methods

As for western blotting, protein samples are separated by SDS-PAGE and transferred to a nitrocellulose or polyvinylidene fluoride (PVDF) membrane. The blocked membrane is then incubated with a probe followed by appropriate wash, and bound probes are visualized.

Generally any protein samples compatible with western blotting can be used including whole cell lysates, purified proteins, and native or denatured samples. The far-western method described here is a rapid and simple protocol in which the membrane is prepared without denature–renature procedures, the probe is labeled directly, and probing is performed in one step [8]. This protocol has been optimized to detect in vitro interaction between modular binding domain probes and immobilized proteins containing short peptide motifs (*see* **Note 1**). Of note, alternative protocols are available including another example in which proteins containing modular domains on a membrane are probed with labeled binding motifs [4, 11–20]. Below, we present a specific protocol for generating GST-SH2 domain probes and using them to probe tyrosine-phosphorylated whole cell lysates. We have also included a method for aligning and quantifying multiple far-western blots which allows for comparative assessment of SH2 binding. Of course these procedures can be adapted for any modular protein binding domain and its binding partners with minor modifications.

For all far-western blotting methods, detection of specific signal is strongly dependent on the quality of the probe protein. Insolubility, aggregation, or denaturation of the protein tends to cause nonspecific background, and even the native probe may bind nonspecifically to abundant proteins in the sample. Thus, it is important to: (1) confirm purified probe is soluble, folded, and not significantly degraded; (2) evaluate activity of the probe and optimize binding conditions if needed; and (3) always include appropriate positive and negative controls for each experiment to ensure any positive signal is indeed specific. To address these considerations, in the following section we will describe a detailed protocol for generation and evaluation of GST-SH2 fusion probes. Appropriate controls should be prepared considering the intended physiological activity of the probe; as an example, several control blots for specific SH2-phosphotyrosine interaction are presented in Fig. 2b. At minimum, GST alone, or more ideally GST fused to the domain of interest bearing a mutation known or suspected to abolish specific binding activity, should be used as a negative control probe.

3.1 Subcloning of GST-SH2 Construct

1. Retrieve cDNA and protein sequences of a SH2 domain-containing protein of interest, e.g., at NCBI Entrez Gene (*see* **Note 2**).
2. Find location of the SH2 domain using the protein sequence at ScanProsite.
3. Find the nucleotide sequence corresponding to the SH2 domain using the sequence editor program (*see* **Note 3**).
4. Find academic or industry source for corresponding cDNA (*see* **Note 4**), otherwise clone the cDNA by RT-PCR method.

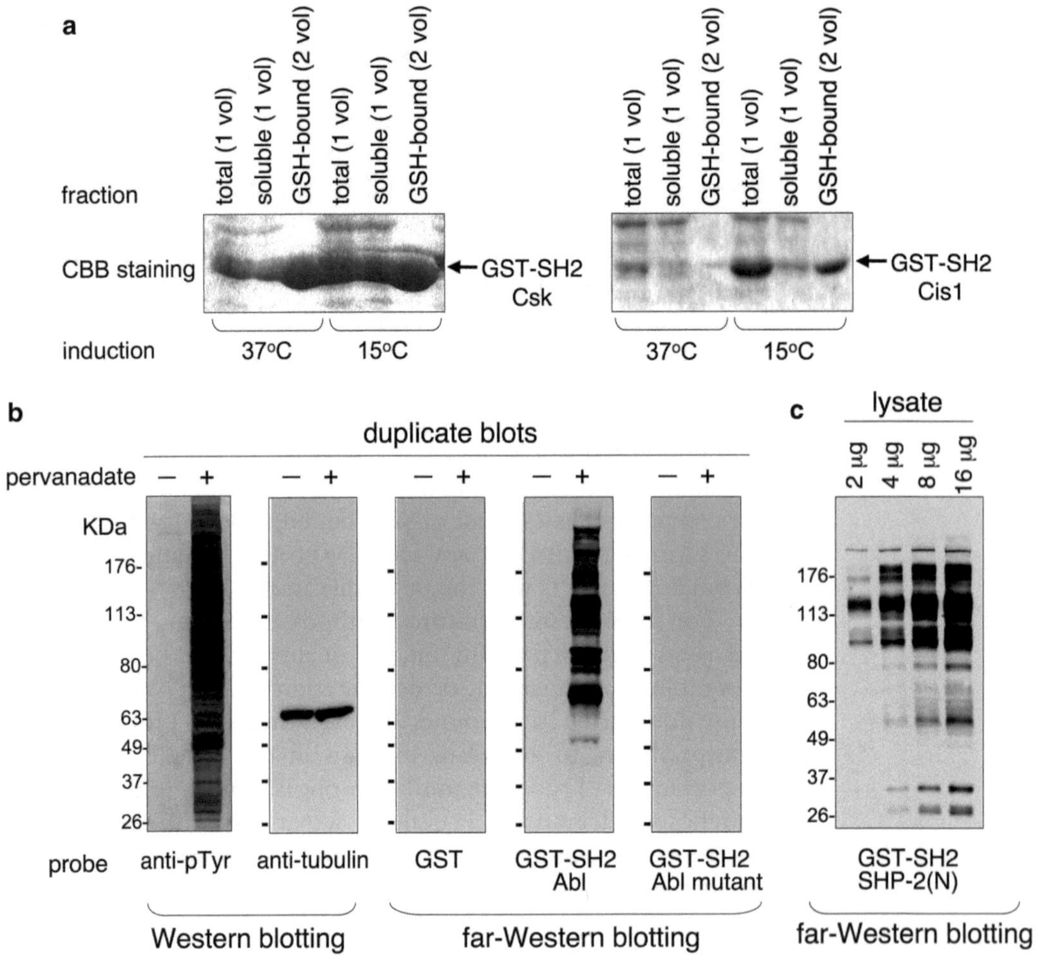

Fig. 2 Generation and evaluation of GST-tagged probes. (**a**) Testing solubility of GST fusion proteins. GST fusions of Csk and Cis1 SH2 domains were expressed in *E. coli* at regular (37 °C, 1 h) and low (15 °C, overnight) culture temperatures. Protein fractions from affinity purification are visualized by Coomassie brilliant blue (CBB) staining: whole cell fraction (total); Triton-soluble fraction (soluble); and fraction bound to glutathione-agarose beads (GSH-bound). The beads fraction represents twice the relative amount of original culture as total and soluble fractions. Cis1-SH2 was much less soluble than Csk-SH2, but its solubility was improved by expression at low temperature. (**b**) Evaluation of GST-SH2 probes. Identical blots of pervanadate-treated lysate (+) and POV-untreated, phosphatase-treated lysate (–), were prepared. To validate specific interaction of a probe with target proteins, different probes were used for following purposes: anti-phosphotyrosine antibody (anti-pTyr), indicator of tyrosine phosphorylation; anti-tubulin antibody, loading control; GST, negative control probe; GST-SH2 domain of Abl, experimental probe; and GST-SH2 Abl mutant, loss-of-function mutant probe. (**c**) Optimization of protein loading. A membrane with various amounts of POV-treated lysate was probed with a GST fusion with N-terminal SH2 domain of SHP-2. High affinity ligand proteins for the SH2 domain are selectively detected with a 2 μg protein per lane, while more proteins are detected with increased protein loading

5. Design primers for PCR (*see* **Note 5**).
6. Amplify the SH2 fragment by PCR using the oligonucleotide primers and the cDNA template.

7. Digest the PCR product, and then purify the fragment from an agarose gel.
8. Insert the purified SH2 fragment into a pGEX vector digested with appropriate restriction enzymes.
9. Transform competent bacteria and grow overnight on LB-ampicillin agar plate (*see* **Note 6**).

3.2 Evaluation of GST-SH2 Clones

Protein expression and solubility of GST-SH2 clones can be tested quickly in small-scale bacterial cultures (*see* **Note 7**), and the activity of the probe can be tested at the same time by a control pull-down assay.

1. Inoculate 0.4 mL dense liquid culture of bacteria into 1.6 mL fresh pre-warmed LB-ampicillin.
2. Shake at 37 °C for 1 h.
3. Add IPTG to 0.1 mM and shake at 37 °C for 1 h to induce protein expression.
4. Transfer 1.5 mL bacteria to microcentrifuge tubes and spin at 10,000 × *g* at 4 °C for 2 min in microfuge.
5. Remove supernatant and resuspend bacterial pellet in 0.4 mL BXB.
6. Vortex to resuspend, then sonicate briefly (e.g., 2–5 s at relatively low power) on ice to break cells, let sit on ice, then repeat. Try to avoid foaming; if foaming occurs, let rest on ice for a few minutes to allow foam to dissipate.
7. Remove 10 μL of total lysate, add 2.5 μL 5× sample buffer for gel (total cell fraction, Fig. 2a).
8. Spin rest of lysate for 5 min in microfuge at 10,000 × *g* at 4 °C, transfer supernatant to a new tube.
9. Take 10 μL of the cleared lysate for gel as above (soluble fraction, Fig. 2a).
10. Take 100 μL of cleared lysate and add to 10 μL glutathione-agarose bead slurry (cut end off pipet tip with razor blade to more accurately pipet beads).
11. Rotate at 4 °C for 30 min.
12. Spin out briefly, and wash beads 3× with 1 mL cold BXB.
13. Resuspend the bead pellet with BXB, take 10 μL, and add 2.5 μL 5× sample buffer for gel (GSH-bound fraction, Fig. 2a) (*see* **Note 8**).
14. Boil all samples and run on 12 % SDS gel.
15. When gel is done, stain for 15 min with Coomassie blue solution, de-stain with fix solution, and then dry (Fig. 2a) (*see* **Note 9**).

Steps 16–19: Evaluate pTyr binding activity by GST pull-down assay (optional).

16. Incubate remaining beads (GSH-bound fractions for pGEX-SH2 clones and a control pGEX clone, if any) with 10 μg cell lysates (*see* **Note 10**).
17. Rotate for 1 h at 4 °C, and wash three times with BXB.
18. Boil all samples and run on 12 % SDS gel (*see* **Note 11**).
19. Perform western blotting with anti-phosphotyrosine antibody (*see* **Note 12**).
20. Subject positive clones to DNA sequencing and store bacteria in 25 % (v/v) sterile glycerol at −70 °C.

3.3 Large-Scale Preparation of GST-SH2 Probe

GST-SH2 probe is purified following the standard protocol for preparation of GST fusion proteins using pGEX series bacterial expression vectors [13] (*see* **Note 13**).

1. Inoculate frozen stock culture of a verified GST-SH2 clone in 50 mL LB-ampicillin overnight.
2. Inoculate 50 mL dense overnight culture to 1 L LB-ampicillin. Shake at 37 °C for 2 h (*see* **Note 14**).
3. Add IPTG to 0.1 mM. Shake at 37 °C for 3 h (*see* **Note 15**).
4. Centrifuge at 5,000 × *g* at 4 °C for 10 min.
5. Resuspend pellet in 5–20 mL ice-cold BXB, transfer to 50 mL tube, and sonicate on ice until cells are broken (*see* **Note 16**).
6. Centrifuge at 5,000 × *g* at 4 °C for 10 min to remove debris (*see* **Note 17**).
7. Add glutathione-agarose to supernatant: 3 mL 50 % (v/v) bead slurry/L original culture.
8. Rotate at 4 °C about 1 h (up to 2 h).
9. Wash beads with TNE buffer: spin out beads at low speed, remove supernatant, and resuspend beads in fresh buffer. Repeat 3–5×.
10. To elute GST-SH2, pour beads into small disposable column, elute with approximately 3 bead volumes of elution buffer (*see* **Note 18**).
11. Change buffer by gel filtration on a Sephadex G-25 PD-10 column according to the supplier's instructions. Briefly, equilibrate column with approximately 25 mL PBS-10 % glycerol. Discard the flow-through. Add sample followed by buffer up to a total volume of 2.5 mL. Discard the flow-through. Elute with 3.5 mL buffer (collect seven 0.5 mL fractions of the eluate in separate tubes (*see* **Note 19**).

12. Estimate relative protein concentration with Bio-Rad Bradford dye reagent, combine top three fractions into one tube (*see* **Note 20**).
13. Determine protein concentration by Bradford assay, take 500 μg of protein and dilute to 0.1 μg/μL with PBS-10 % glycerol, and aliquot 50 μL diluted probe into chilled microcentrifuge tubes. Store the aliquots and undiluted probes at −70 °C (*see* **Note 21**).
14. Evaluate the purification fractions by 12 % SDS gel as in previous Subheading 3.2 (*see* **Note 22**).

3.4 Far-Western Blotting

1. Separate proteins on SDS-polyacrylamide gels (*see* Chapters 11 and 34) and transfer to nitrocellulose or PVDF membranes (*see* Chapters 22 and 34) following general western blotting protocol (*see* **Note 23**).
2. Block membranes in blocking solution for about 1 h at room temperature or at 4 °C overnight (*see* **Notes 24** and **25**).
3. To label probe, thaw the stored probe on ice and add 5 μL GSH-HRP conjugate (0.1 μg/μL) to 50 μL of diluted probe (0.1 μg/μL) (*see* **Note 26**).
4. Incubate on ice for about 1 h. Labeled probes can be stored at 4 °C (*see* **Note 27**).
5. Dilute labeled probe to optimal concentration with blocking buffer and apply to the blocked membrane (*see* **Note 28**).
6. Let probe bind for 1–2 h at room temperature, then wash with multiple changes TBST for 20 min.
7. Visualize signal by enhanced chemiluminescence according to manufacturer's instruction (*see* **Note 29**).
8. Take appropriate exposure using x-ray films or an image analyzer e.g., Kodak Image Station system (*see* **Note 30**).

3.5 Stripping and Reprobing

Generally, fresh membranes are best for far-western blotting; stripping and reprobing of the membrane may result in significant signal loss and increased non-specific background. Nevertheless recycling membranes might be beneficial if sample is limited, or if precise comparison of specific bands is needed within the same membrane.

1. After initial probing, keep membrane wet; it can be stored wet, wrapped in plastic wrap, at 4 °C (up to a week) or −20 °C (for longer period).
2. Rinse the membrane twice with TBST.
3. Immerse membrane in stripping buffer at room temperature for 20 min under gentle rocking agitation (*see* **Note 31**).

4. Wash membrane in large volume of TBST at room temperature for 45 min with frequent buffer changes.
5. Proceed to blocking and reprobing.

3.6 Probing of Replica Membranes

Although a single far-western blot is useful for identifying differences between samples, comparison of probe specificities, such as the phosphosite preference of different SH2 domains, requires parallel probing and quantification of multiple replica membranes. Here we describe a method for the preparation, probing and detection (*see* Subheading 3.6), image adjustment (*see* Subheading 3.7), and batch quantification (*see* Subheading 3.8) of multiple membranes.

1. Prepare multiple nitrocellulose or PVDF membranes of equal size.
2. Draw a frame position mark along the outside edge of each membrane using a permanent marking pen (Fig. 3) (*see* **Note 32**).
3. Run aliquots of lysate samples on multiple protein gels and transfer using identical experimental conditions.
4. For each membrane, follow **steps 2–8** of Subheading 3.4. Keep probing and image acquisition conditions such as blocking, probing, washing stringency, and imager machine settings as constant as possible for all membranes.
5. Following far-western image acquisition, acquire a reference image of the blot under room light before removing it from the imager. Be sure that the frame position mark is clearly visible (*see* **Note 33**).
6. If necessary, strip and reprobe membranes according to **steps 1–5** of Subheading 3.5.
7. After all far-western blotting has been completed, perform western blotting with an anti-phosphotyrosine antibody for all membranes.
8. Visualize signal by enhanced chemiluminescence with appropriate exposures using the same imaging method used for far-western probing (*see* **Note 34**).
9. Acquire a white light reference shot of the blot before removing it from the imager.

3.7 Image Adjustment

For accurate batch quantification, all images must be well aligned. Images from the same physical membrane, e.g., reprobing, can be easily aligned using the frame position marks, while those from different membranes must be carefully aligned using the anti-phosphotyrosine or other reference blot images (Fig. 3).

1. Export imager files in TIFF or high resolution JPEG format, and open as a multi-layer image with image editor software such as Adobe Photoshop.

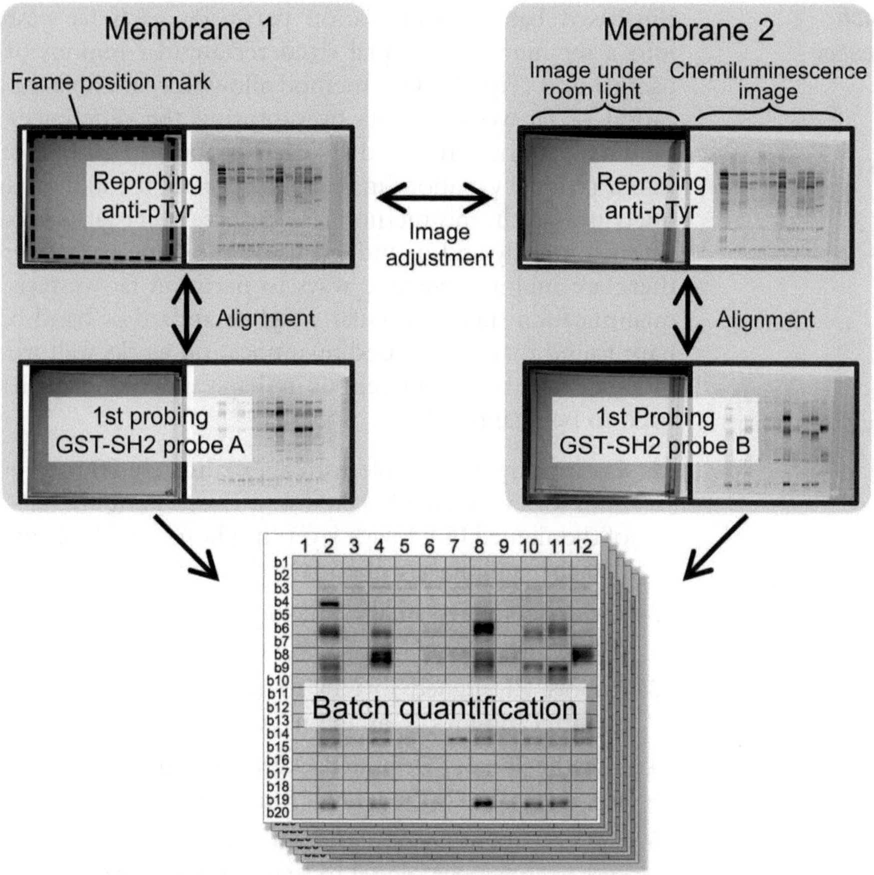

Fig. 3 Image adjustment and batch quantification of replica membranes. To compare multiple far-western results, we use a batch quantification method that requires alignment of images. First, far-western assays are performed under equivalent conditions using membranes marked with a frame position for reference. Following completion of far-western blotting, all membranes are probed with anti-phosphotyrosine. The data images are imported into an image editor and adjusted. Blots derived from the same membrane are typically adjusted with the frame position marks, while those from different membranes have greater variation, and need to be adjusted based on phosphotyrosine band positions. Once aligned, the entire image set is binned and batch quantified using the software tool

2. Align blots derived from the same membrane using the room light images acquired in **step 5** of Subheading 3.6. Utilize the frame position marks to align and overlay membranes using the rotation and scaling functions.
3. For images from different membranes, align the anti-phosphotyrosine bands by utilizing editing functions such as rotation, scaling, and distortion to correct minor inconsistencies in lane and band shape (*see* **Note 35**).

3.8 Batch Quantification

Bin-based batch quantification partitions each far-western lane into a set number of equal sized rectangular regions of interest using a grid (Fig. 3). This method allows for an unbiased quantification of far-western blots by capturing the differences in SH2 binding patterns produced by each probe and compensating for minor residual variations in band position. Since all blots are quantified as a batch, this method also ensures every far-western is partitioned equally and quantified using the same parameters. While there are multiple possible ways to partition far-western data for quantification (i.e., molecular weight standard or band-based) we have found that a bin-based quantification works well when gels/blots are relatively homogenous in shape and have many bands that need to be quantified.

1. Create a mock-up table with approximately 20 rows and a column for every sample lane using Excel. Paste the table on top of the aligned blot image layers in Photoshop. Scale the table so that the columns are aligned with sample lanes (*see* **Note 36**).
2. Export compiled images, including the grid, as ImageJ compatible files (i.e, JPEG, TIFF, etc.).
3. Import all aligned images to ImageJ, or comparable image quantification tools, as an image sequence (*see* **Note 37**).
4. Use the ImageJ default background subtraction tool, invert, and select the Probe Reader plug-in [21].
5. Create a quantification grid with dimensions (size, bin number and arrangement) equal to that of your mock-up grid.
6. Press the green square in the upper left corner of the window to perform batch quantification. The data will be saved to the ImageJ application directory as a single tab-delimited file containing a matrix of bin quantifications.
7. Open the data file with Excel or another spreadsheet software package. Data in the matrix are the background subtracted raw values (*see* **Note 38**).
8. Scale or normalize the data to the image specific maximum bin value or other values as appropriate for comparable band analysis.

4 Notes

1. This approach, termed a reverse-phase assay, maximizes signal to noise when the concentration of analyte is low (e.g., tyrosine-phosphorylated proteins in whole-cell lysates), because efficient binding to the immobilized analyte can be driven by high concentrations of the SH2 domain probe in solution [9].

2. Useful information about modular interaction domains is available at web-based databases:

   Addgene (http://www.addgene.org/browse/).

   NCBI Entrez Gene (http://www.ncbi.nlm.nih.gov/gene).

   ScanProsite (http://prosite.expasy.org/scanprosite/).

   The Human Protein Reference Database (http://www.hprd.org/query).

   UniGene (http://www.ncbi.nlm.nih.gov/unigene/).

   SH2 domain information site (https://sites.google.com/site/sh2domain/).

3. Several sequence editing programs are available, e.g., CLC SEQUENCE VIEWER (http://www.clcbio.com/products/clc-sequence-viewer/).

4. Large collections of full-length or partial cDNAs are now commercially available at reasonable prices, e.g., IMAGE Consortium (http://image.llnl.gov/). To find appropriate cDNA clone, go back and forth between NCBI Entrez Gene site and Unigene sites; referring to the SH2 nucleotides, find IMAGE clones that contain intact SH2 domain. Mayer/Machida Lab GST-SH2 plasmids (complete set of human SH2 domains) can be obtained from Addegene (http://www.addgene.org/Bruce_Mayer/).

5. To ensure maximal binding activity and solubility, we usually include 5–10 amino acids N- and C-terminal to the domain boundary for an SH2 domain probe. In some cases, yield and solubility of the fusion protein are greatly affected by the precise borders of the construct. Structural studies, if any, are useful when planning the borders for any modular domain. PCR primers should have approximately 24 nucleotides of homology to the template; the 5′ end should have five C's (this allows efficient digestion with restriction nucleases) followed by a restriction site to be used for cloning in-frame into the pGEX expression vector, followed by the region with homology to template.

6. We routinely use *E. coli* NB42, which lacks the two major proteases; this may help increase yields by limiting degradation [14]. Similar results can be obtained with typical laboratory strains such as DH5α, though yields may be somewhat lower and bacterial growth is slower. We do not routinely perform restriction endonuclease mapping or DNA sequencing of plasmids at this step, instead we immediately test protein expression of clones as below.

7. Solubility is one of the major determining factors of probe activity in far-western blotting. In our hands about half of the GST-SH2 domains are relatively insoluble, and highly

insoluble domains give poor yields and tend to lack detectable binding activity [9]. Multiple strategies to improve solubility have been reported [15–18]. We have observed that, for about two-thirds of insoluble GST-SH2 proteins, solubility could be significantly improved when protein is expressed in bacteria at lower temperature (e.g., 15 vs. 37 °C, Fig. 2a).

8. If solubility of the GST fusion is unknown, volume used to resuspend beads should be adjusted to load two- to tenfold more of the bead-bound fraction relative to total cell fractions on gel to visualize the band (e.g., twofold, Fig. 2a).
9. A GST-SH2 band of about 40 kDa in size should be easily visible as in Fig. 2a, although the degree of protein expression and solubility may vary depending on the constructs (see Csk SH2 vs. Cis1 SH2 in Fig. 2a). If protein degradation is observed on gel, care must be taken in the large-scale purification.
10. Use positive and negative control lysates for pull-down; as an active SH2 domain should bind to tyrosine-phosphorylated proteins, we use pooled lysates of pervanadate-treated cell lines as a positive control (pervanadate inhibits endogenous protein tyrosine phosphatases, thus strongly enhancing tyrosine phosphorylation in vivo). Corresponding pooled lysates lacking tyrosine phosphorylation (prepared in the absence of phosphatase inhibitors and then treated with phosphatase in vitro) serve as a negative control.
11. The gel lanes should contain: positive control lysate 5 μg; negative control lysate 5 μg; pull-down positive control with GST-bound beads; pull-down negative control with GST-bound beads; pull-down positive control with GST-SH2-bound beads; pull-down negative control with GST-SH2-bound beads.
12. In the pull-down result, a functional GST-SH2 protein should have increased anti-pTyr signal relative to GST control. If GST-SH2 is correct size (and sequence) but the pull-down result is not clearly positive, activity of the probe could be reevaluated by far-western assay (*see* Subheading 3.4 and Fig. 2b).
13. We routinely obtain about 5–10 mg fusion protein per liter of bacteria using the protocol described here (less if the protein is less soluble).
14. Usually an OD of 0.4–0.6 is optimal at this step.
15. If the protein is highly insoluble, consider protein expression at lower temperature: cool down culture with ice before adding IPTG, then add IPTG to 0.1 mM and shake at 30 °C for 4 h or 15 °C about 16 h.
16. Sonication time 2–3 min total with microtip at power setting 3–4; solution should become slightly darker and less turbid-looking. Avoid foaming and overheating the lysate (let solution rest on ice if it warms up detectably).

17. Supernatant does not have to be absolutely clear at this step.
18. If the elution yield is found to be suboptimal, the following conditions may help: shorter incubation time of lysate with glutathione beads (~30 min); higher pH (up to 9.6) and glutathione concentration (up to 50 mM) in elution buffer; iterative batch elution (2–3 × 30 min).
19. Alternatively, the eluate can be dialyzed overnight against several large volumes of PBS-10 % glycerol.
20. To quickly check relative protein concentration, add 2 μL of each PD-10 elution fraction to 50 μL diluted Bradford dye and vortex. Pool the top three fractions with brightest blue color. If color change is not obvious (this occurs when protein concentration is less than 0.2 μg/μL), take third to fifth fractions and proceed to ultrafiltration for concentration (below).
21. If protein concentration is low (<0.5 μg/μL or so), sample can be concentrated by ultrafiltration (YM-10, Amicon) according to manufacturer's instructions.
22. It is important to take aliquots at every step of purification. Take the same fraction of the total at each step (e.g., 1/5,000 of total prep) to monitor percent recoveries at each step. Retain the pellet after sonication and the supernatant after bead binding at −70 °C in case the majority of the fusion protein is there. If final product is highly degraded, we recommended reevaluating parameters of purification such as prep scale, temperature of induction, protease inhibitors, EDTA concentration, and sonication strength.
23. Amount of protein loading on gel is important for an optimal result, and thus should be carefully considered (Fig. 2c). We usually load 10–50 μg of whole-cell protein lysate per lane for far-western blotting with GST-SH2 domains. We routinely use a nitrocellulose membrane with 0.2 μm pore size. We find the advantages of nitrocellulose compared to PVDF are ease of use (no pretreatment is needed) and better signal-to-noise ratio (our observation).
24. Composition of blocking solution is important to maximize signal to noise. Strong blocking buffer lowers background but may decrease or eliminate specific signal. We obtained best results with 10 % nonfat milk-based buffer for GST-SH2 domain probes. For other probes, optimal conditions should be determined empirically; blocking buffer containing 5–10 % non-fat milk (strongest), 1–5 % bovine serum albumin (BSA) (moderate), or 1–5 % ovalbumin (moderate) can be tested.
25. Transferred membrane can be stored at −20 °C for later use: rinse with TBST once and wrap in plastic wrap. Care should be taken to avoid damaging the membrane, which causes undesirable background signal.

26. GSH-HRP (1 μg/μL in PBS-10 % glycerol) can be stored at –70 °C. Recently Sigma has discontinued production of this conjugate. If GSH conjugate is unavailable, we have found that anti-GST-HRP conjugate can be used in the same way, with some decrease in signal strength. To label GST fusion probe, add 0.2 μL (~3 μg) antibody to 50 μL (5 μg) of diluted probe (0.1 μg/μL). This direct labeling method presumably promotes oligomerization of probe and HRP, increasing the avidity of binding to the target and enhancing signal. In addition, this one-step probing and washing procedure is time-saving and may avoid dissociation of probe during washes if probe binding is weak.
27. Purification of the labeled probe is not needed. Labeled probe can be stored at 4 °C with protection from light up to a month.
28. Optimal concentration (maximum signal and low nonspecific background) should be determined empirically in pilot experiments. For GST-SH2 domains, optimal concentration ranges from 0.01 to 0.15 μM (most typically 0.05 μM) [9]. Probing can be done either in a sealed nylon bag or in an open tray with a sufficient volume of solution (1–10 mL/10 cm^2 membrane) with occasional agitation. Do not allow the membrane to dry at any point of the probing procedure through detection.
29. Select chemiluminescence kit with appropriate sensitivity (*see* Subheading 2.4). It is important to cover whole blot area evenly during incubation, and to drain off excess chemiluminescence reagent well (retained solution causes high background).
30. Like western blotting, multiple parameters of the assay influence the level of signal and non-specific background, such as blocking buffer, probe concentration, incubation time and temperature, washing stringency, and detection system. Thus, optimal conditions for a given GST-SH2 or other GST fusion probe should be customized empirically. In general, weak signal can be enhanced by decreased concentration of non-fat dry milk in blocking buffer, increased probe concentration, milder washing, more sensitive chemiluminescence kit, and longer exposure times. Alternatively, domain concatenation may help.
31. If stripping is not sufficient with this method, try a more stringent condition using SDS-ME stripping buffer: briefly, immerse membrane at 50 °C for 30 min under vigorous agitation, and then wash with TBST for 60 min with frequent shaking.
32. For quantitative far-western blotting, it is best to obtain blot images using a digital imager. Having image files of your blots will significantly improve the ease of data analysis.

33. The reference markings on the room light image will be used as a reference for aligning multiple images obtained using a single membrane.
34. Anti-phosphotyrosine blots will be used as a reference to align and adjust far-westerns images. Depending on the bait protein, different types of reference blots including general protein staining should be considered.
35. For accurate comparison of SH2 domain binding patterns in different blots, gel-to-gel variations due to bended lanes and bands, smiles, bubbles, nonuniform background, etc, should be diminished experimentally or computationally. In terms of experimental reproducibility, we routinely use NuPAGE Bis-Tris Precast Gels for their higher protein band resolution and relatively minor gel distortion.
36. Although the number of rows and columns is user customizable, empirically a 20 row grid is sufficient to distinguish bands with slightly different molecular weights on a 4–12 % mini gel. The number of rows can be determined by comparing quantification data from replicated blots, as this should result in nearly identical band profiles if optimal.
37. For ImageJ, we created a plug-in (probe_reader2.jar) for batch quantification of multiple Western/far-Western blots (http://sites.imagej.net/Kazy/plugins/). It is best to name each image file sequentially using numbers, as ImageJ and the Probe Reader will sort images in numerical order based on file name.
38. Data in each cell should correspond to the bin position on your grid. Bin quantifications from each image are stacked vertically on the spreadsheet according to their image sequence order.

Acknowledgment

We are grateful to P. Nollau for his effort for the early development of the method, Z. Haimin and D.-G. Shin for help developing the ImageJ plug-in, C. Thompson for technical assistance, K. Claffey for breast cancer samples and A. Das for a bacterial culture system. The original work for this chapter was partially supported by grants from Breast Cancer Alliance and Connecticut Breast Health Initiative (to K.M.) and NIH grant CA107785 (to B.J.M.). The updated version was partly supported by NIH grant CA1154966 (to B.J.M.) and Quest for CURES (QFC) grant from the Leukemia and Lymphoma Society (to K.M.).

References

1. Edmondson DG, Dent SY (2001) Identification of protein interactions by far Western analysis. In: Coligen JE (ed) Current protocols in protein science, vol 2. Wiley, Hoboken, NJ, pp 1–10
2. Hoeffler JP, Lustbader JW, Chen CY (1991) Identification of multiple nuclear factors that interact with cyclic adenosine 3′,5′-monophosphate response element-binding protein and activating transcription factor-2 by protein-protein interactions. Mol Endocrinol 5: 256–266
3. Akiyama T, Ohuchi T, Sumida S, Xu SQ, Toyoshima K (1992) Phosphorylation of the anti-oncogene products and control of the cell cycle. Tohoku J Exp Med 168:153–157
4. Hall RA (2004) Studying protein-protein interactions via blot overlay or Far Western blot. Methods Mol Biol 261:167–174
5. Cicchetti P, Mayer BJ, Thiel G, Baltimore D (1992) Identification of a protein that binds to the SH3 region of Abl and is similar to Bcr and GAP-rho. Science 257:803–806
6. Macgregor PF, Abate C, Curran T (1990) Direct cloning of leucine zipper proteins: Jun binds cooperatively to the CRE with CRE-BP1. Oncogene 5:451–458
7. Blackwood EM, Eisenman RN (1991) Max: a helix-loop-helix zipper protein that forms a sequence-specific DNA-binding complex with Myc. Science 251:1211–1217
8. Nollau P, Mayer BJ (2001) Profiling the global tyrosine phosphorylation state by Src homology 2 domain binding. Proc Natl Acad Sci U S A 98:13531–13536
9. Machida K, Thompson CM, Dierck K, Jablonowski K, Karkkainen S, Liu B, Zhang H, Nash PD, Newman DK, Nollau P, Pawson T, Renkema GH, Saksela K, Schiller MR, Shin DG, Mayer BJ (2007) High-throughput phosphotyrosine profiling using SH2 domains. Mol Cell 26:899–915
10. Moorhead G, MacKintosh C (2004) Affinity methods for phosphorylation-dependent interactions. Methods Mol Biol 261:469–478
11. Burnham MR, DeBerry R, Bouton AH (2001) Detection of phosphorylation-dependent interactions by far-western gel overlay. Methods Mol Biol 124:209–220
12. Einarson MB, Orlinick JR (2002) Identification of protein-protein interactions with glutathione-S-transferase fusion proteins. In: Golemis E (ed) Protein-protein interactions. Cold Spring Harbor Press, Cold Spring Harbor, NY, pp 37–57
13. Smith DB, Johnson KS (1988) Single-step purification of polypeptides expressed in Escherichia coli as fusions with glutathione S-transferase. Gene 67:31–40
14. Mayer BJ, Jackson PK, Van Etten RA, Baltimore D (1992) Point mutations in the abl SH2 domain coordinately impair phosphotyrosine binding in vitro and transforming activity in vivo. Mol Cell Biol 12:609–618
15. Babon JJ, McManus EJ, Yao S, DeSouza DP, Mielke LA, Sprigg NS, Willson TA, Hilton DJ, Nicola NA, Baca M, Nicholson SE, Norton RS (2006) The structure of SOCS3 reveals the basis of the extended SH2 domain function and identifies an unstructured insertion that regulates stability. Mol Cell 22: 205–216
16. LamLa T, Hoerer S, Bauer MM (2006) Screening for soluble expression constructs using cell-free protein synthesis. Int J Biol Macromol 39(1–3):111–121
17. Welsh M, Mares J, Karlsson T, Lavergne C, Breant B, Claesson-Welsh L (1994) Shb is a ubiquitously expressed Src homology 2 protein. Oncogene 9:19–27
18. Jones RB, Gordus A, Krall JA, MacBeath G (2006) A quantitative protein interaction network for the ErbB receptors using protein microarrays. Nature 439:168–174
19. Sato Y, Kameya M, Arai H, Ishii M, Igarashi Y (2011) Detecting weak protein-protein interactions by modified far-western blotting. J Biosci Bioeng 112(3):304–307. doi:10.1016/j.jbiosc.2011.05.011
20. Wu Y, Li Q, Chen XZ (2007) Detecting protein-protein interactions by Far western blotting. Nat Protoc 2(12):3278–3284. doi:10.1038/nprot.2007.459
21. Machida K, Eschrich S, Li J, Bai Y, Koomen J, Mayer BJ, Haura EB (2010) Characterizing tyrosine phosphorylation signaling in lung cancer using SH2 profiling. PLoS One 5 (10):e13470. doi:10.1371/journal.pone.0013470

Chapter 39

Western Blot Analysis of Adhesive Interactions Under Fluid Shear Conditions: The Blot Rolling Assay

Robert Sackstein and Robert Fuhlbrigge

Abstract

Western blotting has proven to be an important technique in the analysis of receptor-ligand interactions (i.e., by ligand blotting) and for identifying molecules mediating cell attachment (i.e., by cell blotting). Conventional ligand blotting and cell blotting methods employ non-dynamic (static) incubation conditions, whereby molecules or cells of interest are placed in suspension and overlaid on membranes. However, many cell-cell and cell-matrix adhesive interactions occur under fluid shear conditions, and shear stress itself mediates and/or facilitates the engagement of these physiologically appropriate receptors and ligands. Notably, shear forces critically influence the adhesion of circulating cells and platelets to vessel walls in physiologic cell migration and hemostasis, as well as in inflammatory and thrombotic disorders, cancer metastasis, and atherosclerosis. Use of non-dynamic blotting conditions to analyze such interactions can introduce bias, overtly missing relevant effectors and/or exaggerating the relative role(s) of non-physiologic adhesion molecules. To address this shortfall, we have developed a new technique for identifying binding interactions under fluid shear conditions, the "blot rolling assay." Using this method, molecules in a complex mixture are resolved by gel electrophoresis, transferred to a membrane that is rendered semitransparent, and the membrane is then incorporated into a parallel-plate flow chamber apparatus. Under controlled flow conditions, cells or particles bearing adhesion proteins of interest are then introduced into the chamber and interactions with individual immobilized molecules (bands) can be visualized in real time. The substrate molecule(s) supporting adhesion under fluid shear can then be identified by staining with specific antibodies or by excising the relevant band(s) and performing mass spectrometry or microsequencing of the isolated material. This method thus allows for the identification, within a complex mixture and without prior isolation or purification, of both known and previously uncharacterized adhesion molecules operational under dynamic conditions.

Key words Adhesion molecules, Shear stress, Shear conditions, Fluid shear, Flow conditions, Parallel-plate flow chamber, Western blot, Selectins, Selectin ligands, Cell blotting, Blot rolling assay, Gel electrophoresis

1 Introduction

Adhesive interactions of cells with other cells and/or extracellular matrix under fluid shear conditions are critical to a variety of physiologic and pathobiologic processes, including hemostasis, leukocyte trafficking, tumor metastasis, and atherosclerosis.

Biji T. Kurien and R. Hal Scofield (eds.), *Western Blotting: Methods and Protocols*, Methods in Molecular Biology, vol. 1312, DOI 10.1007/978-1-4939-2694-7_39, © Springer Science+Business Media New York 2015

Although various techniques for analyzing cell adhesion have been described, most involve binding assays under non-dynamic (static) conditions. Importantly, certain adhesive receptor-ligand interactions occur preferentially if not solely under physiologic shear stress (e.g., binding of L-selectin to its ligands) or depend on low-affinity and rapidly reversible interactions to serve their functions (e.g., rolling of leukocytes via selectins). Under static binding assay conditions, effectors mediating these types of interactions may be overtly neglected or overshadowed by molecules specialized to form more stable adhesions.

Methods for in vitro study of adhesion under dynamic conditions have been described, including the Stamper-Woodruff [1] and the parallel-plate flow chamber [2] assays. However, these methods are limited in that they require the availability of reagents (e.g., antibodies) that can specifically interfere with the activity of relevant receptors and/or ligands or require isolated substrate materials that can be immobilized on the chamber surface. Thus, the applicability of these methods to examine the structure or function of previously unrecognized/uncharacterized ligands, especially within a complex mixture, is highly constrained.

To address these issues, we developed a method for direct real-time observation of adhesive interactions between cells or particles in flow and proteins separated by SDS-PAGE and immobilized on membranes [3]. This technique, which we have termed the "blot rolling assay," has allowed for the identification of new glycoprotein ligands [4–6], as well as a glycolipid ligand [7] for selectins. This method allows real-time assessment and measurement of interaction parameters (e.g., rolling vs. firm attachment, specificity, and reversibility with inhibitors) in both physiologic and non-physiologic shear conditions, thus permitting a unique user interface for the observation of adhesive events on membrane-immobilized materials. Ligands under investigation can be immobilized directly or segregated by gel electrophoresis (e.g., SDS-PAGE, isoelectric focusing) or other methods prior to transfer to the membrane. This method also provides for real-time manipulation of interaction conditions including wall shear stress, ion requirements, temperature, influence of metabolic inhibitors, and presence of activating agents or inhibitors of cell function. Blot-immobilized substrates can be used repeatedly, allowing in situ manipulation of the substrate under continuous direct visualization or direct comparison of different conditions or different cell populations in shear flow. The capacity to observe sequential experimental and control conditions on a single substrate, and to observe physiologic behaviors and responses to manipulations in real time, provides particularly powerful advantages of this method over conventional static binding assays. Similarly, the ability to observe interaction with individual components of a complex mixture without requiring prior purification or knowledge of the nature of the components provides advantages over conventional flow-based assays. In this chapter, we describe a detailed protocol for performing blot rolling assays of membrane proteins. A video recording of a representative experiment is presented to illustrate the utility of this new analytic technique.

2 Materials

The required components for performing blot rolling assays include an inverted microscope and a parallel-plate flow chamber apparatus. Figure 1 shows the setup utilized in our laboratories. The parallel-plate device we use has internal chamber diameters of 2 cm × 0.5 cm × 0.025 cm. A commercial product with similar properties is available from GlycoTech Corp., Rockville, Maryland, USA (Circular Flow Chamber, Product #31-001). Flow is adjusted by negative pressure regulated by a high-precision, programmable syringe pump (Harvard Apparatus). Wall shear stress (dyn/cm^2) is calculated according to the formula wall shear stress $(\tau) = 6\mu Q/wh^2$, where w = chamber width (cm), h = chamber height (cm), and Q = volumetric flow rate (ml/s). μ = viscosity (poise, or dyn × s/cm^2; for water at 25 °C, μ = 0.009), which is adjusted for contribution of 10 % glycerol (see below), according to the equation μ = 1.37 × medium (μ = 0.0123 at 25 °C) (see below for calculations based on our chamber dimensions). Interaction events can be captured on videotape by using a standard CCD camera and video recording assembly (*see* Fig. 1), for later off-line analysis.

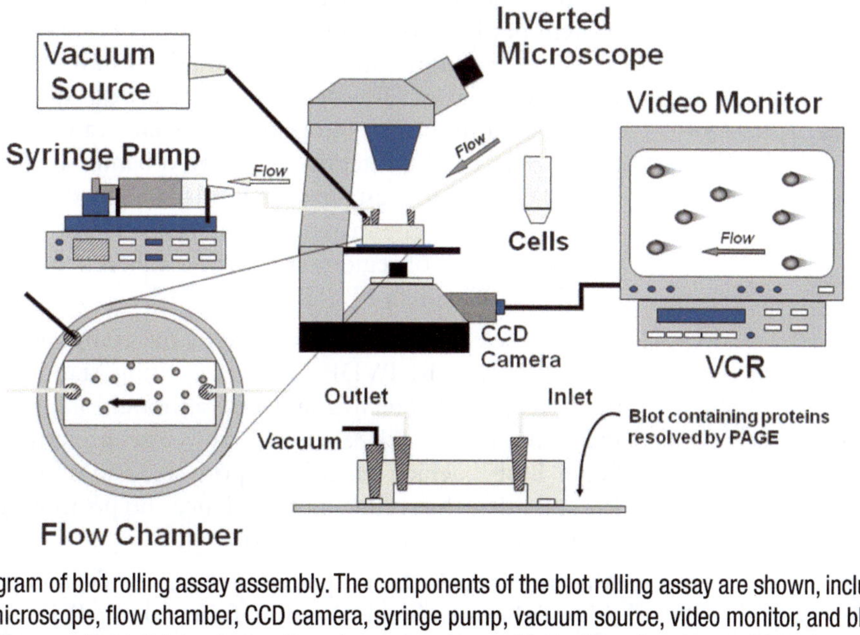

Fig. 1 Diagram of blot rolling assay assembly. The components of the blot rolling assay are shown, including the inverted microscope, flow chamber, CCD camera, syringe pump, vacuum source, video monitor, and blot membrane (resting on a Petri dish located on the microscope stage, with the flow chamber positioned on top of it)

1. PAGE gel system.
2. Sequi-Blot™ PVDF membrane.
3. Fetal bovine serum.
4. Phosphate-buffered saline/0.1 % Tween-20 (PBST).
5. Prestained molecular weight markers.
6. Western Blue stabilized substrate for alkaline phosphatase.
7. Blot rolling medium (H/H medium): Hanks' Balanced Salt Solution with 10 mM HEPES, pH 7.4. Add 2 mM $CaCl_2$, 5 mM Na_2EDTA, and 10 % glycerol to H/H medium.
8. Monoclonal or polyclonal antibody reagents for identifying any epitope(s) of interest (*see* **Note 1**).
9. Acrylamide gradient gels.
10. 1 % *n*-Octylglucoside.

3 Methods

3.1 Preparation of Western Blots/ Visualization of Protein Bands

1. Material of interest can be prepared for PAGE (with or without SDS) using standard methods. In general, we prepare lysates of cells at 4×10^8 cell equivalents per ml of lysate buffer. A number of standard detergents have been utilized for preparation of lysates, though we have found 1 % n-octylglucoside to provide low nonspecific background binding. Acrylamide gradient gels provide improved resolution in higher molecular weight ranges. The amount of protein required will vary with the material under study, but we have found 25–100 μg of protein per lane to be sufficient for most of our studies. Both reducing and nonreducing gel conditions have been utilized with good results. Omission of detergent from the gel and/or non-denaturing gel electrophoresis systems can be utilized to optimize structural integrity of resolved bands. In all gels, it is helpful to include prestained molecular weight standards in the lanes adjacent to the material of interest to assist in alignment and localization of regions supporting adherence (*see* **Notes 2** and **3**) (*see* Chapters 11 and 34).
2. When gel electrophoresis is complete, the component proteins are transferred to PVDF membrane using standard transfer methods (*see* Chapters 22 and 34); although other membrane materials may work as well, we have preferred BioRad Sequi-Blot™, which has 0.2 μm pore size and minimal surface irregularities, thus reducing turbulence and promoting laminar flow over the membrane (*see* **Note 4**).
3. After transfer, the blot is placed protein side facing up (i.e., the side of the membrane that was facing the gel) in a blocking solution (50 % FBS in PBST and 100 % newborn calf serum

have been used with good results) to reduce nonspecific interactions and incubated with gentle agitation (on a rocker platform or rotating platform) for 1 h at room temperature. The blot is then washed twice in PBST for 5 min each with gentle agitation.

4. Immunostaining, if desired, can be performed using standard procedures. Membranes are typically stained with primary antibody for 1 h under gentle agitation, washed 2× in PBS/0.1 % Tween 20 for 5 min, and then incubated with secondary alkaline phosphatase-conjugated antibody in TBS (Tris-buffered saline) for 1 h under gentle agitation. The blot is again washed 2× in PBST for 5 min, then rinsed in PBS followed by TBS (to remove residual PBST), and then developed with alkaline phosphatase substrate (e.g., Western Blue™). After developing blots to an appropriate signal-to-background perspective, reactions are stopped with PBS and the blot is washed twice with PBS. Blots to be used for flow experiments may be stored in buffer at 4 °C until use. Blots will maintain binding fidelity for several days, though best results are obtained with freshly prepared materials (*see* **Note 5**).
5. The membranes are again blocked by incubation in 50 % newborn calf serum or FBS in binding media (H/H with Ca^{2+}/10 % glycerol) for at least 1 h at 4 °C and equilibrated in binding media prior to use in the flow assay.

3.2 Preparation of Selectin-Expressing Cells

To assay blots for selectin ligand activity, we typically employ lymphocytes expressing L-selectin or Chinese hamster ovary (CHO) cells stably transfected to express P-selectin or E-selectin.

1. Human peripheral blood lymphocytes (PBL) can be prepared from whole blood by Ficoll density gradient separation. For analysis of L-selectin function on lymphocytes, monocytes may be removed by plastic adherence or other methods to reduce nonspecific binding (e.g., magnetic bead separation). Murine spleen and lymph node T cells and rat thoracic duct lymphocytes have also been utilized in the same fashion. E-selectin- and P-selectin-transfected cell lines are typically split the day prior to use to achieve log-phase growth and are harvested by replacing the media with H/H- 5 mM EDTA. The cells can then be released from the flask by manual agitation or scraping. We do not use trypsin to release the cells as this may alter the function of some surface adhesion molecules.
2. After counting, the cells are suspended at 10× the concentration desired for use in the flow chamber (e.g., $10–20 \times 10^6$/mL in H/H without Ca^{2+} or Mg^{2+} and without glycerol) and stored on ice. Clumping of adherent cell populations (i.e., CHO cells) may develop after prolonged storage on ice. This

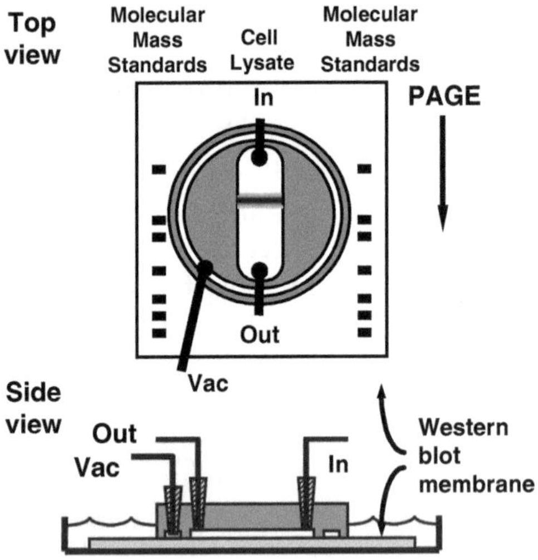

Fig. 2 Blot rolling apparatus. Western blots stained for components of interest are equilibrated in binding media containing 10 % glycerol and placed in a Petri dish. A parallel-plate flow chamber mounted over the area of interest is held in place by low continuous vacuum pressure (Vac). The inlet (In) is connected to a buffer reservoir and the outlet (Out) to a high-precision syringe pump. The apparatus is mounted on the stage of an inverted phase-contrast microscope equipped for video microscopy. Molecular mass(es) of interacting protein(s) resolved in the polyacrylamide gel (PAG) are estimated by alignment with colored mass standards in adjacent lanes (from Fuhlbrigge et al. [3]; reproduced with permission, Copyright 2002, The American Association of Immunologists, Inc.)

can be remedied by washing again in H/H 5 mM EDTA before use (*see* **Note 6**).

3. Controls to confirm specificity for selectin binding to blots include use of mock-transfected CHO cells and/or preincubation of cells with function-blocking anti-selectin mAb.

3.3 Inverted Microscope Setup

1. Setup will vary with the type of microscope. In general, we do not use any filters.
2. Figure 2 shows the placement of the parallel-plate chamber on the blot. Begin visualization at low power to align the chamber and identify the area of interest. Many observations are best performed at relatively low magnification (e.g., 10× objective) to allow observation of larger areas and numbers of cells. Observation under increased magnification can be performed to observe individual cell characteristics. Adjust the shutter arm so that the light narrows specifically over the area of interest on the blot.

3.4 Blot Rolling Assay

1. All media should contain 10 % glycerol. Equilibration in dilute glycerol alters the opacity of the PVDF membrane sufficiently to allow transmission of light and the direct visualization of cells interacting with the surface of the blot by standard light microscopy. Assessment of some functions (e.g., integrin activation) requires maintaining the chamber and media at 37 °C. This can be achieved with a stage warming device and placement of supply media in water baths adjacent to the apparatus (*see* **Note 7**).
2. The chamber should be mounted onto the blot so that the flow channel is aligned over the lane of interest. The flow chamber is secured by application of low vacuum pressure in the same fashion as is used for attachment to glass or plastic surfaces. We have not found it necessary to perform any special maneuvers to maintain an adequate seal on Sequi-Blot PVDF membrane (*see* **Note 8**).
3. The flow field is rinsed with media introduced through the 20 mL syringe, attached to the three-way stopcock, to remove bubbles and prepare the blot surface for cell input (*see* **Note 9**).
4. Immediately before use, aliquots of cells (e.g., PBL or transfected CHO cells) are diluted tenfold into binding medium (H/H w/ Ca^{2+} with 10 % glycerol; final concentration is typically $1–2 \times 10^6$ cells/ml) at room temperature (or 37 °C as indicated) (*see* **Note 10**). This minimizes exposure to glycerol and divalent cations that may affect cell viability and/or promote the formation of cell aggregates.
5. The cell input line is flushed with assay medium and placed into the tube containing cells. Flow is regulated by function of a downstream syringe pump. Use of a programmable pump allows for automation of the assay technique and more reproducible results. We typically initiate flow at a high rate of flow (2 ml/min or ~7.5 dyn/cm^2) to bring cells through the sample tubing and into the chamber. Upon arrival of cells in the field of view, the flow rate is reduced (typically to 0.5–1 dyn/cm^2) to allow cells to interact with the blot surface. The flow rate can be adjusted up or down to increase or decrease, respectively, the stringency of binding interactions. One can program stepwise increases in flow (shear stress) without interruption. Maintaining continuous flow while cells are in the chamber minimizes nonspecific binding to the blot surface. This is critical with adherent cell populations such as CHO cells or monocytes, though lymphocytes typically do not bind when flow is stopped for brief periods (<30 s).
6. Observation of cell interactions with substrate molecule(s) is made in real time and recorded via the camera/VCR for later analysis. Typically, two types of analyses are performed. First, a

"scanning" analysis is performed by moving the microscope stage to view the entire length of the lane where proteins have been resolved while maintaining a constant shear rate. Tethering and rolling interactions are usually most prominent at the leading edge of a "band" on the blot, with rolling cells evident across the band and discontinuation of rolling interactions observed at the downstream edge. Once bands or areas of the blot of interest are identified, we typically perform additional studies comparing control cell preparations (mock-transfected or cells incubated with function blocking antibodies) and varying shear conditions while viewing a fixed area or band on the blot. In this analysis, tethering is observed at low physiologic wall shear stresses (0.5–1.5 dyn/cm^2) for short periods (one to several minutes). Bound cells are then subjected to timed stepwise increases in wall shear stress to assess shear resistance as an estimate of relative strength of binding. The dependence of interactions on the presence of calcium (typical of selectin-mediated binding) can be confirmed by perfusion of the chamber with H/H with 10 % glycerol and 5 mM EDTA and observing release of bound cells, or by repeating the analysis using H/H with 10 % glycerol and 5 mM EDTA for both dilution of the cells and perfusion through the chamber.

7. Antibody inhibition experiments utilizing antibodies to the membrane-bound ligand can be performed in situ by first observing interactions on an area of defined interest, then perfusing the chamber with a solution of antibody (e.g., 1–10 μg/mL for 30 min), and repeating the observation on the same site. If the antibody blocks binding interactions, the blot is unusable for further studies of that ligand. However, if the applied antibody does not block adhesive interactions, the blot may be reused; that is, successive rounds of antibodies or reagents may be screened in this fashion.
8. For selectin-mediated binding, cells must tether and roll in shear flow to be considered specific. Tethering in our studies is defined as reduction of forward motion below the hydrodynamic velocity lasting for a minimum of two video frames (0.07 s), and rolling is defined as >5 cell diameters of lateral translation below the hydrodynamic velocity. The majority of tethered cells in such studies are observed to roll smoothly across the entire band or field of view. Nonspecific interactions (i.e., cellular collisions with the substrate that did not lead to tethering and/or rolling) are defined as interactions lasting for less than 0.07 s and are not included in the analysis. Nonspecifically bound cells (not rolling and/or not released by perfusion with EDTA) are also discounted from analysis. In general, there is a time-dependent increase in nonspecific

(non-rolling) attachments in the absence of flow, though it is our experience that very few cells form nonspecific attachments to the blot in continuous shear flow. Tethering rate is calculated as the number of cells that tether per field per time at a defined shear stress and adjusted to per minute values. As stated above, wall shear stress (τ) values are calculated according to the formula τ (dyn/cm^2) = $6\mu Q/wh^2$ where μ is the viscosity of the solution in the chamber (poise), Q is the volumetric flow rate (ml/s), w is the channel width (e.g., 0.5 cm for chamber we employ), and h is the channel height (e.g., 0.0127 cm for chamber we employ) [8]. A value of 0.009 poise is used for the viscosity (μ) of water at 25 °C [9] and a value of 0.0123 poise is used for the viscosity of 10 % glycerol (v/v) in water at 25 °C [10]. By these values, wall shear stress in 10 % glycerol at 25 °C is approximately 1.37-fold greater than in water at the same temperature.

9. Regions (bands) of interest can be aligned with molecular weight markers run in adjacent lanes to estimate molecular weight. Additionally, one may subsequently probe the blot with antibody to specific proteins of interest to determine whether band(s) supporting binding represent known proteins. This approach has been used to confirm PSGL-1 and identify CD43 as selectin ligands [3, 6]. One can also excise regions of blot supporting binding and submit for mass spectrometry and/or protein microsequencing. This approach was used to identify HCELL, a glycoform of CD44 that functions as a high-affinity L-selectin and E-selectin ligand [4, 5]. Figure 3 is a representative histogram of a blot rolling assay of T cell lysates probed with stably transfected CHO cells expressing P-selectin (CHO-P) and E-selectin (CHO-E). A videotape of these studies is available for viewing on the Harvard Skin Disease Research Center web page, under the Leukocyte Migration Core subheading (http://dermatology.bwh.harvard.edu/leukocyte.html).

4 Notes

1. We typically avoid immunoblotting with function blocking antibodies, since these antibodies may interfere with cell binding to the membrane (*see* Subheading 3.4, **step** 7).
2. Any uniform percent gel or gradient gel can be used so long as sufficient resolution of protein bands is achieved in the molecular weight range of interest.
3. If one wishes to measure the relative strength of interactions between component bands of samples loaded in different lanes of the same gel, it is important to load samples in lanes an

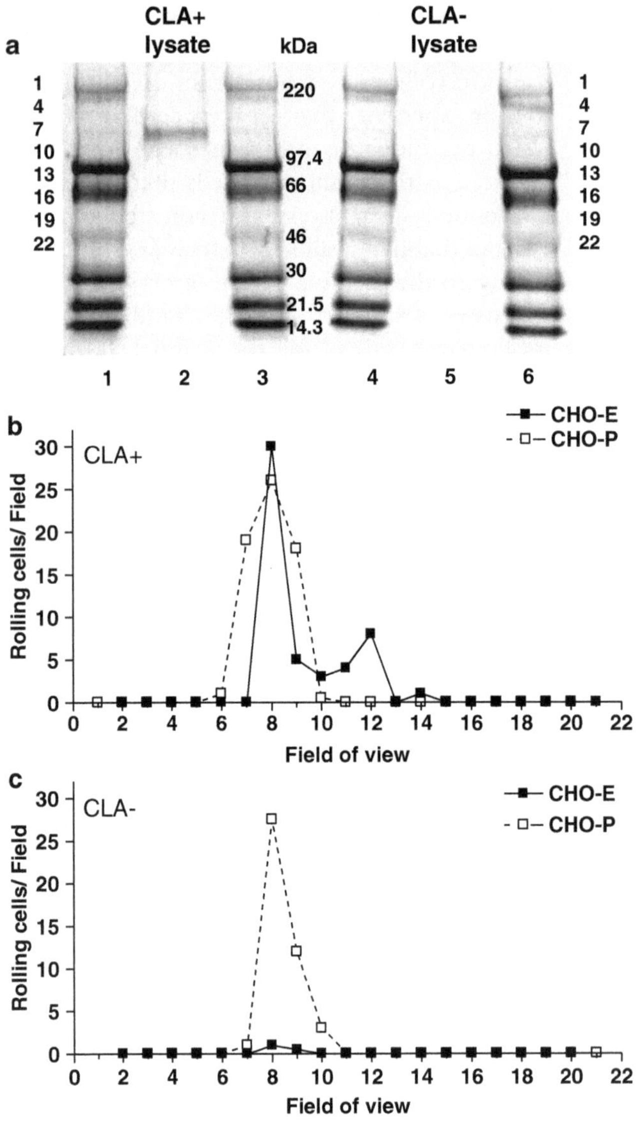

Fig. 3 Cutaneous lymphocyte antigen (CLA) immobilized on western blots supports both E-selectin- and P-selectin-mediated rolling. (**a**) Western blots of T cell lysates stained for CLA expression. 50 μg of cell lysate protein from CLA-positive (*left panel*) or CLA-negative (*right panel*) T cells were subjected to SDS-PAGE, electroblotted onto PVDF membranes, and immunostained with HECA-452 mAb. The single major reactive glycoprotein identified at approximately 140 kDa in CLA-positive cells (*lane 2*) corresponds to the monomer form of CLA/PSGL-1. Similar blots stained with anti-PSGL-1 mAb show approximately equal quantities of 140 kDa PSGL-1 protein in each sample (data not shown). (**b** and **c**) Rolling cells per visual field (10× objective) at 1.75 dyn/cm^2 wall shear stress. The number of cells observed to bind and roll was observed in overlapping fields of view extending from approximately 250 to 40 kDa (identified alongside blot images in *panel* **a**). Both CHO-E and CHO-P cells are observed to tether and roll across the area of the western blot corresponding to CLA/PSGl-1 from CLA-positive T cells (*panel* **b**) while only CHO-P cells are noted to form significant rolling adhesions on PSGL-1 from CLA-negative T cells (*panel* **c**). Both CHO-E and CHO-P cells were noted to tether primarily over the area corresponding to CLA/PSGL-1. Mock-transfected CHO cells did not form rolling tethers on any areas of the blots observed (data not shown). Results presented are the means of two independent experiments on a single substrate and are representative of observations on numerous blots of CLA-positive and CLA-negative T cell lysates (from Fuhlbrigge et al. [3]; reproduced with permission, Copyright 2002, The American Association of Immunologists, Inc.)

appropriate distance apart such that the membrane-attached flow chamber does not cover another sample lane in order to avoid damaging the latter. For our flow chamber and the Criterion precast gel system, we load samples in every third lane, using molecular weight markers in between.

4. Surface irregularities can result in disruption of cell interactions on the membrane. Some membrane surface irregularities due to manufacture are readily visible when the membrane is immersed in methanol prior to the transfer step, while others are only observable when the membrane is immersed in 10 % glycerol. Avoid the use of visibly scored or scratched membranes, as these defects may prevent a proper vacuum seal of the chamber to the membrane or provide a "trough" in which perfused cells may collect and subsequently adhere nonspecifically.
5. In all incubations with mAb or serum, it is critical to avoid using bacterially contaminated solutions in order to prevent digestion of binding epitopes by bacterial proteases and/or glycosidases.
6. It is important that cells are not stored in glycerol, as a 10 % glycerol solution will significantly decrease cell (e.g., CHO) viability within 15 min.
7. Viscosity is temperature dependent. If the assay temperature is adjusted from 25 to 37 °C, make sure to recalculate flow rates using the appropriate viscosity values for cell perfusion at specific shear stresses (*see* Subheadings 2 and 3).
8. We use Dow Corning High-Vacuum Grease to seal the silicon rubber gasket to the flow chamber to create the flow channel. In order to prevent the gasket from slipping after placement, we allow the grease to set for 24 h before use in experiments. It is critical to remove all excess vacuum grease from the channel, gasket surface, and other flow chamber surfaces completely (using a cotton swab and 70 % ethanol) before use in blot rolling assays to prevent the transfer of grease onto the membrane, which will damage the lanes and render them unusable.
9. All bubbles and debris from the channel must be flushed to clear all potential for flow disturbances.
10. The cell concentration can be adjusted to increase or decrease assay sensitivity as desired (i.e., higher cell concentration increases sensitivity).

Acknowledgment

This work was supported by National Institutes of Health grants HL60528 (RS), CA84156 (RS), HL073714 (RS), and AR02115 (RCF). The authors thank Drs. Sandra L. King and Monica M. Burdick for critical review of the manuscript.

References

1. Stamper HB, Woodruff JJ (1976) Lymphocyte homing into lymph nodes: *in vitro* demonstration of the selective affinity of recirculating lymphocytes for high endothelial venules. J Exp Med 144:828–833
2. Lawrence MB, Springer TA (1991) Leukocytes roll on P-selectin at physiologic flow rates: distinction from and prerequisite for adhesion through integrins. Cell 65:859–873
3. Fuhlbrigge RC, King SL, Dimitroff CJ, Sackstein R (2002) Direct, real-time observation of E- and P- selectin mediated rolling on cutaneous lymphocyte-associated antigen immobilized on Western blots. J Immunol 168:5645–5651
4. Dimitroff CJ, Lee JY, Fuhlbrigge RC, Sackstein R (2000) A distinct glycoform of CD44 is an L-selectin ligand on human hematopoietic cells. Proc Natl Acad Sci U S A 97:13841–13846
5. Dimitroff CJ, Lee JY, Rafii S, Fuhlbrigge RC, Sackstein R (2001) CD44 is a major E-selectin ligand on human hematopoietic progenitor cells. J Cell Biol 153:1277–1286
6. Fuhlbrigge RC, King SL, Sackstein R, Kupper TS (2006) CD43 is a ligand for E-selectin on human skin homing T cells. Blood 107:1421–1426
7. Hanley WD, Burdick MM, Konstantopoulos K, Sackstein R (2005) CD44 on LS174T colon carcinoma cells possesses E-selectin ligand activity. Cancer Res 65:5812–5817
8. Glasby MA (1995) Principles of haemodynamics. In: Glasby MA, Huang CL-H (eds) Applied physiology for surgery and critical care. Boston, MA, Oxford, p 131
9. Sengers JV, Watson JTR (1986) Improved international formulations for the viscosity and thermal conductivity of water substance. J Phys Chem Ref Data 15:1291
10. Cheng N-S (2008) Formula for the viscesity of a glycerol-water mixture. Ind Eng Chem Res 47:3285–3288

Chapter 40

Centrifuge Blotting

Jinny Paul

Abstract

Centrifuge blotting is an efficient and convenient method for elution and transfer of proteins from a polyacrylamide gel onto a polyvinylidene difluoride (PVDF) membrane by centrifugation.

Key words Centrifugation, PVDF membrane, Blotting

1 Introduction

Centrifuge immunoblotting is a technique developed by Hermansen et al. [1] to overcome the disadvantages with the downstream steps associated with sodium dodecyl-polyacrylamide gel electrophoresis (SDS-PAGE). This technique alleviates problem, such as low protein recovery, presence of contaminating substances from polyacrylamide gel, fixation and denaturation or blockage of N terminus due to the staining and destaining steps involved, to name a few.

Briefly, the protein samples are prepared and separated by SDS-PAGE as described by Laemmli [2]. Following electrophoresis the gel is soaked in 1 M KCl for 2 min to visualize the protein bands (clear zone against an opaque background). The gel pieces were then used for centrifuge blotting onto the PVDF membranes. PVDF membranes with immobilized proteins could be used for N-terminal sequence analysis or vacuum-dried and stored at −20 °C until further analysis.

Some advantages of centrifuge blotting over the existing techniques are the following: protein concentrations in subnanomolar range can easily be eluted within 1–2 h for Edman degradation. Lack of staining and destaining steps prevents N-terminal blockage, the small size of the PVDF membrane makes it suitable for direct application for microsequencer, and positioning of the dialysis membrane directly beneath the PVDF membrane retains the non-immobilized proteins.

Biji T. Kurien and R. Hal Scofield (eds.), *Western Blotting: Methods and Protocols*, Methods in Molecular Biology, vol. 1312, DOI 10.1007/978-1-4939-2694-7_40, © Springer Science+Business Media New York 2015

2 Materials

2.1 Detection of Protein Bands in the Gel

1. 1 M KCl.
2. Distilled water.

2.2 Centrifuge Blotting

1. PVDF membrane (12 mm diameter) (*see* **Note 8**).
2. 3–10 kDa dialysis membrane discs cut into 12 mm diameter (depending on the size of the protein).
3. 100 % methanol.
4. Eluant solution (0.05 % SDS/5 % methanol/0.05 % DTT) (*see* **Note 1**).

2.3 Centrifuge Receptacle Assembly

1. Outer cylinder constructed of polyoxymethylene (supports whole structure).
2. Inner cylinder constructed of tetrafluoroethylene (reservoir for the eluent).
3. 12 mm polyethene sinter base support (holds dialysis and PVDF membranes in place).
4. O-ring silicone gasket (prevents leakage in receptacle).
5. Polystyrene tube (collection of eluate).

2.4 Centrifugation

Hettich EBA 3S tabletop centrifuge or Sorvall RC5C automatic superspeed refrigerated centrifuge with fixed-angle SS-34 rotor.

3 Methods

3.1 SDS-PAGE

1. Prepare and separate the protein samples according to the method described by Laemmli [2] (*see* Chapters 11 and 34).
2. The dimension of the separation gel is 120 × 120 × 1.5 mm.
3. The acrylamide N,N′-methylenebisacrylamide concentration of the separation gel is about 12 % T, 3 % C.

3.2 Detection of Protein Bands on the Gel

1. Soak SDS-polyacrylamide slab gel in 1 M KCl for 2-min bands in gel.
2. Excise the protein band (clear zone against an opaque background) and soak in distilled water for 3–5 min (can store the moist gel in an Eppendorf tube at −20 °C) to remove excess KCl.

3.3 Centrifuge Blotting

1. Cut the PVDF and dialysis membrane (10 kD molecular weight cutoff) into 12 mm discs.

2. Soak the PVDF membrane in 100 % methanol for 5 min for activation.
3. Soak the gel pieces in the eluant solution for 15 min at 37 °C (*see* **Note 2**).
4. Soak the PVDF and dialysis membranes in the eluant solution for 3–5 min prior to centrifugation.
5. Place the dialysis membrane on the base support of the receptacle followed by the O-ring.
6. Place the PVDF membrane on top of the O-ring after adding 100 μL eluant to the dialysis membrane. This prevents the vacuum formation between the two membranes.
7. Screw in the inner cylinder over the receptacle.
8. Place the gel piece over the PVDF membrane and pipette 300 μL of eluant solution into the receptacle (*see* **Note 3**).
9. Ensure that the assembly is complete as illustrated in Fig. 1.
10. Centrifuge at 3,000 × *g* at 20 °C for 1 h (*see* **Notes 4–7**).

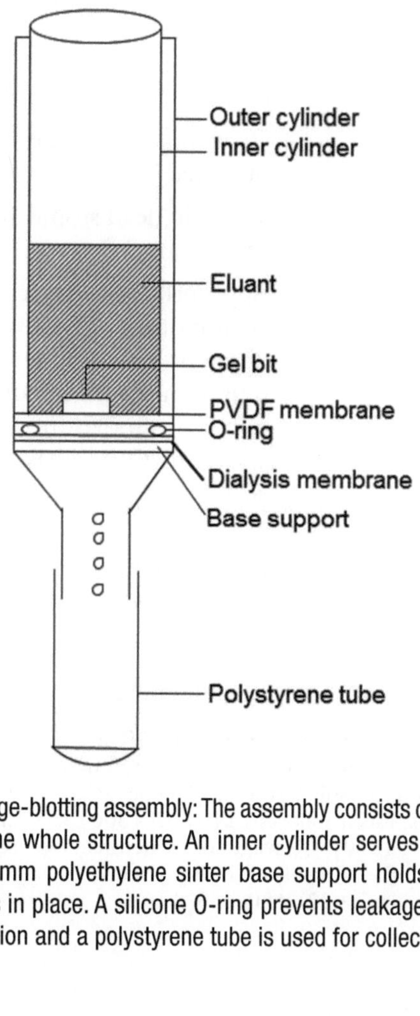

Fig. 1 The centrifuge-blotting assembly: The assembly consists of an outer cylinder which supports the whole structure. An inner cylinder serves as a reservoir for the eluant. A 12 mm polyethylene sinter base support holds the dialysis and PVDF membranes in place. A silicone O-ring prevents leakage in the receptacle during centrifugation and a polystyrene tube is used for collection of the eluate

11. After centrifugation is done, remove the PVDF membrane with the adhering protein, vacuum dry, and store at −20 °C until further analyses (*see* **Note 8**).

4 Notes

1. The use of the eluent solution (0.05 % SDS/5 % methanol/0.05 % DTT) improved the protein yield and secured better adherence of proteins onto the PVDF membranes (methanol weakens the protein-SDS interaction, hence allowing improved adsorption of the protein onto the PVDF membrane).
2. Equilibrating the gel pieces in the eluent solution improves the overall yield.
3. It is essential to have direct contact between the gel piece and the PVDF membrane, one gel piece with max size of 10×6×1.5 mm. The direct contact between the gel piece and the PVDF membrane avoids the endosmosis of protein towards the gel.
4. Polypeptides less than 30 kDa can be blotted by 1-h centrifugation. Higher molecular weight proteins require longer centrifugation time (>3 h). Greater elution of proteins is achieved by extending the centrifugation time up to 6 h or more [3].
5. Centrifuge for longer time at 3,000 ×*g* if dead stop volume is observed.
6. To avoid dead stop volume, use swinging-bucket rotor instead of fixed-bucket rotor.
7. Appropriate flow rate is essential for good desalting and efficient blotting. To eliminate the variability in flow rate between different runs, puncture a hole in a dialysis membrane with 0.05 mm thick wire and using two PVDF membrane, the first one right above the dialysis membrane and the second one above the O-ring gasket [3].
8. PVDF (Immobilon-P) membrane is resistant to the mechanical strain experienced during centrifugation unlike siliconized glass fiber "glassybond" membrane and Whatman glass fiber impregnated with Polybrene.

References

1. Hermansen LF, Pedersen O, Sletten K (1993) Centrifuge-blotting of proteins after separation by sodium dodecyl sulfate-polyacrylamide gel electrophoresis. Electrophoresis 14:1302–1306
2. Laemmli UK (1970) Cleavage of structural proteins during the assembly of the head of bacteriophage t4. Nature 227:680–685
3. Hermansen LF, Juul J, Sletten K (1996) Sodium dodecyl sulfate-polyacrylamide gel electrophoresis and centrifuge blotting: preparation of polypeptides for amino-terminal sequence analysis. Electrophoresis 17: 720–722

Chapter 41

Blotting of Coomassie Blue-Stained Proteins from PAGE Gels to Transparencies

Jinny Paul

Abstract

A simple, convenient, and inexpensive method for long-term non-photographic storage of information present in electrophoresis gel, based on protein blotting patterns, is presented here.

Key words Coomassie Blue, Transparencies, Non-photographic storage

1 Introduction

Blotting patterns on PAGE are treated in one of the two ways for storage purposes. They are either stored in sealed plastic bags containing 7 % aqueous acetic acid or more commonly the gels are photographed and dried. Chu and Whitesides [1] describe an economical transfer method to preserve data from slab gel electrophoresis to transparencies. Such transfer of the blotting patterns from PAGE gels to transparencies is also a convenient method to display data using an overhead projector. The details of the original electropherogram were retained on transfer and did not fade for a period of 3 years. Proteins associated with the blotting pattern transferred along with the dye, but failed to transfer in the absence of the dye. Some other advantages of transferring blotting patterns to transparencies are that they are convenient to store and are excellent substrates for gel scanners, as they are flat, dimensionally stable, and have no color themselves.

2 Materials

2.1 Organic Dyes

1. Coomassie Brilliant Blue (CBB).
2. Fast Green FCF.

Biji T. Kurien and R. Hal Scofield (eds.), *Western Blotting: Methods and Protocols*, Methods in Molecular Biology, vol. 1312, DOI 10.1007/978-1-4939-2694-7_41, © Springer Science+Business Media New York 2015

3. Uniblue A.
4. Procion Blue MX-R.

2.2 Staining Solution

1. Methanol.
2. Ethanol.
3. Acetic acid.
4. Water.

2.3 Transparency Films

The Arkwright PPC-IX and Polaroid 731 (*see* **Note 1**).

3 Methods

3.1 Gel Electrophoresis

Prepare and separate the proteins by Laemmli's procedure for SDS-PAGE [2] (*see* Chapters 11 and 34), Margolis procedure for non-denaturing PAGE [3], and procedure described by Hames and Rickwood for isoelectric focusing (IEF)-PAGE [4] (*see* **Note 2**).

3.2 Staining and Destaining the Gels

1. Prepare the staining solution at suitable concentration (*see* Table 1) and destaining solution (solution without dyes).
2. Stain the gels with dyes 1–4 prepared at suitable concentration (*see* Table 1) (*see* **Notes 3–5**).
3. Destain the gels using the respective solutions of methanol, acetic acid, and water (*see* Table 1) (*see* **Note 6**).

3.3 Transfer Blotting

1. Remove gels from the destaining solution.
2. Immediately place the gels between two sheets of transparency.
3. Leave the setup for 2 h at room temperature or at 40 °C for 30 min (*see* **Note 7**).
4. Disassemble the sandwich and air-dry the transparency.
5. Store the transparency.

Table 1
Preparation of CBB, fast green FCF, Uniblue A, or Procion Blue MX-R used for staining protein gels

Dye	Concentration (wt/vol) (%)	Solvent (vol/vol/vol)
CBB	0.1	Methanol/water/acetic acid (4:5:1)
Fast green	0.1	Methanol/water/acetic acid (3:6:1)
Uniblue A	0.6	Methanol/water/acetic acid (5:4:1)
Procion Blue	0.05	Methanol/water/acetic acid (4:5:1)

4 Notes

1. Dye-protein complex will not transfer to uncoated Mylar film and cellulose films.
2. Neither the gel matrix (polyacrylamide or agarose) nor the type of electrophoresis (denaturing, non-denaturing, or isoelectric focusing) influences the protein blotting from the gel to the transparency.
3. Transfer of color from the PAGE gel is accompanied by transfer of proteins when stained with Coomassie Blue. However proteins would not transfer in the absence of the dye.
4. Protein patterns formed by staining PAGE gels with dyes 2, 3, and 4 (dyes that are anionic and hydrophobic) will also transfer well to the transparencies.
5. Dyes that only form covalent bond with the proteins and do not react chemically, such as silver stain, gold stain, or Stains-all dye, will not transfer successfully to the transparency.
6. Pretreatment of the transparency with reagents such as the gel destaining solution, organic solvents (acetone, dimethylformamide, and dimethyl sulfoxide), and protein-denaturing agents (6 M guanidine hydrochloride, 1 % SDS) will not interfere with the transfer of the protein-dye complexes. However, pretreatment with reagents, such as hot acetic acid, which separates the thin coating films on which the image forms from the structural body of the transparency (the Mylar backing), will inhibit the transfer.
7. Dye-protein complexes cannot be electrophoretically blotted from PAGE gels to transparencies. This could be potentially due to the fact that the transparencies are poreless and not conductive to electric field.

References

1. Chu Y, Whitesides GM (1993) A convenient procedure for transfer blotting of Coomassie blue stained proteins from PAGE-gels to transparencies. Biotechniques 14(6):925–930
2. Laemmli UK (1970) Cleavage of structural proteins during the assembly of the head of bacteriophage t4. Nature 227:680–685
3. Margolis J, Kendrick KG (1968) Polyacrylamide gel electrophoresis in a continuous molecular sieve gradient. Anal Biochem 25:341–362
4. Rickwood D, Hames BD (1990) Gel electrophoresis of nucleic acids: a practical approach, 2nd edn. IRL Press, Oxford

Chapter 42

B-Cell ELISPOT: For the Identification of Antigen-Specific Antibody-Secreting Cells

Hemangi B. Shah and Kristi A. Koelsch

Abstract

The B-cell ELISPOT assay is a sensitive tool that can be utilized to measure total immunoglobulin (Ig) and antigen-specific antibody-secreting cells. Typically, membrane-bound antigen enables binding of antibody secreted by B-cells. Bound antibody is then detected by using an anti-Ig antibody and a colorimetric substrate, resulting in colored spots on the membrane that can be easily enumerated. Here we have described a method to quantitate antigen-specific antibody-secreting cells from the spleen or bone marrow of a vaccinated mouse.

Key words ELISPOT, B-cells, ASC, Antibody-secreting cell, Memory, Humoral response, Antibody, Antigen, Antigen-specific, Antibody response, Antibody-secreting cells, Vaccination

1 Introduction

The B-cell ELISPOT assay was first described more than 30 years ago [1] and yet the basic methodology has undergone only minor modifications since then. The predominant application of the B-cell ELISPOT has been in the detection of B-cell responses to infections and vaccinations [2–4]. B-cell ELISPOT assays have been instrumental in demonstrating the presence and frequency of antigen-specific plasmablasts and long-term memory B-cells in the peripheral blood of vaccinated individuals [5, 6]. Similarly, the ELISPOT assay has been used to quantitate virus-specific memory B-cells in the spleen of adult mice with acute viral infection [7]. The utility of this assay has been important in these kinds of studies as it provides information about antigen-specific B-cell responses that is otherwise difficult to glean using other less sensitive methods.

The source and activation status of the B-cells determine the specific experimental design required for a successful ELISPOT assay. For example, after infection or vaccination, in vivo-activated antibody-secreting cells (ASCs) are briefly detectable in the peripheral blood [6, 8] and can be directly added to ELISPOT

Biji T. Kurien and R. Hal Scofield (eds.), *Western Blotting: Methods and Protocols*, Methods in Molecular Biology, vol. 1312, DOI 10.1007/978-1-4939-2694-7_42,

plates for quantitation. However, "resting" memory B-cells require ex vivo polyclonal activation over 3–8 days before they secrete detectable amounts of antibody [9, 10]. The protocol detailed below describes quantitation of antigen-specific ASCs derived from the spleen and bone marrow of adult mice immunized with a hapten-conjugated model antigen. The B-cell ELISPOT method described here allows one to compare the humoral responses across different strains of mice and to explore the role of adjuvants and of specific cell subsets in enabling this response [11, 12]. The following method has been optimized for a specific antigen and murine B-cell populations. However, this protocol provides the basic methodology, which can be modified for use with other antigens or with B-cell populations from other species.

2 Materials

2.1 Vaccination

1. NP (4-hydroxy-3-nitrophenylacetyl) hapten-conjugated KLH (keyhole limpet hemocyanin) and Imject-Alum (aluminium hydroxide and magnesium hydroxide) adjuvant (*see* **Note 1**).
2. Sterile endotoxin-free PBS (1×).
3. 1 mL Tuberculin syringe with 25 × 5/8 in. needle.
4. Round-bottom polypropylene tubes (14 mL) with snap cap.

2.2 Harvesting Tissue and Cell Isolation

1. Mice: Usually, 6–8-week-old mice are used. A typical experiment would include at least five mice immunized with the antigen of interest with/without adjuvant and five mice injected with PBS only to serve as controls. Include five mice per group for each strain if multiple strains are compared.
2. 70 % ethanol.
3. Dissection board: Pins are optional and may be used to stabilize the animal during dissection.
4. Dissection scissor and forceps (plain and toothed forceps) (*see* **Note 2**).
5. Sterile 60 × 15 mm petri dishes.
6. Microscope glass slides with frosted ends.
7. 27-gauge needle and 1 mL syringe.
8. Sterile 70 μm nylon mesh/cell strainer.
9. RPMI 1640.
10. Ammonium chloride (ACT) lysis buffer: 0.16 M NH_4Cl and 0.17 M Tris–HCl, pH 7.4.
11. Water bath (37 °C).
12. Cell viability and count: Trypan blue, hemocytometer, or automated cell counter.

2.3 ELISPOT

1. 96-Well polyvinylidene fluoride (PVDF) membrane plates (Multiscreen$_{HTS}$ plates, Millipore, Bedford, MA).
2. 35 % ethanol (*see* **Note 3**).
3. 1× PBS.
4. Blocking buffer: RPMI 1640, 10 % bovine calf serum.
5. Cell resuspension media: RPMI 1640, 1 % bovine calf serum.
6. Wash buffer: 1× PBS, 0.05 % Tween 20.
7. Detection antibody diluent: 1× PBS, 5 % bovine calf serum.
8. Substrate: 3-Amino-9-ethyl-carbazole (AEC) tablets containing 20 mg of AEC substrate, *N,N*-dimethylformamide (DMF), 0.1 N glacial acetic acid, 0.1 M sodium acetate, and hydrogen peroxide solution (30 % [w/w] in H_2O).
9. 0.45 μm syringe filter.
10. Capture antigen: 4-Hydroxy-3-nitrophenylacetyl hapten-conjugated bovine serum albumin (NP-BSA) (*see* **Note 4**).
11. Detection antibody: HRP-conjugated goat anti-mouse IgG1 (*see* **Note 5**).
12. Incubator (37 °C, 5 % CO_2).
13. Plate washing: Multichannel pipette, manual plate washer, or automated plate washer can be used.
14. Microscope for manual counting or automated ELISPOT plate reader (e.g., KS ELISPOT 4.10 software, Carl Zeiss, Inc. Thornwood, NY).

3 Methods

Only trained and authorized personnel should handle experimental animals. All aseptic precautions and institutional guidelines should be followed during animal handling.

3.1 Vaccination

1. Prepare immunization solution in a biological safety cabinet by mixing NP-KLH (10 μg/mouse) with 100 μL 1× PBS and 100 μL Imject-Alum adjuvant in a 14 mL round-bottom polypropylene tube with cap. Place tube with stir bar in a beaker, which is placed on a stir plate and the immunization solution is stirred at room temperature for 30 min.
2. Prepare a dose of 200 μL per mouse and inject 100 μL subcutaneously over each flank (*see* **Note 6**).
3. Administer a booster vaccine consisting of NP-KLH (10 μg/mouse) alone (no adjuvant) in 200 μL PBS (100 μL over each flank) subcutaneously 28 days later (*see* **Note 7**).
4. At both vaccination and boost, control mice are injected with 200 μL sterile PBS only.

3.2 Harvesting Tissue and Cell Isolation

1. Euthanize mice 7 days after the booster vaccine and clean the skin over the dissection area by spraying with 70 % ethanol.
2. Place the animal on the dissection board (optional: pin the extremities for stability during dissection) and incise the abdominal cavity to remove the spleen located on the left side below the stomach.
3. To access the long bones (femur and tibia), incise the skin over the legs and use toothed forceps to dislocate the femur from the hip joint. The tibia and femur are also separated to allow complete removal of bone marrow. Clip the ends of the femur and tibia to enable flushing of the bones with RPMI (*see* **Note 8**).
4. Store the harvested spleen and bone marrow (femur and tibia) in 50 mL conical tubes containing 5 mL RPMI 1640. Place tubes containing harvested tissues on ice. To begin tissue processing, transfer tissues from conical tubes into petri dishes containing 5 mL of fresh RPMI 1640.
5. Single-cell suspensions from the spleen are obtained by mechanical disruption. Place the spleen between the frosted ends of two slides. Apply gentle pressure and shear the tissue between the two slides using a circular motion. The cells from the sheared spleen are washed off the slides into the petri dish containing 5 mL fresh RPMI 1640. To remove debris, pass the cell suspension through a 70 μm strainer placed over a new 50 mL conical tube (*see* **Note 9**).
6. To isolate the bone marrow, flush the femurs and tibias with RPMI 1640 using a 27-gauge needle and 1 mL syringe. First, approximately 1 mL of RPMI 1640 is drawn into the 1 mL syringe with needle attached. The needle is then inserted into one end of the femur/tibia, and held by toothed forceps, and the plunger is pushed to flush the bone marrow. To remove debris, pass the cell suspension through a 70 μm strainer placed over a new 50 mL conical tube.
7. Wash cells (spleen and bone marrow) through the 70 μm strainer with an additional 5 mL of fresh RPMI 1640. Bring the total volumes in the new 50 mL conical tubes to 20 mL. Centrifuge the cell suspensions for 5 min at 250 × *g*. Discard supernatants. Continue using the same 50 mL conical tubes.
8. For lysis of red blood cells, gently resuspend spleen and bone marrow cell pellets with 2 mL ACT lysis buffer and incubate for 2 min in water bath at 37 °C. Add cold RPMI to stop the reaction and bring the total volume up to 20 mL. Centrifuge for 5 min at 250 × *g*. Discard supernatant.
9. Resuspend cells in 5 mL RPMI 1640 with 1 % fetal bovine serum and check cell viability by the trypan blue exclusion method.

Appropriate tissue processing should result in >95 % live cells. Perform a manual or automated live cell count using hemocytometer or Cellometer Auto T4 cell counter, respectively.

10. Adjust cell density to 10^7 live cells/mL.

3.3 ELISPOT

1. Pre-wet ELISPOT plates with 15 μL/well of 35 % ethanol for ≤1 min at room temperature (*see* **Note 10**).
2. Wash plates twice by adding 250 μL/well of 1× PBS and tipping the plates to discard the PBS. Gently tap plate upside down on paper towel to remove excess PBS (*see* **Note 11**).
3. Coat plates with 100 μL/well of NP-BSA (10 μg/mL) in 1× PBS. Cover plates to prevent evaporation and incubate overnight at 4 °C (*see* **Note 12**).
4. Wash plates twice with 1× PBS (*see* **step 2**), and then block for 2 h at room temperature by adding 200 μL/well of blocking buffer.
5. Tip the plates to discard blocking buffer and add 200 μL/well of RPMI 1640 to rows B through D and rows F through H.
6. To wells in rows A and E: Add 300 μL/well of appropriate cell suspension (3×10^6 cells/well) from bone marrow and spleen in triplicate. This allows plating of eight samples in one plate (four in row A and four in row E) (*see* **Note 13**).
7. Using a multichannel pipette, threefold serial dilutions of the cells are performed by transferring 100 μL from the top row into the next row (row A to row B and row E to row F) and mixing carefully and thoroughly, repeating through the fourth row. Discard the final 100 μL from the fourth row so that each well has a final volume of 200 μL (*see* **Note 14**).
8. Place the plates in the incubator at 37 °C for 5 h (*see* **Note 15**).
9. Wash plates three times with 250 μL/well wash buffer. Tip plates to discard wash buffer and tap gently on a paper towel to remove excess buffer from the bottom of the well (*see* **Note 16**).
10. Prepare detection antibody by adding HRP-conjugated anti-IgG1 (0.125 μg/mL) or -IgG (0.5 μg/mL) in diluent. Add 100 μL/well to plates (*see* **Note 17**).
11. Cover plates and incubate overnight at 4 °C.
12. Prepare substrate solution 15 min before use. Dissolve one tablet of AEC in 2.5 mL of DMF (about 5 min). Prepare acetate buffer (pH 5.0) by mixing 46.9 mL double-distilled water with 4.6 mL 0.1 N acetic acid and 11 mL 0.1 M sodium acetate. Add 47.5 mL acetate buffer to AEC/DMF solution. Mix thoroughly and filter with sterile 45 μm syringe filter. Just before using the substrate solution add 25 μL H_2O_2 and mix. Wash plates three times with wash buffer (*see* **step 9**) and

develop with 100 μL/well of substrate solution. Allow plates to develop in the dark for 10 min and then wash 10–20 times with double-distilled water. The underdrain (plastic backing on the bottom of the plate) is removed using toothed forceps to allow thorough washing.

13. Store plates upside down in a dark place and allow to dry. Red-colored spots, corresponding to individual ASCs, on the dried plates are enumerated manually or using an automated plate reader (*see* **Note 18**).

4 Notes

1. Other antigens can be used but the optimal dose for immunization and detection via ELISPOT would have to be empirically determined.
2. The toothed forceps will help with dissection of femur and tibia for bone marrow, specifically dislocation of femur from the hip joint and from the tibia.
3. Ethanol evaporates, so prepare fresh working solution (35 %) from stock solutions (95–99.5 %).
4. If antigen-specific ASCs are enumerated, capture antigen will vary depending on the antigen used for vaccination. To measure the frequency of *total* ASCs in the sample, anti-mouse/human Ig antibody can be used.
5. Different subclasses of antibody can be measured using subclass-specific detection antibody. If enzyme-conjugated antibody is unavailable, biotin-conjugated antibody can be used followed by an additional step of streptavidin-HRP (*see* **Note 17**).
6. Depending on the experimental set-up, animals may be anesthetized before vaccination using an inhalational anesthetic such as isoflurane.
7. Booster vaccination will recruit memory B-cells and subsequently result in a robust antibody response. If this step is omitted, the ELISPOT will have to be performed within a week of the primary vaccination.
8. Dissecting and removing the leg muscles will allow better access to the bones.
9. Alternatively, the back end of a syringe plunger can be used to disrupt the spleen through a cell strainer. Avoid using excessive force, which will decrease cell viability. Gentle pressure is enough to disintegrate the spleen into a cell suspension. Filtering the cells will reduce nonspecific spots from debris/artifacts.

10. PVDF membranes are hydrophobic and need to be activated. Ethanol makes the membrane hydrophilic and enhances its protein-binding capacity. Do not let the membranes dry after this step. Incubation with ethanol should be less than 1 min. If multiple plates need to be coated, work with only two plates at a time. If using hydrophilic membranes such as mixed cellulose ester, pre-wetting with ethanol is not required.
11. During wash steps do not allow pipette tips to touch the membrane. The tips may scratch or perforate the membranes. If using an automated washer, make sure that it is compatible with ELISPOT plates. Always press the washer into the corner, not into the middle of the wells.
12. Antigens and antibodies vary in their optimal coating concentration. Generally 10–15 μg/mL is a good starting point, but titration testing should be done for optimization of the reagent used.
13. The number of cells to be added per well depends on the type of cells involved and the expected frequency of ASCs. At least 0.5×10^6 cells/well should be used. The goal is to end up with distinct, well-defined single spots after developing the plate. Too many cells will cause the cells to form layers and prevent accurate spot formation. Typically, 1 week after booster vaccination in mice, frequency of Ig-secreting cells is higher in spleen as compared to bone marrow. Remember to include appropriate negative controls. Cells from mice injected with only PBS or a non-relevant antigen will suffice. To control for background signal due to reagents, assign wells that will be treated with all reagents except cells. If wells cannot hold >200 μL volume, the dilutions can be done in a separate tissue culture plate and transferred to ELISPOT plates.
14. Serial dilutions will help determine the optimal number of cells required per well and performing triplicates will improve accuracy and reproducibility.
15. Incubation times may have to be optimized for specific experiments. Place plates in a single layer in the incubator. The incubator should not be disturbed for the duration of the incubation. Shaking or vibration of plates during incubation can result in streaks and ambiguous/doubled spots.
16. Wash buffer includes Tween-20, which is helpful for reducing background and therefore washing instructions must be followed carefully. Additional steps such as doing additional washes or soaking and then washing wells with wash buffer can remedy high background problems.
17. Concentrations of antibodies to detect other subclasses will have to be determined empirically. Usually ≤1 μg/mL is adequate. If required biotinylated antibodies (1 μg/mL) followed

by streptavidin-HRP (1/1,000) can be utilized to amplify the signal and sensitivity, which would be particularly useful when using human cells.

18. Allow plates to dry completely for accurate detection of spots. Spots on dried plates do not fade away for years if they are kept protected from light and in a dry place.

References

1. Czerkinsky CC, Nilsson LA, Nygren H, Ouchterlony O, Tarkowski A (1983) A solid-phase enzyme-linked immunospot (ELISPOT) assay for enumeration of specific antibody-secreting cells. J Immunol Methods 65(1–2): 109–121
2. Doria-Rose NA, Klein RM, Manion MM, O'Dell S, Phogat A, Chakrabarti B, Hallahan CW, Migueles SA, Wrammert J, Ahmed R, Nason M, Wyatt RT, Mascola JR, Connors M (2009) Frequency and phenotype of human immunodeficiency virus envelope-specific B cells from patients with broadly cross-neutralizing antibodies. J Virol 83(1):188–199. doi:10.1128/JVI. 01583-08
3. Kelly DF, Snape MD, Clutterbuck EA, Green S, Snowden C, Diggle L, Yu LM, Borkowski A, Moxon ER, Pollard AJ (2006) CRM197-conjugated serogroup C meningococcal capsular polysaccharide, but not the native polysaccharide, induces persistent antigen-specific memory B cells. Blood 108(8):2642–2647. doi:10.1182/blood-2006-01-009282
4. Malaspina A, Moir S, Orsega SM, Vasquez J, Miller NJ, Donoghue ET, Kottilil S, Gezmu M, Follmann D, Vodeiko GM, Levandowski RA, Mican JM, Fauci AS (2005) Compromised B cell responses to influenza vaccination in HIV-infected individuals. J Infect Dis 191(9): 1442–1450. doi:10.1086/429298
5. Crotty S, Felgner P, Davies H, Glidewell J, Villarreal L, Ahmed R (2003) Cutting edge: long-term B cell memory in humans after smallpox vaccination. J Immunol 171(10): 4969–4973
6. Smith K, Garman L, Wrammert J, Zheng NY, Capra JD, Ahmed R, Wilson PC (2009) Rapid generation of fully human monoclonal antibodies specific to a vaccinating antigen. Nat Protoc 4(3):372–384. doi:10.1038/nprot.2009.3
7. Slifka MK, Ahmed R (1996) Limiting dilution analysis of virus-specific memory B cells by an ELISPOT assay. J Immunol Methods 199(1): 37–46
8. Vallerskog T, Gaines H, Feldman A, Culbert E, Klareskog L, Malmstrom V, Trollmo C (2008) Serial re-challenge with influenza vaccine as a tool to study individual immune responses. J Immunol Methods 339(2): 165–174. doi:10.1016/j.jim.2008.09.008
9. Jahnmatz M, Kesa G, Netterlid E, Buisman AM, Thorstensson R, Ahlborg N (2013) Optimization of a human IgG B-cell ELISpot assay for the analysis of vaccine-induced B-cell responses. J Immunol Methods 391(1–2):50–59. doi:10.1016/j.jim.2013.02.009
10. Walsh PN, Friedrich DP, Williams JA, Smith RJ, Stewart TL, Carter DK, Liao HX, McElrath MJ, Frahm N (2013) Optimization and qualification of a memory B-cell ELISpot for the detection of vaccine-induced memory responses in HIV vaccine trials. J Immunol Methods 394(1–2):84–93. doi:10.1016/j.jim.2013.05.007
11. Devera TS, Shah HB, Lang GA, Lang ML (2008) Glycolipid-activated NKT cells support the induction of persistent plasma cell responses and antibody titers. Eur J Immunol 38(4): 1001–1011. doi:10.1002/eji.200738000
12. Shah HB, Joshi SK, Rampuria P, Devera TS, Lang GA, Stohl W, Lang ML (2013) BAFF- and APRIL-dependent maintenance of antibody titers after immunization with T-dependent antigen and CD1d-binding ligand. J Immunol 191(3):1154–1163. doi:10.4049/jimmunol.1300263

Chapter 43

T Cell ELISPOT: For the Identification of Specific Cytokine-Secreting T Cells

Kerry M. Leehan and Kristi A. Koelsch

Abstract

The ELISPOT is a powerful functional assay used to detect biological activity and immunological secretions from immune cells. In this chapter, we specifically discuss T cell ELISPOT methods for the detection of secreted cytokines. A detailed protocol is given enabling the detection of interferon gamma-secreting $CD8^{+}$ T cells and/or peripheral blood mononuclear cells following their isolation and polyclonal activation. Included is a brief discussion on choosing the activation method for your T cell ELISPOT assay, as well as additional instructions for the adaptation of this protocol for the study of memory and antigen-specific T cell responses.

Key words ELISPOT, T cell, T cell activation, PBMC, Polyclonal activation, Cytokine, Cytokine-secreting cells, Interferon gamma, IFN-γ, Specific peptide/antigen response, Spot-forming-cell

1 Introduction

The enzyme-linked immunospot (ELISPOT) assay is an extremely sensitive membrane-based assay that is useful for detecting secreted proteins in cellular immune responses. The assay is based on the same theory as the enzyme-linked immunosorbent assay (ELISA), but is useful for the detection of protein secretion from single cells. The assay is both qualitative (detects specific proteins, such as cytokines or antigen-specific antibodies) and quantitative (it is useful for identifying the frequency of secreting cells within a given population).

There are many permutations of the assay design depending on the cell type and protein being detected. The basic principle is as follows: an analyte-specific antibody is bound to the ELISPOT membrane and cells with specific or nonspecific (control) stimulants are incubated in a monolayer in individual wells. Cells responding to the antigen (stimulant) secrete cytokines, which are captured by the membrane-bound antibody. After incubation, the cells are removed, and the analyte of interest is detected by an

Biji T. Kurien and R. Hal Scofield (eds.), *Western Blotting: Methods and Protocols*, Methods in Molecular Biology, vol. 1312, DOI 10.1007/978-1-4939-2694-7_43, © Springer Science+Business Media New York 2015

enzyme-conjugated detection antibody. A chromogenic substrate is then added, causing the precipitation of colored dots where the specific stimulation occurred.

T cell ELISPOTS have been utilized to great effect in the detection of antigen-specific T cells [1], memory T cell responses [2], and identification of T cell cancer epitopes [3, 4]. The advantages of the ELISPOT over supernatant-based assays are the ability to enumerate cells within a given population that respond to stimulation, as well as the strength of that response (in spot size and intensity), and the ability to identify rare populations. The following protocol gives instruction for the isolation and polyclonal stimulation of human $CD8^+$ T cells and or peripheral blood mononuclear cells (PBMCs), followed by the detection of interferon gamma (IFN-γ) secretion by responder cells using a single-color ELISPOT assay. Additional information is included to guide the adaptation of this protocol for the study of memory and antigen-specific T cell responses.

2 Materials

2.1 Cell Isolation

2.1.1 For Whole PBMC

1. Fresh human blood (whole blood 30–40 mL) or buffy coat.
2. Lympholyte H Cell Separation Media.
3. Sterile PBS (1×), pH 7.2.
4. PBMC wash buffer: Sterile PBS (1×) containing 1 % bovine serum albumin (BSA).
5. 50 mL conical tubes.

2.2 For CD8+ T Cells

1. Anti-human CD8 magnetic isolation beads (Miltenyi Biotec, San Diego, CA).
2. LS MACS column (Miltenyi Biotec).
3. MACS buffer: 1× sterile PBS + 5 % FCS, degassed.
4. Catch tubes: 15 mL conical tubes (sterile).

2.3 Cell Stimulation

Use PMA/ionomycin, OR PHA stimulation (*see* **Note 1**).

1. Phorbol 12-myristate 13-acetate (PMA).
2. Ionomycin.
3. Phytohemagglutinin (PHA).

2.4 ELISPOT

1. EMD-Millipore Multiscreen Immobilon-PVDF 96-well assay plate (EMD Millipore, Danvers, MA).
2. Capture and detection antibodies: Human IFN-γ ELISPOT pair (Becton-Dickinson, Franklin Lakes, NJ).
3. 35 % ethanol.
4. ELISPOT coating and wash buffer: Sterile PBS (1×).
5. Dilution buffer: 1× sterile PBS, 10 % FBS, sterile filtered.

6. ELISPOT medium and blocking buffer: RPMI 1640 supplemented with 2 mM L-glutamine, 1 mM HEPES, 10 % FCS, 1× MEM NEAA, 1 mM sodium pyruvate, and 1× penicillin/streptomycin.
7. Enzyme conjugate: Steptavidin HRP (Becton Dickinson #557630, Franklin Lakes, NJ).
8. Substrate solution: AEC reagent set (Becton Dickinson, Franklin Lakes, NJ).
9. Complete human T cell medium (CTCM): RPMI 1640 supplemented with 2 mM L-glutamine, 1 mM HEPES, 10 % human AB serum, 1× MEM NEAA, 1 mM sodium pyruvate, and 1× penicillin/streptomycin.

2.5 Miscellaneous

1. Tissue culture incubator, 37 °C, 5 % CO_2.
2. Plate washer or multichannel pipette.
3. Sterile filter tips.
4. Biological safety cabinet.
5. Hemocytometer and trypan blue for cell counting.
6. Microscope for cell counting.
7. Dissection microscope or ELISPOT plate reader for visualization and analysis of spots.

3 Methods

Only those appropriately trained in the handling of human blood and infectious substances should complete this work. Human blood should always be treated as infectious, and researchers should use correct personal protective equipment.

All work requires aseptic technique and should be performed in a biosafety cabinet until the live cells are removed from the ELISPOT plate in Subheading 3.5, **step 1**.

This protocol can be adapted for memory and antigenic T cell responses (*see* **Note 2**).

Experimental Time Line

- Day 1: Isolate and purify PBMC/T cells, wet out and coat ELISPOT plate (steps A–C).
- Day 2: Block and load ELISPOT plate (steps D and E).
- Day 3–5: Develop ELISPOT plate (step F, IFN-γ requires 24–48 h).

3.1 Cell Isolation

3.1.1 Isolation of Human PBMCs

1. Bring Lympholyte H Cell Separation Media (1.077 g), 1× PBS (sterile), and whole blood or buffy coat to room temperature.
2. Dilute whole blood or buffy coat 1:1 with 1× PBS.

3. Pipette 10 mL Lympholyte H Cell Separation Media into a 50 mL conical tube.
4. Carefully layer 30 mL blood/PBS mixture on top of Lympholyte H Cell Separation Media.
5. Centrifuge at 800 × *g* for 20 min, with NO BRAKE.
6. Carefully remove the PBMC layer.
7. Wash PBMCs with wash buffer (1× PBS, 1 % BSA). Centrifuge at 800 × *g* for 15 min.
8. Repeat PBMC wash and spin, and decant supernatant.
9. Resuspend cells at 1×10^5 in CTCM, overnight at 37 °C (T25 flasks).

3.1.2 Isolation of CD8+ T Cells

1. Determine the total number of PBMCs.
2. Centrifuge at 300 × *g* for 10 min, and then aspirate supernatant completely.
3. Resuspend cell pellets in 80 μL MACS buffer, per 10^7 cells.
4. Add 20 μL anti-CD8 magnetic beads per 10^7 cells.
5. Mix well and incubate for 15 min at 4 °C.
6. Wash with 1.5 mL MACS buffer, per 10^7 cells.
7. Centrifuge at 300 × *g* for 10 min, and then aspirate supernatant completely.
8. Resuspend up to 10^8 cells in 500 μL MACS buffer.
9. Wash LS MACS column three times with 3 mL MACS buffer by applying buffer to column and allowing it to flow through into catch tube. Empty flow-through from catch tube and reposition catch tube under column.
10. Apply cell suspension to column, allowing negative cells to flow through into catch tube.
11. Wash column three times with 3 mL MACS buffer, allowing flow-through to drip into catch tube, and then place on ice. This fraction will represent the "negative" fraction ($CD8^-$ T cells) and will serve as single-color controls for FACS analysis (*see* **Note 3**).
12. Remove column from magnet, add 5 mL MACS buffer, and firmly flush out positive ($CD8^+$) cells into clean catch tube.
13. Centrifuge $CD8^+$ and $CD8^-$ fractions at 200 × *g* for 10 min at 22 °C.
14. Resuspend $CD8^+$ and $CD8^-$ fractions in 5 mL CTCM and count cells using the trypan blue exclusion method.
15. Resuspend $CD8^+$ and $CD8^-$ fractions at 2×10^6 cells/mL in CTCM for overnight culture at 37 °C.

3.2 Pre-wetting and Coating ELISPOT Plates

1. To wet ELISPOT plate, add 15 μL 35 % ethanol to each well, and let it sit for 1 min.
2. Flick plate to remove ethanol.
3. Wash four times with 200 μL sterile ddH_2O.
4. Wash one time with 200 μL 1× PBS (sterile, pH = 7.4).
5. Dilute capture antibody into 1× PBS (1 μL antibody into 200 μL PBS, 5.0 μg/mL final concentration scale up as needed).
6. Add 100 μL of dilute antibody solution to each well.
7. Incubate overnight at 4 °C.

3.3 Blocking ELISPOT Plate

1. Discard coating antibody solution (flick out gently).
2. Wash wells five times with 200 μL/well, 1× PBS.
3. Add 200 μL/well ELISPOT media/blocking buffer. Incubate for 2 h, covered, at room temperature.

3.4 Loading ELISPOT Plate

See plate layout for plate map (Fig. 1).

1. Wash T cells (or PBMC) in ELISPOT media, and centrifuge at 600 × *g* for 7 min. Carefully aspirate supernatant. Repeat.
2. Resuspend T cells at 1×10^6/mL in a minimum of 2 mL (this is tube 1).

3.4.1 T Cell/PBMC Dilutions and Plating

See plate layout for plate map (Fig. 1).

1. Add 1 mL of ELISPOT media to three sterile Eppendorf tubes, labeled 2–4.

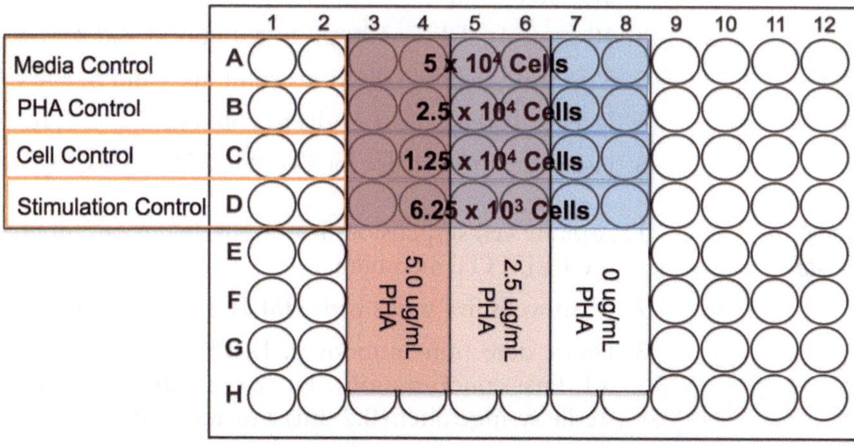

Fig. 1 ELISPOT plate layout for T cell polyclonal stimulation. Control wells are plated in duplicate rows *A–D*, columns *1* and *2*. Controls include media control (no cells or stimulation), PHA control (no cells), cell control (cells with no stimulation), and optional stimulation control (cells with alternate stimulation—PMA and ionomycin). For the assay, four decreasing concentrations of cells are loaded in rows *A–D*, columns *3–8*, and stimulated with three different concentrations of PHA (5.0 μg/mL, columns *3* and *4*; 2.5 μg/mL, columns *5* and *6*; 0 μg/mL, columns *7* and *8*). All conditions are loaded in duplicate sets

2. Take 1 mL of cell suspension from tube 1, add to tube 2, and mix well.
3. Take 1 mL of cell suspension from tube 2, add to tube 3, and mix well.
4. Remove 1 mL of cells from tube 3, and move to tube 4. These are your T cell/PBMC dilutions (*see* **Note 4**).
5. Load 100 μL of tube 1 into cells in Row A, columns 3–8.
6. Load 100 μL of tube 2 into cells in Row B, columns 3–8.
7. Load 100 μL of tube 3 into cells in Row C, columns 3–8.
8. Load 100 μL of tube 4 into cells in row D, columns 3–8.

3.4.2 Control Well Plating

See plate layout for plate map (Fig. 1).

1. Cells A1/A2: Add 200 μL ELISPOT media (no cells, media control).
2. Cells B1/B2: Add 100 μL PHA (5 μg/mL) solution, and 150 μL ELISPOT media (PHA control).
3. Cells C1/C2: Add T cells/PBMC alone (100 μL of a 5×10^5/mL dilution), and 100 μL ELISPOT media (baseline response, cell control).
4. Cells D1/D2: (100 μL of a 5×10^5/mL dilution) T cells/PBMC + 2 ng/mL PMA, 300 ng/mL ionomycin (optional, stimulation control) (*see* **Notes 1**, **5**, and **6**).

3.4.3 Loading Stimulants

1. Take stock PHA (20 mg/mL) and mix 1 mL of solution at each of the following concentrations: deliver 100 μL to the indicated well.
2. 10 μg/mL, deliver 100 μL to columns 3 and 4, rows A–D.
3. 5 μg/mL, deliver 100 μL to columns 5 and 6, rows A–D.
4. Deliver 100 μL plain media to columns 7 and 8, rows A–D.
5. Place plate in incubator for 24–48 h, 37 °C, 5 % CO_2 (*see* **Note 7**).

3.5 Develop ELISPOT Plate

1. Aspirate cell suspension from wells, being careful not to touch the ELISPOT membrane.
2. Wash wells five times with 200 μL sterile 1× PBS.
3. Dilute detection antibody in 1× PBS (1:250 dilution 2.0 μg/mL final concentration). Filter the antibody solution using a sterile syringe filter (0.2 μm) and add 100 μL per well.
4. Replace lid, and incubate for 2 h at room temperature.
5. Discard detection antibody solution.
6. Wash wells five times with 200 μL sterile PBS (1-min soak, pre-filtered).
7. Dilute SAV-HRP enzyme conjugate in dilution buffer (1:100 dilution).

8. Cover and incubate for 1 h at room temperature.
9. Discard enzyme conjugate solution from wells.
10. Wash wells five times with 200 μL sterile PBS (let each wash sit for 1 min before proceeding to the next wash).
11. Remove plate underdrain, and prop plate so the membranes do not touch anything.
12. Add 100 μL substrate solution to each well.
13. Monitor spot development for 5–60 min (do not allow to overdevelop).
14. Stop reaction by washing extensively with DI tap water (minimum of ten fill-and-flicks).
15. Air-dry plate overnight in the dark, upside down.
16. When plate is completely dry, spots can be counted on a dissection microscope or using an automated ELISPOT reader.

4 Notes

1. Special considerations should be given to the method of stimulation and activation in T cells. The approaches to nonspecific T cell stimulation are numerous and beyond the scope of this chapter to discuss in detail. However, two alternative stimulation reagents (PMA/ionomycin and PHA) are given in this protocol. Both approaches will give nonspecific activation of T cells, but it is important for one to consider the different mechanisms that drive activation by each method and decide which would be most appropriate for the intended assay. PHA is a lectin that activates by cross-linking the T cell receptor (TCR) and co-stimulatory molecules, resulting in a TCR signaling cascade and subsequent activation. Alternatively, PMA and ionomycin both cross the cell membrane and directly activate protein kinase C (PKC) and NFAT pathways, bypassing the requirement for TCR signaling, causing activation and calcium flux via the ionophore, ionomycin. PMA/ionomycin stimulation is robust and rapid, but, for incubations longer than 24 h, will result in loss of cell viability, which is an important factor in successful ELISPOT assays. In those cases requiring longer incubation periods, PHA stimulation would be preferred over PMA/ionomycin.
2. To adjust protocol for memory and antigen-specific responses: When testing T cells of unknown specificity against peptide antigens, the antigen-presenting cell (APC) chosen must be capable of surface loading of antigenic peptides. Adequate time between peptide introduction and co-culture of T cells and APCs must elapse to ensure that peptides are loaded into MHC. When testing whole antigen, APCs must be capable of uptake, processing, and presentation of antigen and antigenic peptides.

MHC restriction of both T cells and APCs must match. When testing peptide or whole antigen, controls with irrelevant peptide, no peptide, irrelevant protein, and no protein must be included on the ELISPOT plate. Chosen APC and T cells must be plated alone, with and without stimulants, to determine background activity. This is particularly important when using human donors, as lymphocyte activation status varies widely from person to person.

3. Prior to running the ELISPOT assay, FACS analysis can be used to verify cell isolation purity. The negative population from this step can serve as single-color controls for compensation analysis.
4. Optimal cell density will be dependent on the cell type and the anticipated frequency of responders within the population being assayed. Total PBMCs should be plated at a density of 50–100,000 per well. Purified CD8+ T cells should be plated at a density of 6–50,000 depending on the specificity of the stimulation. To determine optimal density, cells should be titrated using PMA + ionomycin stimulation.
5. Appropriate controls must be included in every plate: stimulated cells, unstimulated cells, and each individual component of the experiment on its own. This helps to determine sources of background, which can be problematic with this assay.
6. For optimal reproducibility, run each sample in duplicate, or triplicate. In addition, total well volume should never be more than 200 μL per well. Increased volumes can cause the membrane to come into contact with the underdrain, leading to artifacts during development.
7. It is very important not to disturb the plate while incubating. Even slight vibrations, from a centrifuge adjacent to the incubator, for example, can lead to cell vibration, causing double-spotting and diffuse, difficult-to-read spots.

References

1. Hecht TT, Longo DL, Matis LA (1983) The relationship between immune interferon production and proliferation in antigen-specific, MHC-restricted T cell lines and clones. J Immunol 131(3):1049–1055
2. Heeger PS, Greenspan NS, Kuhlenschmidt S, Dejelo C, Hricik DE, Schulak JA, Tary-Lehmann M (1999) Pretransplant frequency of donor-specific, IFN-gamma-producing lymphocytes is a manifestation of immunologic memory and correlates with the risk of posttransplant rejection episodes. J Immunol 163(4):2267–2275
3. Pass HA, Schwarz SL, Wunderlich JR, Rosenberg SA (1998) Immunization of patients with melanoma peptide vaccines: immunologic assessment using the ELISPOT assay. Cancer J Sci Am 4(5):316–323
4. Scheibenbogen C, Lee KH, Stevanovic S, Witzens M, Willhauck M, Waldmann V, Naeher H, Rammensee HG, Keilholz U (1997) Analysis of the T cell response to tumor and viral peptide antigens by an IFNgamma-ELISPOT assay. Int J Cancer 71(6):932–936

Chapter 44

Membrane Microplates for One- and Two-Color ELISPOT and FLUOROSPOT Assays

Alexander E. Kalyuzhny

Abstract

Membranes are widely used as protein blotting matrices for a large variety of research applications including western blotting and enzyme-linked immunospot assay (ELISPOT). The largest advantage of using membranes versus solid plastic support is the porosity of membranes allowing for immobilization of high concentrations of proteins and antibodies which, in turn, increases the sensitivity of detection. Similar to plastic surfaces, polyvinylidene difluoride (PVDF) and nitrocellulose membranes create good microenvironment for live cells cultured in vitro and do not interfere with cellular physiology. It appears that PVDF-backed microplates are a golden standard for ELISPOT assays: such plates are inexpensive, easy to use and after assay development, membranes can be removed from the plates and archived. Given the convenience and reliability of membrane microplates, they are widely used in ELISPOT assays for basic research and clinical trials. The ELISPOT assay is an antibody "sandwich" technique aimed at trapping cell-secreted molecules between capture and detection antibodies, followed by either chromogenic enzymatic or fluorescence detection. This review covers the principles of the ELISPOT assay on membrane microplates including single-color and two-color detection techniques with the emphasis on assay design, choosing membrane microplates, and troubleshooting protocols.

Key words ELISPOT-FLUOROSPOT, Membrane microplates, Detection antibodies, Capture antibodies, Chromogenic substrates, Single-color ELISPOT, Two-color ELISPOT, Spot forming cells, Quantification of spots, ELISPOT readers, Membrane removal device

1 Introduction

1.1 ELISPOT Assay Historic Background

In 1983 Sedgwick and Holt [1, 2] published a paper describing a novel technique to quantify the number of antibody-secreting B cells in vitro. The novel assay has a very simple and elegant design utilizing a concept of counting "footprints" left by protein-secreting cells cultured in vitro. Such "footprints" were the product of the reaction between the immobilized antigen on the bottom of the plate with B cell-secreted antibodies. Unlike ELISA, the readout in the novel assay was distinct and easy to count color spots: the higher the number of spots, the more antibody-secreting B cells. Later in 1983, another paper describing a similar antibody

Biji T. Kurien and R. Hal Scofield (eds.), *Western Blotting: Methods and Protocols*, Methods in Molecular Biology, vol. 1312, DOI 10.1007/978-1-4939-2694-7_44, © Springer Science+Business Media New York 2015

detection technique was published by Czerkinsky and colleagues [3], who coined the name for this assay: Enzyme-Linked Immunospot, or ELISPOT. A fascinating historic overview of ELISPOT assay development has been recently published by its original inventor Sedgwick [2]. Currently ELISPOT assays can be used either for quantification of antibody-secreting B cells, or cells secreting protein antigens (e.g., T-cells secreting cytokines or glial cells and macrophages secreting growth factors). ELISPOT is a very sensitive assay and some researchers report that it is much more sensitive than ELISA for the detection of cell-secreted cytokines. Whereas ELISA requires a minimum of 10^4 cells to detect secreted cytokines, ELISPOT allows detection of cytokines secreted from as few as 10–100 cells in a well [4]. Such a high sensitivity makes ELISPOT a technique of choice for the detection of spontaneous and antigen-induced secretion of cytokines from peripheral blood lymphocytes [5, 6]. Using ELISPOT and ELISA in parallel allows for measuring the average amount of cytokines secreted by a single cell [7]. ELISPOT can be used for both research [8–17] and human diagnostic application such as testing for tuberculosis in HIV-positive individuals [18] and allergies to metals [19].

A comprehensive collection of articles on various applications of ELISPOT assay can be found in recently published first and second editions of *Handbook of ELISPOT* [20, 21].

1.2 ELISPOT Assay Protocols

Like ELISA, ELISPOT is a sandwich immunoassay but ELISPOT readout is different: it does not measure the amount of secreted cytokines [22] but rather detects the number of cytokine-secreting cells [2, 3]. Another big difference between ELISA and ELISPOT is that the latter is a combination of immunoassay and bioassay. The quality of the readout depends on the viability of cells cultured in the microplate. ELISA is just an immunoassay which uses cell-free samples like cell culture conditioned media or/and cell lysates. Its performance entirely depends on the quality of capture and detection antibodies, enzyme conjugates, and color reagents. In addition to viability of cultured cells, specificity of capture and detection antibodies, enzyme/fluorescent conjugates play a critical role in the performance of the ELISPOT assay as well. Key components of the chromogenic ELISPOT assay include (1) antibodies (capture and detection), (2) enzyme conjugates (e.g., streptavidin-HRP, streptavidin-AP), (3) chromogenic substrates (e.g., (3-amino-9-ethylcarbazole, $C_{14}H_{14}N_2$) (AEC), 5-bromo-4-chloro-3-indolylphosphate *p*-toluidine salt, $C_{15}H_{15}N_2O_4PClBr$ (BCIP) and nitroblue tetrazolium chloride, $C_{40}H_{30}N_{10}O_6Cl_2$ (NBT)), and (4) synthetic membranes used for immobilization of capture antibodies and cell support. When developing an ELISPOT assay, all four components mentioned above should be optimized to generate spots with a high signal-to-noise ratio, enabling recognition by a machine vision ELISPOT reader system. Ideally, spots should have a strong staining intensity and have a high contrast [23]. It is also desirable that spots have a small diameter so they don't merge with each other because merged spots will be recognized by ELISPOT reader as single spots, affecting the accuracy of spot quantification.

2 Antibodies

In sandwich immunoassays, capture and detection antibodies are utilized to trap protein molecules, allowing their subsequent measurement. Capture (immobilized on the solid support) and detection antibodies (in solution) represent a so-called match pair. To avoid competition between these antibodies for binding to the protein molecule, it is necessary to use capture and detection antibodies raised against nonoverlapping epitopes of the target protein [22]. Therefore, it may not be recommended to use the same monoclonal antibody for both capture and detection in the same ELISPOT assay unless the protein of interest has multiple identical epitopes in its structure. Host species are not important and both monoclonal and polyclonal antibodies can be successfully used in ELISPOT assays as capture and detection antibodies. These antibodies can be raised either against the large antigen molecule (e.g., anti-recombinant protein antibodies) or against a small fragment of the antigen (e.g., anti-peptide antibodies). Interestingly, antibodies that produce good staining in such applications as immunohistochemistry, western blotting, and even ELISA may not work for ELISPOT assay. Usually capture antibodies work within a wide range of concentration (1–30 μg/mL), whereas detection antibodies work at much lower concentrations (0.1–0.5 μg/mL). Detection antibodies should be conjugated to biotin to enable interaction with streptavidin-conjugated enzymes in chromogenic assays [24] and to fluorescent tags in FLUOROSPOT assays. Biotinylated detection antibodies allow the utilization of enzyme conjugated streptavidin rather than a third layer of anti-detection antibodies which can sometimes cross-react with capture antibodies, especially when both capture and detection antibodies are raised in the same species (e.g., mouse). Antibodies (e.g., anti-mouse) conjugated to enzyme may bind to both capture and detection antibodies rather than binding to detection antibodies only. Another approach would be to use detection antibodies directly conjugated to HRP and AP enzymes.

2.1 Enzyme Conjugates in Chromogenic Assay

In chromogenic ELISPOT assay streptavidin conjugates of horseradish peroxidase (HRP) and alkaline phosphatase (AP) are frequently used for spot development in single-color assays (Fig. 1) [24]. HRP (optimum pH 7.6) in the presence of hydrogen peroxide (H_2O_2) catalyzes the oxidation of substrates, causing them to change color with the loss of electrons. The advantage of using

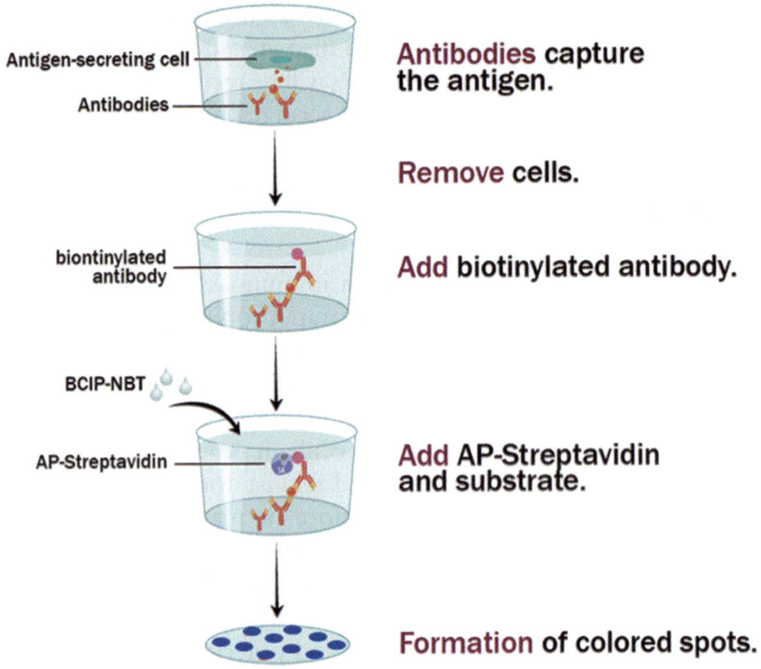

Fig. 1 Schematics of single-color ELISPOT assay

HRP is its high turnover rate (spots develop faster), whereas the drawback is increased background. Unlike HRP, AP (optimum pH 9.0–9.6) has a linear reaction rate (spots develop slower) allowing for longer incubations with chromogenic substrates [24] without a risk of developing background staining. Longer incubation may be done if it is necessary to increase the sensitivity of AP-based assay. By combining HRP and AP, the ELISPOT assay can be used for simultaneous detection of two different cell-secreted molecules (Fig. 2) [25–27]. The major drawback of multianalyte systems is the loss of sensitivity for each of the antigens. In a previous study we found that a number of spots formed by IL-2 and IFN-gamma secreted from peripheral blood mononuclear cells in the plate coated with anti-interleukin-2 (IL-2) and anti-interferon (IFN)-gamma antibodies were noticeably lower in comparison to corresponding single-cytokine assays. We also reported that the drop in sensitivity becomes even more profound if the ELISPOT plate is coated with more than two capture antibodies [28]. The mechanism underlying this phenomenon is not clear and we hypothesize that the binding of secreted cytokines to capture antibodies has a negative impact on feedback regulation and affects production and secretion of other cytokines in the cells. If this is the case, then increasing the sensitivity of the ELISPOT assay by using better capture antibodies, more efficient enzyme conjugates and chromogenic substrates could become a solution in multiplex ELISPOT assays.

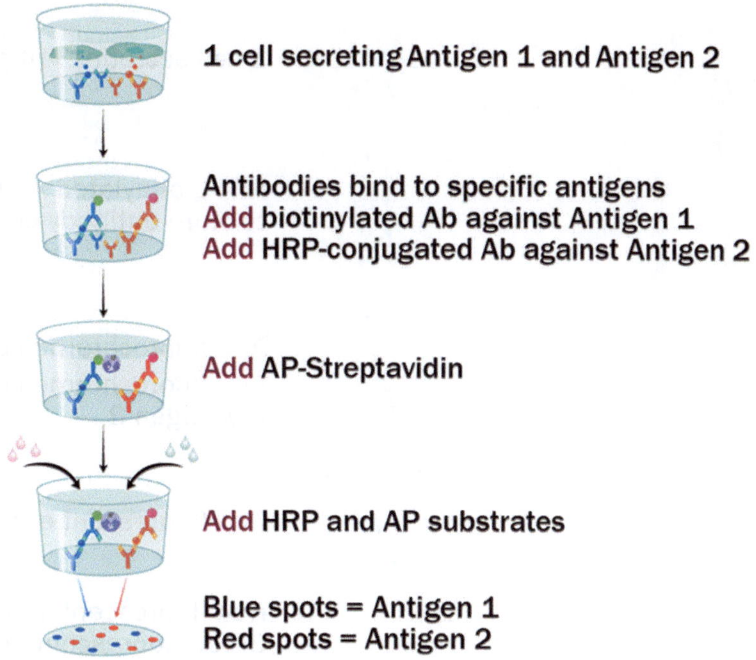

Fig. 2 Schematics of two-color ELISPOT assay

2.2 Enzyme Substrates

Regardless of which enzyme conjugate is used, their corresponding substrates should produce intense and stable colors, and minimal nonspecific background staining. The popular HRP substrate AEC can produce intense red/brown color spots [24]. However AEC is unstable and spots may decolorize over short period of time, which in turn results in loss of primary experimental data. Another HRP substrate, DAB (3,3′-diaminobenzidine, $C_{12}H_{14}N_4$), produces dark brown color spots, which, unlike AEC spots, have a lower contrast [24]. Unlike AEC, spots formed by DAB are stable, but DAB is a carcinogen, which may be a major drawback in using this chromogen. It appears that the most frequently used substrate for AP is a mixture of BCIP with NBT. BCIP/NBT is converted by AP into insoluble black-blue spots of high contrast and intensity [24]. Spots formed by BCIP/NBT do not fade over long periods of time allowing repeat quantification of spots on archived ELISPOT plates or membranes removed from them.

2.3 FLUOROSPOT: Fluorescence Assay

Detection of cell-secreted proteins can be also done using fluorescent probes. This modification of ELISPOT was given a name FLUOROSPOT and employs anti-cytokine detection antibodies conjugated to fluorescent probes (Fig. 3). Like ELISPOT, FLUOROSPOT allows for the simultaneous detection of two different cytokines secreted by the same pool of cultured immune system cells [29, 30]. The advantage of FLUOROSPOT over

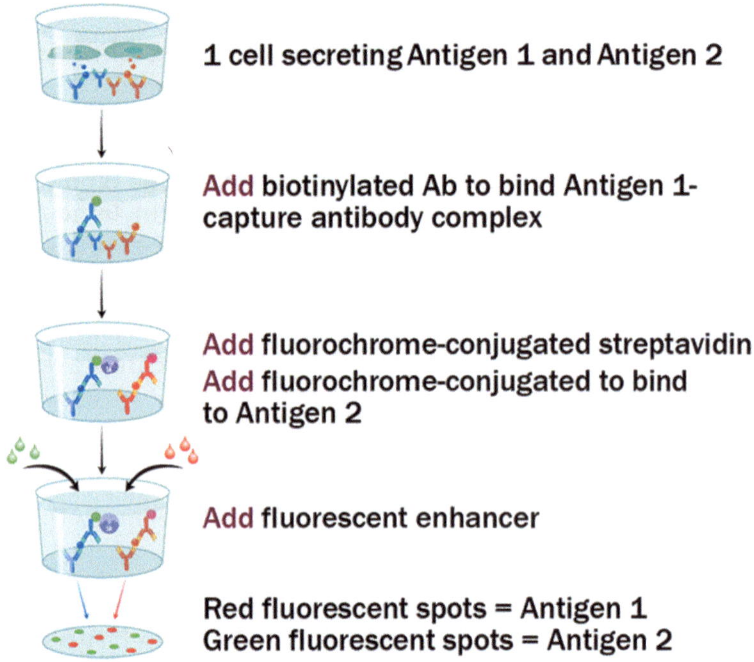

Fig. 3 Schematics of two-color FLUOROSPOT assay

ELISPOT is that the former is better suited for detecting overlapping spots produced by different cytokines by visualizing spots using different color channels. In ELISPOT, this is always a challenge because overlapping red and blue spots obscure each other making it difficult for machine vision-based ELISPOT readers to recognize such spots. On the other hand, the advantage of ELISPOT is that it is more sensitive than FLUOROSPOT due to a "snowball" character of chromogen precipitation via enzymatic reaction. ELISPOT and FLUOROSPOT assays use the same protocols with regard to coating plates with capture antibodies, cell culture, and stimulation and washing steps.

2.4 Cell Culture in ELISPOT Assay

To study protein-secreting capacity of the cells, they can be processed in two ways: (1) cells are cultured in a culture flask and then transferred into ELISPOT plates [31–33], or (2) cells are plated directly into ELISPOT plate in which they are cultured and stimulated with mitogens [34, 35]. If cells are cultured and stimulated outside ELISPOT plate, they need to be washed with fresh culture media before plating them into ELISPOT plate to avoid background staining (*see* **Notes 1** and **2**).

2.5 Membrane Microplates

Although ELISPOT assays can be performed using all-plastic plates [4, 36], plastic plates backed with either PVDF [37, 38] or nitrocellulose membranes [36, 39] are considered to be plates of choice for this assay. In spite of the fact that membrane backed plates were originally designed for filtration purposes, it turned out that membranes provide a very good support for cultured cells and don't affect their viability. What is more important is that due to their porous structure membranes have a much higher binding capacity for capture antibodies than all-plastic plates, and white color of membranes on the bottom of the plates provides an excellent contrast background for colored spots (characteristics and attributes of membranes are reviewed in detail by Weiss [40]. In the ELISPOT assay the flow of reagents through or across the membrane is not required, but rather a diffusion of cell-secreted molecules towards capture antibodies immobilized on the membrane. It appears that PVDF-backed plates developed by Millipore is a golden plate standard for ELISPOT and FLUOROSPOT assays: for one- and two-color ELISPOT, it is recommended to use MAIPN0B50 plates and for FLUOROSPOT—S5EJ104I07 plates.

2.6 Archiving Primary ELISPOT Data

After finishing the experiment, ELISPOT plates are considered primary experimental data and it may be required to store them in a safe place. Unfortunately 96-well plates are bulky and their storage requires a lot of space. However, Millipore plates are designed in such a way that after finishing the assay, membranes with colored spots can be removed from the plate for archiving purposes: removed membranes can be laminated and bar-coded, and compactly stored in a binder using a membrane removal device (www.mvspacific.com), which allows quick removal of all membranes from entire ELISPOT plate.

2.7 ELISPOT Assay Formats

There are two major commercial formats of ELISPOT assay: (1) fully developed and optimized ready-to-use kits and (2) a set of capture and detection antibodies, and color developing reagents which researcher can use to develop his or her assay. ELISPOT assay standardization and validation techniques (*see* **Note 3**) are covered in detail by Janetzki et al. [35]. Ready-to-use ELISPOT kit formats are ideal for large-scale clinical trial projects which require convenience and a high degree of accuracy and reliability [41]. R&D Systems, Inc. (www.rndsystems.com) was the first company which designed and introduced ready-to-use ELISPOT kits which include precoated PVDF membrane microplates, wash buffers, detection antibody, enzyme conjugate, and a color reagent.

3 ELISPOT Readout

In the ELISA assay the concentration of the molecules in the sample is determined by measuring the optical density of the color substrate solution filling the wells [22], whereas in ELISPOT, cell-secretion capacity is measured by counting colored spots on the bottom of the well (Fig. 4) [2, 3, 35]. The term "spot forming cells," or SFC,

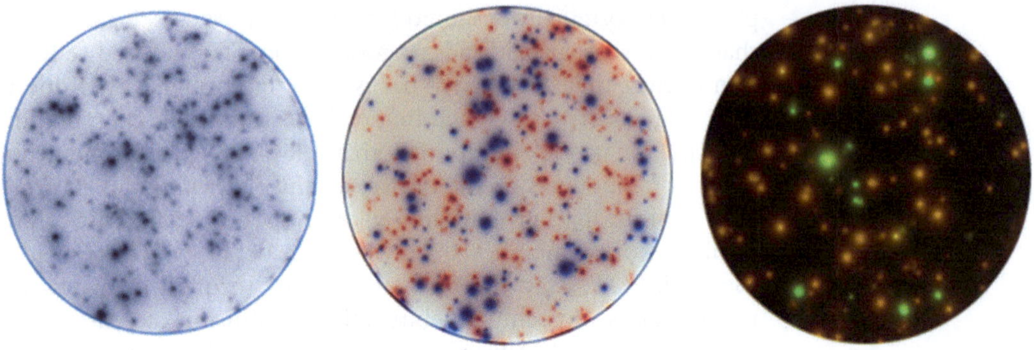

Fig. 4 Appearance of spots in ELISPOT and FLUOROSPOT assays

is used as a quantitative measure of the cell-secretion activity in the ELISPOT assay [42, 43]. After finishing the assay, spots can be counted either manually or by using computer-aided image analysis [44]. Manual counting can be done using, for example, a dissecting microscope but it is very labor-intense, inaccurate, and time-consuming. Computer-aided quantification of spots using machine vision technology appears to be more reliable than the manual counting [44, 45], and currently several automated ELISPOT readers are available from different vendors (e.g., C.T.L., MVS Pacific, and AID): Fig. 5 shows a screenshot of a two-color ELISPOT plate analysis using QuantiHub Color 5.1 ELISPOT reader from MVS Pacific (www.mvspacific.com).

4 Notes

1. *Quality of spots.* The quality of colored spots in ELISPOT assay is a prerequisite for their accurate quantification. It is required that stained spots have high contrast which is a function of strong color of spots and a low background [46]. Background staining may be caused by using low affinity capture antibodies: cytokine molecules secreted from cells dissociate from capture antibodies surrounding the releasing cell diffuse in the well and bind to capture antibodies in the cell-free zone. Background staining can be also caused by disturbing ELISPOT plate with cells during the incubation: frequent opening and closing of the incubator's door will shake the ELISPOT plate and facilitate diffusion of secreted cytokines away from the secreting cell. Another source of background is the adsorption of detection antibodies, enzyme conjugate, and precipitating substrate onto the membrane: preliminary experiments need to be done to determine whether reagents can bind to membranes in ELISPOT plates even without cells. We have found that wrapping ELISPOT

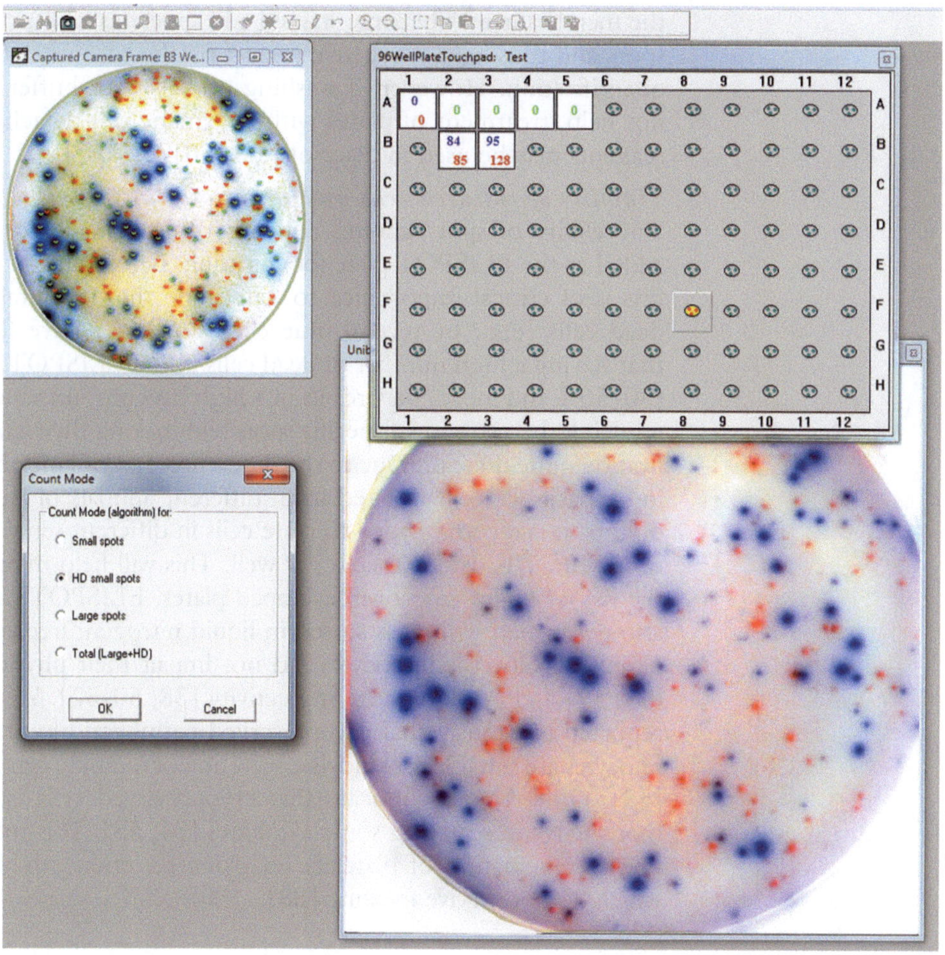

Fig. 5 Screenshot of the graphic-user interface of QuantiHub Color 5.1 software illustrating quantification of two-color ELISPOT images. *Lower right* is the "live" image of the membrane with *colored spots*; *upper right* is the touchpad resembling the layout of the 96-well ELISPOT plate with spot counts in corresponding wells; *Lower left* is the drop-down window allowing to choose a spot quantification mode depending on the size of the spots and their density; *upper left* is the captured image of the membrane with spots which have been already counted: *red spots* are marked with *red labels* and *blue spots* with the *turquoise labels*

plates into aluminum foil reduces background staining, improves contrast of stained spots across the entire membrane, and improves well-to-well reproducibility [47]. It is not clear how such a simple trick as wrapping plates into aluminum foil works, and it is tempting to speculate that aluminum foil facilitates even distribution of heat across the bottom of ELISPOT plate in CO_2 incubator.

2. *ELISPOT plate washing*: After finishing cell incubation, ELISPOT plates must be thoroughly washed to remove cells which may contribute to background staining. It is necessary to wash of as many cells as possible because cells which stick to

the membrane may become stained and appear as false-positive spots and thus have a negative impact on the accuracy of spot quantification. However, if washing of cells with buffer does not help, treatment of plates with enzymatic cell-detachment reagents may be used in place of PBS [56].

3. *Viability of cultured cells and their removal from the plate.* The clarity of spot staining depends on the quality of cells added to the ELISPOT plate for culturing. Before plating cells, this is of critical importance to determine the percentage of dead cells (e.g., by trypan blue exclusion). We have found that having a high number of dead cells in the ELISPOT plate (30–50 % and more) may result in a high background staining, reducing the contrast of specific spots and making their quantification quite difficult. Due to the differences in cytokine-secreting capacity of cells isolated from different animals or human donors, it is recommended to plate cells in different serial dilutions: 10^3, 10^4, 10^5, 10^6 cells per well. This will help minimize the risk of under- and overdeveloped plates. ELISPOT can be also performed with cells stored in liquid nitrogen: freezing of peripheral blood lymphocytes did not impair their physiology [48, 49] and cytokine-secreting activity [38, 50–52]. We have reported previously that cryopreserved peripheral blood lymphocytes are suitable for studying secretion of multiple cytokines [53]. Other reports indicate that cryopreserved cells can be more active in secreting some cytokines [54, 55]. This may be due to elimination of platelets (cytokine secretion inhibitors) which do not survive freezing [38].

References

1. Sedgwick JD, Holt PG (1983) A solid-phase immunoenzymatic technique for the enumeration of specific antibody-secreting cells. J Immunol Methods 57:301–309
2. Sedgwick JD (2005) ELISPOT assay: a personal retrospective. Methods Mol Biol 302: 3–14
3. Czerkinsky CC, Nilsson LA, Nygren H, Ouchterlony O, Tarkowski A (1983) A solid-phase enzyme-linked immunospot (ELISPOT) assay for enumeration of specific antibody-secreting cells. J Immunol Methods 65: 109–121
4. Tanguay S, Killion JJ (1994) Direct comparison of ELISPOT and ELISA-based assays for detection of individual cytokine-secreting cells. Lymphokine Cytokine Res 13:259–263
5. Bienvenu J, Monneret G, Fabien N, Revillard JP (2000) The clinical usefulness of the measurement of cytokines. Clin Chem Lab Med 38:267–285
6. Mashishi T, Gray CM (2002) The ELISPOT assay: an easily transferable method for measuring cellular responses and identifying T cell epitopes. Clin Chem Lab Med 40:903–910
7. Hagen J, Houchins JP, Kalyuzhny AE (2012) Combining ELISPOT and ELISA to measure amounts of cytokines secreted by a single cell. Methods Mol Biol 792:115–122
8. Pass HA, Schwarz SL, Wunderlich JR, Rosenberg SA (1998) Immunization of patients with melanoma peptide vaccines: immunologic assessment using the ELISPOT assay. Cancer J Sci Am 4:316–323
9. Asai T, Storkus WJ, Whiteside TL (2000) Evaluation of the modified ELISPOT assay for gamma interferon production in cancer patients receiving antitumor vaccines. Clin Diagn Lab Immunol 7:145–154
10. Kamath AT, Groat NL, Bean AG, Britton WJ (2000) Protective effect of DNA immunization against mycobacterial infection is associated

with the early emergence of interferon-gamma (IFN-gamma)-secreting lymphocytes. Clin Exp Immunol 120:476–482

11. Eriksson K, Nordstrom I, Horal P, Jeansson S, Svennerholm B, Vahlne A et al (1992) Amplified ELISPOT assay for the detection of HIV-specific antibody-secreting cells in subhuman primates. J Immunol Methods 153: 107–113
12. Howell DM, Feldman SB, Kloser P, Fitzgerald-Bocarsly P (1994) Decreased frequency of functional natural interferon-producing cells in peripheral blood of patients with the acquired immune deficiency syndrome. Clin Immunol Immunopathol 71:223–230
13. Schmittel A, Keilholz U, Thiel E, Scheibenbogen C (2000) Quantification of tumor-specific T lymphocytes with the ELISPOT assay. J Immunother 23:289–295
14. Smith SM, Brookes R, Klein MR, Malin AS, Lukey PT, King AS et al (2000) Human CD8+ CTL specific for the mycobacterial major secreted antigen 85A. J Immunol 165: 7088–7095
15. Pelfrey CM, Cotleur AC, Lee JC, Rudick RA (2002) Sex differences in cytokine responses to myelin peptides in multiple sclerosis. J Neuroimmunol 130:211–223
16. Schmid-Grendelmeier P, Altznauer F, Fischer B, Bizer C, Straumann A, Menz G et al (2002) Eosinophils express functional IL-13 in eosinophilic inflammatory diseases. J Immunol 169: 1021–1027
17. Groth R, Aanonsen L (2002) Spinal brain-derived neurotrophic factor (BDNF) produces hyperalgesia in normal mice while antisense directed against either BDNF or trkB, prevent inflammation-induced hyperalgesia. Pain 100:171–181
18. Chapman AL, Munkanta M, Wilkinson KA, Pathan AA, Ewer K, Ayles H et al (2002) Rapid detection of active and latent tuberculosis infection in HIV-positive individuals by enumeration of Mycobacterium tuberculosis-specific T cells. AIDS 16:2285–2293
19. Jakobson E, Masjedi K, Ahlborg N, Lundeberg L, Karlberg AT, Scheynius A et al (2002) Cytokine production in nickel-sensitized individuals analysed with enzyme-linked immunospot assay: possible implication for diagnosis. Br J Dermatol 147:442–449
20. Kalyuzhny A.E. (2005) Handbook of Elispot: Methods and Protocols. 1 ed. Methods in Molecular Biology, ed. J.M. Walker. Vol. 302. Totowa, New Jersey: Humana Press.
21. Kalyuzhny AE (2011) Handbook of ELISPOT: methods and protocols, 2nd edn. In: Walker JM (ed) Methods in molecular biology, vol 792. Humana, Totowa, NJ
22. Kemeny DM (1997) Enzyme-linked immunoassays. In: Johnstone AP, Turner MW (eds) Immunochemistry 1. Oxford University Press, Oxford, pp 147–175
23. Karulin AY, Lehmann PV (2012) How ELISPOT morphology reflects on the productivity and kinetics of cells' secretory activity. Methods Mol Biol 792:125–143
24. Savage MD, Mattson G, Desai S, Nielander GW, Morgensen S, Conklin EJ (1992) Avidin-biotin chemistry: a handbook. Pierce Chemical Company, Rockford, IL, p 467
25. Okamoto Y, Abe T, Niwa T, Mizuhashi S, Nishida M (1998) Development of a dual color enzyme-linked immunospot assay for simultaneous detection of murine T helper type 1- and T helper type 2-cells. Immunopharmacology 39:107–116
26. Okamoto Y, Gotoh Y, Tokui H, Mizuno A, Kobayashi Y, Nishida M (2000) Characterization of the cytokine network at a single cell level in mice with collagen-induced arthritis using a dual color ELISPOT assay. J Interferon Cytokine Res 20:55–61
27. Okamoto Y, Nishida M (2005) Dual-color ELISPOT assay for analyzing cytokine balance. Methods Mol Biol 302:263–272
28. Palzer S, Bailey T, Hartnett C, Grant A, Tsang M, Kalyuzhny AE (2005) Simultaneous detection of multiple cytokines in ELISPOT assays. Methods Mol Biol 302:273–288
29. Gazagne A, Malkusch W, Vingert B, Fridman WH, Tartour E (2005) Fluorospot assay: methodological analysis. Methods Mol Biol 302:289–296
30. Malkusch W (2005) High resolution as a key feature to perform accurate ELISPOT measurements using Zeiss KS ELISPOT readers. Methods Mol Biol 302:133–150
31. Favre N, Bordmann G, Rudin W (1997) Comparison of cytokine measurements using ELISA, ELISPOT and semi-quantitative RT-PCR. J Immunol Methods 204:57–66
32. Herr W, Schneider J, Lohse AW, Meyer zum Buschenfelde KH, Wolfel T (1996) Detection and quantification of blood-derived CD8+ T lymphocytes secreting tumor necrosis factor alpha in response to HLA-A2.1-binding melanoma and viral peptide antigens. J Immunol Methods 191:131–142
33. Arlen P, Tsang KY, Marshall JL, Chen A, Steinberg SM, Poole D et al (2000) The use of a rapid ELISPOT assay to analyze peptide-specific immune responses in carcinoma patients to peptide vs. recombinant poxvirus

vaccines. Cancer Immunol Immunother 49: 517–529

34. Janetzki S, Song P, Gupta V, Lewis JJ, Houghton AN (2000) Insect cells as HLA-restricted antigen-presenting cells for the IFN-gamma elispot assay. J Immunol Methods 234:1–12
35. Janetzki S, Cox JH, Oden N, Ferrari G (2005) Standardization and validation issues of the ELISPOT assay. Methods Mol Biol 302:51–86
36. Ronnelid J, Klareskog L (1997) A comparison between ELISPOT methods for the detection of cytokine producing cells: greater sensitivity and specificity using ELISA plates as compared to nitrocellulose membranes. J Immunol Methods 200:17–26
37. Schielen P, van Rodijnen W, Tekstra J, Albers R, Seinen W (1995) Quantification of natural antibody producing B cells in rats by an improved ELISPOT technique using the polyvinylidene difluoride membrane as the solid support. J Immunol Methods 188:33–41
38. McCutcheon M, Wehner N, Wensky A, Kushner M, Doan S, Hsiao L et al (1997) A sensitive ELISPOT assay to detect low-frequency human T lymphocytes. J Immunol Methods 210:149–166
39. Taguchi T, McGhee JR, Coffman RL, Beagley KW, Eldridge JH, Takatsu K et al (1990) Detection of individual mouse splenic T cells producing IFN-gamma and IL-5 using the enzyme-linked immunospot (ELISPOT) assay. J Immunol Methods 128:65–73
40. Weiss AJ (2005) Membranes and membrane plates used in ELISPOT. Methods Mol Biol 302:33–50
41. Klencke B, Matijevic M, Urban RG, Lathey JL, Hedley ML, Berry M et al (2002) Encapsulated plasmid DNA treatment for human papillomavirus 16-associated anal dysplasia: a Phase I study of ZYC101. Clin Cancer Res 8: 1028–1037
42. McGhee ML, Ogawa T, Pitts AM, Moldoveanu Z, Mestecky J, McGhee JR et al (1989) Cellular analysis of functional mononuclear cells from chronically inflamed gingival tissue. Reg Immunol 2:103–110
43. Merville P, Pouteil-Noble C, Wijdenes J, Potaux L, Touraine JL, Banchereau J (1993) Detection of single cells secreting IFN-gamma, IL-6, and IL-10 in irreversibly rejected human kidney allografts, and their modulation by IL-2 and IL-4. Transplantation 55:639–646
44. Herr W, Linn B, Leister N, Wandel E, Meyer zum Buschenfelde KH, Wolfel T (1997) The use of computer-assisted video image analysis for the quantification of CD8+ T lymphocytes producing tumor necrosis factor alpha spots in response to peptide antigens. J Immunol Methods 203:141–152
45. Lehmann PV (2005) Image analysis and data management of ELISPOT assay results. Methods Mol Biol 302:117–132
46. Kalyuzhny AE (2005) Chemistry and biology of the ELISPOT assay. Methods Mol Biol 302:15–31
47. Kalyuzhny A, Stark S (2001) A simple method to reduce the background and improve well-to-well reproducibility of staining in ELISPOT assays. J Immunol Methods 257:93–97
48. Merker R, Check I, Hunter RL (1979) Use of cryopreserved cells in quality control of human lymphocyte assays: analysis of variation and limits of reproducibility in long-term replicate studies. Clin Exp Immunol 38:116–126
49. Lewalle P, Rouas R, Lehmann F, Martiat P (2000) Freezing of dendritic cells, generated from cryopreserved leukaphereses, does not influence their ability to induce antigen-specific immune responses or functionally react to maturation stimuli. J Immunol Methods 240: 69–78
50. Keane NM, Price P, Stone SF, John M, Murray RJ, French MA (2000) Assessment of immune function by lymphoproliferation underestimates lymphocyte functional capacity in HIV patients treated with highly active antiretroviral therapy. AIDS Res Hum Retroviruses 16: 1991–1996
51. Smith JG, Liu X, Kaufhold RM, Clair J, Caulfield MJ (2001) Development and validation of a gamma interferon elispot assay for quantitation of cellular immune responses to varicella-zoster virus. Clin Diagn Lab Immunol 8:871–879
52. Sobota V, Bubenik J, Indrova M, Vlk V, Jakoubkova J (1997) Use of cryopreserved lymphocytes for assessment of the immunological effects of interferon therapy in renal cell carcinoma patients. J Immunol Methods 203:1–10
53. Bailey T, Stark S, Grant A, Hartnett C, Tsang M, Kalyuzhny A (2002) A multidonor ELISPOT study of IL-1beta, IL-2, IL-4, IL-6, IL-13, IFN-gamma and TNF-alpha release by cryopreserved human peripheral blood mononuclear cells. J Immunol Methods 270:171–182
54. Venkataraman M (1994) Effects of cryopreservation on immune responses: VII. Freezing induced enhancement of IL-6 production in human peripheral blood mononuclear cells. Cryobiology 31:468–477

55. Venkataraman M (1995) Effects of cryopreservation on immune responses. VIII. Enhanced secretion of interferon-gamma by frozen human peripheral blood mononuclear cells. Cryobiology 32:528–534

56. Grant A, Palzer S, Hartnett C, Bailey T, Tsang M, Kalyuzhny AE (2005) A cell-detachment solution can reduce background staining in the ELISPOT assay. Methods Mol Biol 302:87–94

Chapter 45

SDS-PAGE to Immunoblot in One Hour

Biji T. Kurien, Debashish Danda, Michael P. Bachmann, and R. Hal Scofield

Abstract

An ultra-rapid method for electrophoresing proteins on sodium dodecyl sulfate-polyacrylamide gel electrophoresis, transfer of proteins to nitrocellulose membranes, and immunoblotting is described here. Electrophoresis of the autoantigens La and Ro60, as well as molecular weight standards on a 4–20 % gradient gel, was performed in about 10 min using heated (70–75 °C) normal Laemmli running buffer. Electrophoretic transfer of these proteins was achieved in 7 min using a semidry transfer method. Finally, immunoblotting of La and Ro60 was carried out in 30 min. Thus, the entire process of electrophoresis, electrotransfer, and immunoblotting could be carried out in 1 h.

Key words SDS-PAGE, Western blotting, Nitrocellulose, Molecular weight standards

1 Introduction

Proteins are first separated on the basis of size in sodium dodecyl sulfate-polyacrylamide gel electrophoresis (SDS-PAGE) [1, 2]. Proteins separated on SDS-PAGE can be visualized by either staining with various protein stains or immunoblotting. Electrophoresis of proteins on a mini-gel takes at least 2 h to complete to ensure ideal separation without "smiling" artifacts. However, Haeberle demonstrated electrophoresis of proteins in 5 min with a special gel and a special buffer that was heated to 70 °C [3].

For immunoblotting, the separated proteins are transferred to nitrocellulose or polyvinylidene difluoride membranes. The transfer to membranes has been achieved by (a) simple diffusion [4]; (b) vacuum-assisted solvent flow [5]; and (c) electrophoretic elution [6]. Electrophoretic transfer of proteins, resolved by SDS-PAGE, to nitrocellulose is a fundamental step prior to detection of specific proteins with specific antibodies [7–9]. The protein transfer procedure normally takes about 2–4 h at about 70 V or an overnight transfer at 30 V. High-molecular-weight proteins are often stubbornly resistant to transfer [7] in spite of prolonged runs and this

Biji T. Kurien and R. Hal Scofield (eds.), *Western Blotting: Methods and Protocols*, Methods in Molecular Biology, vol. 1312, DOI 10.1007/978-1-4939-2694-7_45, © Springer Science+Business Media New York 2015

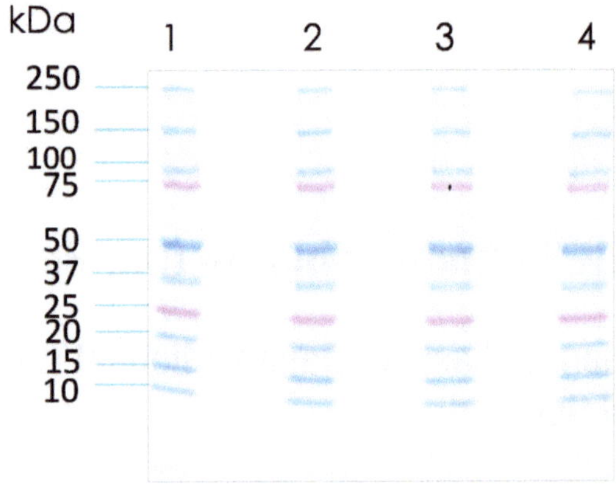

Fig. 1 Ultrafast SDS-PAGE electrophoresis and electrotransfer of proteins to nitrocellulose membrane. Prestained high-molecular-weight proteins were electrophoresed on a 4–20 % gradient gel in 10 min and transferred to nitrocellulose membrane in 7 min. *Lanes 1–4* show prestained high-molecular-weight protein standards

problem is accentuated when higher percentage gels are used. Prolonged electrotransfer (16–20 h) at high current density coupled with inclusion of 0.01 % sodium dodecyl sulfate, to enhance protein elution, has been used to transfer high-molecular-weight proteins in an efficient manner [9, 10].

Trans-blot turbo system employs a semidry method of transfer of proteins from gel to membranes. Proteins of varying sizes (10–250 kDa) can be transferred to membranes in 3–7 min very efficiently [9]. We have shown that both low- and high-molecular-weight proteins can be transferred very efficiently to nitrocellulose membranes in a very short time using heated transfer buffer without methanol [10, 11].

Traditional immunoblotting normally takes about 4–5 h, with 1 h for blocking, 2 h for incubation with primary antibody, 1 h with secondary antibody, 30 min for washing between primary and secondary antibody incubation, and finally development of protein bands with substrate.

Here we show that the entire process of electrophoresis of autoantigens Ro60 and La (actual gel running), western transfer, and immunoblotting with specific autoantibodies can be carried out in 1 h (Figs. 1 and 2), a process normally carried out in a period of up to 2 days.

2 Materials

Prepare all solutions using ultrapure water (prepared by purifying deionized water, to attain a sensitivity of 18 M-Ω cm at 25 °C) and analytical grade reagents. Prepare and store all reagents at room

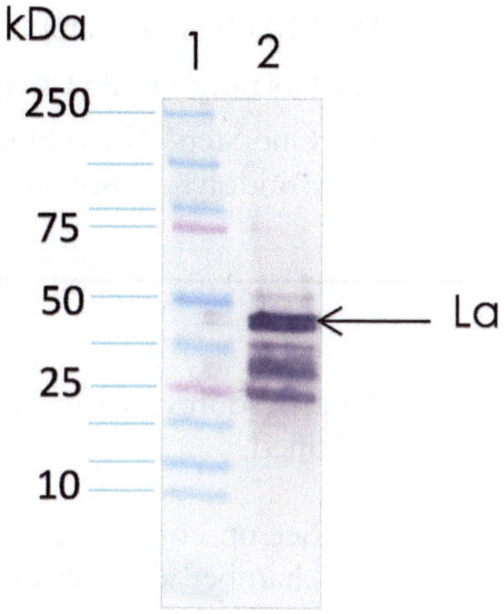

Fig. 2 Ultrafast immunoblotting. Autoantigens La and Ro were electrophoresed in 10 min, transferred in 7 min, and immunoblotted in 35 min. *Lane 1*: Prestained molecular weight standards; *Lane 2*: bovine La autoantigen

temperature (unless indicated otherwise). Diligently follow all waste disposal regulations when disposing waste materials. We do not add sodium azide to our reagents.

1. 10 % SDS-PAGE precast gradient gels (10-well).
2. SDS lysis buffer (5×): 0.3 M Tris–HCl (pH 6.8), 10 % SDS, 25 % β-mercaptoethanol, 0.1 % bromophenol blue, 45 % glycerol. Leave one aliquot at 4 °C for current use and store remaining aliquots at −20 °C (*see* **Note 1**).
3. SDS-PAGE running buffer: 10× running buffer from BioRad (*see* **Note 2**).
4. Phosphate-buffered saline (PBS), pH 7.4.
5. Purified bovine Ro60 (SS-A) [12, 13] and La (SS-B) autoantigen [14] were from Immunovision, Springdale, AK, USA.
6. BenchMark prestained molecular weight standards.
7. Nitrocellulose membranes.
8. Transfer buffer: 0.025 mM Tris–HCl, 192 mM glycine, 20 % methanol.
9. BioRad TransBlot® Turbo™ transfer system.
10. BioRad TransBlot® Turbo™ mini nitrocellulose transfer pack.
11. TBS containing 0.05 % Tween-20 (TBST).
12. Blocking solution: 0.5 % milk in TBS, pH 7.4. Store at 4 °C (*see* **Note 3**).
13. Diluent solution: 0.5 % milk in TBS, pH 7.4 containing 0.05 % Tween. Store at 4 °C (*see* **Note 3**).
14. Mini PROTEAN® 3 System Glass plates.

15. FB300 power supply.
16. Corning PC-351 magnetic stirrer.
17. One Step® Nitro blue tetrazolium (NBT)/5-bromo-4-chloro-3-indolyl phosphate (BCIP) reagent.

3 Methods

All procedures are carried out at room temperature unless otherwise specified.

3.1 SDS-PAGE

1. Set up commercial (4–20 %) gradient gels in electrophoresis chamber for electrophoresis.
2. Heat running buffer to about 90 °C (*see* **Notes 4** and **5**).
3. Add molecular weight marker and the autoantigens to the wells of the gel.
4. Transfer the heated buffer very carefully into the electrophoresis chamber without disturbing the samples in the wells (*see* **Notes 6** and 7).
5. Add couple of drops of 0.1 % bromophenol blue to the running buffer and mix gently.
6. Electrophorese the sample at top settings of the FB power supply.
7. The samples can be immediately seen to start migrating from the wells into the gel. The sample dye reaches the bottom of the gradient gel in about 10 min (*see* **Note 8**).

3.2 Semidry Electrophoretic Transfer

1. Set up the sandwich for electrotransfer using the transfer kit according to manufacturer's instructions.
2. Pre-warm all reagents (blocking, diluent, TBST, substrate) in a walk-in 37 °C incubation room.
3. Insert the membrane gel sandwich into a cassette of the BioRad turbo transfer system.
4. Transfer at 2.5 A (25 V) for 7 min (use Turbo Midi setting).
5. Remove the nitrocellulose membrane from the sandwich and incubate for 10 min with 0.5 % milk blocking that has been pre-warmed to 37 °C (*see* **Notes 9** and **10**).
6. Incubate membrane for 10 min with primary antibody diluted in diluent pre-warmed to 37 °C (*see* **Notes 11** and **12**).
7. Wash three times with deionized water and three times with pre-warmed TBST, all in 3 min.
8. Incubate with appropriate conjugate for 10 min in diluent pre-warmed to 37 °C (*see* **Note 13**).

9. Wash as in **step 7**.
10. Incubate with pre-warmed substrate (*see* **Note 14**).
11. Scan and save results.

4 Notes

1. SDS precipitates at 4 °C. Therefore, the lysis buffer needs to be warmed prior to use.
2. Simple method of preparing running buffer (0.025 M Tris–HCl, pH 8.3, 0.192 M glycine, 0.1 % SDS): Prepare 10× native buffer (0.25 M Tris–HCl, 1.92 M glycine). Weigh 30.3 g Tris–HCl and 144 g glycine, mix, and make it to 1 L with water. Dilute 100 mL of 10× native buffer to 990 mL with water and add 10 mL of 10 % SDS. Care should be taken to add SDS solution last, since it makes bubbles.
3. Add 100 mL of 10× TBS to a 1 L graduated cylinder and make it to about 800 mL with water. Transfer 50 g skim milk powder into the cylinder and mix stir until dissolved. Make to 1 L with water. Separate 500 mL as the blocking solution. To the remaining 500 mL add 250 μL of Tween-20 (cut the end of blue tip to aspirate Tween-20 easily), dissolve, and use it as the diluent.
4. If using l liter of buffer, use at least a 2 L glass beaker to heat. This is because the SDS bubbles. We used Bio-Rad's running buffer with SDS already in it. If making own running buffer, add SDS after heating.
5. Cover beaker with Saran wrap or some kind of cling wrap. Use extreme caution when heating buffer.
6. Exercise caution in handling the hot contents. Use thermo gloves to hold the hot beaker and pour carefully.
7. Transfer hot buffer directly into outer electrophoresis chamber. However, when transferring into inner chamber the hot buffer comes directly in contact with the antigens in the well. Therefore, transfer the buffer gently into chamber along a 10 mL pipette so that the antigens in the wells are not disturbed significantly. Alternatively, the molecular weight marker and other antigens can be added as done normally, after filling up the electrophoresis chamber with buffer. Samples can also be loaded in the wells after filling inner chamber with non-heated buffer. The outer chamber can then be filled with heated buffer. Running the gel in this manner will, however, increase running time to 15–20 min.
8. Using the Haeberle gel electrophoresis system [3], we were able to electrophorese proteins in about 5 min (data not shown).

9. The membrane can be stained with Fast Green to see the effectiveness of protein transfer. However, transferring for 7 min with the BioRad TransBlot system has consistently transferred low- and high-molecular-weight protein standards almost completely. This can be directly observed on the gel from which the proteins have been transferred. Also, we have stained the gel post-transfer with Coomassie brilliant blue and have observed that most of the proteins have been transferred to membrane.
10. We used only 0.5 % milk, since a report showed that 5 % milk blocking can remove antigens from membrane [15].
11. We diluted anti-Ro60 human sera 1:100 and the anti-La monoclonal 1:10.
12. Exercise universal precaution when handling human sera. Treat each serum sample as potentially dangerous.
13. We typically use Jackson Immunochemicals conjugates at 1:5,000 dilution.
14. With the Ro60 and La antigens the signals could be observed in 30 s or less.

References

1. Laemmli UK (1970) Cleavage of structural proteins during assembly of the head of bacteriophage T4. Nature 227:680–685
2. Srinivas PR (2012) Introduction to protein electrophoresis. Methods Mol Biol 869:23–28
3. Haeberle JR (1997) High-temperature sodium dodecyl sulfate polyacrylamide gel electrophoresis. Biotechniques 23:638–640
4. Kurien BT, Scofield RH (1997) Multiple immunoblots after non-electrophoretic bidirectional transfer of a single SDS-PAGE gel with multiple antigens. J Immunol Methods 205:91–94
5. Peferoen M, Huybrechts R, De Loof A (1982) Vacuum-blotting: a new simple and efficient transfer of proteins from sodium dodecyl sulfate-polyacrylamide gels to NC. FEBS Lett 145:369–372
6. Towbin H, Staehelin T, Gordon J (1979) Electrophoretic transfer of proteins from polyacrylamide gels to NC sheets: procedure and applications. Proc Natl Acad Sci U S A 76: 4350–4354
7. Bolt MW, Mahoney PA (1997) High efficiency blotting of proteins of diverse sizes following sodium dodecyl sulfate-polyacrylamide gel electrophoresis. Anal Biochem 247:185–192
8. MacPhee DJ (2010) Methodological considerations for improving Western blot analysis. J Pharmacol Toxicol Methods 61:171–177, Review
9. Biorad transblot turbo transfer system. http://www.bio-rad.com/en-us/product/semi-dry-rapid-blotting-systems/trans-blot-turbo-transfer-system
10. Kurien BT, Scofield RH (2000) Ultrarapid electrophoretic transfer of high and low molecular weight proteins using heat. Methods Mol Biol 536:181–190
11. Kurien BT, Scofield RH (2002) Heat-mediated, ultra-rapid electrophoretic transfer of high and low molecular weight proteins to nitrocellulose membranes. J Immunol Methods 266:127–133
12. Huang SC, Pan Z, Kurien BT, James JA, Harley JB, Scofield RH (1995) Immunization with vesicular stomatitis virus nucleocapsid protein induces autoantibodies to the 60 kD Ro ribonucleoprotein particle. J Investig Med 43:151–158
13. Kurien BT, Asfa S, Li C, Dorri Y, Jonsson R, Scofield RH (2005) Induction of oral tolerance in experimental Sjogren's syndrome autoimmunity. Scand J Immunol 61:418–425
14. Tröster H, Bartsch H, Klein R, Metzger TE, Pollak G, Semsei I, Schwemmle M, Pruijn GJ, van Venrooij WJ, Bachmann M (1995) Activation of a murine autoreactive B cell by immunization with human recombinant autoantigen La/SS-B: characterization of the autoepitope. J Autoimmun 8:825–842
15. DenHollander N, Befus D (1989) Loss of antigens from immunoblotting membranes. J Immunol Methods 122:129–135

Chapter 46

Single-Cell Western Blotting

Syed M.S. Quadri

Abstract

Cell heterogeneity is a variation in cellular processes in functionally similar cells. Cells from the same tissue which are considered genetically identical may have difference in size, structure, and level of protein expression which can lead to major impact on the functions of cell leading to difference in physiological consequences. Single-cell proteome-wide studies are used to detect cell heterogeneity. Flow cytometry and immunocytochemistry do play an important role in evaluating cell heterogeneity. However, these methods are based on separation by antibodies with limited specificity. Cross-reactivity can occur leading to bias in result. Western blot is done to separate the proteins according to molecular weight. Therefore, off-target and on-target signals can be discriminated. Detection of protein expression from a tissue can be done with the help of western blot. However, it is unable to differentiate protein expression of individual cells. For detection of this cell-to-cell variation, a highly advanced technique termed "single-cell western blotting" is carried out. Single-cell western blot has enabled us to detect protein expression at cellular level at a fairly advanced high resolution using a western blot designed to assess cell heterogeneity.

Key words Single-cell western blotting, Single-cell immune-blotting, Cell heterogeneity, Cell-to-cell variation

1 Introduction

Western blot commonly called as immunoblot is one of the common techniques used to study the presence and or comparison of protein levels found in minute levels in body. Protein samples are loaded in a polyacrylamide gel and the proteins are separated based on their molecular weight using electric current in a process called electrophoresis. These proteins are then transferred to a polyvinylidene difluoride or nitrocellulose membrane. Monoclonal or polyclonal primary antibodies against specific proteins are applied to the membrane which is then exposed to secondary antibodies. Finally, the membrane is exposed to substrate for detection of bands of proteins if they are present.

Generally, the cells of the same type are considered to give a homogenous response when stimulated. However, there are differences in cell responses which results in cell-to-cell variation.

Biji T. Kurien and R. Hal Scofield (eds.), *Western Blotting: Methods and Protocols*, Methods in Molecular Biology, vol. 1312, DOI 10.1007/978-1-4939-2694-7_46, © Springer Science+Business Media New York 2015

This variation results in difference in protein expression in cells (*see* **Notes 1** and **2**). Conventional western blot is able to detect the changes in protein level to a good extent; however it does not help us in determination of individual cell protein expression.

To detect protein expression in different cells, an advanced stage process utilizing the same technology as western blotting is done that is termed as "single-cell western blotting" [1]. In this process a microscope slide having a thin film of photoactive polyacrylamide gel is micropatterned with an array of 6,720 microwells in 16 blocks. The gel is constructed against a silicon wafer augmented with SU-8 micropost. Each microwell is approximately the size of a single cell (20 μm in diameter and 30 μm in depth approximately). The solution containing cells to be studied is applied to these slides and allowed for the settlement of cells. After appropriate settlement of cells in the wells, the slides are washed with PBS for removal of excess cells from the gel surface. The settling of cells and microwell occupancy is observed under bright-field microscope. The cells are then lysed using RIPA buffer and proteins are separated by applying electric field across the gel for a very short period of time. The slide is then exposed to UV light for immobilization of proteins. Unlike the conventional immunoblotting the gel is not transferred to a membrane. Since the gel layer is extremely thin (30 μm), diffusion can easily occur when antibodies are applied. Primary antibodies are directly applied to slides, followed by application of fluorescently labeled secondary antibodies. Finally, the fluorescence is detected by fluorescence imaging and the data is analyzed. Next, stripping buffer is applied to the slide for removal of antibodies followed by washing with TBST. Thus, the slide can be reused for another set of antibodies.

2 Materials

1. Regular microscope slide.
2. Photoactive polyacrylamide gel.
3. SU-8 2025 micropost (Microchem Corp, Westborough, MA).
4. 2× 8-well microarray hybridization cassette.
5. Radio immunoprecipitation assay (RIPA) lysis/electrophoresis buffer—0.5 % SDS, 96 mM glycine, 0.1 % v/v Triton X-100, 0.25 % sodium deoxycholate 0.25 % sodium deoxycholate in 12.5 mM Tris, bring to pH 8.3. RIPA buffer is denaturing but nonreducing.
6. UV mercury arc lamp (Lightningcure LC5 Hamamatsu).
7. Gel precursor solution—8 % T (w/v total acrylamide), 2.7 % C (w/w of the cross-linker *N*,*N*-methylene bisacrylamide) from 30 % T, 2.7 % C stock, 3 mM BPMAC from 100 mM stock in

DMSO, 0.1 % SDS, 0.1 % Triton X-100, 0.0006 % riboflavin 5′ phosphate, 0.015 % ammonium persulfate (APS), 0.05 % tetramethylethylenediamine (TEMED) in 75 mM Tris buffer. Titrate with HCl to a pH of 8.8.

8. Dichlorodimethylsilane (DCDMS).
9. Phosphate-buffered saline (PBS)—pH 7.4.
10. SU-8 developer solution (Microchem Corp).
11. TBST—Titrate 100 mM Tris, containing 150 mM NaCl, with HCl to pH 7.5, followed by 0.1 % Tween 20.
12. Bovine serum albumin (BSA).
13. Stripping buffer—2.5 % SDS, 1 % β-mercaptoethanol, 62.5 mM Tris. Titrate to a pH of 6.8 with HCl.
14. Platinum wire electrodes (0.5 nm diameter, Sigma-Aldrich).

3 Methods

3.1 Method of Culturing Cells (See Note 3)

1. Neural stem cells are extracted from hippocampus of an adult rat.
2. Coat the culture-treated polystyrene plates with polyornithine 10 μg/mL and laminin 5 μg/mL.
3. Culture the cell in required media.
   Preparation of media: Take 250 mL of DMEM-F12 and mix with 250 mL N-2. Add 10 μg of recombinant human FGF-2 (Peprotech). Culture up to 80 % of confluency.
4. For cell detachment use accutase.

3.2 Preparation of Protein Standard (Ladder)

For making protein standard or ladder for single-cell western some of the fluorescent antibodies are prepared, and others are obtained in premade condition. The ladder is made by purified proteins having molecular weight of 27–132 kDa (*see* Fig. 1). The following proteins can be used:

1. Purified Dronpa—27 kDa
2. OVA—ovalbumin—45 kDa
3. BSA—66 kDa
4. OVA dimer—90 kDa
5. BSA dimer—132 kDa

Purified albumin and BSA were Alexa Fluor 488 labeled and can be obtained from Life Technologies. Purified dronpa can be prepared by expression in Rosetta-competent cells transformed with tobacco etch virus (TEV) ligase-independent cloning vector and then affinity purified.

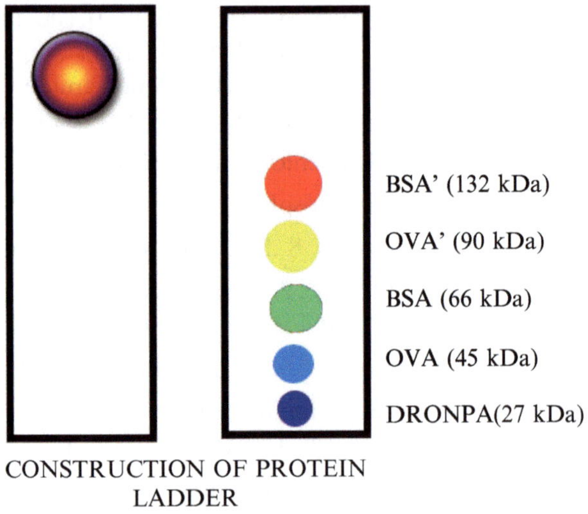

Fig. 1 Construction of protein ladder

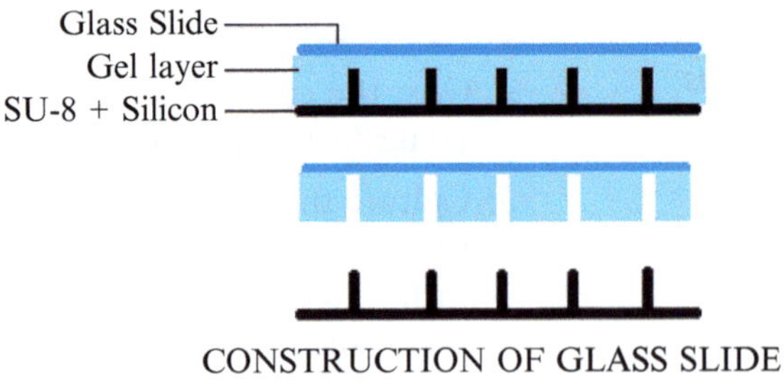

Fig. 2 Construction of glass slide

The fluorescence at different distance is recorded and the distance measured. The fluorescence within 10 % of the target region is considered as positive.

3.3 Preparation of Fabricated Slides

1. Microposts (SU-8) are fabricated on silicon wafer by standard lithography method (*see* Fig. 2).
2. Spin SU-8 2025 according to the manufacturer's guidelines to reach a thickness of 30 μm.
3. Expose it to UV light (365 nm) at around 40 mW/cm^2 (use Myler mask with 20 μm circular features at 20,000 d.p.i.).
4. Arrange the features in square configuration. Keep the pitch as follows:
   (a) 500 μm in direction of separation.
   (b) 190 μm in transverse direction.

(c) 2×8 blocks of 14×30 are separated by space of 9 mm in order to accurately match the dimensions of 2×8-well microarray hybridization cassette.

5. Optical profilometry can be done to analyze the uniformity of thickness using SU-8 developer solution.
6. Silanize the wafer by using vapor deposition of DCDMS for approximately 1 h in vacuum.
7. Wash the wafer with deionized water and dry it using nitrogen stream.
8. In the next step, take a microscope glass slide and silanize it in order to create a surface layer of methacrylate functional group.
9. Silanized slides are placed on micropost wafer and aligned to SU-8 rail and micropost features.
10. Prepare gel precursor solution (as described in materials).
11. Sonicate the precursor solution and degas it for 1 min in vacuum before addition of Triton X-100, SDS, APS, riboflavin, and TEMED.
12. Add the precursor solution in the area between glass slide and silicon wafer with the help of pipette.
13. Wait for 30 s for the solution to settle.
14. Expose the glass slide to blue light for 7.5 min at 4,701× from a collimated 470 nm LED.
15. Leave it for 10–11 min for polymerization of gel.
16. Wet the edges of the fabricated slide with 1–2 mL PBS.
17. Carefully separate the slide from wafer with the help of blade.
18. Fabricated slides are now ready for experiment (*see* Fig. 3). These slides can be stored at 4 °C in PBS for 1–2 weeks.

3.4 Single-Cell Western Blot

1. Remove the slides from PBS solution.
2. Tilt the slide to one side and drain the excess liquid from the corner of the slide. Absorb it with the help of Kimwipe.

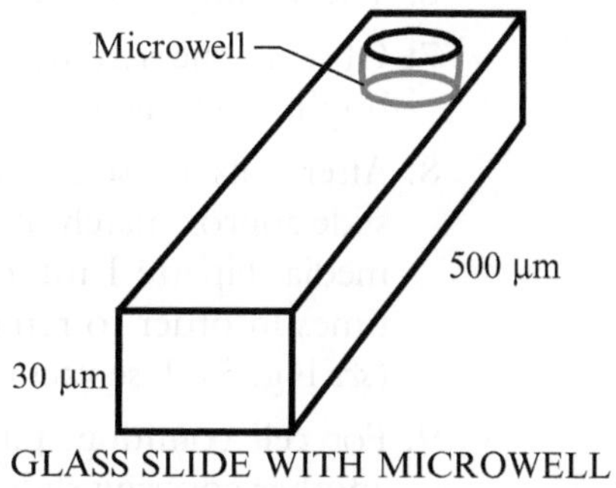

Fig. 3 Glass slide showing a typical microwell

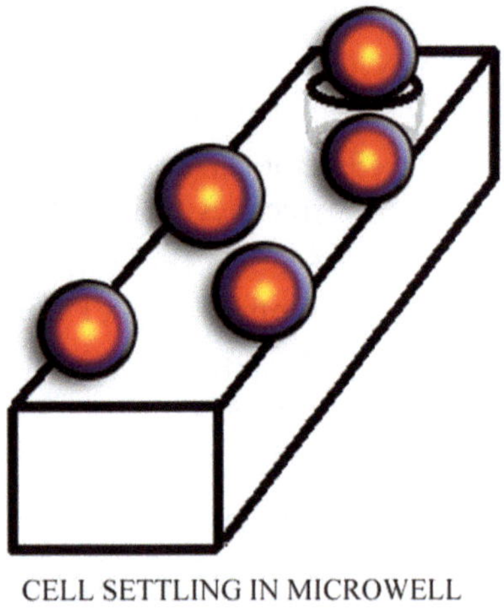

Fig. 4 Cell settling in microwell

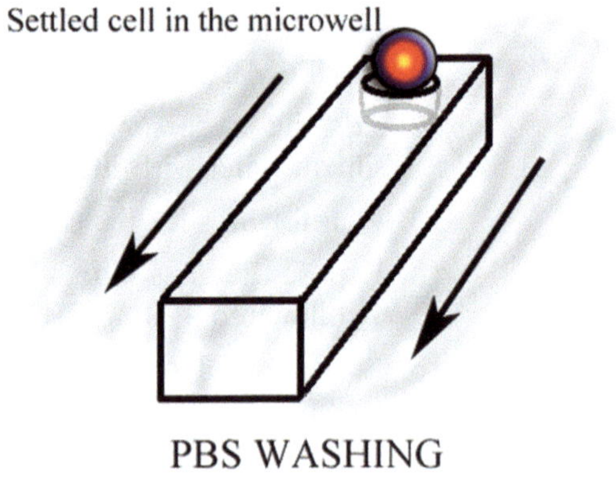

Fig. 5 PBS washing

3. Apply 1–2 mL of cell suspension solution and wait for settling of cells (*see* Fig. 4).
4. Place the slide in petri dish 100 × 100 mm.
5. After every 2–5 min gently shake the petri dish for 10 s.
6. The settling time ranges from 5 to 30 min.
7. Observe the settling of cells and microwell occupancy under bright-field microscope.
8. After cells are settled down in microwells, lift one end of the slide approximately at an angle of 10–20° for removal of excess media. Pipette 1 mL of PBS to the raised side of the slide 4–5 times in order to remove the cells on the surface of the slide (*see* Fig. 5). Use vacuum at low pressure if necessary.
9. For cell counting, put 1 mL of PBS on the slide and place another covering slide to prevent bubble formation.
10. Image under bright-field microscopy at 4× magnification controlled by MetaMorph software.

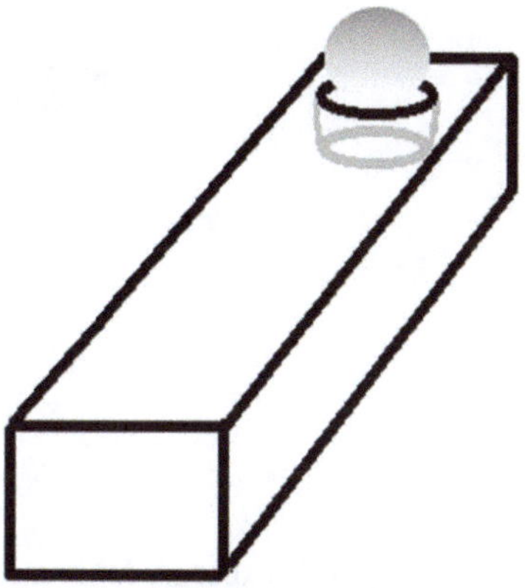

Fig. 6 Lysis of cell by RIPA buffer

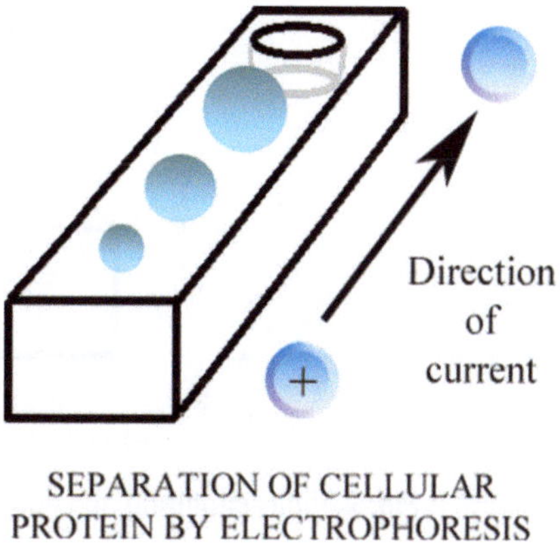

Fig. 7 Separation of cellular protein by electrophoresis

11. After the counting is done, remove the upper glass slide gently across the gel layer.
12. Transfer the slide with settled cell to 60 × 100 mm customized horizontal electrophoresis chamber fabricated from 3 mm plastic (Perspex).
13. Place platinum wire electrode along the long edge of the chamber and connect it to standard electrophoresis power supply with the help of alligator clips.
14. Attach the slides to the bottom of the chamber using petroleum jelly.
15. Add 10 mL of RIPA lysis buffer to the slide to lyse the cells and wait for 10 s (*see* Fig. 6).
16. Turn on the current with setting of 200 V (E = 40 V/cm) for about 30 s (*see* Fig. 7).

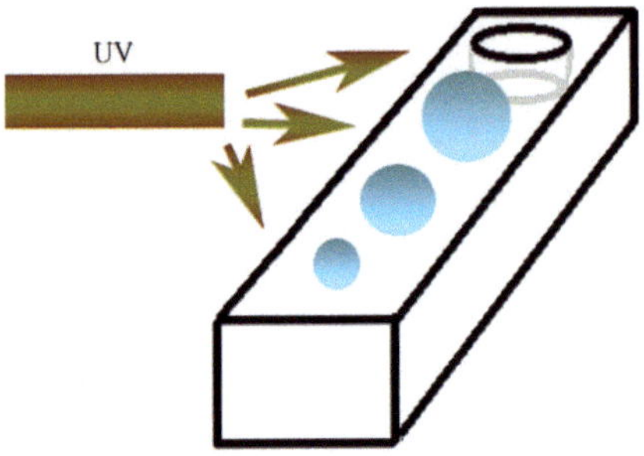

Fig. 8 UV exposure for immobilization of proteins

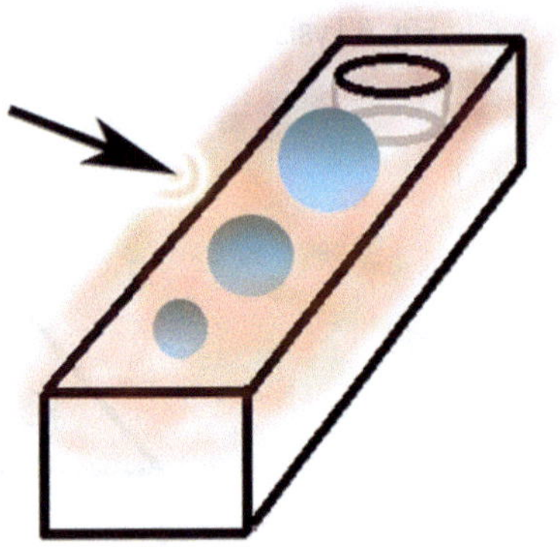

Fig. 9 Photo-capturing and addition of primary antibodies

17. Monitor the separation of proteins from a single cell in real time with a magnification of 10×. The filters are optimized for particular cells and may vary in different cell types.
18. After separation is done, expose to UV light using UV mercury arc lamp at a distance of 10 cm above the slide with a UW power of 40 mW/cm^2. The exposure helps in immobilization of proteins (*see* Fig. 8).
19. Wash the slides with 10 mL of denaturing RIPA buffer for about 10 min.
20. Wash the slides with 10 mL of TBST for about 10 min (at this point the slides can be stored up to a week at 4 °C in TBST).
21. Incubate each block of separation with 40 μL of primary antibody solution diluted in TBST and 2 % BSA (*see* Fig. 9).
22. Keep the slide on rotator for about an hour under gentle shaking mode.

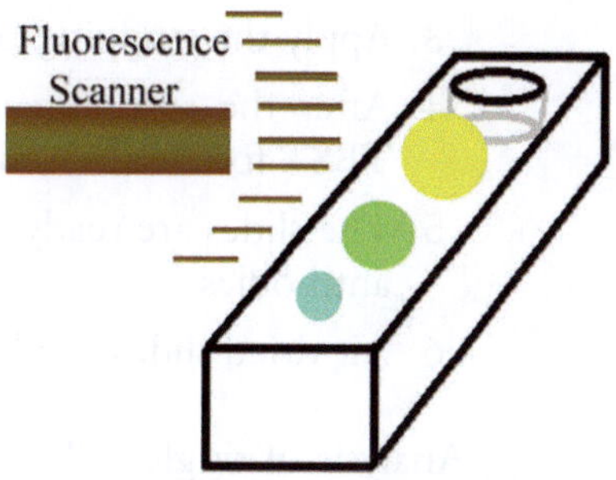

Fig. 10 Addition of fluorescent secondary antibodies and fluorescent scanning

23. Remove the slides and wash the slide adding 10 mL of TBST and place on the rotator for 15 min under gentle shaking mode.
24. Repeat the above step two times for a total of three washes or 45 min.
25. Incubate each block of separation with 40 μL of fluorescently labeled secondary antibody solution diluted in TBST and 2 % BSA.
26. Remove the slides and wash the slide adding 10 mL of TBST and place on the rotator for 15 min under gentle shaking mode. Repeat the same steps two times for a total of three washes or 45 min (*see* Fig. 10).
27. Finally wash the slide with 10 mL of deionized water for 5 min and dry under nitrogen.
28. The slide is ready for imaging studies. Use microarray scanner for imaging studies (*see* Fig. 10).

3.5 Single-Cell Western Blot for Purified Protein

1. The same protocol is used for purified proteins with few different steps.
2. Slides are incubated in RIPA denaturing buffer for 30 min and then submerged in fresh RIPA buffer for about 5 s.
3. A second glass slide is placed to trap the proteins.
4. The sandwiched slide is then subjected to electrophoresis and photo-capture using UV light. Second glass cover can then be removed.
5. The remaining steps are similar.

3.6 Reprobing with Another Antibody

1. Unlike conventional western blotting, the slide of single-cell western blot can be reused again for detection of another protein.
2. Heat the stripping buffer to 50 °C.

3. Apply the stripping buffer to slide and incubate for 3 h.
4. After the stripping is done, wash the slides with 10 mL of TBST for about 5 min three times.
5. The slides are ready for the next set of primary and secondary antibodies.
6. Air-dried slides can be reused again when preserved at 4 °C.

3.7 Analysis of Single-Cell Western Blot

Analysis of single-cell western blot involves confirmation of presence of single cell in a well. Well containing no cell or more than one cells is not included. The distance between the fluorescent purified protein and well is determined. The fluorescence activity within 10 % of the region can be considered on-target.

4 Notes

1. Understanding cell heterogeneity helps in determining the percentage of cells responding to stimulus as compared to non-responding cells.
2. Causative factor responsible for decrease in cell response can be better studied under high-resolution single-cell western blotting. Individual cells not responding to stimulus may have similar or different cause. These can only be studied using single-cell western blot.
3. Any cells can be cultured for this purpose.

Reference

1. Hughes AJ, Spelke DP, Xu Z, Schaffer DV, Herr AE (2014) Single cell western blotting. Nat Methods 11(7):749–755

Chapter 47

Protein Detection by Simple Western™ Analysis

Valerie M. Harris

Abstract

Protein Simple© has taken a well-known protein detection method, the western blot, and revolutionized it. The Simple Western™ system uses capillary electrophoresis to identify and quantitate a protein of interest. Protein Simple© provides multiple detection apparatuses (*Wes*, *Sally Sue*, or *Peggy Sue*) that are suggested to save scientists valuable time by allowing the researcher to prepare the protein sample, load it along with necessary antibodies and substrates, and walk away. Within 3–5 h the protein will be separated by size, or charge, immuno-detection of target protein will be accurately quantitated, and results will be immediately made available. Using the *Peggy Sue* instrument, one study recently examined changes in MAPK signaling proteins in the sex-determining stage of gonadal development. Here the methodology is described.

Key words Protein Simple©, Simple Western™, Protein

1 Introduction

This chapter discusses the recent development of the Simple Western™ by Protein Simple©. This technology allows the researcher to perform a western blot [1] without using gels or blots resulting in minimal effort by the researcher and saving valuable research time for additional experiments. The methodology behind Simple Western™ involves a one-time addition of the provided capillaries with the stacking matrix, and a separation matrix which allows for size- or charge-based separation of a protein of interest. The initial step also requires the addition of an extracted protein sample, primary antibody, secondary horseradish peroxidase-conjugated secondary antibody, and substrate for detection. After loading the samples into any of the commercially available Simple Western™ detection instrumentation (*Wes*, *Peggy Sue*, *Nano Pro 1000*, and *Sally Sue* (*see* **Note 1**)) the researcher can simply walk away and quantitated results are available within 3–5 h [2, 3].

Biji T. Kurien and R. Hal Scofield (eds.), *Western Blotting: Methods and Protocols*, Methods in Molecular Biology, vol. 1312, DOI 10.1007/978-1-4939-2694-7_47, © Springer Science+Business Media New York 2015

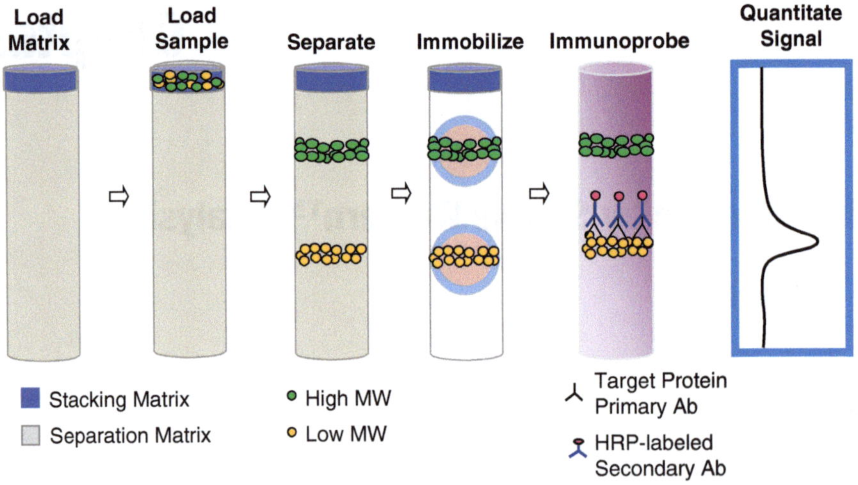

Fig. 1 Simple Western™ automated procedure. The method uses an assay plate setup by the researcher to load the stacking and separation matrix followed by the protein samples. The proteins are separated by charge or size and immobilized on the capillary walls. After extrusion of the gel material, the proteins are probed with primary antibody followed by secondary antibody. Substrate is added and the signal is quantitated

By automating the steps that are manually performed during a western blot experiment, the researcher only has to extract the protein, and load the samples and necessary reagents on to an assay plate, and the instrument will do the rest. The instrument of choice can (a) perform protein separation (size or charge), (b) immobilize the separated protein onto the capillary wall, (c) immuno-probe the target protein using a primary and secondary antibody, (d) perform washes between antibody addition, and (e) develop, identify, and quantitate the signal (*see* Fig. 1) Pam Sigger et al. [4] used the *Peggy Sue* system to detect mitogen-activated protein kinase (MAPK) signaling proteins, the methods and materials of which are described below.

2 Materials (*See* Note 2)

1. RIPA buffer: 150 mM NaCl, 1 % NP40, 0.5 % sodium deoxycholate, 0.1 % SDS, 50 mM Tris, pH 7.5.
2. Protein Simple© Aqueous and DMSO inhibitor cocktails.
3. Simple Western™ sample dilution buffer and fluorescent standards.
4. Primary antibody (applicable dilution).
5. Secondary horseradish peroxidase-conjugated antibody (applicable dilution).

6. Separation and stacking matrix.
7. Chemiluminescent substrate.
8. 384-Well assay plate.

3 Methods

1. Lyse tissue using 50 μL RIPA buffer supplemented with Protein Simple© Aqueous and DMSO inhibitor cocktails.
2. Remove cell debris by centrifugation by spinning for 10 min at 16,000 ×*g*.
3. Mix lysates (7.5 μL) with 2.5 μL of Simple Western™ sample dilution buffer containing reducing agent and fluorescent standards, and denature at 95 °C for 5 min (*see* **Note 3**).
4. Dispense lysates (10 μL), primary antibodies, horseradish peroxidase-conjugated secondary antibodies, separation matrix, stacking matrix, and substrate into designated wells in a 384-well assay plate.
5. Place Simple Western™ assay buffers, capillaries, and the prepared assay plate in *Peggy Sue* instrument (*see* **Note 4**).
6. Assay steps were carried out automatically for up to eight cycles.

4 Notes

1. *Wes* is a size separation assay that holds 25 samples per experiment; *Sally Sue* is a size separation assay that holds 96 samples per experiment; and *Peggy Sue* is a size and charge separation assay that holds 96 samples per experiment. *Nano Pro 1000* is a charge separation assay that holds 96 samples per experiment, but is also able to give detailed information on posttranslational modifications of the protein.
2. Separation Master Kits or individual reagents can be purchased from Protein Simple. The kits are particular to each separation assay (*Peggy Sue*, *Wes*, or *Sally Sue*), and contains the consumables and reagents necessary to perform the assay. Species-specific (rabbit or mouse) or total protein kits contain assay plates, capillaries, buffers, chemiluminescent substrates, separation and stacking matrices, and fluorescent standards with system control protein. These kits do not include primary antibodies or protein lysis buffers.
3. 5 μL of this sample will generate eight data points per run.
4. Instrumentation (*Peggy Sue*, *Wes*, or *Sally Sue*) and Compass Software are provided with purchase of the instrument. Compass Software is used to process and analyze all data results.

References

1. Towbin H, Staehelin T, Gordon J (1979) Electrophoretic transfer of proteins from polyacrylamide gels to NC sheets: procedure and applications. Proc Natl Acad Sci U S A 76: 4350–4354
2. Erik G, Thayer W, John P (2013) Peggy™: size- or charge-based western blotting at the push of a button. Nature Methods 10:i–ii
3. Protein Simple©. "The Simple Western Overview™". Protein Simple© 2011–2014. www.proteinsimple.com
4. Siggers P, Carre' G-A, Bogani D, Warr N, Wells S et al (2014) A novel mouse Fgfr2 mutant, hobbyhorse (hob), exhibits complete XY gonadal sex reversal. PLoS One 9(6): e100447

Chapter 48

Western Blotting Using Microfluidics

Pothur R. Srinivas

Abstract

Together with polyacrylamide gel electrophoresis, the western blot has been an invaluable research technique in biological sciences. It continues to serve as an important diagnostic tool in medical laboratories. The procedure, however, involves multiple steps that are often time and resource intensive in addition to being of low throughput. Using advances in microfluidics, Hughes and Herr et al. [1] initially developed a microfluidic western blot approach that significantly optimizes resources and assay times. More recent developments have enabled multiplexing to facilitate probing of multiple proteins.

Key words Western blot, Microfluidics, Electrophoresis

1 Introduction

The western blot in combination with polyacrylamide gel electrophoresis has been a valuable research tool for life sciences research. It provides qualitative information on predominantly low-abundance proteins in a biological sample. There are, however, multiple steps in the process involving separation of the proteins on a gel matrix, transfer to a membrane, and probing of the membrane with antibodies specific to the protein of interest. Additional time- and resource-intense steps are required to minimize nonspecific binding and enhance the specificity of antibody binding to the protein of interest. Given the recent biological insights enabled through high-throughput proteomic approaches it would be extremely beneficial were an automated, miniaturized, and rapid western blot system available to validate the findings from the high-throughput proteomic approaches and otherwise. Recent advances have led to the incorporation of microfluidics into western blotting approaches. One approach, described by Pan et al. [2], involves a combination of conventional polyacrylamide gel electrophoresis, transfer of separated proteins onto membranes followed by microfluidic based manipulation to facilitate detection of

Biji T. Kurien and R. Hal Scofield (eds.), *Western Blotting: Methods and Protocols*, Methods in Molecular Biology, vol. 1312, DOI 10.1007/978-1-4939-2694-7_48, © Springer Science+Business Media New York 2015

multiple proteins. A more recent approach developed by Gerver and Herr [3] combines all steps in a single platform. The method detailed below is based on Pan et al.'s [2] published approach.

2 Materials

1. Typhoon Imager (Amersham Biosciences, now acquired by GE Healthcare Life Sciences): For capture of fluorescent signals generated from secondary antibodies.
2. ImageQuant TL8.1 software (GE Healthcare Life Sciences): Used to quantify the fluorescent signals captured by the imager.
3. The microfluidic template: The template and microchannels were fabricated in-house using soft lithography on a silicon wafer as described by Pan et al. [2]. The template can be made in any fabrication facility that has a "clean" room.

3 Methods

1. Use a mini-gel system to resolve proteins from the biological sample of interest. A polyacrylamide gel made up of 5 % stacking gel and 10 % running gel will work well for low-molecular-weight proteins. The percentage of acrylamide in the running gels can be altered depending on the protein/s of interest. Generally, a higher percent running gel offers better separation of low-molecular-weight proteins.
2. Perform electrophoresis at room temperature, first at 80 V for 20 min and then at 120 V for 65 min, for example using Bio-Rad's Mini-PROTEAN system.
3. Post-electrophoresis, transfer the separated proteins from the gel to a polyvinylidene difluoride membrane (PVDF) at 260 mA for 1 h, for example using Bio-Rad's mini Trans-Blot system.
4. After transfer incubate the PVDF membrane at 37 °C for 1 h to facilitate protein absorption to the membrane and minimize diffusion in subsequent microfluidic assay phases.
5. Place the membrane in 0.1 % Ponceau S stain for about 5 min and wash in 5 % acetic acid for 5 min. Repeat the wash two more times. Place the membrane in distilled water and wash twice more, 5 min each time or until background becomes clear (*see* **Note 1**). The limit of detection for Ponceau S is about 250 ng of protein (*see* ref. 4).
6. Place the fabricated microfluidic template (*see* **Note 2** and also **Note 7**) with the channels perpendicular to the stained protein bands and seal by gently pressing the template against the membrane.

7. Fill the microchannels with 0.05 % Tween-20 solution prepared in PBS and let stand for about 1 min to dampen the membrane (*see* **Note 3**) followed by 5 % BSA prepared in PBS for 20 min to prevent nonspecific binding.
8. Apply 1–2 μL of the primary antibody diluted in PBS (1:20; *see* **Note 4**) to the protein of interest and let incubate for 20 min. Use different channels to probe for different proteins using the respective primary antibodies.
9. Remove the microfluidic template and rinse the membrane in distilled water.
10. Soak in isopropanol for 30 s, remove, and place in blocking buffer containing 5 % BSA, prepared in PBS, for 20 min.
11. Wash the membrane in PBS containing 0.05 % Tween-20 twice, 5 min each time with fresh PBS.
12. Incubate the membrane with fluorescence-tagged secondary antibody (1:1,000) diluted in PBS containing 5 % BSA and 0.05 % Tween-20 for 30 min with mild agitation.
13. Rinse the membrane in washing buffer (*see* **Note 5**) and wash in the same buffer twice, 5 min each time with fresh buffer.
14. Scan the bands using a fluorescent imager such as a Typhoon Imager (*see* **Note 6**) and quantify the signals using software such as ImageQuant TL8.1.

4 Notes

1. Care should be taken not to wash the membrane after Ponceau S staining with excess distilled water to avoid washing away the stain from the protein bands.
2. The fabricated network template is soft and gentle pressure is enough to seal it to the membrane.
3. The solutions are introduced into the microchannels as droplets at the microchannel inlet and applying vacuum to the outlet.
4. Primary antibodies should be diluted in PBS containing 5 % BSA and 0.25 % Tween-20.
5. Washing buffer used after incubation with secondary antibody contains 0.05 % Tween-20 in PBS.
6. The recommended setting for the Typhoon Imager are as follows: Laser—488 nm; emission filter—520 nm; PMT—480 V 7; and pixel size—25 μm.
7. Dimensions of the microfluidic network as described by Pan et al. are as follows: channel height—100 μm; channel width—150 μm; channel length—3.5 cm; gap between channels—1,000 μm; and number of channels—10.

References

1. He H, Herr HE (2010) Automated microfluidic protein immunoblotting. Nat Protoc 5: 1844–1856
2. Pan W, Chen W, Jiang X (2010) Microfluidic western blot. Anal Chem 82:3974–3976
3. Gerver R, Herr HE (2014) Microfluidic western blotting of low-molecular- mass proteins. Anal Chem 86:10625–10632
4. Salinovich O, Montelaro RC (1986) Reversible staining and peptide mapping of proteins transferred to nitrocellulose after separation by sodium dodecylsulfate polyacrylamide gel electrophoresis. Anal Biochem 156:341–347

Chapter 49

Two-Dimensional Gel-Based Protein Standardization Verified by Western Blot Analysis

Hisao Haniu, Daisuke Watanabe, Yusuke Kawashima, and Hiroyuki Matsumoto

Abstract

In data presentation of biochemical investigation the amount of a target protein is shown in the *y*-axis against the *x*-axis representing time, concentrations of various agents, or other parameters. Western blot is a versatile and convenient tool in such an analysis to quantify and display the amount of proteins. In western blot, so-called housekeeping gene product(s), or "housekeeping proteins," are widely used as internal standards. The rationale of using housekeeping proteins for standardization of western blot is based on the assumption that the expression of chosen housekeeping gene is always constant, which could be false under certain physiological or pathological conditions. We have devised a two-dimensional gel-based standardization method in which the protein content of each sample is determined by scanning the total protein density of two-dimensional gels and the expression of each protein is quantified as the density ratio of each protein divided by the density of the total proteins on the two-dimensional gel. The advantage of this standardization method is that it is not based on any presumed "housekeeping proteins" that are supposed to be being expressed constantly under all physiological conditions. We will show that the total density of a two-dimensional gel can render a reliable protein standardization parameter by running western blot analysis on one of the proteins analyzed by two-dimensional gels.

Key words Protein standardization, Total protein density of a two-dimensional gel, Ratio of a protein against the total gel density, Quantitative western blot

1 Introduction

Western blot analysis of proteins existing in biological specimens is one of the most frequently used experimental protocols in biochemistry and molecular biology. Results will be interpreted based on the amount of expressed protein on the *y*-axis against variables on the *x*-axis. There has been always a concern in western blot analysis with regard to how to standardize the measurement. Traditionally, so-called housekeeping proteins that are presumably expressed constantly are used for internal standards. Recently, however, the notion that expression of housekeeping proteins is

Biji T. Kurien and R. Hal Scofield (eds.), *Western Blotting: Methods and Protocols*, Methods in Molecular Biology, vol. 1312, DOI 10.1007/978-1-4939-2694-7_49, © Springer Science+Business Media New York 2015

constant is challenged [1, 2]. We have devised a method to quantitate multiple proteins displayed by two-dimensional gel electrophoresis [3]. After scanning the gels we quantified each protein standardized against the total protein density of each gel. By utilizing this protein quantification method we showed that proteins expressed in a neonatal mouse retina can be mapped along the developmental time axis and that we can classify proteins according to the expression profile (proteomic trajectory mapping). The standardization principle used in this technique can be applied to western blot on one-dimensional gel electrophoresis. In this chapter we briefly explain the protein stain-based quantitation performed on two-dimensional gel electrophoresis, which we designate "2-DE density method," and the correspondence of the result of 2-DE density method to that of one-dimensional gel-based quantitative western blot designated "western blot method."

2 Materials

1. Two-dimensional gel electrophoresis (2DE) apparatus for a glass tube-based IEF system followed by SDS-PAGE: The apparatus was described previously in a chapter of this protocol series [4].
2. Dissect developing retinas from neonatal C57BL/6 mice at postnatal (P) day at 1, 3, 5, 7, 9, 14, 21, and adult (>28 days). Flash-freeze retinas in liquid nitrogen and store at −80 °C until use.
3. IEF lysis solution containing 9 M urea, 4 % CHAPS, 2 % Bio-Lyte pH 3–10 (BioRad, Hercules, CA), 40 mM Tris, and 5 % β-mercaptoethanol.
4. A transmission scanner (PowerLook III; UMAX Technologies, Dallas, TX) at 14-bit depth and a resolution of 300 dpi.
5. Kodak Photographic Step Tablet No. 2 for calibration of the scanner.
6. Progenesis Workstation software (Nonlinear Dynamics, Newcastle upon Tyne, UK).
7. Li-Cor Odyssey Infrared Gel Scanner.
8. Rabbit-anti-mouse IgG against Nrdg-1 (Santa Cruz, polyclonal goat IgG).
9. IR fluorescent dye conjugated with donkey IgG: Li-Cor IRDye® 800CW anti-goat IgG (H+L).
10. BCA™ Protein Assay Kit.
11. SDS-PAGE apparatus: NuPAGE® Novex Midi Gel.
12. Electroblotting apparatus: iBlot™.

3 Methods

1. Carry out two-dimensional gel electrophoresis (2-DE) by a glass tube-based IEF system followed by SDS-PAGE. The protocol was described previously in a chapter of this protocol series [4].
2. Dissect developing retinas from neonatal C57BL/6 mice at postnatal (P) day at 1, 3, 5, 7, 9, 14, 21, and adult (>28 days). Homogenize 6–16 retinas harvested at each developmental stage with a lysis solution containing 9 M urea, 4 % CHAPS, 2 % Bio-Lyte pH 3–10, 40 mM Tris, and 5 % β-mercaptoethanol in a water bath sonicator at 10 °C. After centrifugation at 16,000 × *g* for 15 min at 4 °C, collect the supernatants and assay the protein contents by Coomassie Plus protein assay and store at −85 °C until use.
3. One set of 2-DE analysis consists of total proteins extracted from P1, P3, P5, P7, P9, P14, P21, and adult retinas. Since the volume of retinal tissue increases as it matures, the number of retinas at each developmental stage was adjusted accordingly; the number of retinas varied from 16 to 6.
4. Carry out 2-DE of each retinal sample containing 400 μg proteins four times ($n = 4$).
5. After focusing, equilibrate IEF gels and apply to the second dimension on 11 % SDS-PAGE.
6. Stain the 2-DE gels by CBB R-250 and then destain.
7. Sandwich the gels in between cellophane membranes and air-dry.
8. Conduct four sets of 2-DE analyses ($n = 4$) in order to statistically evaluate the numeric data.

3.1 Quantitation by 2-DE Density Method

1. Stain each 2-DE gel with CBB and scan by a transmission scanner at 14-bit depth and a resolution of 300 dpi.
2. Calibrate the scanner up to the transmission OD of 3.1 using a Kodak Photographic Step Tablet No. 2, ensuring linearity.
3. Analyze the scanned gel images by Progenesis Workstation software to quantify the protein spots. Figure 1 represents a quadruple set of eight 2-DE gels showing the whole image of each gel of total retinal proteins at P1, P3, P5, P7, P9, P14, P21, and adult stage, respectively. In this illustration, a part of each 2-DE gel that contains a protein of interest, in this case Ndrg1, is shown.
4. The Progenesis software automatically defines the boundaries of 2-DE gel spots for each gel and also assigns the set of corresponding protein spots on the eight gels.

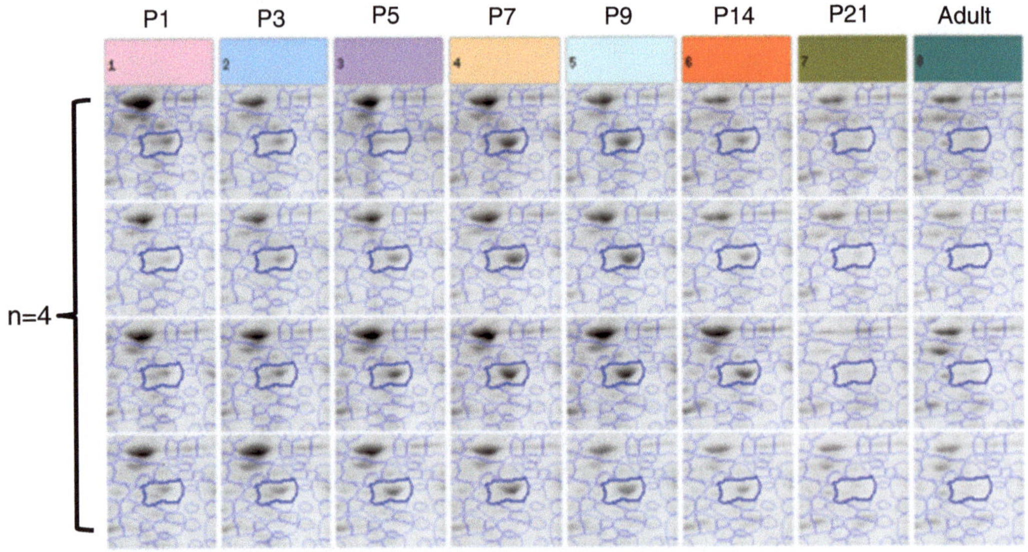

Fig. 1 Two-dimensional gel profiles of mouse retinal proteins during neonatal development P1 through adult. The gels were stained with CBB R-250. The region of each gel representing Ndrg1 (N-myc downstream regulated gene 1; 45 kDa, p*I* = 5.7) is shown. A *thicker border* among other borders illustrates the Ndrg1 protein. For each developmental stage, four independent mouse groups ($n = 4$) were analyzed. The protein expression kinetics (proteomic trajectory map) is illustrated in Fig. 2, "2-DE Density Method." These portions of 2-DE gel images were automatically extracted from the original 2-DE gels by Progenesis software

5. Inspect the assignment of the 2-DE gel spot boundaries visually and corrected, when necessary.
6. Visually inspect also the correspondence of each 2-DE gel spot between eight gels and correction made, when necessary.
7. Express the percentage of each protein spot after the normalization as percentile (%).
8. Conduct cluster analysis of these retinal proteins and group them into four groups, i.e., juvenile type, transient type, adult type, and constant-type, depending on their expression patterns [3].

3.2 Quantitation by Western Blot Method

1. Using a separate batch of developing C57BL/6 retinas, carry out SDS-PAGE in order to evaluate the protein expression profile of one of the transient-type proteins Ndrg1, which was shown to be expressed transiently higher around stages P5–P14.
2. For each developmental stage, run 10 μg of total retinal proteins on an SDS-PAGE by using NuPAGE® Novex Midi Gel System according to the manufacturer's instructions.
3. Transfer the gels electrophoretically onto a PVDF membrane using an electroblotting apparatus according to the manufacturer's instruction.

4. Block background proteins by Odyssey Blocking Buffer solution.
5. Incubate PVDF membrane with goat primary antibody against mouse Ndrg1 and wash twice.
6. Incubate PVDF membrane with secondary donkey antibody against goat IgG tagged with IR dye, Li-Cor IRDye® 800CW anti-goat IgG (H+L), and wash three times.
7. Scan the gel image with Li-CoR IR scanner and quantify. The overall protocol for the quantitative western blot is illustrated in Fig. 2.
8. In Fig. 3 the developmental expression profiles of Ndrg1 obtained from two independent methods, i.e., the 2-DE density method and the western blot method, are shown (*see* **Notes 1** and **2**).
9. The results indicate that the 2-DE density method and the western blot method project a consistent protein expression profile during the development of neonatal retina, assuring the reliability of using total protein density for standardization.

4 Notes

1. The data shown in Fig. 3 illustrates (a) that the total protein density on a 2-DE can be a reliable measure for standardization of a protein expressed in the system, and (b) that quantitative western blot using an infrared fluorescent dye-tagged second antibody to detect the primary antibody bound with the antigen protein separated on SDS-PAGE would be a simpler protocol to quantitate a protein compared to that by the 2-DE density method.
2. It should be noted that the 2-DE density method is required in order to profile the two-dimensional protein map with each protein identified by peptide mass fingerprinting [3]. After the protein map is established, however, the western blot method will serve as a more efficient and less time-consuming protocol for the quantification of protein in a set of biological samples.

Acknowledgment

The projects described in this chapter were partially supported by Oklahoma Center for Advancement of Science and Technology (OCAST) Health Research Program Grant HR10-120 and NIH R21 EY017888.

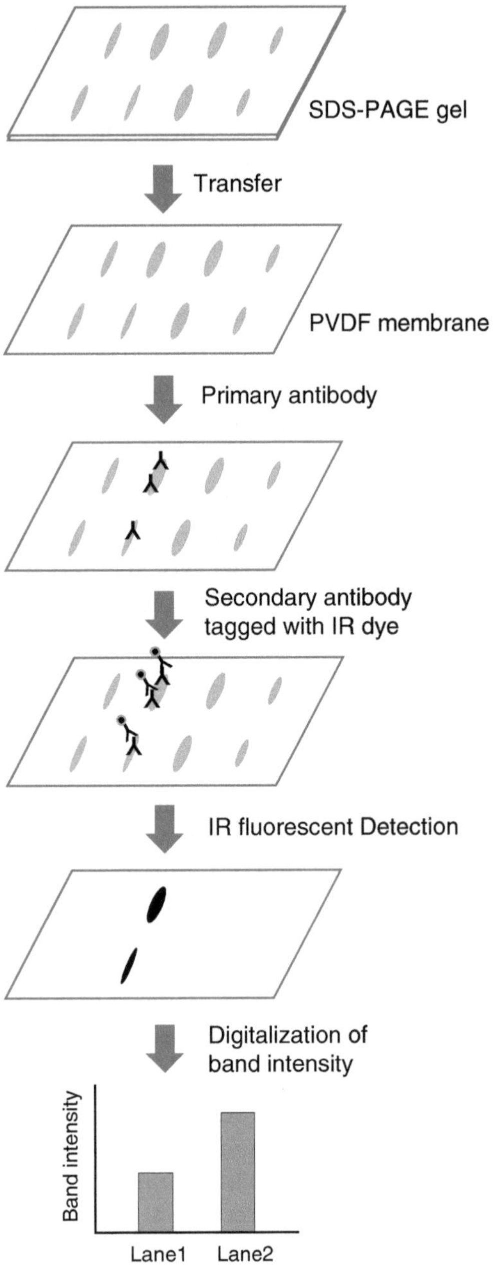

Fig. 2 Illustrated workflows for quantitative western blot using infrared fluorescent-labeled secondary antibody detected by Odyssey Infrared Imaging System (Li-Cor). After electroblotting proteins separated on SDS-PAGE onto a PVDF membrane the proteins were visualized by a primary antibody specific to the protein of interest (e.g., goat anti-Ndrg1 antibody in this case), followed by the secondary donkey antibody against goat IgG tagged with IR dye, Li-Cor IRDye® 800CW. The PVDF membrane was scanned by Odyssey Infrared Scanner (Li-Cor) and the software of the Odyssey scanner analyzed the images (http://www.licor.com/bio/applications/quantitative_western_blots/)

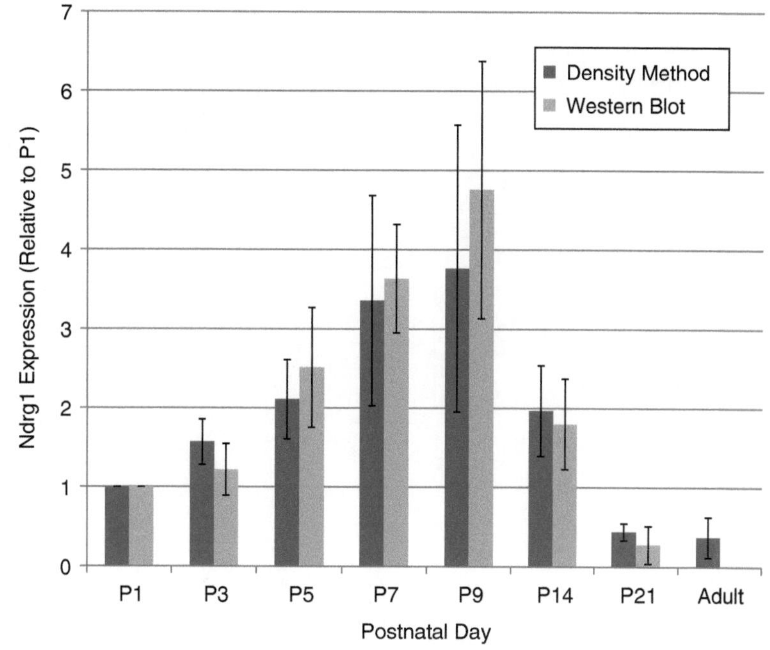

Fig. 3 "(2-DE) Density Method" and "Western Blot (Method)" characterize the proteomic trajectory map of the target protein Ndrg1 consistently. The error bars indicate standard deviations

References

1. Li R, Shen Y (2013) An old method facing a new challenge: re-visiting housekeeping proteins as internal reference control for neuroscience research. Life Sci 92:747–751
2. Kim HJ, Na JI, Min BW, Na JY, Lee KH, Lee JH, Lee YJ, Kim HS, Park JT (2014) Evaluation of protein expression in housekeeping genes across multiple tissues in rats. Korean J Pathol 48:193–200
3. Haniu H, Komori N, Takemori N, Singh A, Ash JD, Matsumoto H (2006) Proteomic trajectory mapping of biological transformation: application to developmental mouse retina. Proteomics 6:3251–3261
4. Matsumoto H, Haniu H, Kurien BT, Komori N (2012) Two-dimensional gel electrophoresis: glass tube-based IEF followed by SDS-PAGE. Methods Mol Biol 869:267–273

Chapter 50

Fingerprint Deposition on Nitrocellulose and Polyvinylidene Difluoride Membranes Using Alkaline Phosphatase

Biji T. Kurien, Debashish Danda, and R. Hal Scofield

Abstract

Dactyloscopy or fingerprint identification is a vital part of forensic evidence. Identification with fingerprints has been known since the finding of finger impressions on the clay surface of Babylonian legal contracts almost 4,000 years ago. The skin on the fingers and palms appears as grooves and ridges when observed under a microscope. A unique fingerprint is produced by the patterns of these friction skin ridges. Visible fingerprints can be deposited on solid surfaces. Colored inks have been used to deposit fingermarks on documents. Herein, we show that alkaline phosphatase can be used to transfer prints from fingers or palm to nitrocellulose or polyvinylidene difluoride membranes. The prints can be detected by using the nitro blue tetrazolium/5-bromo-4-chloro-3-indolyl phosphate method of detection.

Key words Dactyloscopy, Fingermarks, Fingerprints, Forensic science, Identification, Alkaline phosphatase

1 Introduction

Dactyloscopy or fingerprint identification, the singling out of individuals from indentations made by their fingers, is the mainstay of forensic evidence even today. Identification with fingerprints has been known since the finding of finger impressions on the clay surface of Babylonian legal contracts almost 4,000 years ago [1]. Reports in the nineteenth century have suggested that impressions on clay/glass or bloody fingermarks could be used for the scientific identification of criminals.

Fingerprints form a vital part of forensic evidence and have been used in a number of criminal investigations and as a signature for preventing impersonation or repudiation of a handwritten signature [2, 3]. Fingerprints are currently useful only when police or other security agencies obtain a positive match with those prints present on databases.

Biji T. Kurien and R. Hal Scofield (eds.), *Western Blotting: Methods and Protocols*, Methods in Molecular Biology, vol. 1312, DOI 10.1007/978-1-4939-2694-7_50, © Springer Science+Business Media New York 2015

The skin on the fingers and palms appears as grooves and ridges when observed under a microscope. A unique fingerprint is produced by the patterns of these friction skin ridges [4].

Fingerprints are classified into three groups depending on the kind of surface on which they occur and whether they are visible or not. Fingerprints found on hard surfaces are either latent (invisible) or patent (visible) prints, while those on soft surfaces like wet paint, fresh caulk, soap, and wax are three-dimensional plastic prints. When paint, color, dirt, blood, or ink are transferred from a thumb or a finger to a solid surface, it leads to the deposition of visible prints (http://www.forensicsciencesimplified.org/prints/how.html) [5–7]. These three types of fingermarks can be encountered during criminal investigations.

Visible fingerprints can also be deposited on solid surfaces and maintained in a database for future identification purposes. Colored inks have been used to deposit fingerprints on documents [2]. Herein, we show that alkaline phosphatase can be used to transfer prints from fingers or palm to nitrocellulose. The prints can be detected by using the nitro blue tetrazolium/5-bromo-4-chloro-3-indolyl phosphate (NBT/BCIP) method of detection [8].

2 Materials

1. Nitrocellulose membrane.
2. Polyvinylidene difluoride membrane.
3. Anti-human IgG alkaline phosphatase conjugate.
4. Anti-rabbit IgG alkaline phosphatase conjugate.
5. NBT/BCIP.

3 Methods

3.1 Depositing Fingerprint on Nitrocellulose/PVDF Membranes

1. Cut a nitrocellulose or PVDF membrane to the required size.
2. Remove the top flap covering the membrane.
3. Dilute alkaline phosphatase conjugate 1:500 with required volume of water into an appropriate container (*see* **Note 1**).
4. Dip finger, thumb, or palm into the diluted conjugate (*see* **Note 1**).
5. Gently make the finger, thumb, or palm impression on the membrane (*see* **Notes 2** and **3**).
6. Let the membrane dry. It should dry within 5 min on nitrocellulose membrane.

Fig. 1 Thumbprint using alkaline phosphatase and detection with NBT/BCIP

Fig. 2 Index fingerprint using alkaline phosphatase and NBT/BCIP detection system

3.2 Detection with NBT/BCIP

1. Once the impression of the thumb or finger is made on the membrane, allow it to dry. It should take less than 2 min for this.
2. There is no need to wash the dry membrane containing the finger, thumb, or palm impression.
3. Add 10 mL of NBT/BCIP mixture and swish it over the membrane (*see* **Note 4**).

4. The fingerprint or palm print should appear in a few minutes. Wait till the development is complete and then rinse with deionized water.
5. Scan to save image (*see* Figs. 1 and 2).

4 Notes

1. The conjugates were diluted 1:500 with water. Shake of excess conjugate from finger before making impressions on membrane. If the fingerprint appears smudged, make one more finger, thumb, or palm impression on the membrane using the residual alkaline phosphatase. Alternatively, a paper pad soaked in conjugate would also work for transferring fingerprints to membrane. Conjugates from Jackson ImmunoResearch contain 0.54-0.64 % sodium azide (NaN_3). However, since we are using 1:500 dilution of the conjugate, the NaN_3 level will be 0.001 %. Even though NaN_3 is a hazardous chemical as such, the diluted conjugate is not considered to be a hazardous substance according to the OSHA Hazard Communications Standard (CFR 1910.1200), which states that, if a mixture contains less than 1 % of a hazardous chemical or 0.1 % of a carcinogen, the mixture shall not be considered hazardous (https://jacksonimmuno.com/msds/14). The preparation used in the study will have only 0.001 % NaN_3, a very negligible amount that will not pose any significant risk. However, it is recommended to wash finger or palm immediately after transferring prints to membrane. No buffers were checked as diluent.
2. It takes much longer for the conjugate to dry on the PVDF membrane. It may be possible to make the conjugate to diffuse into the PVDF membrane. Diluting the conjugate in a 20 % methanol solution may work in making the alkaline phosphatase to diffuse. However, studies should be undertaken to find out the level of methanol that allows the alkaline phosphatase to remain active.
3. Fingerprints obtained on PVDF with alkaline phosphatase appear fuzzy and smudged. Nitrocellulose would be the membrane of choice for depositing fingerprints with alkaline phosphatase.
4. The NBT/BCIP mixture is stored at 4 °C. It would accelerate the reaction if the required volume is brought to room temperature prior to applying on the membrane.

References

1. Ashbaugh DR (1999) Quantitative–qualitative friction ridge analysis: an introduction to basic and advanced ridgeology. CRC Press, Boca Raton, FL, USA
2. Faulds H (1880) Skin furrows of the hand. Nature 22:605
3. Herschel WJ (1880) Skin furrows of the hand. Nature 23:76
4. Leggett R, Lee-Smith EE, Jickells SM, Russell DA (2007) "Intelligent" fingerprinting: simultaneous identification of drug metabolites and individuals by using antibody-functionalized nanoparticles. Angew Chem Int Ed Engl 46:4100–4103
5. Lee HC, Gaensslen RE (2001) Methods of latent fingerprint development. In: Lee HC, Gaensslen RE (eds) Advances in fingerprint technology, 2nd rev. ed. CRC Press, Danvers, MA, USA, pp 105–175
6. Mink T, Voorhaar A, Stoel R, de Puit M (2013) Determination of efficacy of fingermark enhancement reagents; the use of propyl chloroformate for the derivatization of fingerprint amino acids extracted from paper. Sci Justice 53:301–308
7. Templeton JE, Linacre A (2014) DNA profiles from fingermarks. Biotechniques 57:259–266
8. Kurien BT, Scofield RH (2009) Nonelectrophoretic bidirectional transfer of a single SDS-PAGE gel with multiple antigens to obtain 12 immunoblots. Methods Mol Biol 536:55–65

Chapter 51

Other Notable Protein Blotting Methods: A Brief Review

Biji T. Kurien and R. Hal Scofield

Abstract

Proteins have been transferred from the gel to the membrane by a variety of methods. These include vacuum blotting, centrifuge blotting, electroblotting of proteins to Teflon tape and membranes for N- and C-terminal sequence analysis, multiple tissue blotting, a two-step transfer of low- and high-molecular-weight proteins, acid electroblotting onto activated glass, membrane-array method for the detection of human intestinal bacteria in fecal samples, protein microarray using a new black cellulose nitrate support, electrotransfer using square wave alternating voltage for enhanced protein recovery, polyethylene glycol-mediated significant enhancement of the immunoblotting transfer, parallel protein chemical processing before and during western blot and the molecular scanner concept, electronic western blot of matrix-assisted laser desorption/ionization mass spectrometric-identified polypeptides from parallel processed gel-separated proteins, semidry electroblotting of peptides and proteins from acid-urea polyacrylamide gels, transfer of silver-stained proteins from polyacrylamide gels to polyvinylidene difluoride (PVDF) membranes, and the display of K^+ channel proteins on a solid nitrocellulose support for assaying toxin binding. The quantification of proteins bound to PVDF membranes by elution of CBB, clarification of immunoblots on PVDF for transmission densitometry, gold coating of nonconductive membranes before matrix-assisted laser desorption/ionization tandem mass spectrometric analysis to prevent charging effect for analysis of peptides from PVDF membranes, and a simple method for coating native polysaccharides onto nitrocellulose are some of the methods involving either the manipulation of membranes with transferred proteins or just a passive transfer of antigens to membranes. All these methods are briefly reviewed in this chapter.

Key words Centrifuge blotting, Black cellulose nitrate, Protein microarray, Square wave alternating voltage, Acid electroblotting, Electronic western blotting, Gold coating of membranes

1 Vacuum Blotting

Renart et al. first used a capillary blotting system to passively transfer proteins from polyacrylamide gels onto nitrocellulose membrane. However, transfer was slow and inefficient using this method. Electroblotting, described later by Towbin and Burnette, was much more efficient. However, Peferoen et al., owing to serious financial constraints in their department, developed another way of transferring proteins. They decided to speed up the transfer process (compared to that obtained by the capillary transfer method) by using suction force to draw proteins out of the gel. The authors

Biji T. Kurien and R. Hal Scofield (eds.), *Western Blotting: Methods and Protocols*, Methods in Molecular Biology, vol. 1312, DOI 10.1007/978-1-4939-2694-7_51, © Springer Science+Business Media New York 2015

used the suction power of a pump that was connected to a slab dryer system to drive proteins from a gel (polyacrylamide or sodium dodecyl polyacrylamide gels) to a nitrocellulose membrane. Using this method, Peferoen et al. efficiently blotted *Sarcophaga bullata* egg homogenate and *Leptinotarsa decemlineata* hemolymph proteins [1].

2 Centrifuge Blotting

Hermansen et al. [2] describe a method to elute and transfer proteins, detected in the gel by visualization with 1 M potassium chloride (KCl), to a polyvinylidene difluoride (PVDF) (Immobilon-P, BioRad Laboratories, Hercules, CA, USA) membrane by centrifugation instead of electroblotting. The SDS-polyacrylamide gel was immersed in 1 M KCl for 2 min. Protein bands that appear as a clear zone against an opaque background were cut out and immersed in distilled water to remove excess KCl (3–5 min). The gel pieces were then used for centrifuge blotting or stored moist in a microcentrifuge tube at −20 °C.

PVDF and dialysis membrane (10 kDa cut off) were excised into 12 mm disks. The PVDF disk was wetted in methanol for 5 min for activation. The gel bit was soaked in an eluant for 15 min at 37 °C. The dialysis and PVDF membranes were soaked in the eluant (3–5 min) prior to centrifugation.

The centrifuge receptacle assembly consists of an outer chamber that supported the entire structure, an inner cylinder (serving as the reservoir for the eluant), a 12-mm sinter base support made of polyethene which holds the dialysis and PVDF membrane in place, a polystyrene tube for eluate collection, and a flat O-ring gasket made of silicon to prevent leakage in the receptacle during centrifugation.

The dialysis membrane was positioned on the base support of the receptacle followed by the O-ring. Eluant was added on the dialysis membrane prior to placing the PVDF membrane. This prevents formation of vacuum between the PVDF and dialysis membranes. The inner cylinder is screwed in the receptacle after the PVDF membrane is placed on top of the O-ring. Eluant was added into the receptacle after placing the gel bit on top of the PVDF membrane. Centrifugation was carried out at 3,000 × *g* at 20 °C for 1 h in a tabletop centrifuge. Following centrifugation the PVDF membrane with the transferred protein adhering to it was vacuum dried and stored at −20 °C for N-terminal sequence analysis or other analyses. The dialysis membrane placed beneath the PVDF membranes helps retain the non-immobilized proteins [2].

3 Electroblotting of Proteins to Teflon Tape and Membranes for N- and C-Terminal Sequence Analysis

Several membranes have been used in electroblotting since the inception of the electrotransfer of proteins from polyacrylamide gels to membranes and the subsequent characterization of the blotted proteins [3]. The chemical inertness of a membrane has been one of the critical properties that determined the usefulness of a particular membrane in electroblotting. Proteins eletrotransferred to nitrocellulose cannot be subjected to Edman sequencing but are amenable to in situ proteolytic digestion. PVDF (polyvinylidene difluoride) membranes have been widely used for N-terminal sequencing applications on account of the fact that it is inert to Edman degradation chemistry.

This study, to evaluate the electrotransfer of proteins to Teflon tape and membranes, was initiated on account of the fact that PVDF membranes were not inert to the chemistry used on the Hewlett-Packard G1009A C-terminal sequencer. Burkhart et al. [4] report the optimized conditions for electrotransfer of proteins from polyacrylamide gels to low-density Teflon tape and GORE-TEX expanded polytetrafluorethylene membranes that were discovered to provide performance similar to that obtained with PVDF. Teflon was found to be a suitable Edman sequencing support in an earlier study [5]. Teflon blots were found to be amenable to amino acid analysis, in situ proteolytic digestion, and a combination of N-terminal sequencing followed by C-terminal sequencing in addition to being inert to the chemistry used on the Hewlett-Packard G1009A C-terminal sequencer.

In this procedure, electrophoretically separated proteins were electrotransferred onto Teflon membrane. The Teflon tape or GORE-TEX was moistened thoroughly with absolute ethanol prior to assembling the blotting sandwich. The proteins were stained with 0.005 % sulforhodamine B in 30 % methanol for 10 min following transfer. The blots were washed several times with distilled water, prior to drying, to remove excess stain [4].

Blotted protein bands were cut out from dry blots and placed directly into the reaction cartridge without additional washing for automated sequencing purposes. They were found to be amenable to both N- and C-terminal sequencing. As long as the described wetting procedures (it was important to moisten the membranes with absolute ethanol or isopropanol and not methanol and to keep the membrane wet with ethanol and not equilibrated with transfer buffer prior to being placed next to the gel on the blotting sandwich) were followed, the ease of preparing samples on Teflon blots was similar to PVDF membranes. However, Teflon tapes were not useful for western blotting owing to their low porosity.

4 Multiple Tissue Western Blot

An important part of protein blotting is the study of comparative protein levels in various tissues. Analysis of this kind permits characterization of tissue-specific protein isoforms, detection of immunologically related proteins, and examination of posttranslational modifications leading to changes in molecular weight. Immunoblot analysis of this kind requires considerable expenditure of money, time, and energy in terms of obtaining hard-to-get human tissues, protein content normalization, preparation of protein samples, electrophoretic separation, and transfer. The availability of prefabricated immunoblots would allow the investigator to proceed directly to the antigen detection phase of the experiment.

Human multiple tissue western (MTW) blots provide a new immunological tool for the investigation of tissue-specific protein expression. MTW blots are premade immunoblots [6] and are prepared utilizing proteins isolated from adult human tissue. The proteins are obtained from whole tissue under conditions of minimal proteolysis and ensuring maximal representation of tissue-specific proteins. Proteins are solubilized in sodium dodecyl sulfate (SDS) and fractionated by SDS-polyacrylamide gels. They are then electroblotted onto polyvinylidene fluoride membranes to generate blots ready for incubation with specific antibodies. Following detection of antigen, the blots can be reused several times using a stripping protocol that permits the selective removal of both primary and secondary antibodies in a single incubation. Multiple reprobings make this protocol very useful to study human tissue-specific proteins.

5 A Two-Step Transfer of Low- and High-Molecular-Weight Proteins

Transfer of high-molecular-weight proteins has been the bane of investigators everywhere [7]. When higher percentage gels are used, this problem is accentuated. Interestingly investigators have sought to increase the transfer of large proteins by enhancing the degree of protein migration out of the gel during the transfer. Researchers have used prolonged electrotransfer (16–21 h) at high current density along with inclusion of sodium dodecyl sulfate (SDS) in the transfer buffer, to enhance protein elution and efficiently transfer high-molecular-weight proteins [8, 9].

Adding SDS to a final concentration of 0.1 % in the transfer buffer and transferring for 21 h have helped to quantitatively transfer proteins [8]. Some investigators have used novel gels and blotting buffers [7] to efficiently transfer high-molecular-weight proteins. These investigators made gels using a 50:1 ratio of acrylamide:bisacrylamide in all experiments, since they were mainly interested in transferring high-molecular-weight proteins. In addition, they also used a different blotting buffer. Others [10] have cleaved high-molecular-weight proteins with periodate or alkali, before transfer, to facilitate transfer of proteins. Partial proteolytic digestion of high-molecular-weight proteins prior to transfer [11] has also been attempted. Bigger pore sizes provided by agarose gels have been used to advantage in [12] a composite agarose-polyacrylamide gel containing SDS and urea for this same purpose.

Otter et al. [9, 13] used a two-step procedure to electrotransfer both high-molecular-weight (greater than 400,000) and low-molecular-weight (less than 20,000) proteins from polyacrylamide gels to nitrocellulose sheets. The salient features of this method included a two-step electrotransfer. The low-molecular-weight polypeptides were eluted at a low current density (approximately 1 mA/cm^2) for 1 h. This was followed by prolonged electrotransfer (16–20 h) at high current density (approximately 3.5–7.5 mA/cm^2) under conditions that favored the transfer of high-molecular-weight proteins. SDS (0.01 %) was added to the transfer buffer to enhance protein transfer. The nitrocellulose was air-dried following the transfer to eliminate protein loss during subsequent processing. This transfer procedure was found to work well with all polyacrylamide gel systems tested and with proteins prepared from many different cell types.

6 Toxin Binding to Chimeric K^+ Channels Immobilized on a Solid Nitrocellulose Support

Chip technology is playing an important role in pharmaceutical and biological research. Microarrays that are used currently involve DNA. Peptides are also used to some extent. Large proteins, however, have not been used extensively mainly owing to difficulties associated with purification and structural integrity [14, 15]. It is even more problematic in the case of membrane proteins that are often stabilized by the lipid moiety. There have been reports of the functional immobilization on gold/glass [16] or sensor surface of BIAcore chip [17] of a ligand-activated G-protein-coupled receptor. K^+ channels are membrane proteins making up the largest and most ubiquitous family of ion channels that control excitability in a number of cell types. Several neurological diseases have been thought to be involved owing to the dysfunction of these channels and as a result are potential targets for therapeutic drugs [15]. Therefore, displaying K^+ channel proteins on a solid surface is valuable, potentially leading to clinical applications. Such an approach also has the potential for drug screening methodologies.

7 Development of a Membrane-Array Method for the Detection of Human Intestinal Bacteria in Fecal Samples

All mammals, including human beings, are adapted to life in a world of microbes. For every gram weight of human intestinal contents, there are 10^{10}–10^{12} colony forming units of bacteria (this is 10–20 times the total number of tissue cells in the whole body) [18]. At least 400 different species of bacteria (of which 30–40 species account for 99 % of the total microflora) colonize the human gastrointestinal tract (GIT) [19, 20]. The indigenous intestinal bacteria play important roles in food digestion, production of vitamins and other essential nutrients, metabolism of endogenous and exogenous compounds, and in preventing pathogenic microflora from colonizing the GIT. Therefore the knowledge relating to the numbers and bacterial species found in the human GIT is important [21]. The safety implications of antibiotic resistant bacteria in foods, contamination of foods by fecal material, the effect of diet, food additives, veterinary drug residues on the ecosystem of the intestine, and the use of probiotics in the prevention and treatment of GI disorders have stimulated the interest in intestinal microflora [20].

The population of anaerobic bacteria in the human GIT has been characterized traditionally by biochemical, microscopic, physiological, and selective culture plating methods of fecal samples from human beings. A number of molecular techniques have been used in recent years to analyze the bacterial flora in human fecal samples. Such analysis can detect changes in the human GIT flora rapidly and precisely [19, 20].

Wang et al. [20] have used a nitrocellulose membrane-array method to detect human GIT bacteria in fecal samples without the use of expensive microarray-arrayer and laser scanner. Three 40-mer oligonucleotides specific for each of 20 important human GIT bacterial species (total 60 probes) were designed and synthesized, based on comparison of the 16S rDNA sequences in the GenBank data library.

The oligos were diluted and xylene cyanol was added as an indicator. The diluted oligos were heated in a boiling water bath for 2 min and then cooled immediately in ice-water for 1 min. The cooled oligos were applied to the nitrocellulose membrane with a micropipetman in a 6×10 array. The membranes were heated at 80 °C for 2 h after air drying and UV-crosslinking. Digoxigenin (DIG)-labeled 16S rDNAs were amplified by polymerase chain reaction from human fecal samples or pure cultured bacteria with the help of two universal primers and hybridized to the membrane array. Hybridization signals were read by NBT (nitro blue tetrazolium)/BCIP (5-bromo-4-chloro-3-indolyl phosphate) color development. The two universal primers were found to amplify full size 16S rDNA from all the 20 bacterial species that were tested.

The membrane-array method was thus found to be a reliable method for the detection of important human intestinal bacteria in the fecal samples [20].

8 A New Black Cellulose Nitrate Support for Protein Microarray

Protein microarray is still in an early stage of development compared to the well-established DNA microarray technology. This is due to the different physical and chemical properties of nucleic acids and proteins. There is no comparable amplification method for proteins like PCR used for nucleic acids and also proteins are much more complicated to purify. The irreversible denaturation of proteins during the process of immobilization and the insufficient stability of purified proteins are still problematic [22, 23]. An optimal surface for all proteins has not been discovered yet owing to the varying properties of different proteins. Reck et al. [24] have used a modified nitrocellulose membrane for the optimization of a microarray sandwich-enzyme-linked immunosorbent assay (ELISA) against hINF (human interferon)-γ. This membrane was found to provide an excellent signal-to-noise ratio (SNR) and very little autofluorescence. The novel microarray slide used by these investigators was a self-made prototype coated with a black microporous cellulose nitrate substrate. The porous nitrocellulose substrate was produced by Sartorius Stedim Biotech (GmbH, Gottingen, Germany) using an evaporation technology. Porous nitrocellulose substrates that are available commercially tend to produce a high background fluorescence through their overall thickness even though they provide high protein binding capacity. Such background fluorescence may be caused by the substrate itself or by the adhesives used to attach the cellulose nitrate to glass. Sartorius developed a novel recipe and adhesive-free manufacturing procedure to overcome this problem. The white substrate made by Sartorius was already optimized for high binding capacity and low autofluorescence (made possible by the choice of various cellulose nitrate grades). The black substrate was made to provide an additional benefit by adsorbing background fluorescence originating in the depth of the structure. The added coloring material is essential for the low autofluorescence of the black substrate.

Reck et al. [24] used this self-made black nitrocellulose membrane with a high SNR and low autofluorescence as a microarray substrate. For spotting automation, an affymetrix 417 contact printer was used. Probes were spotted using a spotting buffer containing phosphate buffered saline and 0.5 % trehalose to which 40 % glycerol was added to prevent denaturation and improving the stability of protein probes. Using this black nitrocellulose membrane the authors have optimized a microarray sandwich-ELISA against hINF-γ [24].

9 Quantification of Proteins Bound to PVDF Membranes by Elution of Coomassie Brilliant Blue

Proteins transferred to PVDF membranes can be stained with a variety of stains including Coomassie Brilliant Blue (CBB), amido black, and colloidal gold [25]. It becomes essential to determine the amount of protein present on the membrane in order to accurately determine the sensitivity of immunostaining, initial yields from protein microsequencing, and the specific activity of enzymes bound to PVDF. Since the extent and rate of electroblotting is likely to change between samples, such values cannot be obtained from the amount of loaded proteins. Therefore, Kain and Henry [26] developed a method to quantify proteins bound to PVDF membrane by elution of CBB, in order to quantify proteins for microsequencing. Proteins were resolved on a 13 % gel and transferred to PVDF.

The membrane was stained with CBB and rapidly destained. The stained membrane was air-dried and individual protein bands were excised with a scalpel. The protein from each PVDF piece was eluted with 0.1 % SDS/50 % isopropanol. The colored liquid was removed and read at 595 nm using a Beckman Model DU-40 spectrophotometer after calibrating with CBB eluted from a PVDF fragment derived from a region of the blot that did not have a protein band. The proteins are not eluted along with the dye. The authors eluted CBB from ^{14}C labeled proteins electrotransferred to PVDF membranes and subjected the eluted solution and the PVDF membrane containing the radiolabeled protein to liquid scintillation counting. Virtually all of the protein was found to be retained on the PVDF membrane fragment during CBB elution. The authors also found that each of the protein that they tested gave a linear response with respect to the change in absorbance in the protein range of 500 ng to 10 μg. The slope of each curve was found to differ considerably, confirming the well-known fact regarding the variance of CBB bound by different proteins. Thus, in any assay of protein content a standard curve with the protein of interest needs to be obtained to get the most accurate quantitative results. This protocol is useful for the analysis of multiple samples since the absorbance readings obtained from the eluted dye stays stable for up to 1 h. This procedure is not dependent on bandshape since it could be useful in the quantitation of proteins separated by two-dimensional gel electrophoresis. Densitometric scanning of 2-D gels is complicated by the lack of uniformity in the shape of protein spots. On account of the fact that the proteins are not extracted from the membrane, this procedure is useful for protein quantitation before protein microsequencing, for analysis of amino acid composition, and for immunological procedures.

10 Enhanced Protein Recovery After Electrotransfer Using Square Wave Alternating Voltage

Bienvenut et al. [27] studied the efficiency of protein transfer using the conventional continuous current (as in Towbin transfer system) and the use of an unusual square wave alternative voltage (SWAV).

In this procedure, immediately following SDS-PAGE, the gels were soaked in deionized water for 5 min, and then equilibrated two times (5 min each time) in the cathodic blotting buffer. Trans-blot PVDF membranes were equilibrated with the anodic buffer for 5 min. The standard blotting technique used a continuous current (1 mA/cm^2) using 3-(cyclohexylamino)-1-propane-sulfonique acid (CAPS) buffer. The voltage used in the SWAV method of transfer was an asymmetrical alternating voltage also using CAPS buffer. This method delivered +12 V for 125 ms followed by −5 V for 125 ms repetitively. This corresponded to a 4 Hz frequency signal and an average tension of 3.5 V. Following transfer the gels were stained with Coomassie blue and the membranes were stained with amido black and destained by repeatedly washing with deionized water. An average 65 % increase of protein recovery was observed using the SWAV technology in combination with CAPS buffer compared to that obtained with standard immunoblotting conditions [27].

11 Polyethylene Glycol-Mediated Significant Enhancement of the Immunoblotting Transfer

Zeng et al. [28] have studied the effects of polyethylene glycol (PEG) on protein fixation, electrotransfer from SDS-polyacrylamide gels onto PVDF membranes, and immunoblotting. Serum proteins were resolved by SDS-PAGE and the gel was then immersed in a 30 % PEG 2000 buffer for 2 h for reversibly fixing the proteins (unlike trichloroacetic acid-sulfosalicylic acid or acetic acid-methanol systems that irreversibly fix proteins). PEG 1000, 1500, or 2000 (at 30 % level) were found to have almost equal ability to preserve the protein bands in the gels very well. PEG less than 1000 was not found to have significant effect. The proteins in the gel were electrotransferred to PVDF for 24–48 h (in a −20 °C freezer at 200 mA/120 V) using the same buffer and visualized finally by the indirect HRP-antibody method. The total protein detected (lowest level) per lane was 25 pg. It was found that when the transfer was carried out in the absence of PEG, a similar immunostain revealed a significantly lower sensitivity and the bands obtained were found to be slightly blurred. PEG 1000–2000 was capable of increasing the sensitivity of immunoblotting by 10–100-fold. This increase is thought to be brought about by its nondenaturing hydrophobicity with (a) self-association of protein molecules that could diminish protein blow-through (through the PVDF membrane), (b) miniaturization caused by PEG, possibly increasing the intraband protein density, and (c) possible enhancement of the interaction between free antigen and antibody by PEG treatment.

12 Acid Electroblotting onto Activated Glass

Data obtained from partial amino acid sequencing has been used frequently in assisting in isolation of a gene coding for a specific protein and also for confirming that the right gene has been isolated. In order to establish the structure of the mature gene translation product and to cross-check sequences determined at the nucleic acid level, more extensive amino acid sequence data can be employed. Aeberold et al. [29] have used a procedure to electroblot proteins to activated glass to isolate subpicomolar levels of proteins for microsequencing. They considered nitrocellulose and nylon membranes for this purpose. However, nitrocellulose was found to dissolve during the sequencing chemistry while the charge modified nylon membranes were found to collapse into a solid pellet during the procedure. The glass fiber paper support was found to be completely stable to the sequencing conditions. However, untreated glass fiber sheets possess a very limited ability to adsorb proteins. However, the authors found that the glass fiber sheets developed considerable capacity to adsorb proteins (7–10 μg/cm^2) following treatment with trifluoroacetic acid (TFA). The mechanism of protein adsorption was a result of ionic interaction of the net positively charged proteins (owing to the acidic transfer buffer) to the negative charges on the glass fiber sheet. The TFA treatment of the glass sheet apparently unveiled these negative charges on the sheet.

12.1 Activation of Glass Fiber Sheets (Acid Etching)

Whatman GF/C or GF/F circles or sheets were placed inside neat TFA in a covered Petri dish and incubated for 1 h at room temperature. Extreme care was taken to avoid air bubbles if more than one sheet was used. The glass fiber sheets were dried completely until there was no trace of TFA.

12.2 Acid Blotting

The SDS from the proteins, following SDS-PAGE, was displaced by immersing the gels in 0.5 % (v/v) acetic acid containing 0.5 % Nonidet P-40 for 10 min at room temperature. Before assembling the sandwich for blotting, each of the blotting component was incubated with the blotting buffer. However, in order to minimize acid-catalyzed protein cleavage, acid concentration was kept as low as possible. Blotting was carried out. At this low pH of protein transfer, the proteins have a net positive charge and migrate toward the cathode onto the glass fiber paper.

Protein bands, following detection with Coomassie or fluorescent staining, were excised out of the glass fiber sheets and placed in the cartridge of a gas-phase sequenator without further treatment [29].

13 Clarification of Immunoblots on PVDF for Transmission Densitometry

For protein quantitation, methods such as radioimmunoassay or enzyme-linked immunosorbent assay have been used. However, these assays cannot provide much information regarding the characteristics of the protein, nor can they distinguish between cross-reactive species including different protein components that react with the probe. Consequently, SDS-PAGE followed by immunoblotting is required to quantify events such as cleavage of polypeptide fragments utilizing polyclonal sera or the determination of enzyme activation status consequent to a loss of pro-enzyme domains [30].

While results are compromised by the limited linearity of photographic methods, immunoblot quantification has been done indirectly by densitometry carried out on the autoradiograph of the blot [31] as well as by making a transparent copy of the blot [30]. Tagami et al. [32] has performed densitometry of a dry, color-stained blot with a laser on account of the intensity and collimation of the light source. However, it was found that the absorbance of even the unstained parts of the membrane was approximately 2 and also leading to variable baselines since they were found to be very sensitive to warping [33]. However, it was found that such densitometry was possible if the membranes could be rendered transparent like a polyacrylamide gel. Nitrocellulose could be made transparent by treating with three-in-one lubricating oil [33], thus allowing the blot to be quantitated using a conventional densitometer, analogous to the manner in which gels are scanned. However, PVDF (on account of its more hydrophilic character) was not made transparent by treatment with oil.

Tarlton and Knight tried several solvents differing in refractive index to make PVDF transparent [30]. Their idea was that if PVDF could be made transparent like nitrocellulose they could be used for several experiments that were not possible with nitrocellulose, such as adsorbing lipids for the detection of anti-cholesterol antibodies, to be used in conjunction with transmission densitometry. They found that ethylene glycol was the most effective. It was found that ethylene glycol/glycerol mixture (9:1; v/v; refractive index of 1.433) was found to bring about the most optimal clarification of PVDF membranes. For this procedure dry PVDF membrane was first moistened in methanol and then immersed in solvent mixture with a minimum of one change and optical absorbance of PVDF was measured with a laser densitometer. Immunoblots were taken from storage in water and immersed in the clarification mixture with two changes.

The authors found that the PVDF immunoblots could be examined over long periods of time without loss of transparency, owing to the low volatility of both ethylene glycol and glycerol. The authors found no fading of bands after repeated clarification of dried PVDF blots by ethylene glycol/glycerol over a period of 2 months. However, clarification of nitrocellulose for the purpose of quantitative densitometry was found to produce some fading when oil was used.

14 Parallel Protein Chemical Processing Before and During Western Blot and the Molecular Scanner Concept

Following gel electrophoresis, the proteins are stained with either Coomassie Brilliant Blue or silver and spots containing the protein of interest are excised either manually or using robotic methods. Following several rounds of washing, the gel slices are subjected to in-gel tryptic digestion. The tryptic peptides are extracted from each gel slice and concentrated, purified, and subjected to mass spectrometry. The robots that carry out the initial sample preparation are expensive. Furthermore, the investigator has to keep track of how each sample was related to the original gel during all of these steps, thus introducing opportunities for confusion. Matrix-assisted laser desorption/ionization (MALDI) mass spectrometry directly on a polymeric surface has been performed as early as 1990 [34]. Specifically, proteins transferred to a capture membrane can be analyzed by scanning the surface with the laser beam of a MALDI machine. The concept of molecular scanning [35] involves the use of enzymatic digestion during the blotting from an SDS-PAGE gel or two-dimensional gel to a PVDF capture membrane (a trypsin-coated membrane being placed between gel and the PVDF membrane). This procedure does not require the use of gel staining, spot excision, or extraction. These authors found that a combination of in-gel digestion (prior to western blotting) together with the positioning of a trypsin-coated membrane between the gel and the PVDF capture membrane (during western blotting) resulted in obtaining the best digestion efficiency (digestion of high-molecular-weight and basic proteins without losses of low-molecular-weight polypeptides due to diffusion) compared to either in-gel digestion or the use of trypsin-coated membrane alone during transfer. The capture membrane absorbs the peptide fragments from the digested proteins, following which the membrane is treated with a MALDI ionizing matrix and mass spectrometry analysis is carried out directly on the membrane. All the proteins that were originally present in the gel are processed at the same time using methods compatible with mass spectrometry. In order to obtain maximum digestion and transfer to the capture membrane, the authors used a pulsed or alternating electric current (an unusual square wave alternative voltage—*see* ref. 27). The time of contact between the proteins migrating out of the gel, with the trypsin immobilized on the membrane, was maximized by this oscillating current. This present study highlighted a positive influence of the "shaking" effect of the asymmetric alternative voltage on gel protein extraction.

15 Electronic Western Blot of Matrix-Assisted Laser Desorption/Ionization Mass Spectrometric-Identified Polypeptides from Parallel Processed Gel-Separated Proteins

As described above, proteins separated by one-dimensional or two-dimensional gel electrophoresis are transferred directly through a trypsin-coated membrane onto a membrane ready for matrix-assisted laser desorption/ionization (MALDI) [36]. The authors found that the protein transfer and efficiency of digestion was more than 95 %. They identified 110 unique proteins obtained from an *Escherichia coli* extract and 149 proteins from a mouse liver homogenate resolved by one-dimensional (1-D) sodium dodecyl sulfate polyacrylamide gel electrophoresis (SDS-PAGE). Furthermore, they used a Visual Basic Program to plot the identified proteins according to where they were found on the gel. Thus, the presence and distribution of any of the identified proteins could be visualized as in a western blot without the use of an antibody.

Nadler et al. [36] used a special membrane and procedure to covalently couple trypsin to obtain a high activity, compared to that used by Bienvenut et al. [35]. Thus, they could avoid starting the digestion process in the separating gel before electroblotting and also the use of the special oscillating electroblotting apparatus as done earlier [35]. The aldehyde-activated polyethersulfone membrane, Gelman US450, reacts with nucleophiles (such as primary amines) to form nucleophiles that are then reduced with sodium cyanoborohydride. The authors determined that the amount of trypsin covalently attached to this membrane ranged from 14 to 22 μg/cm^2 of membrane. This amount of trypsin had sufficiently high activity to bring about a quick digestion of proteins in a single pass [36].

16 Gold Coating of Nonconductive Membranes Before Matrix-Assisted Laser Desorption/Ionization Tandem Mass Spectrometric Analysis Prevents Charging Effect

As seen earlier, it is possible to perform matrix-assisted laser desorption/ionization (MALDI)-mass spectrometry directly on PVDF (polyvinylidene difluoride) membrane. However, it is not possible to carry out tandem mass spectrometry (MS/MS) directly on this polymeric surface. The reason is that the isolating material is not able to dissipate the charge made by the MALDI process [34, 37]. This charging effect has been postulated to create local perturbations in the electric field between the acceleration plate and the sample. To overcome this problem and allow MS/MS analysis on tandem TOF instruments, the sample has to be conductive to dissipate the charges. By depositing a thin gold layer on the surface of a nonconductive membrane like PVDF, Scherl et al. [34] describe a positive effect. The thin gold layer is applied by anodic vaporization following matrix deposition. The conductive material permits the dissipation of the charges, resulting in the first successful MS/MS analysis of peptides from PVDF membranes using a MALDI-TOF/TOF instrument under normal operating conditions [34].

17 Semidry Electroblotting of Peptides and Proteins from Acid-Urea Polyacrylamide Gels

Electroblotting protocols have been mostly designed for protein transfer from SDS-containing gels, using tank-type apparatus typically requiring 10–18 h for transfer. Semidry transfer methods have been developed for SDS-PAGE not long ago [38]. Owing to its speed and convenience this procedure has been adopted for DNA and RNA electroblotting.

Polyacrylamide gels without SDS provide an important avenue for separating proteins under partially denaturing or native conditions. Separation of proteins and peptides as a function of their combined size and charge has been made possible with the use of low pH PAGE systems. The acidic, urea-containing (AU)-PAGE system [39] enables excellent resolution of several proteins and peptides that cannot be resolved by SDS-PAGE.

Model polypeptides like calf thymus histone (21.5 kDa), ribonuclease A (13.7 kDa), human lysozyme (14 kDa), pancreatic trypsin inhibitor (6.5 kDa), and others with molecular weight ranging from 6.5 to 3.3 kDa were used to test transfer parameters. Using a power setting of 115 mA and 5 V, a transfer solution of 5 % acetic acid (the same solution was used for electrophoresis), and a transfer time of 15 min, it was possible to transfer the polypeptides almost completely. PVDF (0.2 μm) was found to be a superior membrane, compared to nitrocellulose (0.2 μm), for efficient transfer.

The authors studied the effect of adding methanol to the transfer buffer. Methanol addition has been recommended for increasing protein binding capacity of nitrocellulose membrane [40]. The authors [41] found that the addition of methanol (10 %) brought about precipitation of protein on the gel, impeding the electrophoretic transfer. Also, they found no increase in protein binding to either PVDF or nitrocellulose membranes upon addition of methanol to the buffer.

18 Transfer of Silver-Stained Proteins from Polyacrylamide Gels to Polyvinylidene Difluoride Membranes

Wise and Lin [42] showed that they could transfer almost all the silver-stained proteins from a polyacrylamide gel to PVDF by rinsing the gel in 2× SDS Laemmli sample before transfer. Some silver-stained proteins were also found to be directly transferred without a rinse with the Laemmli buffer. The antigenicity of the transferred proteins was found to be retained when transferred either way.

19 A Simple Method for Coating Native Polysaccharides onto Nitrocellulose

Lipid-free or protein-free polysaccharides (PS) have been reported to have very low affinity for plastic and polystyrene. Carbohydrate molecules have been modified in an effort to increase their binding affinity. These modifications have involved the covalent attachment of poly-l-lysine biotin or tyramine to the PS. The modification has the possibilities of being limited by high background, loss of specificity, loss of antigenicity, and lack of reproducibility.

Feng et al. describe a method to immobilize PS to nitrocellulose without using any modification and permitting antibody analysis by enzyme-linked immunoassays. Bacterial levan (a $\beta(2 \rightarrow 6)$-polyfructosan) and dextran (a polymer of $\alpha(1 \rightarrow 6)$ glucose), both neutral PS and a highly charged PS (*Neisseria meningitides* group C polysaccharide; a polymer of $\alpha(2 \rightarrow 9)$ sialic acid) have been coated onto nitrocellulose through filtration devices. Various blotting assays can be used with the PS-coated membrane to investigate specific antibodies [43, 44].

20 Fabrication of Electrospun PVDF Nanofiber Membrane for Western Blot with High Sensitivity

The microphase separation-based porous PVDF membrane normally used in western blotting methods is extremely hydrophobic. Unless pre-wetted with methanol, the membrane does not wet in aqueous solutions. In addition, the membrane's protein binding capacity and mechanical strength deteriorates following long-term immersion of the PVDF membrane in methanol. Cho et al. [45] studied a mechanically strong PVDF membrane with uniform and opened pores. This membrane, prepared by electrospinning and subsequent calendering, contained long nanofibers. Cho et al. showed that this membrane did not require a methanol pre-wet step and displayed high sensitivity, low background, and high protein binding capacity.

References

1. Peferoen M, Huybrechts R, De Loof A (1982) Vacuum-blotting: a new simple and efficient transfer of proteins from sodium dodecyl sulfate-polyacrylamide gels to nitrocellulose. FEBS Lett 145:369–372
2. Hermansen LF, Pedersen O, Sletten K (1993) Centrifuge-blotting of proteins after separation by sodium dodecyl sulfate-polyacrylamide gel electrophoresis. Electrophoresis 14:1302–1306
3. Patterson SD (1994) From electrophoretically separated protein to identification: strategies for sequence and mass analysis. Anal Biochem 221:1–15, Review
4. Burkhart WA, Moyer MB, Bailey JM, Miller CG (1996) Electroblotting of proteins to Teflon tape and membranes for N- and C-terminal sequence analysis. Anal Biochem 236:364–367
5. Burkhart W, Moyer M, Bodnar W, Everson A, Valladares V et al (1995) Direct collection into Zitex and PVDF for Edman sequencing: elimination of polybyrene. In: Crabb J (ed) Techniques in protein chemistry VI. Academic, San Diego, pp 169–176
6. Kain SR, Mai K, Sinai P (1994) Human multiple tissue western blots: a new immunological tool for the analysis of tissue-specific protein expression. Biotechniques 17:982–987
7. Bolt MW, Mahoney PA (1997) High efficiency blotting of proteins of diverse sizes following sodium dodecyl sulfate-polyacrylamide gel electrophoresis. Anal Biochem 247:185–192
8. Erickson PF, Minier LN, Lasher RS (1982) Quantitative electrophoretic transfer of polypeptides from SDS polyacrylamide gels to nitrocellulose sheets: a method for their re-use in immunoautoradiographic detection of antigens. J Immunol Methods 51:241–249
9. Otter T, King SM, Witman GB (1987) A two-step procedure for efficient electro transfer of both high-molecular weight (greater than 400,000) and low-molecular weight (less than 20,000) proteins. Anal Biochem 162:370–377
10. Renart J, Reiser J, Stark GR (1979) Transfer of proteins from gels to diazobenzyloxymethyl paper and detection with and detection with anti-sera: a method for studying antibody specificity and antigen structure. Proc Natl Acad Sci U S A 76:3116–3120
11. Gibson W (1981) Protease-facilitated transfer of high-molecular-weight proteins during electrotransfer to nitrocellulose. Anal Biochem 118:1–3
12. Elkon KB, Jankowski PW, Chu JL (1984) Blotting intact immunoglobulins and other high-molecular-weight proteins after composite agarose-polyacrylamide gel electrophoresis. Anal Biochem 140:208–213
13. Kurien BT, Scofield RH (2006) Western blotting. Methods 38:283–293
14. Tomizaki KY, Usui K, Mihara H (2005) Protein-detecting microarrays: current accomplishments and requirements. Chembiochem 6:782–799, Review
15. Legros C, Martin-Eauclaire MF, Pongs O, Bougis PE (2007) Toxin binding to chimeric K^+ channels immobilised on a solid nitrocellulose support. Biochem Biophys Res Commun 353:1086–1090
16. Neumann L, Wohland T, Whelan RJ, Zare RN, Kobilka BK (2002) Functional immobilization of a ligand-activated G-protein-coupled receptor. Chembiochem 3:993–998
17. Bieri C, Ernst OP, Heyse S, Hofmann KP, Vogel H (1999) Micropatterned immobilization of a G protein-coupled receptor and direct detection of G protein activation. Nat Biotechnol 17:1105–1108
18. Suau A, Bonnet R, Sutren M, Godon JJ, Gibson GR, Collins MD et al (1999) Direct analysis of genes encoding 16S rRNA from complex communities reveals many novel molecular species within the human gut. Appl Environ Microbiol 65:4799–4807
19. Moore WE, Moore LH (1995) Intestinal floras of populations that have a high risk of colon cancer. Appl Environ Microbiol 61: 3202–3207
20. Wang RF, Kim SJ, Robertson LH, Cerniglia CE (2002) Development of a membrane-array method for the detection of human intestinal bacteria in fecal samples. Mol Cell Probes 16:341–350
21. Falk PG, Hooper LV, Midtvedt T, Gordon JI (1998) Creating and maintaining the gastrointestinal ecosystem: what we know and need to know from gnotobiology. Microbiol Mol Biol Rev 62:1157–1170, Review
22. Glökler J, Angenendt P (2003) Protein and antibody microarray technology. J Chromatogr B Analyt Technol Biomed Life Sci 797: 229–240, Review
23. Kusnezow W, Hoheisel JD (2003) Solid supports for microarray immunoassays. J Mol Recognit 16:165–176, Review
24. Reck M, Stahl F, Walter JG, Hollas M, Melzner D, Scheper T (2007) Optimization of a microarray sandwich-ELISA against hINF-gamma on a modified nitrocellulose membrane. Biotechnol Prog 23:1498–1505

25. LeGendre N, Matsudaira P (1988) Direct protein microsequencing from Immobilon-P transfer membrane. Biotechniques 6:154–159
26. Kain SR, Henry HL (1990) Quantitation of proteins bound to polyvinylidene difluoride membranes by elution of Coomassie brilliant blue R-250. Anal Biochem 189:169–172
27. Bienvenut WV, Déon C, Sanchez JC, Hochstrasser DF (2002) Enhanced protein recovery after electrotransfer using square wave alternating voltage. Anal Biochem 307: 297–303
28. Zeng C, Suzuki Y, Alpert E (1990) Polyethylene glycol significantly enhances the transfer of membrane immunoblotting. Anal Biochem 189:197–201
29. Aebersold RH, Teplow DB, Hood LE, Kent SB (1986) Electroblotting onto activated glass. High efficiency preparation of proteins from analytical sodium dodecyl sulfate-polyacrylamide gels for direct sequence analysis. J Biol Chem 261:4229–4238
30. Tarlton JF, Knight PJ (1996) Clarification of immunoblots on polyvinylidene difluoride (PVDF) membranes for transmission densitometry. J Immunol Methods 191:65–69
31. Likavcanova E, Lagacé J (1992) Quantitative analysis of immunoglobulin G subclass responses to Pseudomonas aeruginosa antigens in cystic fibrosis. J Med Microbiol 36: 437–444
32. Tagami T, Nakamura H, Sasaki S, Miyoshi Y, Imura H (1993) Estimation of the protein content of thyroid hormone receptor alpha 1 and beta 1 in rat tissues by Western blotting. Endocrinology 132:275–279
33. Maciewicz RA, Knight PJ (1988) Transmission densitometry of stained nitrocellulose paper. Anal Biochem 175:85–90
34. Scherl A, Zimmermann-Ivol CG, Di Dio J, Vaezzadeh AR, Binz PA, Amez-Droz M et al (2005) Gold coating of non-conductive membranes before matrix-assisted laser desorption/ionization tandem mass spectrometric analysis prevents charging effect. Rapid Commun Mass Spectrom 19:605–610
35. Bienvenut WV, Sanchez JC, Karmime A, Rouge V, Rose K, Binz PA et al (1999) Toward a clinical molecular scanner for proteome research: parallel protein chemical processing before and during western blot. Anal Chem 71:4800–4807
36. Nadler TK, Wagenfeld BG, Huang Y, Lotti RJ, Parker KC, Vella GJ (2004) Electronic Western blot of matrix-assisted laser desorption/ionization mass spectrometric-identified polypeptides from parallel processed gel-separated proteins. Anal Biochem 332:337–348
37. Knochenmuss R (2004) Photoionization pathways and free electrons in UV-MALDI. Anal Chem 76:3179–3184
38. Jacobson G, Kårsnäs P (1990) Important parameters in semi-dry electrophoretic transfer. Electrophoresis 11:46–52
39. Panyim S, Chalkley R (1969) High resolution acrylamide gel electrophoresis of histones. Arch Biochem Biophys 130:337–346
40. Johnson TK, Yuen KC, Denell RE, Consigli RA (1983) Efficient transfer of proteins from acetic acid-urea and isoelectric-focusing gels to nitrocellulose membrane filters with retention of protein antigenicity. Anal Biochem 133: 126–131
41. Wang MS, Pang JS, Selsted ME (1997) Semidry electroblotting of peptides and proteins from acid-urea polyacrylamide gels. Anal Biochem 253:225–320
42. Wise GE, Lin F (1991) Transfer of silver-stained proteins from polyacrylamide gels to polyvinylidene difluoride membranes. J Biochem Biophys Methods 22:223–231
43. Feng SH, Rubinstein LJ, Stein KE (1991) A simple method for coating native polysaccharides onto nitrocellulose. J Immunol Methods 137:261–266
44. Kurien BT, Scofield RH (2009) A brief review of other notable protein blotting methods. Methods Mol Biol 536:367–384, Review
45. Cho E, Kim C, Kook J-K, Jeong Y II, Kim JH, Kim YA et al (2012) Fabrication of electrospun PVDF nanofiber membrane for Western blot with high sensitivity. J Memb Sci 389:349–354

INDEX

A

Acid blotting 496–497
Acid etching 496
Acid-urea-PAGE 500
Activated glass 496–497
Activity gel electrophoresis 26, 61–71
Adhesive interactions under fluid shear 399–409
Affinity immunoblotting 109–116, 269
Affinity purification 39, 168–169, 257–267, 386
Alkaline phosphatase buffer. *See* Buffer
Assay
  blot rolling 399–409
  dot-immunobinding 105–108
  ECL (*see* Chemiluminescence)
  electrophoretic mobility shift assay (EMSA) 61, 62, 356–364, 369, 370
  ELISPOT
    B cell 419–426
    fluorescence ELISPOT 439–440
    T cell 427–434
  immunochromatographic 301
  multiple antigen blot 301–318
Autoantibody
  anti-centromere antibody 257
  anti-La (SS-B) 251
  anti-Ro 60 110
  clonotype analysis 109–116
  multiple nuclear dot anti-nuclear antibody 258
Autoantigens. *See* Systemic lupus erythematosus
Automation 31–39, 405, 493

B

B Cell Elispot 419–426
Blotting. *See* Protein blotting
Buffer
  alkaline phosphatase 79, 111, 252, 262, 348, 352
  antibody-binding 330, 340
  cell lysis 199
  cetyltrimethylammonium bromide (CTAB)
    sample 152, 153
  Eastern blotting 154, 156
  immunoprecipitation 330, 340, 344
  isoelectric focusing (IEF) equilibration 67
  isoelectric focusing (IEF) sample 154, 157
  Laemmli
    modified 334
    5x Lysis buffer 79
  SDS-PAGE sample 169, 170, 191, 228
  SDS-PAGE stacking 228
  SDS-PAGE 5x sample 451
  Towbin buffer 23
    CAPS buffer 23
    Towbin system buffer 23
Burnette, W. Neal 15, 20, 487
  western blotting 9–12

C

Capillary blotting 123, 125, 487
Cationic electrophoresis 151–161
CBB. *See* Coomassie Brilliant Blue (CBB)
CBB-stained proteins
  elution of CBB for quantification 494
  transfer to transparencies 252, 253, 415–417
Centrifuge blotting 411–414, 488
Chemiluminescence 36, 56, 134, 154, 161, 221, 235, 274, 280, 368, 378, 383, 389, 390, 396
Cleveland peptide mapping 230–233
Concentrating proteins with membranes 21
Coomassie Brilliant Blue (CBB) 21, 22, 26, 43, 52, 65, 74, 134, 156, 160, 189, 191, 192, 230, 233, 235, 252, 253, 270, 386, 415, 416, 454, 475, 476, 494, 498
CTAB-PAGE 152, 155, 160

D

Densitometry 497–498
Diffusion blotting 25–27, 61, 62, 65, 69, 70, 73–76, 322, 325
Dot blotting 18
Dot-immunobinding assay (Dot-Iba) 105–108
Double-blotting 277–282

E

Eastern blotting 151–161
ECL. *See* Enhanced chemiluminescence (ECL)
Edman sequencing on PVDF 228, 489

Biji T. Kurien and R. Hal Scofield (eds.), *Western Blotting: Methods and Protocols*, Methods in Molecular Biology, vol. 1312, DOI 10.1007/978-1-4939-2694-7, © Springer Science+Business Media New York 2015

Electroblotting of very large proteins 285
Electrophoresis
activity gel.. 26, 61–71
cationic 151–161
electrophoretic mobility.. 207
isoelectric focusing (IEF) (*see* Gel)
native 343–352
sodium dodecyl sulfate-polyacrylamide gel electrophoresis (SDS-PAGE)........................ 450
two-dimensional.. 75
vertical agarose.. 360, 366
Electrophoretic destainer.. 6, 15
Electrophoretic mobility shift assay (EMSA). *See* Assay
ELISPOT. *See* Enzyme-linked immunospot assay (ELISPOT)
Enhanced chemiluminescence (ECL)................ 55, 89, 205, 216, 219, 235, 274, 289, 368, 389, 390
Enzyme-linked immunospot assay (ELISPOT).. 419–444
Enzyme substrate
AEC 421, 423, 429, 436, 439
DAB
NBT/BCIP.. 439
ethylene glycol 497
luminol and peroxide containing reagents 154
PVDF clarification 497, 498

F

Far-western-blotting 379–397
Fast green 21, 23, 249, 258, 263, 266, 269, 273, 275, 332, 336, 337, 415, 416
Fingerprinting 477

G

Gel
agrose-gel electrophoresis 121–125
CTAB-PAGE 155
immobilized pH gradient 321–326
isoelectric focusing (IEF)
flatbed non-denaturing IEF 114
phast gel.. 237–244, 325
sodium dodecyl sulfate-polyacrylamide gel electrophoresis (SDS-PAGE)............ 17, 18, 25, 27, 56, 75, 77–85, 143, 144, 223, 232, 247–249, 251, 383, 498
Gene screen nylon 22
Glycosaminoglycan blotting................................... 119–126
Grid-immunoblotting 375–378

H

Heat-mediated blotting... 248
High-molecular weight protein
immobilized on solid nitrocellulose 491
transfer with heat... 490, 491
two-step transfer.. 490–491
History of blotting.. 301
Horseradish peroxidase
sodium azide... 161
thimerosal... 161
Horseradish peroxidase substrate
3-Amino-9-ethyl-carbazole (AEC) 439
3,3-diaminobenzidine (DAB) 154
luminol and peroxide containing reagents 154

I

Immobilization of large proteins on solid nitrocellulose.. 491
Immunochromatographic... 301
Immunodetection of proteins 63
Immunodominant bands 148–149
Immunoprecipitation western (IP)
electrophoresis 328
immunoprecipitation 327–342
India ink 22, 23, 97–99, 102, 161
Isoelectric focusing (IEF) 25, 37, 58, 61, 64, 66–67, 78, 109, 110, 112–115, 152, 154, 157, 229, 277, 321, 323–326, 400, 416, 417, 474, 475
flatbed non-denaturing IEF 114
Isoelectric point (pI) 63, 110, 321, 324, 344

M

Mass spectrometry
MALDI TOF
electronic western blot 499
(LC)-ESI-MS/MS...................................... 499–500
gold coated membrane and charging effect.. 499–500
Membrane blocking
blotto 154, 157
for calcium detection 406
for carbohydrate detection 298
fish gelatin 161
gelatin and BSA 97
goat serum 91
heat inactivated BSA 161
Ig depleted BSA 10
milk and Blotto.. 154, 157
milk and Tween-20............................. 79, 190, 376, 384
milk, EDTA and orthovanadate 384
non-fat milk.. 395
Membranes
activated paper 22
black 341
microplate.. 435–444
nitrocellulose
disadvantages 21

immobilization mechanism.......... 20–21
synthesis.......... 20
polyvinylidene difluoride (PVDF)
advantages.......... 21
clarification for transmission
densitometry.......... 497–498
edman sequencing.......... 228, 489
immobilization mechanism.......... 21–22
silver stained proteins.......... 501
Microblot.......... 16–168, 170–172
Microfluidics.......... 469–471
Miniaturized blotting.......... 165–172
Multiple antigen blot.......... 301–318
Multiple antigenic peptide
immunoblotting.......... 269–275
staining with Coomassie.......... 270
transfer to nitrocellulose.......... 279
Multiple immunoblots.......... 62, 63, 77–85, 247
Multiplex immunoblotting of paraffin embedded
tissue.......... 175–183
Multistrip western blotting.......... 197–224

N

Native western blot.......... 344–346, 348, 349, 351
NBT/BCIP substrate.......... 273
Nitrocellulose
black.......... 493
disadvantages.......... 21
immobilization mechanism.......... 20–21
oligonucleotide array method.......... 358, 492
synthesis.......... 20
Non-electrophoretic blotting.......... 70, 80–82
Non-electrophoretic transfer.......... 25, 80–83
Nonspecific binding.......... 10, 22, 59, 85, 116, 183, 199, 275, 277–282, 363, 368, 405, 469
Northern blotting.......... 10, 17
Nylon
disadvantages.......... 22–23
gene screen.......... 22
Zetabind.......... 22

O

On-membrane tryptic digestion of proteins.......... 499
Origins of protein blotting: 1.......... 5, 498

P

Parallel protein chemical processing
molecular scanner concept.......... 498–499
western blot.......... 499
Passive diffusion.......... 77–85
Pas-staining.......... 160
Peptide purification.......... 41, 269
Phenol red.......... 156, 160, 201
Phosphate specific fluorophore.......... 352
Polyethylene glycol
protein fixation.......... 495
protein transfer.......... 495
Polyvinylidene difluoride (PVDF)
advantages.......... 22
calcium-induced renaturation.......... 406
clarification for transmission densitometry.......... 497–498
edman sequencing.......... 489
immobilization mechanism.......... 20–21
silver stained proteins.......... 501
Ponceau S.......... 21, 23, 154, 156, 157, 160, 161, 167, 170, 470, 471
Protein blotting
acid.......... 496–497
acid-urea-PAGE.......... 500
activated glass.......... 496
adhesive interactions under fluid shear.......... 399–409
affinity immunoblotting.......... 109–116
affinity purification of antibodies from NC
strip.......... 258
blotting by ultrasound.......... 242
blotting from immobilized pH gradient
gels.......... 321, 322
capillary blotting.......... 123, 125, 487
CBB-stained proteins to transparencies.......... 415–417
centrifuge blotting.......... 411, 488
diffusion blotting.......... 26, 69
dot blotting.......... 18
dot-immunobinding (Dot-Iba)
advantages.......... 106
limitations.......... 106
double-blotting.......... 278
eastern blotting.......... 151–161
electroblotting to Teflon tape.......... 489
ELISPOT
B Cell.......... 419–425
fluorescence ELISPOT.......... 499–500
T Cell.......... 427–434
far-western-blotting.......... 379–397
fluorescence ELISPOT.......... 439–440
grid-immunoblotting.......... 375
heat-mediated.......... 249, 251, 252
history.......... 5–7
immunoprecipitation-blotting of low abundance
proteins.......... 368
intermittent microwave irradiation.......... 33
large protein blotting.......... 285
microblot.......... 167, 168
miniaturized blotting system.......... 165–172
molecular scanner concept.......... 498–499
multiple antigen blot.......... 301–318
multiple tissue blot.......... 490
multiplex immunoblotting of paraffin
embedded tissue.......... 175–183
multistrip western blotting.......... 197–224

Protein blotting (*cont.*)
native western blot 343–352
non-electrophoretic transfer 25, 80–82
polyethylene glycol mediated 495–496
quantitative computerized western blotting
immunodominant bands 148–149
net band intensity 147–148
total lane intensity 142, 149
semi-dry protein transfer 295
shift-western blotting 355–375
silver stained proteins 501
simple diffusion 25–26, 77, 247, 449
slice blotting 87–92
square wave alternating voltage 495
strip blotting 65
tissue printing 490
vacuum blotting 487–488
Protein concentration 35, 41–47, 53, 68, 131–135, 183, 323, 364, 366, 389, 395, 411
Protein fixation with glutaraldehyde 495
Protein marker 52, 62, 80, 85, 211, 240, 2499
non-electrophoretic transfer 25, 81, 82
Protein microarray 140, 493
Protein–protein interactions by far-western 379–394
Protein stains
Coomassie Brilliant Blue (CBB) 52, 189, 192, 230, 253
fast green 250, 266
India ink 97, 98
pas-staining 160
phenol red 160
Ponceau S 167, 170, 470
silver staining 160

Q

Quantification of proteins on PVDF 494
coomassie elution 494
Quantitative computerized blotting .... 141–150. *See also* Protein blotting

R

Reminiscences 12
Renaturation of proteins on blots 69
Reverse-phase protein array (RPPA) 129–148, 392

S

Semi-dry blotting 295
Sequential use of immunoblots 198
Shift-western blotting 62, 355–371
Silver staining 62, 63, 70, 74, 160, 270, 294
Simian Virus 40 (SV40) 14, 15
Simple western 465–467
Single cell western 455–464
Slice blotting 87–92
Sodium azide (NaN_3) 78, 97, 111, 131, 161, 224, 250, 259, 270, 451, 484
horseradish peroxidase (HRP) 161
Sodium dodecyl sulfate-polyacrylamide gel electrophoresis (SDS-PAGE) 10, 11, 17–19, 25, 27, 35, 36, 50–54, 62–71, 73–75, 77–85, 102, 134, 137, 143, 144, 151, 152, 160, 168–170, 172, 186–188, 191, 194, 223, 228–230, 232, 235, 238, 239, 247–252, 261, 262, 269–271, 274, 294, 295, 297, 321, 369, 383, 384, 400, 408, 411, 412, 416, 449–453, 474–478, 495–500
Solubilization of proteins 49–59
Southern EM 1–3, 6, 7, 14, 15
Spectrometric methods 26
Square wave alternating voltage 495
protein transfer 495
Strip immunoblotting 269–275
Systemic lupus erythematosus
La autoantigen (SS-B) 451
Ro 60 autoantigen (SS-A) 110–112, 248, 250, 251, 270

T

Teflon tape for N-and C-terminal sequencing 489
Tissue printing 93–102
Tissue slice preparation 888–90
brain 88–90
Total cell lysates 219, 321–326
Towbin, Harry
protein blotting 5–28
Towbin system 23
Transfer (protein)
bidirectional transfer 77, 247
electrophoretic 15, 28, 33, 77, 191, 192, 198, 248, 263, 273, 325, 449, 452–453, 500
electrophoretic transfer (heat mediated) 252
non-electrophoretic 80–83
semi-dry protein transfer 341
setting 24–25
simple diffusion 25
Transfer of paraffin embedded tissue 178
Transmission densitometry 437–438
Trypsin
coated membrane 498, 499, 501
digestion during transfer 498
Tryptic peptide purification 498
Two dimensional gel electrophoresis 75, 357, 474, 475, 499

U

Ultrasound blotting 244

V

Vacuum blotting 27, 487–488
V8 protease digestion 232–233

W

Western blot stripping 389

X

X-ray film 384, 389

Z

Zetabind nylon 22
Zymography 26, 61, 62, 64–67, 69–70

MIX
Papier aus verantwortungsvollen Quellen
Paper from responsible sources
FSC® C105338

If you have any concerns about our products,
you can contact us on
ProductSafety@springernature.com

In case Publisher is established outside the EU,
the EU authorized representative is:
Springer Nature Customer Service Center GmbH
Europaplatz 3, 69115 Heidelberg, Germany

Printed by Libri Plureos GmbH
in Hamburg, Germany